JURAN'S QUALITY MANAGEMENT AND ANALYSIS

SIXTH EDITION

JURAN'S QUALITY MANAGEMENT AND ANALYSIS

SIXTH EDITION

Joseph A. De Feo
CEO, Juran Institute, Inc.

Frank M. Gryna, Ph.D. (late)
Distinguished Professor of Industrial Engineering Emeritus
Bradley University

Earlier editions of this book were written by
J. M. Juran, Chairman Emeritus of the Juran Institute, Inc.

JURAN'S QUALITY MANAGEMENT AND ANALYSIS, SIXTH EDITION
International Edition 2015

Published by McGraw-Hill Education, 2 Penn Plaza, New York, NY 10121. Copyright © 2015 by McGraw-Hill Education. All rights reserved. Previous editions © 2007, and 2001. No part of this publication may be reproduced or distributed in any form or by any means, or stored in a database or retrieval system, without the prior written consent of the publisher, including, but not limited to, in any network or other electronic storage or transmission, or broadcast for distance learning.

Some ancillaries, including electronic and print components, may not be available to customers outside the United States.

This book cannot be re-exported from the country to which it is sold by McGraw-Hill. This International Edition is not to be sold or purchased in North America and contains content that is different from its North American version.

10 09 08 07 06 05 04 03 02 01
20 15 14
CTP   SLP

All credits appearing on page or at the end of the book are considered to be extension of the copyright page.

When ordering this title, use ISBN 978-1-259-25530-4 or MHID 1-259-25530-1

The Internet addresses listed in the text were accurate at the time of publication. The inclusion of a website does not indicate an endorsement by the authors or McGraw-Hill Education, and McGraw-Hill Education does not guarantee the accuracy of the information presented at these sites.

Printed and bound by CPI Group (UK) Ltd, Croydon, CR0 4YY

www.mhhe.com

Preface

We have entered the century of quality; we leave a century of productivity. The first decade of this century has resulted in dramatic changes in the field of managing for quality. There has been a surge of growth in performance excellence initiatives. The traditional quality management activities have become integrated with other functions to meet the changing needs of our customers. Professional societies, the many websites devoted to quality, publication of books and papers, awards, and consultancies have all grown exponentially. Much of this growth can be attributed to the need of our organizations to meet the new globally driven customer and marketplace. Organizations must excel in a competitive world that features a service-driven economy, global and virtual supply chains, mass customization, and technology that seems to change in a nanosecond.

Our customers demand perfection in safety, quality, reliability, and cost, and timely performance and delivering large quantities are now the norm. Failure to meet such needs and high-performing goals will produce declines in business performance. At times, we have seen rare and critical failures that make headlines and cause significant declines in brand reputation. We refer to this as poor quality. It is the reason the quality practitioners exist—to prevent these events from happening.

This edition has been rewritten for the practitioners who search to find the right methods to create an enterprise quality management system to make quality happen. This new edition is not about meeting product specifications; it's not about statistical techniques. It is about how to use them to attain superiority in quality and sustain business results. This book is about "Big Q," as Dr. Juran referred to it. In this book, "Big Q" is about what it takes for an entire organization to manage for superiority in quality. It is for goods and service organizations. It is for improving the quality of processes, systems, organizations, and leaders. In other words, this book is about enterprise transformation using quality as the driver.

This book was written for:

- Business and engineering students, both as an introductory text to the subject of managing for quality and as an advanced text in quality management for those specializing in operations management or industrial engineering.
- Managers and practitioners of ongoing improvement initiatives such as Performance Excellence, Operational Excellence, Six Sigma, and Lean.
- Quality directors, managers, and engineers, including those preparing for the certification examinations such as the Certified Quality Engineer, Quality Manager, Software Quality Engineer, and Reliability Engineer.

- Black Belts and Green Belts—this book serves well as the primary text for Black Belts and Green Belts as they undergo training and pursue certification. Master Black Belts will also find this book useful in providing a solid grounding in the concepts, tools, and techniques for attaining enterprise quality leadership.

New features of this edition include:

- Developing an integrated business management system to attain superior and sustainable performance.
- New and expanded topics, including Delivering Quality with Lean; Six Sigma Improvement; Design for Quality; Design for Six Sigma; Lean; Quality Risk Assessment; Employee Engagement; and Enterprise Assurance: A Look Ahead.
- Updated and showcased real business project examples and presentations.
- A case study showcasing the application of Lean Six Sigma to error-proof process documentation in an FDA-regulated biotech environment.
- A more extensive treatment of hypothesis testing, including a hypothesis testing road map to guide practitioners in the choice of statistical tests. This can add value to Six Sigma Experts, Green Belts, and Black Belts.
- Use of MINITAB software to analyze data and solve problems. Real-world examples using MINTAB are featured.
- A website for instructors and students to access data sets, homework assignments, and web links for further study and reference.

In addition, a full Instructor's Manual is available, containing solutions to all problems and additional case studies and their solutions. Enlargements of key figures from the book are on the website.

We hope you will agree with the changes and find this textbook useful.

—J. A. De Feo

ACKNOWLEDGMENTS FOR THE SIXTH EDITION

We would like to thank the following contributors to this edition. Without them we would not be able to provide a broader view of quality.

John Early, Kevin Caldwell, Brian Swayne, Charles Aubrey, Joseph M. De Feo, and Brad Wood co-authored some of the chapters.

Tina Pietraszkiewicz and Michelle Matschke provided needed review and edits.

And, as always, our thanks go to Dr. Juran and Dr. Gryna for providing us the opportunity to learn the methods of managing for quality while allowing us the leeway to move to new methods and the means to apply it.

About the Authors

JOSEPH A. De Feo is President and Chief Executive Officer of Juran Global. Mr. De Feo is an author, executive coach, and performance improvement practitioner. Among his areas of expertise are the Management of Quality, Lean, and Six Sigma Deployment; Strategic Planning; and Business Process Management methodologies.

Mr. De Feo's 35 years of experience comes from a wide range of industries, including consumer products, retail, electronics, chemical processing, aerospace, automotive, semiconductors, and both secondary and higher education.

Mr. De Feo has published more than 100 papers in national and international publications, including many web-based publications. He serves on the advisory boards of *Six Sigma Forum* magazine and FarmAid.

Mr. De Feo is the co-author of *Juran Institute's Six Sigma: Breakthrough and Beyond* with Dr. William W. Barnard, past senior vice president of the Institute. Published in October 2003, the book was one of the first management texts to present the full philosophy and methodology of Juran Institute. Mr. De Feo also co-authored the 2010 publication of *Juran's Quality Handbook 6th Edition, The Complete Guide to Performance Excellence*; the "go to" resource for deployment leaders. With the goal of helping practitioners and leaders learn to move beyond their initial Six Sigma efforts to achieve sustainable bottom line results, the book features performance-improving applications based on the Juran Trilogy of planning processes, controlling processes, and systematically achieving breakthrough improvement in processes.

FRANK M. GRYNA, Ph.D., had degrees in industrial engineering and more than 50 years' experience in the managerial, technological, and statistical aspects of quality activities.

From 1991 to 1999 he served first as director of the Center for Quality and then as Distinguished University Professor of Management at the University of Tampa. From 1982 to 1991, he was with the Juran Institute as senior vice president. Prior to 1982, Dr. Gryna was based at Bradley University, where he taught industrial engineering and served as acting dean of the College of Engineering and Technology. Prior to his passing, he was Distinguished Professor of Industrial Engineering Emeritus. In addition, he had been a consultant for many companies on all aspects of quality and reliability programs from initial design through field use.

Dr. Gryna also served in the U.S. Army Signal Corps Engineering Labs and the Esso Research and Engineering Company. At the Space Systems Division of the Martin Company, he was manager of reliability and quality assurance.

He co-authored *Quality Planning and Analysis* with J. M. Juran and was associate editor of the second, third, and fourth editions of *Juran's Quality Handbook*. His research project, *Quality Circles,* received the Book of the Year Award sponsored by various publishers and the Institute of Industrial Engineers. He received recognitions as a Fellow of the American Society for Quality, a Fellow of the Institute of Industrial Engineers, a Certified Quality Engineer, a Certified Reliability Engineer, and a Professional Engineer (Quality Engineering). He also received various awards, including the E. L. Grant Award of the American Society for Quality, Engineer of the Year Award of the Peoria Engineering Council, teaching and professional excellence awards, and the Award of Excellence of the Quality Control and Reliability Engineering Division of the Institute of Industrial Engineers. Dr. Gryna was also the recipient of the Ott Foundation Award, presented by the Metropolitan Section of the American Society for Quality.

Contents

Introduction   1

1 The Bottom Line: Quality and Business Performance   5
- 1.1 Why Quality Management? A Brief History Lesson   5
- 1.2 Quality and Its Impact on Financial Performance   7
- 1.3 *Integrated Quality System* Defined   9
- 1.4 The Quality Function   11
- 1.5 Relationship of Quality to Productivity, Costs, Cycle Time, and Value   13
- 1.6 Universal Principles for Managing for Quality   15
- 1.7 Quality Disciplines and Other Disciplines   17
- 1.8 The Way Forward—Developing a System Plan for Success   18
- Summary   19
- Problems   19
- References   20

2 Integrating Quality into the Enterprise Strategic Plan   21
- 2.1 Quality and the Strategic Plan   21
- 2.2 A Strategic Planning Model   23
- 2.3 Developing a Mission and Vision   25
- 2.4 Developing Long-Term Strategies   28
- 2.5 Development of Short-Term Annual Goals   30
- 2.6 Deployment of Goals—Hoshin Kanri   34
- 2.7 Leadership to Implement the Strategies   36
- 2.8 Review Progress Assessments, Audits, and a Balanced Scorecard   39
- 2.9 Review and Refresh the Plan   39
- 2.10 Obstacles to Achieving Goals   41
- Summary   43
- Problems   44
- References   44

3 Organizationwide Assessment of Quality   47
- 3.1 Why Assess Performance?   47
- 3.2 Quality Risk Assessment   47
- 3.3 Strategic Alignment, Deployment, and Assessment   48
- 3.4 Plan the Assessment   51

	3.5	Cost of Poor Quality	53
	3.6	Categories of Quality Costs	55
	3.7	Relating the Cost of Poor Quality to Business Measures	61
	3.8	Optimum Cost of Quality	62
	3.9	Assessment and Standing in the Marketplace	64
	3.10	Assessing Using National Performance Standards and Awards	67
	3.11	Baldrige Scoring System	72
	3.12	Using Quality Awards as a System Assessment Tool	73
	3.13	Assessing to the International Standards	75
	3.14	Quality System Certification/Registration	78
	3.15	Industry-Specific Adoptions and Extensions of ISO 9000 Standards	79
	3.16	Benchmarking Best Practices: Moving Toward Sustainability	85
	3.17	Benchmarking: What it is and What it is Not	89
	3.18	Internal and External Competitive and Noncompetitive Benchmarking (Who)	93
	3.19	Data and Information Sources (How)	95
	3.20	Data Normalization	98
	3.21	Analysis and Identification of Best Practices	100
		Summary	101
		Problems	102
		References	103
4	**Improving Quality While Decreasing Cost**		**105**
	4.1	Lean Six Sigma as a Breakthrough Improvement Model	105
	4.2	Breakthrough Improvement: Project-By-Project Approach	107
	4.3	Six Sigma Improvement Project Example	107
	4.4	The Business Case for Quality Improvement	110
	4.5	Lessons Learned with the Project-By-Project Approach	111
	4.6	Lean and Six Sigma Improvement	112
	4.7	Define Phase	114
	4.8	Measure Phase	118
	4.9	Analyze Phase	125
	4.10	Improve Phase	140
	4.11	Control Phase	147
	4.12	Lean Six Sigma Summary and Project Example	148
	4.13	Maintaining a Focus on Continuous Improvement	155
		Summary	160
		Problems	160
		References	163
5	**Quality by Design to Increase Sales**		**165**
	5.1	Contribution of Quality to Sales Income	165
	5.2	Quality and Financial Performance	166
	5.3	Achieving Marketplace Superiority	167
	5.4	Customer Satisfaction Versus Customer Loyalty	169
	5.5	Customer Loyalty and Retention	169
	5.6	Economic Worth of a Loyal Customer	173

	5.7	Impact of Poor Quality on Lost Sales	174
	5.8	Level of Satisfaction to Retain Present Customers	175
	5.9	Life-Cycle Costs	176
	5.10	Spectrum of Customers	177
	5.11	The Quality by Design Road Map	178
	5.12	Quality by Design (QbD) and Design for Six Sigma (DFSS)	184
		Summary	190
		Problems	191
		References	191
6	**Control of Quality to Maintain Superior Performance**		193
	6.1	Compliance and Control Systems	193
	6.2	The Importance of Information and Measurement	194
	6.3	The Ultimate Inspector: Self-Managed Processes	196
	6.4	Understanding What Is Critical to Control	198
	6.5	Establish Measurement and Establish Standards of Performance	199
	6.6	Measure Actual Performance	206
	6.7	Compare to Standards	207
	6.8	Take Action on the Difference	211
	6.9	Developing Process Control	213
		Summary	214
		Problems	214
		References	215
7	**Business Process Management**		217
	7.1	Functional Versus Business Process Management	217
	7.2	Business Process Management	218
	7.3	Selection of Processes	219
	7.4	Organize the Process Team	221
	7.5	Example of Business Process Management	221
	7.6	The Planning Phase of Business Process Management	222
	7.7	Transferring and Managing the New Process	232
	7.8	Impact of Business Process Management on an Organization	232
		Summary	234
		Problems	234
		References	235
8	**Organization Roles to Support a Quality Culture**		237
	8.1	Evolution of Organization for Quality	237
	8.2	Role of the Quality Office	238
	8.3	Organizing the Quality Office of the Future	241
	8.4	Role of Upper Management	249
	8.5	Role of the Quality Director	251
	8.6	Role of Management	255
	8.7	Role of the Workforce	256

8.8	Role of Teams	256
8.9	Training and Certification of Personnel	258
	Summary	261
	Problems	262
	References	262

9 Creating a Quality Culture — 265

9.1	Corporate Culture	265
9.2	Quality Culture	266
9.3	Creating a Culture of Quality	267
9.4	The Performance Excellence Change Model	268
9.5	The Roles of Leaders to Attain a Breakthrough in Leadership	275
9.6	Resistance to Change	285
9.7	Policies and Cultural Norms	287
9.8	Human Resources and Cultural Patterns	287
9.9	A Transformation Roadmap	294
	Summary	296
	Problems	297
	References	298

10 Inspection, Test, and Measurement — 299

10.1	The Terminology of Inspection	299
10.2	Conformance to Specification and Fitness for Use	300
10.3	Disposition of Nonconforming Product	303
10.4	Inspection Planning	305
10.5	Seriousness Classification	309
10.6	Automated Inspection	310
10.7	Inspection Accuracy	312
10.8	Errors of Measurement	313
10.9	How Much Inspection is Necessary?	325
10.10	The Concept of Acceptance Sampling	328
10.11	Sampling Risks: The Operating Characteristic Curve	329
10.12	Quality Indexes for Acceptance Sampling Plans	332
10.13	Types of Sampling Plans	334
10.14	Characteristics of a Good Acceptance Plan	335
10.15	ANSI/ASQC Z1.4	336
10.16	Selection of a Numerical Value of the Quality Index	340
10.17	How to Select the Proper Sampling Procedures	342
	Summary	344
	Problems	345
	References	349

11 Understanding Customer Needs — 351

11.1	Quality and Competitive Advantage	351
11.2	Identify the Customers	351
11.3	Customer Behavior	353

11.4	Scope of Human Needs and Expectations	353
11.5	Sources of Market Quality Information	354
11.6	Market Research in Quality ("Voice of the Customer")	355
11.7	Needs Related to Product Features	357
11.8	Needs Related to Product Deficiencies	363
11.9	Measuring Customer Satisfaction	365
11.10	Market Research for Internal Customers	366
	Summary	367
	Problems	367
	References	368

12 New-Product Development's Role in Designing for Quality — 371

12.1	Opportunities for Improvement in Design	371
12.2	Design and Development as a Process	372
12.3	Phases of Product Development and the Early Warning Concept	373
12.4	Designing for Basic Functional Requirements; Quality by Design Spreadsheets	374
12.5	Designing for Time-Oriented Performance (Reliability)	379
12.6	Availability	392
12.7	Designing for Safety	393
12.8	Designing for Manufacturability	397
12.9	Cost and Product Performance	398
12.10	Design Review	399
12.11	Software Development	400
12.12	Measurement in Design	401
12.13	Improving the Effectiveness of Product Development	402
	Summary	404
	Problems	404
	References	406

13 Managing Quality in Operations—Manufacturing — 409

13.1	Quality in Manufacturing in the 21st Century	409
13.2	Lean Manufacturing and Value Stream Management	410
13.3	Initial Planning for Quality	417
13.4	Concept of Controllability: Self-Control	424
13.5	Automated Manufacturing	436
13.6	Overall Review of Manufacturing Planning	438
13.7	Organizing for Quality in Manufacturing Operations	439
13.8	Planning for Evaluation of Product	440
13.9	Conducting Process Quality Audits	443
13.10	Quality Measurement in Manufacturing Operations	444
13.11	Maintaining a Focus on Continuous Improvement	445
13.12	Case Study on Error-Proofing	447
	Summary	456
	Problems	456
	References	457

14 Managing Quality in Operations—Service — 459
- 14.1 The Service Sector — 459
- 14.2 Initial Planning for Quality — 460
- 14.3 Planning for Self-Control — 466
- 14.4 Control of Quality in Service Operations — 471
- 14.5 Process Quality Audits — 475
- 14.6 Frontline Customer Contact — 475
- 14.7 Organizing for Quality in Service Operations — 478
- 14.8 Six Sigma Projects in Service Industries — 479
- 14.9 Quality Measurement in Service Operations — 480
- 14.10 Maintaining a Focus on Continuous Improvement — 483
- Summary — 484
- Problems — 484
- References — 485

15 Managing the Supply Chain — 487
- 15.1 Supplier Relations—A Revolution — 487
- 15.2 Scope of Activities for Supplier Quality — 489
- 15.3 Specification of Quality Requirements for Suppliers — 490
- 15.4 Supplier Selection; Outsourcing — 493
- 15.5 Assessment of Supplier Capability — 495
- 15.6 Supply Chain Quality Planning — 499
- 15.7 Supply Chain Quality Control — 502
- 15.8 Supply Chain Quality Improvement — 509
- Summary — 512
- Problems — 513
- References — 515

16 Quality Assurance and Audits — 519
- 16.1 Definition and Concept of Assurance — 519
- 16.2 Concept of Quality Audits — 520
- 16.3 Principles of a Quality Audit Program — 522
- 16.4 Subject Matter of Audits — 523
- 16.5 Structuring the Audit Program — 525
- 16.6 Planning Audits of Activities — 526
- 16.7 Audit Performance — 530
- 16.8 Audit Reporting — 531
- 16.9 Corrective Action Follow-Up — 533
- 16.10 Human Relations in Auditing — 534
- 16.11 Product Audit — 536
- 16.12 Sampling for Product Audit — 538
- 16.13 Reporting the Results of Product Audit — 538
- Summary — 540
- Problems — 540
- References — 541

17 The Role of Statistics and Probability 543
17.1 Statistical Tools in Quality 543
17.2 The Concept of Variation 543
17.3 Tabular Summarization of Data: Frequency Distribution 544
17.4 Graphical Summarization of Data: The Histogram 546
17.5 Boxplots 548
17.6 Graphical Summarization of Time-Oriented Data: The Run Chart 549
17.7 Methods of Summarizing Data: Numerical 550
17.8 Probability Distributions: General 551
17.9 The Normal Probability Distribution 553
17.10 Probability Curves and Histogram Analysis 555
17.11 The Exponential Probability Distribution 558
17.12 The Weibull Probability Distribution 559
17.13 The Poisson Probability Distribution 563
17.14 The Binomial Probability Distribution 564
17.15 Basic Theorems of Probability 565
17.16 Computer Software for Statistical Analysis 566
Summary 566
Problems 567
References 569

18 Tools for Analyzing Data 571
18.1 Scope of Data Planning and Analysis 571
18.2 Statistical Inference 572
18.3 Sampling Variation and Sampling Distributions 572
18.4 Statistical Estimation: Confidence Limits 575
18.5 Importance of Confidence Limits in Planning Test Programs 578
18.6 Determination of the Sample Size Required for a Specified Accuracy in an Estimate 579
18.7 Test of Hypothesis 579
18.8 Testing a Hypothesis When the Sample Size Is Fixed in Advance 584
18.9 Drawing Conclusions from Tests of Hypotheses 596
18.10 Determining the Sample Size Required for Testing a Hypothesis 597
18.11 The Design of Experiments 598
18.12 An Illustrative Case of Paint Blisters 602
18.13 Contrast Between the Classical and Modern Methods of Experimentation 606
18.14 Analysis of Variance 608
18.15 Fractional Factorial Experimental Design 609
18.16 Planning for Experimental Design 611
18.17 Regression Analysis 611
18.18 Advanced Tools of Data Analysis 615
Worked Examples Using Minitab 616
Summary 620
Problems 620
References 624

19 Statistical Tools to Design for Quality — 625
- 19.1 The Statistical Toolkit for Design — 625
- 19.2 Failure Patterns for Complex Products — 625
- 19.3 The Exponential Formula for Reliability — 628
- 19.4 The Relationship Between Part and System Reliability — 630
- 19.5 Predicting Reliability During Design — 633
- 19.6 Predicting Reliability Based on the Exponential Distribution — 635
- 19.7 Predicting Reliability Based on the Weibull Distribution — 635
- 19.8 Reliability as a Function of Applied Stress and Strength — 636
- 19.9 Availability — 637
- 19.10 Setting Specification Limits — 639
- 19.11 Specification Limits and Functional Needs — 639
- 19.12 Specification Limits and Manufacturing Variability — 640
- 19.13 Specification Limits and Economic Consequences — 644
- 19.14 Specification Limits for Interacting Dimensions — 645
- Summary — 648
- Worked Examples Using Minitab — 649
- Problems — 651
- References — 654

20 Tools to Maintain the Control of Quality — 657
- 20.1 Definition and Importance of Quality Control — 657
- 20.2 Advantages of Decreasing Process Variability — 658
- 20.3 Statistical Control Charts—General — 660
- 20.4 Advantages of Statistical Control — 662
- 20.5 Steps in Developing Control Charts — 663
- 20.6 Control Chart for Variables Data — 665
- 20.7 Pre-Control — 673
- 20.8 Attributes Control Charts — 675
- 20.9 Special Control Charts — 677
- 20.10 Process Capability — 679
- 20.11 Measuring Process Performance — 687
- 20.12 Planning for the Process Capability Study — 689
- 20.13 Six Sigma Concept of Process Capability — 690
- 20.14 Statistical Process Control and Quality Improvement — 692
- 20.15 Software for Statistical Process Control — 693
- 20.16 Worked Example Using Minitab — 694
- Summary — 696
- Problems — 696
- References — 701

Appendix I: Tables — 703

Name Index — 719

Subject Index — 721

Introduction

By the time Dr. Juran published his autobiography in 2003, he was 99 years of age and had not been pleased with the progress toward the United States becoming a global leader in quality. In his own words, "I became increasingly unhappy with the status of quality in the United States."

His major concern was that although there were a few U.S. companies that had become world-class quality leaders by then, there were still too many that did not. Many of those companies in the latter group continued to lose market share to top performers such as Japanese companies. Most organizations and leaders from 1980 to 2000 learned quickly of the negative impact from the Japanese revolution in managing for quality. It had vaulted Japan to superpower status taking with it most of our manufacturing capabilities. By the end of the 2000s, a number of U.S. companies had bounced back by doing what the best companies were doing—managing the quality of their products, services, and processes in a better way. That way has been well documented and we call it the "Juran Management System," or JMS.

We often speak of our system of quality management as the Toyota Production System, the Value Management System (or the Performance Excellence System), the Malcolm Baldrige System, and so on. Many quality professionals and C-suite leaders understand the methods of quality management such as Process Improvement, ISO 9000, and lean and six sigma, but few know where this all originated from, and frankly, they do not need to.

What every expert and leader should know is that the JMS was the first body of knowledge (BOK) that many organizations and associations such as ASQ use today.

The Juran BOK has been researched, practiced, and its methods carried out in many forms since long before the Japanese Quality Revolution. The Juran BOK may even be the reason why many U.S. companies are still with us today and why world-class companies become world-class. Simply put, the Juran BOK creates world-class companies. Not because the Japanese have a culture conducive to this stuff, it is because it works.

The Juran BOK creates a never-ending pursuit of satisfying client needs and being the best in the market at doing that. This pursuit leads to sustainable business and financial performance longer than others.

This revised publication is about the few methods that are needed to be successful and world-class. This publication began when Dr. Juran and Dr. Gryna set out to teach the quality engineer how to manage for quality. When it was first published in 1951, the need for quality control was clear. We were a manufacturing society and control of quality was the first universal process to be mastered. Control of quality requires sometimes sophisticated data analysis to be able to make risk-based decisions and not cause harm or loss of dollars due to poor-performing machines. This led to the certified quality engineer profession. As society grew and the need for nonmanufacturing services flourished, these concepts were transferred to these new businesses beginning the evolution of quality management.

We are now into the 6th edition and the evolution continues. This is no longer a text just for quality engineers. It is not just a text on the science of quality. It is not just a text for manufacturing companies. It is a text for those who want to be informed of the universal methods and organizational system to drive business results. It is the best way we know to sustain business performance.

This is not a prescription. It is a book about what has worked and the set of methods that worked for so many organizations globally. It is to be used by a business to meet its customer needs and stakeholder needs. It includes universal processes to think about the organization and the system needed to be successful.

Global quality leadership can come from any country or industry. What makes them the leader is how they think about quality and its impact on business performance.

This book is to inform our leaders what can be done to improve quality. It is about how to never stop pursuing the needs of society and your customers to keep our business alive and healthy.

The JMS is not the only choice a leader has, but it has been one of the most highly respected choices to attain a great ROI.

This book is being published in my 25th year at Juran Global. I have learned a lot. We at Juran are lifelong learners. I could not have written this book alone and will not take the credit alone. There were many inspiring professionals; some have left us and others still around.

I would like to thank:

Dr. Juran and Dr. Gryna for giving me the chance to continue where they left off although a difficult task.

Our current and past clients. Without you I would not have any lessons learned to share with future generations.

My colleagues and business partners that helped me complete this updated text. Specifically I would like to acknowledge:

John Early for his work revising the tougher parts of the management of quality specifically the statistical sections.

Dr. Richard Chua and the use of Minitab examples to make it easy for us to calculate tough numbers.

Tina Pietraszkiewicz, Jeremy Hopfer, Michelle Matschke, Kaitlin Tyer, and Joseph M. De Feo for their research and assistance in pulling this together.

And most importantly, I want to thank my late wife Monica for her support for 32 years; 25 of those years of which I worked for Juran Institute, her support had enabled me to continue doing what I love to do, which was to improve the performance of an institution. I could not have done it without her.

I hope you benefit from the Juran BOK as I have.

Sincerely,

Joseph A. De Feo
CEO, Juran Institute, Inc.

CHAPTER 1

The Bottom Line: Quality and Business Performance

➔ 1.1 WHY QUALITY MANAGEMENT? A BRIEF HISTORY LESSON

Quality is a word with multiple meanings. It is often used to describe the term a customer uses when they like something they tried or purchased. It is a word that is used synonymously with an organization's mission—to provide superior quality to customers and stakeholders. It is also a word used to describe many methods to improve business performance, such as this text's title, *Quality Management.* A brief look at history will help you better understand where quality management is today and where we may be headed tomorrow.

Practitioners are urged to learn about the different approaches to quality, choose the best methods to fit your business needs, and then customize the framework for your own organizations. In this book we will present the body of knowledge that works. We will refer to examples of the most popular frameworks such as the US Malcolm Baldrige National Award for Excellence (and all similar frameworks around the globe), the Juran Quality Management System (QMS), and the ISO Quality Management System to name a few. Before we do that lets take a look at history. Two dramatic examples illustrate how quality has an impact on both sales revenue and costs.

First, two examples of poor quality. An international organization that claimed to have strong customer service refused to accept sales orders for delivery in less than 48 hours—even though competitors honored delivery requests of 24 hours. Imagine the millions of dollars of sales revenue lost each year because this company did not recognize the need of its customers.

Second, listen to the president of a specialty manufacturing company: "Our scrap and rework costs this year were five times our profit. Because of those costs, we have had to increase our selling price and we subsequently lost market share. Quality is no longer a technical issue; it is a business issue."

Samsung electronics used quality as a mantra, as a method to attain $1.6 billion in savings by focusing on improving the quality of flat-screen TV sets, making them the best practice globally.

Everyone knows about Toyota quality. Why? Because its products have a reputation of being the best.

There are many stories like this. The conclusion is the same. If an organization focuses on the relentless pursuit of quality, it will attain a sustainable business performance and positive global reputation. Our forefathers knew—as we know—that quality is important. Metrology, specifications, inspection—all go back many centuries.

Then came the 20th century. The pace quickened with a lengthy procession of "new" activities and ideas launched under a bewildering array of names: quality control, continuous quality improvement, defect prevention, statistical process control, reliability engineering, quality cost analysis, zero defects, total quality management, supplier certification, quality circles, quality audit, quality assurance, quality function deployment, Taguchi methods, competitive benchmarking, lean and six sigma. This book discusses all these concepts and places them into a context that is needed to sustain business performance.

Following World War II, two major forces emerged that have had a profound impact on quality.

The first force was the Japanese revolution in quality. Prior to World War II, many Japanese products were perceived, throughout the world, as of poor quality. To help sell their products in international markets, the Japanese took some revolutionary steps to improve quality:

1. Upper level managers personally took charge of leading the revolution.
2. All levels and functions received training in the quality disciplines.
3. Quality improvement projects were undertaken on a continuing basis—at a revolutionary pace.

The Japanese success has been almost legendary.

The second major force to affect quality was the prominence of product quality in the public mind. Several trends converged to highlight this prominence: product liability cases; concern about the environment; some major disasters and near disasters; pressure by consumer organizations; and the awareness of the role of quality in trade, weapons, and other areas of international competition. This emphasis on quality has been further accented by the emergence of national awards such as the Baldrige and European Quality Awards.

Quality is not limited to the manufacturing sector. Quality concepts are applied to other sectors such as healthcare, education, not-for-profit organizations, and governments. Product quality is not the only focus. Service quality, process quality, and data quality are now being measured, controlled, and improved.

During the 20th century, a significant body of knowledge emerged on achieving superior quality. Many individuals contributed to this knowledge, and five names deserve particular mention: Juran, Deming, Feigenbaum, Crosby, and Ishikawa.

J. M. Juran emphasizes the importance of a balanced approach using managerial, statistical, and technological concepts of quality. He recommends an operational framework of three quality processes, quality planning, quality control, and quality improvement. The foundation for this book is the Juran approach. This book makes frequent references to Juran's *Quality Handbook,* sixth edition, which is denoted *JQH6,* as well as the fifth edition, *JQH5.*

W. Edwards Deming also had a broad view of quality, which he initially summarized in 14 points aimed at the management of an organization. These 14 points rest on a system of "profound knowledge" that has four parts: the systems approach, understanding of statistical variation, the nature and scope of knowledge, and psychology to understand human behavior.

A. V. Feigenbaum emphasizes the concept of total quality control throughout all functions of an organization. Total quality control really means both planning and control. He urges creating a quality system to provide technical and managerial procedures that ensure customer satisfaction and an economical cost of quality.

Philip Crosby defines quality strictly as "conformance to requirements" and stresses that the only performance standard is zero defects. His activities demonstrated that all levels of employees can be motivated to pursue improvement but that motivation will not succeed unless tools are provided to show people how to improve.

Kaoru Ishikawa showed the Japanese how to integrate the many tools of quality improvement, particularly the simpler tools of analysis and problem solving.

The approaches of these gurus have similarities as well as differences—particularly in the relative emphasis on managerial, statistical, technological, and behavioral elements. This book draws upon the contributions of these and other experts.

Juran (1995) provides a comprehensive history of managing for quality for different time spans (ancient, medieval, modern), geographical areas, products, and political systems.

The major forces affecting managing for quality led to a changing set of business conditions.

➔ 1.2 QUALITY AND ITS IMPACT ON FINANCIAL PERFORMANCE

The prominence of product quality in the public mind has resulted in quality becoming a cardinal priority for most organizations. The identification of quality as a core concern has evolved through a number of changing business conditions. These include

1. *Competition.* In the past, higher quality usually meant higher price. Today, customers can obtain high quality and low price simultaneously. Quality is now a "given," i.e., customers assume that they will receive adequate quality. In the service sector, deregulation in areas such as airlines and utilities created competition that did not exist before, and quality is a key dimension of that competition.
2. *The customer-focused organization.* The impact of quality as a tool of competition has led to viewing quality as customer satisfaction and loyalty rather than conformance to specifications. In the United States, the growth of the service sector (which now employs more than 75% of all workers) with so much direct customer contact has been a driver of customer focus. Also, the concept of "customer" now includes both external and internal customers (see Section 1.3).
3. *Higher levels of customer expectation.* Higher expectations, spawned by competition, take many forms. One example is lower variability around a target value of

a product characteristic, even though all product meets the specification limits. Another form of higher expectation is improved quality of service both before and after the sale.
4. *Performance excellence and improvement.* Quality, cycle time, cost, and profitability have become interdependent. Many organizations now talk of "performance improvement" or "business excellence" rather than quality.
5. *Changes in organization forms.* Most organizations no longer try to be completely self-sufficient with layers of management for various functional activities. What has emerged are concepts like partnering with other organizations, outsourcing of complete functions, process management, and various types of permanent and temporary teams—and all this with fewer layers of management.
6. *Changing workforce.* These changes include a higher level of education for some parts of the workforce, a multilingual workforce, and downsizing.
7. *Information revolution.* The key operating parameters now are labor, material, equipment, capital—and information. The relative ease with which information can be collected and disseminated throughout an organization now makes it feasible to plan and control activities that were unthinkable a few decades ago. Automated real-time access to business data is now common. Internets and intranets have become a way of life.
8. *Electronic commerce.* Organizations now use the Internet for many activities—to provide customers with a wealth of information, enable customers to order products, collect information on customer needs and purchasing behavior, customize products based on customer needs, link suppliers, and link dealers. Electronic commerce applies to companies and their consumer customers and also to commerce between companies. The impact on functional activities—marketing, purchasing, product development, operations, and customer service—will be pervasive. Evolving issues concerning quality include ensuring the quality of e-commerce transactions, measuring the quality of information provided to customers, and gathering and analyzing information on customer needs and problems.
9. *Role of a "quality department."* In previous decades, many organizations (particularly in the manufacturing sector) had a quality department that performed various roles involving formal evaluation of products and assistance to line departments in planning for quality. The recent emphasis on quality and the enormous training provided within an organization have resulted in transferring some activities from the staff quality department to the line departments. The integration of quality into line departments has decreased the size of (or even eliminated) the quality department. A reality is that in some organizations the integration was wishful thinking, and the reduction of the department was done prematurely—but the process has forced quality departments to review the services they provide to internal and external customers.

Each of these changing business conditions must be thoroughly understood if organizations are to survive in competitive world markets. The role of quality standards (such as ISO 9000, TS-16949, FDA cGMPs) as a prerequisite for doing business is also changing the way companies view quality. This book addresses these changing conditions and describes the approaches needed for future quality leadership.

Before examining these approaches, we need to define some terms.

❖ 1.3 INTEGRATED QUALITY SYSTEM DEFINED

The dictionary provides numerous definitions for the word *quality*. One short definition of *quality* is "customer satisfaction and loyalty." "Fitness for use" was used by Juran when business was typically manufacturing-focused. Today we use "fit for purpose." This means that no matter what product or service your organization provides, it must be "fit for the purpose the customer intended to use it for." Internally, all processes must be fit for purpose—meaning every step in a process must meet the next internal customer step. Although there are many short, easy-to-remember definitions, we must develop it further to provide a basis for action—to be able to manage it.

Unfolding the meaning starts with defining the word *customer*. A *customer* is "anyone who is impacted by the service, product, or process."

1. *External customers* include ultimate users (current and potential) and also intermediate processors, as well as retailers. Other customers who are not purchasers have some connection to the product, e.g., government regulatory bodies, shareholders, suppliers, partners, investors, the media, and the general public. External customers clearly are of primary importance.
2. *Internal customers or functions and processors* include other divisions of a company that are provided with information or components for an assembly and also departments or persons that supply products to each other. Thus when a purchasing department receives a specification from an engineering department for a procurement, purchasing is an internal customer of engineering. When the procurement is provided, then engineering is the internal customer of purchasing. Similarly, at a bank, the payroll department and operations department are internal customers of each other.

These external and internal customers are sometimes called "stakeholders."

A product is the output of any process. Three categories can be identified:

1. *Goods:* e.g., automobiles, circuit boards, reagent chemicals.
2. *Software:* e.g., a computer program, a report, an instruction.
3. *Service:* e.g., banking, insurance, transportation. Service also includes support activities within companies, e.g., employee benefits, plant maintenance, secretarial support.

Throughout this book *product* means "goods, software, or services." At times, we will interchange the word *service* for *product* to balance between a service-focused organization and a goods producer.

Customer satisfaction and loyalty are achieved through two dimensions: features and freedom from deficiencies. Examples of the main categories of these dimensions are shown in Table 1.1 for manufacturing and service industries. We see dramatic differences within manufacturing industries (assembly versus chemicals) and within services (restaurants versus banking). Each organization must identify the dimensions of quality that are important to its customers.

A closer examination of what quality means to customers follows two dimensions:

1. The *features of a product* have a major effect on *sales income* (through market share, premium prices, etc.). In many industries, the customer population can be

TABLE 1.1
Two dimensions of quality

Manufacturing industries	Service industries
Features	
Performance	Accuracy
Reliability	Timeliness
Durability	Completeness
Ease of use	Friendliness and courtesy
Serviceability	Anticipating customer needs
Esthetics	Knowledge of server
Availability of options and expandability	Appearance of facilities and personnel
Reputation	Reputation
Freedom from deficiencies (failures or non-value-added costs)	
Product free of defects and errors at delivery, during use, and during servicing	Service free of errors during original and future service transactions
All processes free of rework loops, redundancy, and other waste	All processes free of rework loops, redundancy, and other waste

segmented by the level or "grade" of quality desired. Thus the spectrum of customers leads to a demand for luxury hotels and budget hotels and to a demand for inexpensive automobiles and expensive ones with many special features as well as for those with basic cooling capabilities. The features refer to the aspects of the product that happen through what is called *quality of design*. The designer intended to create a product that had features that meet the needs of customers. In this case, increasing the quality of the design generally leads to higher costs and higher sales!

2. All products must be *free from as many deficiencies as possible*. If there are *too many deficiencies*, it has a major effect on *costs* through reductions in rework, customer complaint handling, scrap, and other results of deficiencies. "Deficiencies" are typically stated in different units, e.g., errors, defects, failures, off-specification. Freedom from deficiencies refers also to *quality of conformance*. Increasing the quality of conformance usually results in lower costs. In addition, higher conformance means fewer complaints and therefore decreased customer dissatisfaction and less cost.

When talking about quality to leaders, we must be able to speak in the language of leaders—the language of money. How features and freedom from deficiencies interrelate and lead to higher profits is shown in Figure 1.1.

To summarize, quality is defined by the customer. Features and freedom from deficiencies are the main determinants of satisfaction. For example, an external customer of an automobile desires certain performance features along with a record of few defects and breakdowns. The manufacturing department, as an internal customer

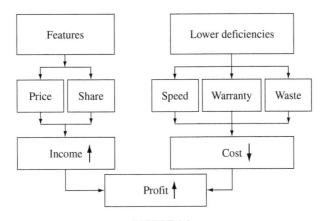

FIGURE 1.1
Quality, market share, and return on investment. (*From Juran's Quality Handbook: The Complete Guide to Performance Excellence 6e, 2010*)

of the product development department, wants an engineering specification that is producible in the shop and is free of errors or omissions. Both of these customers want "the right product right." Design projects improve features, and improvement projects reduce deficiencies. Different techniques are directed for each. For example, quality by design (QbD), or design for six sigma (DMADV), focuses on improving features, whereas lean and six sigma (DMAIC) focus on lowering deficiencies. Quality as used here is not limited to just quality performance metrics. In addition to quality-based metrics, time and cost dimensions of performance are also relevant to both features and deficiencies. For example, customers may demand accurate and responsive service, available 24/7, 365 days a year.

Finally, we note that quality experts offer different shorthand definitions of quality—"fitness for use" (Juran), "conformance to specifications" (Crosby), "loss to society" (Taguchi), "predictable degree of uniformity" (Deming). These definitions are complementary and provide operational meaning at different phases of quality activities. For the record, the International Organization for Standardization (ISO) defines quality as the "totality of characteristics of an entity that bear on its ability to satisfy stated and implied needs."

➔ 1.4 THE QUALITY FUNCTION

Attainment of quality requires performing a wide variety of identifiable activities or quality tasks. Obvious examples are the study of customers' quality needs, design review, product tests, and field-complaint analysis. In a tiny enterprise, all these tasks (sometimes called work elements) may be performed by a few persons. As the enterprise grows, however, specific tasks may become so time-consuming that we must create specialized departments to perform them. Corporations have created departments such as product design, operations, and customer service, which are

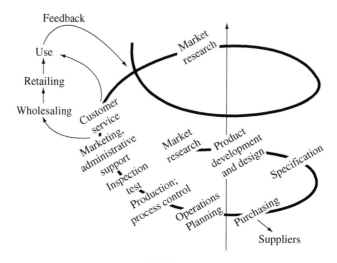

FIGURE 1.2
Spiral of progress in quality.

essential to launching any new or changed product. These functions follow a relatively unvarying sequence of events (see the spiral in Figure 1.2). In addition to the main "line" activities in the spiral, we need many administrative and support activities such as finance, human resources, and information technology.

The spiral shows that many activities and tasks must be performed to attain customer satisfaction and loyalty. Some of these are performed within manufacturing or service companies. Others are performed elsewhere—by suppliers, merchants, regulators, etc. A convenient shorthand name for this collection of activities is "quality function." The quality function is the entire collection of activities through which we achieve customer satisfaction and loyalty, no matter where these activities are performed.

Some practitioners look upon the spiral or the quality function as a system, i.e., a network of activities or subsystems. Some of these subsystems correspond to segments of the spiral. Others, although not shown on the spiral, are nevertheless present and active, e.g., data processing, standardization. These subsystems, when well designed and coordinated, become a unified system that carries out the intended quality objectives.

The traditional scope of quality activities is undergoing a radical and exciting change from the historical emphasis on quality of physical products in manufacturing industries ("little q") to what is now emerging as the application of quality concepts to all products, all functional activities, and all industries ("big Q"). Table 1.2 summarizes this change in scope.

Thus quality concepts are now being applied to order entry, inventory management, human resources, and new product development. Two books have even been written that apply quality concepts to personal activities—both on and off the job (see Roberts and Sergesketter, 1993, and Forsha, 1992).

Under this enlarged concept, all jobs encompass three roles for the jobholder: the customer who receives inputs of information and physical goods, the processor who converts these inputs into products (outputs), and the supplier who delivers the resulting

TABLE 1.2
Little q and big Q

Topic	Content of little q	Content of big Q
Products	Manufactured goods	All products, goods, and services, whether or not for sale
Processes	Processes directly related to manufacture of goods	All processes; manufacturing support; business, etc.
Industries	Manufacturing	All industries; manufacturing; service, government, etc., whether or not for profit

products to customers. This concept is called the triple-role concept (see Figure 1.3 for clarification). Thus a product development department has three roles: as an internal customer of the marketing department, the product development department receives information on customer needs; as a processor, product development creates designs for new products; and as a supplier, it furnishes specifications to the operations department to create the new products.

1.5 RELATIONSHIP OF QUALITY TO PRODUCTIVITY, COSTS, CYCLE TIME, AND VALUE

Does an emphasis on quality have a positive or negative impact on productivity, costs, cycle time, and value? The easy answer is "positive," but the reality is that, although these parameters should be mutually compatible, they may be either supportive or opposing forces.

Quality and Productivity

Productivity is the ratio of salable output divided by the resources used. The resources include labor, raw materials, and capital. Any one of these (or the total) can be the denominator in the productivity ratio. McCracken and Kaynak (1996) discuss alternative definitions of productivity and their relationships to quality (a key conclusion: as quality increases, productivity increases).

A common measure of productivity is labor productivity, e.g., number of salable units per hour of direct labor. When quality is improved by identifying and eliminating the causes of errors and rework, more usable output is available for the same amount of labor input. Thus the improvement in quality results directly in an increase in productivity. Productivity is now a key issue in national economic policy decisions, and thus the definition and measurement of productivity is important to any country. For a discussion of five links between national economics and product quality, see Brust and Gryna (1997).

Quality and Costs

As the quality of design (features) increases, costs typically increase. As the quality of conformance increases, the reduction in rework, complaints, scrap, and other deficiencies

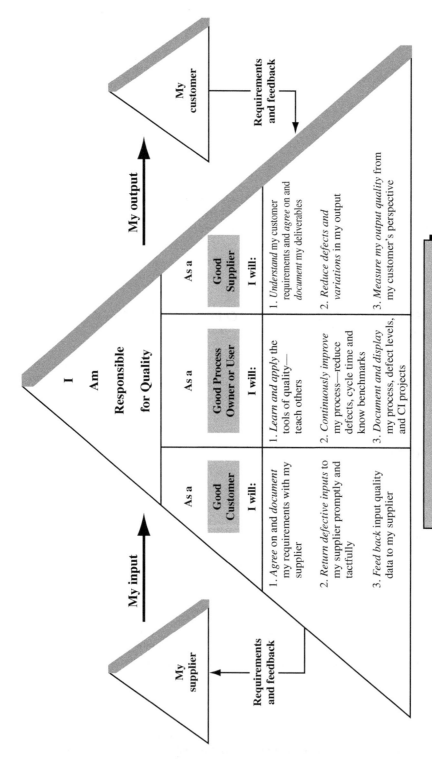

FIGURE 1.3
Triple-role concept. Continuous improvement practices.

results in a significant decrease in costs. An ideal strategy calls for using the savings from reduced deficiencies to pay for any increase in features without increasing the selling price, thus resulting in higher customer satisfaction and increased sales revenue. Many exemplary firms are quietly following this strategy.

Quality and Cycle Time

In both the manufacturing and service sectors, the cycle time to complete the activities required for customers is now a key parameter. In service industries, customers view cycle time to provide a transaction as a quality parameter. Customers simply demand fast response. Thus when a quality improvement effort reduces rework, redundant operations, and other deficiencies, a simultaneous reduction in cycle time occurs.

Quality and Value

Value is quality divided by price. The reality is that customers do not separate quality from price; they consider both parameters simultaneously. Improvements in quality that can be provided to customers without an increase in price result in better "value." As customers compare sources of supply, organizations must evaluate the value they provide relative to the competition. Advertising has trivialized the word *value*, but the concept itself is sound because value is what customers demand.

Quality, productivity, costs, cycle time, and value are interrelated. Quality activities must try to detect quality problems early enough to permit action without requiring a compromise in cost, schedule, or quality. The emphasis must be on prevention rather than on just correction of quality problems. Improving quality can be the driving force to improve results in the other parameters. In an end-of-chapter problem, readers are asked to calculate the benefits that a quality improvement program contributes to cost savings, delivery schedules, and productivity.

Companies that have high product quality typically achieve superior financial performance. For example, recipients of the Malcolm Baldrige National Quality Award (see Section 2.9) saw benefits (for the U.S. economy) to costs ratio of 820 to 1 (Link and Scott, 2011).

➔ 1.6 UNIVERSAL PRINCIPLES FOR MANAGING FOR QUALITY

Quality management is the management of a quality function within an enterprise. The process of identifying and administering the activities needed to achieve the customer-driven objectives of an organization must be carried out to sustain performance. A useful way to introduce the concept of quality management is to relate it to another well-known management concept, i.e., financial management.

Financial management is accomplished by the use of three managerial processes: planning, control, and improvement. Some key elements of these three processes are shown in Table 1.3. The same three processes apply to quality. The three financial processes provide a methodical approach to addressing finance; the three quality processes provide

TABLE 1.3
Financial processes

Process	Some elements
Financial planning	Budgeting
Financial control	Expense measurement
Financial improvement	Cost reduction

a methodical approach to addressing quality. Of particular importance is that all three quality processes can be further defined in a universal sequence of activities. Table 1.4 summarizes these sequences. Later chapters develop further detail for these "road maps."

The three processes of Juran's Trilogy are interrelated. Figure 1.4 shows the interrelationship applied to one of the two components of the quality definition, freedom from deficiencies. This figure is of uncommon importance. For example, note the graphic distinction between the noisy *sporadic* problem and muted *chronic* waste. The sporadic problem is detected and acted upon by the process of *control*. The chronic problem requires a different process, namely, *improvement*. Dr. Joseph M. Juran calls improvement, breakthrough improvement. The key word is "breakthrough." Dr. Juran defined this word in his 1964 book, *Managerial Breakthrough* (Juran, 1964). It was a precursor to six sigma when Motorola Inc. created six sigma from it. He described it as a means to achieve a change in business performance from historical performance. This change does not just happen. It requires a "systematic change process" focused on a systematic approach to achieve breakthroughs and a change in culture. DeFeo and Barnard (2003) identify six types of breakthroughs (leadership, organization, current performance, management, adaptability and culture) as prerequisites for achieving and sustaining improvements in business performance. Such chronic problems are traceable to an inadequate *planning* process. Later chapters develop these concepts in more detail.

The experience gained from improvement activities provides feedback ("lessons learned") for replanning activities.

The planning process is really operational planning directed at product and process planning. This activity is in contrast to strategic planning, which establishes long-term

TABLE 1.4
Universal processes for managing quality

Planning (design)	Control (compliance)	Improvement (breakthrough)
Establish the project	Choose control subjects	Prove the need
Identify customers	Establish measurement	Identify projects
Discover customer needs	Establish standards of performance	Organize project teams
Develop product		Diagnose the causes
Develop process	Measure actual performance	Provide remedies, prove that the remedies are effective
Develop process controls, transfer to operations	Compare to standards	
	Take action on the difference	Deal with resistance to change
		Control to hold the gains

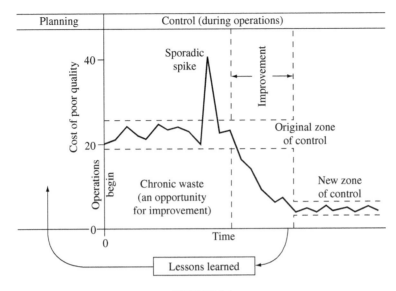

FIGURE 1.4
The Juran Trilogy diagram. (*From Juran's Quality Handbook: The Complete Guide to Performance Excellence 6e, 2010.*)

goals and the approach to meet those goals. In the context of the Juran Trilogy, design for six sigma (DMADV) and other design methodologies are for planning, whereas lean six sigma (DMAIC), Lean, and such methodologies are for improvement. Lean concepts can also be applied in the design or planning efforts. In any case, control is needed upon conclusion of either planning or improvement. This is evident from the Juran Trilogy diagram in Figure 1.4.

For the trilogy of processes to be a successful framework for achieving quality objectives, the processes must occur in an environment of inspirational leadership where the practices are strongly supportive of quality. Without such a quality "culture," the trilogy of processes cannot be fully effective. These elements have an impact on people at all levels. Again, later chapters provide additional explanation.

➔ 1.7 QUALITY DISCIPLINES AND OTHER DISCIPLINES

This book discusses the concepts, techniques, and tools essential to modern competition in quality. *Quality functions* is the term used to denote the body of quality-related knowledge. Originally, this knowledge was applied to the quality of manufacturing processes for physical goods (little q). As the scope of quality activities has expanded to all processes—and to both external and internal customers—knowledge from other disciplines has become useful. The contributions of other disciplines are sometimes unique and sometimes overlap quality disciplines. A summary of the contributions of other disciplines is shown in Table 1.5. People concerned with quality management must draw upon the contributions of all disciplines. Potential applications are identified throughout the book.

TABLE 1.5
Contributions of various disciplines

Discipline	Example of contribution
Finance	Measuring the cost of poor quality
Industrial engineering	Design of integrated systems, measurement, problem solving, work analysis
Information technology	Measurement, analysis, and reporting on quality
Market research	Competitive standing on quality; understand customer desires
Operations management	Management of integrated systems
Operations research	Analyzing product design alternatives for optimization
Organizational behavior	Understanding quality culture; making teams effective
Organizational effectiveness	Satisfying the needs of both internal and external customers
Strategic planning	Quality as a means of achieving a unique competitive advantage
Systems engineering	Translating customer needs into product features and process features
Value engineering	Analysis of essential functions needed by customer

→ 1.8 THE WAY FORWARD—DEVELOPING A SYSTEM PLAN FOR SUCCESS

As a preview of the rest of this book, Table 1.6 shows two views of quality: the conventional internal view still followed by many organizations and the modern external view that many other organizations have found imperative for survival. Stay tuned for some obvious and some more subtle aspects of the entries in this table.

In the 1980s, the emphasis on customer satisfaction and loyalty, application of quality concepts beyond manufacturing to the service and public sectors, and participation of all employees gave rise to a new title—total quality management (TQM). TQM is the system of managerial, statistical, and technological concepts and techniques to achieve quality objectives throughout an organization. It is still in use today in many companies around the world—including Toyota. This system will unfold throughout this book. Although it may go by various names, the authors believe that Enterprise Quality is a more appropriate and descriptive name.

A new approach to quality will always come about, and it will come with a new name (e.g., 5S, TQM, LSS). Inevitably, the enthusiasm for a new approach is the result of not being successful with a prior approach or of a promise of more benefits than are actually delivered, resulting in "quality" not living up to expectations. Further, other approaches will emerge as competitors, e.g., reengineering, performance improvement, six sigma.

The reality is that quality-related problems will always be present—because we cannot always predict the right products and services to meet customer needs; thus, our products and services will have errors and defects. These quality issues must be addressed.

Readers are urged to learn about all approaches to quality, choose the best ideas, and then customize the framework for their own organizations.

In this book, we refer to examples of frameworks used such as the U.S. Malcolm Baldrige National Award for Excellence, The Juran QMS, and the ISO Quality Management System to name a few.

TABLE 1.6
Two views of quality

Internal view	Customer-focused view
Compare product to specification	Compare product to competition and to the best
Get product accepted at inspection	Provide satisfaction over product life
Prevent plant and field defects	Meet customer needs on goods and services
Concentrate on manufacturing	Cover all functions
Use internal quality measures	Use customer-based quality measures
View quality as a technical issue	View quality as a business issue
Efforts coordinated by quality manager	Efforts directed by upper management

Source: Juran's Quality Handbook: The Complete Guide to Performance Excellence 6e, 2010.

SUMMARY

- Quality is customer satisfaction and loyalty.
- Quality has two components: product features and freedom from deficiencies.
- Product features affect sales income.
- Freedom from deficiencies affects costs.
- Attainment of quality requires activities in all functions of an organization.
- Traditional quality activities have concentrated on manufacturing ("little q"); modern quality activities encompass all activities ("big Q").
- All jobs have three roles: customer, processor, supplier.
- We can identify three quality processes: planning, control, improvement. Each process has a defined list of steps.
- Sporadic and chronic quality problems require different approaches.
- Quality, costs, and schedules can be mutually compatible.
- Quality management draws upon the knowledge of many other disciplines.
- Both internal and external views of quality are essential.

PROBLEMS

Note: An instructor's manual is available for instructors using this book as an assigned text.

1.1. For customer satisfaction, computer software must not only be free of errors or "bugs" but also have the necessary product features. List some important features of a word processing software package.

1.2. *Quality* has been defined as "customer satisfaction and loyalty" (or "fitness for use"). *Quality* may also be defined as "conformance to specification." In theory, creating the proper specifications and then manufacturing product that conforms to those specifications should lead to customer satisfaction. Alas, life isn't that simple. Consider these four situations for a product:

- Conforms to specifications; is competitive on fitness.
- Conforms to specifications; is not competitive on fitness.

- Does not conform to specifications; is competitive on fitness.
- Does not conform to specifications; is not competitive on fitness.

Which two of these four situations are theoretically not plausible, but in practice do occur (and cause much confusion and concern)? Can you cite any examples?

1.3. Select one functional department in a manufacturing or service organization, such as the product development department.
   (a) For the selected department, list two internal customers. For example, the manufacturing department is a likely internal customer of product development.
   (b) For each internal customer, describe in one sentence a likely requirement of that customer that the supplier department must meet.
   (c) Propose a measurement that might be used to quantify how well the requirement is being met.

1.4. For each 100 units of product manufactured, a certain process yields 85 conforming units, five that must be scrapped, and 10 that must be reprocessed. Each unit scrapped results in a $50 loss; each reprocessed unit requires an extra half hour of processing time. The resource time of producing the original 100 units is 20 hours.
   (a) Calculate the scrap cost, reprocessing time, and productivity per hour. Productivity should be calculated in terms of conforming units per hour of resource input.
   (b) A quality improvement program has recently been successfully introduced. For each 100 units manufactured, the process now yields 95 conforming units, one to be scrapped, and four for reprocessing. Repeat the calculations of paragraph (a). What are the quantitative benefits of the quality effort to costs, to delivery schedules, and to productivity?

1.5. Based on your experience, identify three examples of chronic or sporadic deficiencies that were the result of either poor work effort or poor quality planning.

REFERENCES

Brust, P. J. and F. M. Gryna (1997). *Product Quality and Macroeconomics—Five Links.* Report No. 904, College of Business Research Paper Series, University of Tampa, Tampa, FL.
DeFeo, J. A. and W. Barnard (2004). *Juran's Six Sigma Breakthrough and Beyond,* McGraw-Hill, New York.
Forsha, H. I. (1992). *The Pursuit of Quality through Change,* Quality Press, American Society for Quality, Milwaukee.
Juran, J. M., ed. (1995). *A History of Managing for Quality,* Quality Press, ASQ, Milwaukee.
Juran, J. M. (1995). *Managerial Breakthrough,* McGraw-Hill, New York.
Juran Institute (1990). *Designs for World Class Quality.* Juran Institute, Wilton, CT.
Juran, J. M. and J. A. DeFeo. (2010). *Juran's Quality Handbook, The Complete Guide to Performance Excellence,* 6e.
Link, Albert N. and J. T. Scott (2011). *Economic Evaluation of the Baldrige Performance Excellence Program.* www.nist.gov/director/planning/upload/report11-2.pdf
McCracken, M. J. and H. Kaynak (1996). "An Empirical Investigation of the Relationship between Quality and Productivity," *Quality Management Journal,* vol. 3, no. 2, pp. 36–51.
Roberts, H. V. and B. F. Sergesketter (1993). *Quality Is Personal,* Free Press, New York.

CHAPTER 2

Integrating Quality into the Enterprise Strategic Plan

→ 2.1 QUALITY AND THE STRATEGIC PLAN

Previous chapters present some specific concepts and techniques for achieving quality excellence. With that knowledge, we are now ready to take a longer range view of quality, i.e., strategies for quality. Strategies provide a road map for the future.

Strategic quality management is the process of establishing long-range customer-focused goals and defining the approach to meeting those goals. Strategic quality management becomes an integral part of the overall strategic plan of an organization and is developed, implemented, and led by upper management. Strategic quality management (this chapter) is performed at the top organizational level. Chapter 5 focuses at the product and process level with middle management.

We will examine the basic elements of strategic management and then address how the quality parameter can be integrated. The following elements provide a widely accepted framework:

- Define the mission and critical success factors.
- Study the internal and external environments, and identify the strengths, weaknesses, opportunities, and threats to the organization.
- Define a long-term, ultimate goal (a "vision").
- Develop key strategies to achieve the vision.
- Develop strategic goals (long term and short term).
- Subdivide the goals and develop operational plans and projects to achieve the goals throughout the organization and be held accountable.
- Hold executive leadership accountable to implement the strategies.
- Review progress with measurements, assessments, and audits.

Note that these elements cover both the development and the deployment of strategies.

Typically, strategy covers a five-year span in broad terms, with the first year in more detail, and with annual updating of the five-year strategy.

Thompson and Strickland (1998) is a helpful reference that elaborates on the general concept of strategic planning. *JQH6,* Section 13 on strategic deployment, discusses both development and deployment of quality strategies.

Specific approaches to strategic quality management are still evolving, but these are some necessary ingredients:

1. A focus on customer needs.
2. Continuous improvement to all processes in the organization (big Q).
3. Understanding the key customer, market, and operational conditions as input to setting strategic direction. This aspect links directly to four components of quality assessment described in Chapter 3. These components identify strengths, weaknesses, opportunities, and threats—in the language of strategic management, a "SWOT analysis." If a significant difference exists, then strategies, goals, and actions must be identified—call it a "gap analysis." See Figure 2.1 below for an example of a SWOT analysis.
4. Leadership by upper management. Quality needs to be integrated into the eight elements of strategic management identified above. This ingredient of strategic quality management includes actions to set up organizational machinery to carry out improvement, empower the workforce to make improvements, train all levels to execute their quality responsibilities, establish measures and review progress against improvement goals, provide recognition for superior performance, and expand the reward system to reflect changes in job responsibilities.

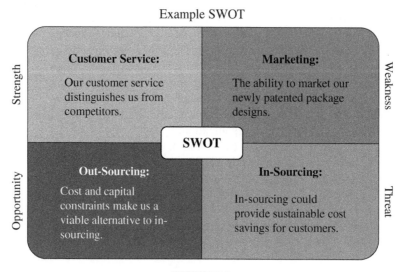

FIGURE 2.1
Example SWOT

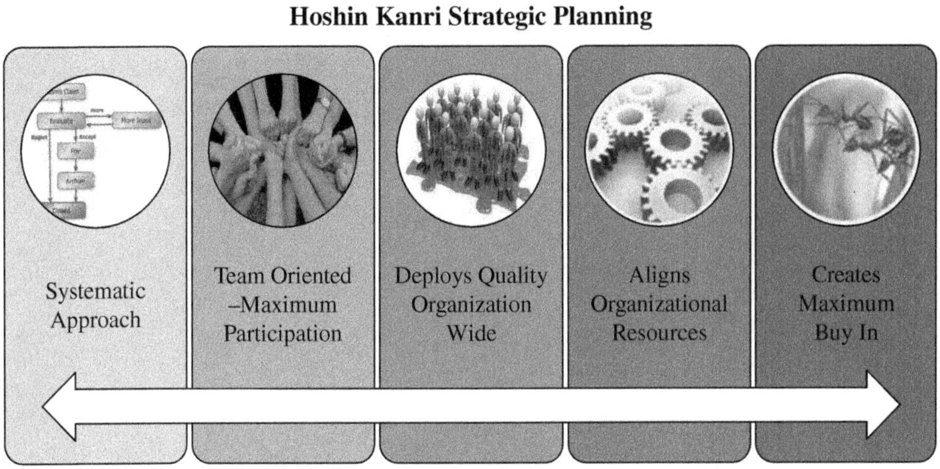

Integrating Quality into the Enterprise Strategic Plan 23

Hoshin Kanri Strategic Planning

| Systematic Approach | Team Oriented –Maximum Participation | Deploys Quality Organization Wide | Aligns Organizational Resources | Creates Maximum Buy In |

FIGURE 2.2
Hoshin Kanri Strategic Planning

5. Translation and deployment of strategies into annual business plans. This aspect requires a structured approach to align activities at three levels: organizational level, key process level, work unit/individual level. Also, emphasis must be on actions by line departments instead of relying on a quality department.
6. Adequate resources "where people have the knowledge, skills, authority, and desire to decide, act, and take responsibility for the results of their actions and for the contribution to the success of the company." Also, certain resource capabilities serve as a source of competitive advantage over rival firms (e.g., product development, marketing). These "core competencies" require special attention and support.

The Japanese use a similar approach to strategic management, called *"hoshin planning"* (or "policy deployment"), which is built around the classic management cycle of plan, execute, and audit (or plan, do, check, act). Key aspects of *hoshin* planning include a focus on the planning process, company targets known by all employees, individual initiative, self-audit, and documentation and communication (See Figure 2.2).

Next we present an overview of the strategic quality management process using four examples, and then we discuss the individual elements of the process.

➔ 2.2 A STRATEGIC PLANNING MODEL

In a classic case, General Electric conducted a full strategic planning study for one of its products. Product quality was one parameter addressed in the study (Utzig, 1980).

In this example, General Electric was the minimum-unit-cost producer, but the company was losing market share. The strategic planning study posed several questions.

TABLE 2.1
Customer-based measurements

Product attributes	Mean importance rating	Competitor performance ratings		
		GE	A	B
Reliable operation	9.7	8.1	9.3	9.1
Efficient performance	9.5	8.3	9.4	9.0
Durability/life	9.3	8.4	9.5	8.9
Easy to inspect and maintain	8.7	8.1	9.0	8.6
Easy to wire and install	8.8	8.3	9.2	8.8
Product service	8.8	8.9	9.4	9.2

1. What are the financial goals for product X?
    The short-range financial goal was a return on investment of at least 25%; the 5-year goal for cumulative net income was $120 million.
2. What is our present quality goal with respect to the competition?
    The goal was to be equal to competitors A and B.
3. What are the key quality factors that influence the purchasing decision of potential customers?
    Market research determined that the factors for product X were reliability, efficient performance, durability, ease of inspection and maintenance, ease of wiring and installation, and product service.
4. How do we compare to the competition on each of the key factors?
    The surprising results are shown in Table 2.1. Note that GE was rated lowest on all six factors. Of course, the results were not believed—until the study was done three times. Can't you just hear the criticisms of the study?
5. Does anyone have a unique competitive advantage on quality?
    Table 2.1 shows that competitor A was best on all six factors.
6. What are the internal results on quality?
    Failure costs were low and complaints were low. Thus a traditional cost-of-poor-quality study would *not* have identified the seriousness of the problem.
7. What are alternative quality goals with respect to competition?
    Several alternatives were examined by studying the benefits and costs. It was decided that the present goal (to be equal in quality to competitors A and B) was to be retained.
8. For the chosen goal, what departmental goals are needed to achieve the changes of level in the key factors?
    New departmental goals were needed in the design, manufacturing, service, and quality assurance departments.
9. What departmental plans must be developed?
    The goals were translated into specifics such as strengthening the magnetic structure to improve reliability, providing more uniform heat treatment, instituting a special training program for service technicians, and performing additional life testing on the product.
10. What resources are required?
    After the usual give-and-take of negotiation, appropriate resources were assigned.

Note that the driving force in this model (item 1) was a profit goal, not a quality goal. The competitive analysis, when viewed in light of the profit goal and regaining market share, was decisive in changing priorities and assigning the necessary resources. Business planning, involving items 8, 9, and 10, glued the quality strategy into place. This action led to a successful result.

Middle managers at Carolina Power and Light were the source of input in developing a strategy (Allen and Bailes, 1988). A series of meetings was held to identify critical issues that were obstacles to quality. More than 100 issues were identified. These were reduced to 22 issues, e.g., enhanced leadership, direction and control by senior management, and further meetings took place to develop goals to address the 22 issues. The result was 297 goals, e.g., sense of unity of purpose, visible commitment to corporate mission and beliefs. Finally, the goals were grouped into four "macrostrategies": communications, changing the culture and management style, recognition and reward, and education and training. For each of these strategies, brainstorming was employed to identify action items. Action items, providing a framework for each strategy, then led to the development of a five-year plan for quality.

Increasingly, organizations are finding that although upper management leadership is essential, the development of strategy should involve both top-down and bottom-up viewpoints.

In another service industry example, Figure 2.3 shows the approach at Kelly Services, Inc. Note how the link between strategic planning and performance management begins with internal and external customer input.

A final example comes from Xerox. Figure 2.4 depicts the major steps in going from developing strategy at the corporate level and the linkage from corporate direction to annual objectives at the division level and then to the operations level.

In the next section, we discuss how quality-related activities can be linked and integrated into the eight elements of strategic planning, resulting in a strategic quality management framework.

➔ 2.3 DEVELOPING A MISSION AND VISION

A *mission* is a statement of the organization's purpose and the scope of its operations, i.e., "the business we are in."

Some examples:

Our mission is to provide any customer a means of moving people and things up, down, and sideways over short distances with higher reliability than any similar enterprise in the world (Otis Elevator).

Our business is renting cars. Our mission is total customer satisfaction (Avis Rent-a-Car).

We exist to create, make, and market useful products and services to satisfy the needs of our customers throughout the world (Texas Instruments).

The mission of the American Red Cross is to improve the quality of human life; to enhance self-reliance and concern for others; and to help people avoid, prepare for, and cope with emergencies.

Note how all of these mission statements recognize quality.

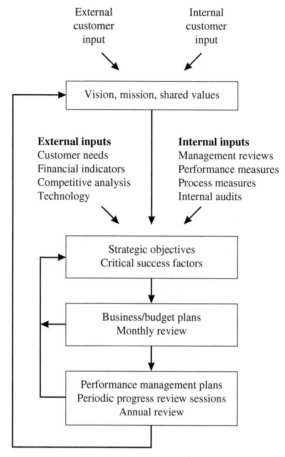

FIGURE 2.3
Kelly Services strategic planning process. (*From McCain, 1995.*)

The mission statement sets forth what an organization is doing today, but we must then examine the external and internal environments and think strategically about the future. With respect to quality, the environmental analysis should focus on four elements: cost of poor quality, market standing on quality, the current quality culture, and the current quality system. These elements, explained in Chapter 3, "Organization-wide Assessment of Quality," are essential because they identify the driving forces that must be addressed to define a vision and develop and deploy strategies.

A *vision statement* defines the desired future state of the organization. A *vision* can be viewed as the ultimate goal that may take five years or more to achieve.

Some examples:

To be the leading supplier of PCs and PC servers in all customer segments (Compaq Computers).

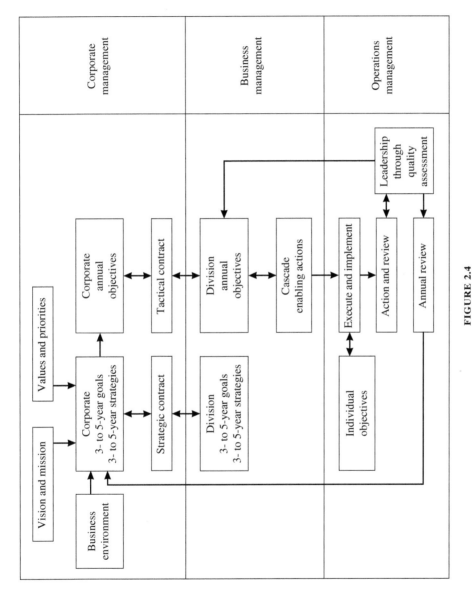

FIGURE 2.4
Xerox managing-for-results process. (*From Leo, 1994.*)

We will lead in delivering affordable, quality health care that exceeds the service and value of our customers' expectations (Kaiser-Permanente).

To engineer, produce, and market the world's finest automobiles (Cadillac Motor Car Division).

To be America's best quick service restaurant chain. We will provide each guest great tasting, healthful, reasonably priced fish, seafood, and chicken in a fast, friendly manner on every visit (Long John Silver's).

Note how all these vision statements recognize quality.

In practice, mission statements focus on "what our business is now"; vision statements emphasize "what our business will be later." Sometimes one statement covers both the present and the future.

2.4 DEVELOPING LONG-TERM STRATEGIES

A *strategy* is a guide on how to pursue the organization's mission and vision. Strategies set direction by identifying the key issues or activities that help develop specific goals and plans. Strategies must contribute significantly to the vision. In developing strategies and goals, we need the leadership of upper management and the participation of middle management. This scope provides a source of valuable ideas and also encourages ownership for implementation.

With respect to quality, there are usually a small number of strategies, say three to five. For example, the Cadillac Motor Car Division identified three strategies: a cultural change where teamwork and employee involvement are considered a competitive advantage, a focus on the customer with customer satisfaction in the master plan, and a more disciplined approach to planning that focuses all employees on quality goals.

Strategies can aim at both operational effectiveness and achieving a competitive advantage (Porter, 1996). *Operational effectiveness* means performing similar activities better than the competition does (e.g., conducting operations with less waste); *achieving a competitive advantage* means not performing the same activities as competitors perform or performing similar activities in different ways (e.g., conducting a formal, intensified product development process to generate unique product features).

Leveraging Core Competencies

The strategic planning process should recognize and leverage your organization's core competencies. Core competencies are your organization's greatest strengths and capabilities. They are of primary competitive advantage, are difficult to duplicate, and may be patented or copyrighted. Often, they are secrets of success, and the organization wants and should keep information about them from its competitors or potential competitors. See Table 2.2 on the next page for examples of core competencies.

These core competencies can be identified or developed during the strategic planning phase of the SWOT analysis. They may be flushed out during that discussion regarding strengths. The process may specifically allow a step or a specific phase to

TABLE 2.2
Examples of core competencies

- Unbridled customer service and focus
- Engineering expertise to create, design, and build highly efficient assembly lines
- Unparalleled regulatory profile and compliance
- Continuous improvement culture and process improvement capabilities to reduce costs and increase quality
- Unique patents on the reinvention of product delivery systems

identify core competencies if they have not been recognized or reviewed. The status of core competencies should be reviewed when the strategic plan is reviewed and refreshed as part of the next strategic planning cycle.

These core competencies are usually critically important to fulfilling your mission, as well as potentially vital to achieving your vision. As such, they are typically highly leveraged in the strategic plan and will have one or more strategies to enhance and further deploy them. In addition, typically multiple strategic goals as well as projects will involve each core competency.

Because your core competencies provide your organization with competitive advantage, your competitors will try to discover them. Core competencies may be unique or may be the particularly high quality of a product or service delivery, such as business or technical acumen or a particular marketplace niche or customer segment.

The distinction between organization effectiveness and competitiveness applies to developing quality strategies. Table 2.3 presents examples of some quality strategies for both operational effectiveness and competitive advantage.

The distinction reminds us that quality activities go far beyond reducing waste and the cost of poor quality. Strategies are the engine for quality success. These strategies must then be supported with goals, operational plans, and projects.

TABLE 2.3
Examples of quality strategies

Operational effectiveness	Competitive advantage
Reduce costs and cycle time through waste reduction using the six-sigma approach	Pursue customer loyalty and retention
	Formalize the product development process
Institute formal cross-functional process management	Pursue the Baldrige Award
Decentralize decision making through employees trained in problem solving	Focus on quality/price ratio
	Build a learning organization
Realign reward and recognition systems to focus on quality	Significantly engage the workforce to drive satisfaction and competency
Emphasize both incremental improvement and major breakthrough improvement	Reinvent packaging capability
Pursue ISO certification	Develop one-stop contact for all customer information and inquiry
Work jointly with suppliers to achieve quality excellence	Design and implement customer feedback mechanisms

→ 2.5 DEVELOPMENT OF SHORT-TERM ANNUAL GOALS

A *goal* is a desired result to be achieved in a specified time. ("A goal is a dream with a deadline"—Anonymous.) Goals often have different names: The overall organization goal is sometimes called a "vision"; long-range company goals (say, five years) are called "strategic goals"; short-range goals (say, one year) are "tactical goals"; goals at various levels may be called "business goals," "objectives," or "targets." The terminology simply is not standardized.

DeFeo (1995) in *JQH5* identifies seven areas in which goals are minimally required:

- Product/service performance
- Competitive performance
- Quality improvement
- Cost of poor quality improvement
- Performance of business processes
- Customer satisfaction
- Customer loyalty and retention

Some examples of corporate quality goals prepared for a health products company for the coming year are

1. Reduce quality costs for the company by _____ percent.
2. Keep material loss for the company under _____.
3. Reduce the average leakage rate for product _____ to _____.
4. Employ _____ certified quality engineers.
5. Determine quality costs for at least one product.
6. Develop and implement a specific in-process quality data analysis technique for at least one product.
7. Define numerical reliability and maintainability objectives for at least one product.
8. Implement a procedure to ensure that all corporate product specifications are reviewed by the plants before plant production planning starts.
9. Implement a procedure to ensure that all specifications for suppliers are agreed to by the supplier before the purchasing contract is finalized.
10. Develop a quality procedures manual.
11. Require the president or senior vice president to make at least _____ visits to customers to review product quality.

Other examples of overall quality goals are

- Create the Taurus/Sable car model at a level of quality that is "best in class."
- Reduce by 50% the time required to resolve customer complaints.
- Increase the percentage of research results that become incorporated into products by _____ %.

Note that these statements include a quantification in terms of a product characteristic or a date (the end of the calendar year). These cover both product characteristics and tasks in the overall company quality program. Quality goals can also be created for individual departments.

TABLE 2.4
Business risk categories

- Uncertainty in financial markets
- Project failures (design, development production, or life cycle)
- Legal liabilities
- Credit risk
- Accidents

Consideration of Risks

An extremely important area that has emerged is the area of business risk. While we typically identify weakness and threats as part of a SWOT analysis, the idea of risk and risk management is required in this "cruel new world." This involves identification, assessment, prioritization, and actions to leverage or mitigate potential risks.

Every organization is vulnerable to business risk. Business risk could be in any of the types or categories listed in Table 2.4.

Formulation of Quality Goals

Quality goals can be identified from several inputs. The most important source is the collection of four studies described in Chapter 3, "Organizationwide Assessment of Quality": cost of poor quality, market standing on quality, quality culture, and the quality system. These studies identify the strengths, weaknesses, opportunities, and threats. Other inputs to help formulate goals include

- Pareto analysis (see Chapter 4) of repetitive external measures (field failures, complaints, returns, etc.)
- Pareto analysis of repetitive internal measures (scrap, rework, sorting, 100% test, etc.)
- Proposals from key insiders—managers, supervisors, professionals, union shop stewards
- Proposals from suggestion schemes
- Surveys, interviews, focus group, and field study of users' needs, costs
- Data on performance of products versus competitors' (from users and from laboratory tests, customer surveys, and competitive benchmarking)
- Comments of key people outside the company (customers, vendors, journalists, trade journals, critics, and competitors' annual reports)
- Findings and comments of government regulators, independent laboratories, reformers, and the Internet

As in formulating policies, analysis of these inputs requires a mechanism that gives managers the opportunity to participate in setting goals without the burden of performing the detailed staff work. Quality engineers and other staff specialists are assigned the job of analyzing the available inputs and of creating any essential missing inputs. These analyses point to potential projects that are then proposed. The proposals are reviewed by managers at progressively higher organizational levels. At each level, there is a summary and consolidation until the corporate level is reached. The foregoing process is similar to that used in preparing the annual financial budget.

The process employed at a chemical plant of Union Carbide to set annual goals is shown in Figure 2.5. Note that this approach includes a vision, a strategy (long range), and tactical plans (short range).

Several alternative criteria may be used to define quality goals: historical performance, engineering analysis, and competition (see "Competitive Benchmarking"), or some absolute value (e.g., six sigma). Companies aspiring to excellence often set goals beyond those that are clearly attainable to encourage people to generate unusual approaches to achieving excellence. These are "stretch goals." Sometimes, the results can be remarkable, indeed extraordinary. All goals—but particularly stretch goals—require a strong follow-through of goal deployment and assignment of resources.

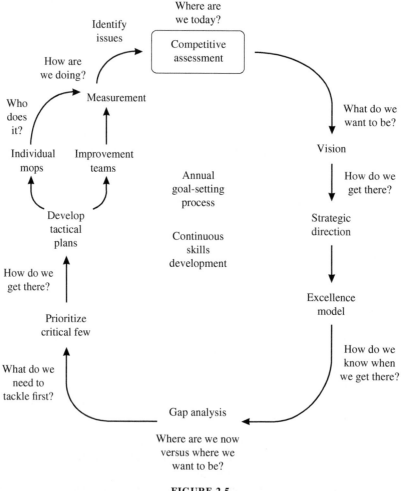

FIGURE 2.5
Annual goal-setting process. (*From Perry, 1989.*)

TABLE 2.5
Benchmarks

The specification
Customer desires
Competition
Best in our industry
Best in any industry

Finally, the nature of quality goals changes with the maturity of an organization's quality initiative, e.g., earlier goals for establishing infrastructure versus later goals that stem from quality assessments and benchmarking.

Competitive Benchmarking

Competitive benchmarking is "the continuous process of measuring products, services, and practices against the company's toughest competitors or those companies renowned as industry leaders" (Camp, 1994).

A benchmark is simply a reference point that is used as a standard of comparison for actual performance. Table 2.5 lists typical benchmarks.

Unfortunately, adherence to a specification may be woefully insufficient to generate sales. Also note that some of the other benchmarks listed in Table 2.5 go beyond competition. In this context, *competition* means other organizations that compete to sell, e.g., office reproduction equipment. But a benchmark organization may be "best in our industry" or "best in any industry." For example, the Xerox Corporation used IBM and Kodak (direct competitors on some products) as benchmark organizations to evaluate many Xerox operations. But for warehousing and distribution activities, Xerox chose the L.L. Bean Company, a catalog sales distributor of clothing and other consumer products, as a benchmark. Benchmarks serve not only as a standard of comparison but also as a means of self-evaluation and subsequent improvement. The concept of searching for the best performer in *any* industry is a valuable contribution of the benchmarking approach.

The benchmarking process applies to subjects such as products, customer services, and internal processes. The steps in the process are

1. Identify the benchmark subjects.
2. Identify benchmark partners (organizations that will serve as benchmarks).
3. Determine the data collection method and collect the data.
4. Determine the competitive gap (comparing our company to the benchmark partners).
5. Project the future performance of the industry and our company.
6. Communicate the results.
7. Establish functional goals.
8. Develop action plans.
9. Implement plans and monitor results.
10. Recalibrate the benchmarks (repeat the benchmarking process, typically every three to five years depending on the subject).

Competitive benchmarking is an essential input to the formulation of goals. But benchmarking also provides sources of ideas for improvement along with the evidence that better practices not only exist but also must be instituted to be competitive. For a complete discussion on benchmarking, see *JQH6,* Chapter 15. Camp (1998) presents case studies of benchmarking in the manufacturing, service, nonprofit, government, and education sectors.

➔ 2.6 DEPLOYMENT OF GOALS—HOSHIN KANRI

Broad goals do not lead directly to results; they must first be "deployed." *Deployment* means subdividing (aligning) the goals and allocating the goals to lower levels for conversion into operational plans and projects (see Figure 2.6 and Table 2.6). Thus goals must be deployed from the organizational level to the process level and to individual jobs.

For example, an organization had a strategy to be "known as the provider of the most responsive customer service." To support this strategy, four goals were set (Juran Institute, 1992):

1. Reduce order lead times to two days by 1993.
2. Reduce by 50% the time interval to reply to customer complaints.
3. Decrease by 90% customer application development times.
4. Cut contract proposal times by 80%.

Note that each goal is specific, observable, and measurable. Specific projects were then set up to achieve these goals.

One hospital methodically defines the linkages among elements of its strategic planning process. The hospital has five long-term goals, e.g., improve customer success

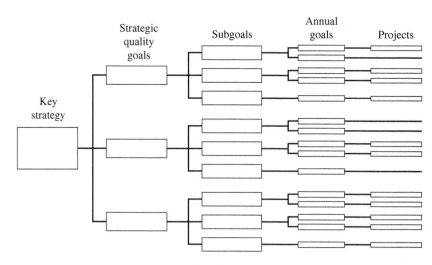

FIGURE 2.6
Deploying the vision. (*Juran Institute, Inc.*)

TABLE 2.6
Strategic Plan Projects with Strategy

- **Value-Driven Leadership**
  - "Develop and implement a proactive associate communication strategy."
  - "Develop and implement a formal succession planning program."
- **Surpass Customer Expectations**
  - "Develop compliance packaging strategy."
- **Superior Technology and Innovation**
  - "Create metrics tree that links all KPI's to departments."
- **World-Class Workforce**
  - "Benchmark, develop, and implement a workforce and workplace engagement strategy."
- **Process and Quality Management**
  - "Develop and implement the visual factory concept to enhance communication and visibility."
  - "Create a maintenance excellence benchmarking system."
  - "Collaborate to develop core competencies and intellectual property vision and approach."

and satisfaction. Each goal is linked (in a matrix) to specific strategic objectives, key performance measures, annual targets, and 11 core processes, e.g., deliver services. The impact of a goal on a core process is further defined as high, medium, or low. Management "champions" then work with process owners to identify and pursue projects that will achieve the performance targets. This cascading process of deploying strategy and goals to lower levels is called *hoshin kanri* in Japanese. Translated it means policy (strategic) deployment.

Here is a key point: vision, strategies, and goals provide direction, but specific projects and other forms of action supply the methods to achieve results. Sometimes the number of projects can be dramatic, e.g., for General Electric's six sigma goal, the number of projects grew from 200 in 1995 to about 47,000 in 1999 (Slater, 1999). For the Ford Taurus/Sable goal of becoming best in its class, more than 400 subgoals were defined, which led to more than 1500 project teams.

Communication and Accountability

Two-way communication and participation during the strategic plan development process is critical between upper management downward (top down) as well as upward (bottom up). Even more important is the two-way communication that takes place during the execution phases of the strategic plan.

One of the most important contributions of a strategic plan is the alignment that must be communicated and achieved throughout the entire organization. Two things are critical. First, upper management—with the help of staff support resources—must get the future direction right. This means the vision, strategies, and strategic goals along with projects and project goals need to be the right ones. But equally important is the alignment of the staff throughout the organization to execute in the direction that they have been given. This requires extremely good communication throughout the organization of the directions, expectations of results and measures, accountability, review and feedback, and, ultimately, reward and recognition. If the organization gets this right, it should be highly successful.

→ 2.7 LEADERSHIP TO IMPLEMENT THE STRATEGIES

The most important factor in implementing quality strategies is the personal leadership of upper management. The organizational mechanism used is the quality council (see Chapter 8 under "Role of Upper Management").

Although upper management leadership is essential, the development of strategy and plans should involve both top-down and bottom-up viewpoints. This process requires two-way communication between top management and lower levels—the Japanese call this concept "catch ball." The key elements are

- Communication of what top management proposes as the key focus areas for strategic planning.
- Nominations by managers at lower levels of additional areas for attention.
- Decisions on strategies, goals, and resources for deployment.

At Fannie Mae, a financial services company, senior executives drafted the original vision, mission, and key strategies. Then directors and middle-level managers provided comments, which led to a draft of strategic goals. Next more than 600 managers and supervisors contributed ideas for deploying the goals to subgoals and projects. Senior management then created the final version of the key strategies and goals. Note the similarity to the approach employed at Carolina Power and Light (see Section 2.2).

To implement the strategies successfully requires upper management to provide an infrastructure for quality. For example,

- Build an organization with the necessary competencies and assign sufficient resources to the activities critical to the strategies.
- Emphasize continuous improvement.
- Install operating and information systems to support personnel, including quality policies and procedures.
- Create a positive quality culture, including rewards and recognition for achievement of superior performance.
- Exert personal leadership to drive implementation of the strategies.

This book explains these elements. One tool in providing direction is the concept of quality policies.

Quality Policies or Values

A *policy* is a broad guide to action. It is a statement of principles or values. A policy or value differs from a procedure, which details how a given activity is to be accomplished. Thus a quality policy might state that quality costs will be measured. The corresponding procedure describes how the costs are to be measured.

Policies or values do not have to be vague. They can be specific enough to provide useful guidance. Here are two dramatic examples from different computer manufacturers:

> A new product must perform better than the product it replaces and better than the competition's, and this must be the case at the time of the first regular customer shipment.

In selecting suppliers, decision makers are responsible for choosing the best source even if this means internal sources are not selected. (This policy rescinded a previous policy that placed a priority on buying from "sister divisions.")

The following corporate quality policies or values were prepared for discussion at a health products company.

1. At both the corporate and plant levels, the quality control department shall be independent of the production function.
2. The company shall place a new product on the market only if the overall quality is superior to the competition.
3. All tasks necessary to achieve superior quality shall be taken, but each task shall be evaluated to ensure that the investment has a tangible effect on quality.
4. Specific quality responsibilities for all company areas, including top management, shall be defined in writing.
5. Quality activities shall emphasize the prevention of quality problems rather than only detection and correction.
6. Quality and reliability shall be defined and measured in quantitative terms.
7. All quality parameters and tests shall reflect customer needs, use conditions, and regulatory requirements.
8. Total company costs for achieving quality objectives shall be periodically measured.
9. Technical assistance shall be provided to suppliers to improve their quality control programs.
10. Each quality task responsibility defined for a functional department shall have a written procedure describing how to perform the task.
11. The company shall propose to regulating agencies or other organizations any additions or changes to industry practice that will ensure minimum acceptable quality of products.
12. Each year, quality objectives shall be defined for corporate, division, and plant activities and shall include both product objectives and objectives for tasks in the company quality program.
13. All levels of management shall have defined quality motivation programs for the employees in their department.

These policies or values were prepared to provide guidelines for (1) planning the overall quality program and (2) defining the action to be taken when personnel request guidance.

Policies may also be needed within a functional department. Policies for use within a quality department might include the following statements:

1. The amount of inspection of incoming parts and materials shall be based on criticality and a quantitative analysis of supplier history.
2. The evaluation of new products for release to production shall include an analysis of data for compliance to performance requirements and shall also include an evaluation of overall fitness for use, including reliability, maintainability, and ease of user operation.
3. The evaluation of new products for compliance with performance requirements shall be made to defined numerical limits of performance.

4. Suppliers shall be supplied with a written statement of all quality requirements before a contract is signed.
5. Burn-in testing is not a cost-effective method of eliminating failures and shall be used only on the first units of a new product type (or a major modification of an existing product) to get rapid knowledge of problems.

Note that these examples of policies or values state (1) a principle to be followed or (2) *what* is to be done but not *how* it is to be done. The how is described in a procedure. Often it is best to have a policy instead of a procedure to provide flexibility for different situations.

A sensitive policy issue may arise as the result of improvement projects that reduce rework or reprocessing to correct errors. The people who have been doing the rework wonder, what will happen to me if this work is no longer necessary? This apprehension should be faced head-on and a policy formulated. Several alternatives are possible:

- Guarantee that no employee will lose employment as a result of the quality effort. A few companies have issued such a policy statement.
- Rely on resignations and retirements as a source of new jobs for those whose jobs have been eliminated. Retrain affected workers to qualify them for the new jobs.
- Reassign affected employees to other areas. This approach can include creating positions for additional quality improvement work.
- Offer early retirement.
- If all else fails, offer termination assistance to help workers locate jobs in other companies.

Planning for this situation in advance can often lead to creative solutions; failure to plan in advance often leads to a minefield of problems, including serious disruptions of the lives of people. A compassionate approach is clearly a hallmark of true leadership.

It should be noted here that the values or policies that we articulate and currently practice are hollow if upper management does not practice them, hold others accountable, and reward and recognize for adherence.

Capacity and Capability Requirements

An important part of the strategic planning process is analyzing the strategic needs for staff capacity and capability. *Capacity* references the amount of work that individuals can be expected to accomplish because the strategic plan usually identifies business growth opportunities (for products, services, customers, markets, geography, etc.).

This means that staff resources must grow accordingly. Not only must they grow to grow the business, they need to be in place before the business can grow. This means recruiting, promoting, and developing the right number of people at the right time to coincide with the ramp-up of the expected sales or market growth.

Capability refers to the appropriate skills that the staff must have to be able to manufacture, deliver, service, sell, manage, support, and maintain the anticipated growth of the business. In other words, the right number of staff (capacity) needs to be in place at the right time along with the required skills (capability) in order for the strategic plan to be successful.

Therefore, as decisions are being made regarding strategies, strategic goals, projects and project goals, staff capacity, and capability, needs must be a clear part of the documentation, process, and results expected to successfully implement the strategic plan.

➔ 2.8 REVIEW PROGRESS ASSESSMENTS, AUDITS, AND A BALANCED SCORECARD

Once goals have been set and deployed into subgoals, business plans, and projects, then key measurements must be established. Various names apply, e.g., key results indicators, performance indicators. Readers are urged to review the basic elements of measurement—see Chapter 6 under "Measurement" and particularly Figure 6.2.

At the strategic level, measurements should be developed for each strategic goal defined in the strategic plan. The measurements usually include areas such as product performance, competitive performance, quality improvement, cost of poor quality, performance of business processes, customer satisfaction, and customer loyalty and retention (see Section 2.5, "Development of Goals"). Luther (1993) describes how Corning Inc. uses more than 200 key results indicators to integrate quality and business objectives.

Some organizations combine their measurements from financial, customer, internal processes, and learning and growth areas into a "balanced scorecard." Figure 2.7 illustrates the concept. The scorecard is used in conjunction with four management processes: translating the vision, communicating and linking strategy to departmental and individual objectives, integrating business and financial plans, and modifying strategies to reflect real-time learning.

For a discussion of broad assessments and audits, see Chapter 3, "Organization-wide Assessment of Quality."

➔ 2.9 REVIEW AND REFRESH THE PLAN

Most organizations refresh or substantially review and update their strategic plan annually or every other year. While most strategic plans are for a three- to five-year period, the rate of change in this world and almost every industry continues to accelerate every year.

This makes the ability to see clearly three to five years out almost impossible. Therefore, upper management must look to see if the plan is on track and makes sense given the changing, visible horizon. This allows for necessary changes along the way rather than waiting three years to find out the organization should have changed direction three years earlier or stopped the effort completely.

Most members of upper management also review progress on strategies and projects quarterly or semi-annually. This involves reviewing both project and strategy

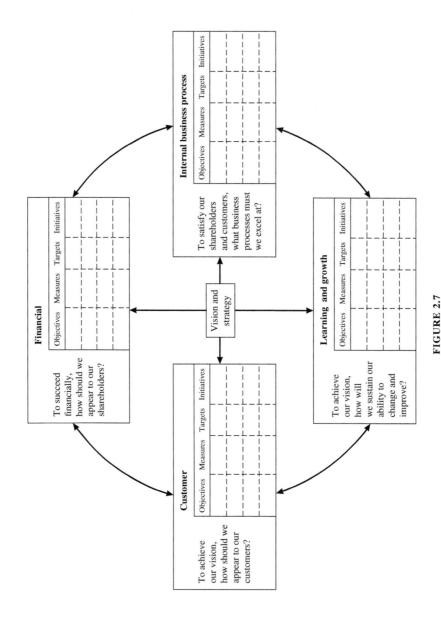

FIGURE 2.7

Translating vision and strategy: four perspectives. From *The Balanced Scorecard* by R. S. Kaplan and D. P. Norton. Boston, MA, 1996. Copyright © 1996 by the Harvard Business School Publishing Corporation; all rights reserved. Reprinted by permission of Harvard Business School Press.

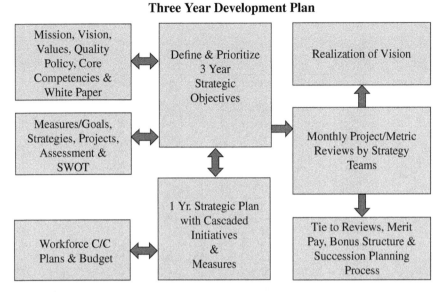

FIGURE 2.8
Three-year development plan

accomplishments, as well as measureable results, against preset targets either by upper management or lower levels of management as projects are assigned and delegated.

Of course, upper management can then change course and provide more or fewer resources. Most important is providing feedback that is motivational to both the strategy and project leaders and their teams. During the year, this feedback should propel enthusiasm and energy around achieving the goals. Upon completion, it is also important that these strategic contributions should be recognized in performance evaluations, merit and promotional decisions, and (if possible) bonus and special award programs, which will provide for even more enthusiasm, commitment, and motivation for future strategic plan assignments.

➔ 2.10 OBSTACLES TO ACHIEVING GOALS

The reasons for failure are many, but eight stand out as important:

1. *Lack of leadership by upper management.* Without this leadership, quality success is likely to be doomed. Some in upper management may be "committed," but lack of *evidence* of this commitment has a damaging result on the rest of the organization. Chapter 8, "Organization Roles to Support a Quality Culture," discusses the leadership role of upper management.
2. *Lack of an infrastructure for quality.* With other major activities, management has successfully delegated responsibility but only after evolving mechanisms that include clear goals, plans, organizational mechanisms for carrying out the plans,

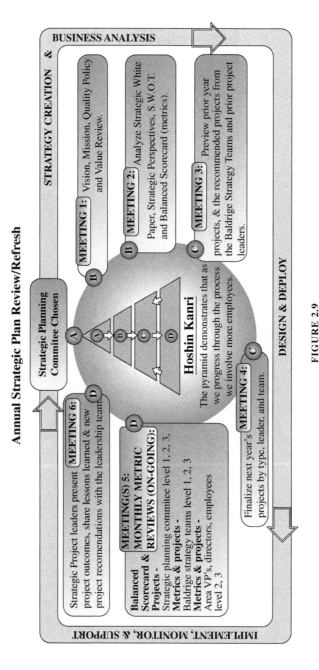

FIGURE 2.9
Annual strategic plan review/refresh

budgets, and provision for recognition and rewards. In contrast, these elements are usually vague or are missing with respect to quality.
3. *Failure to understand the skepticism about the "new quality program."* Many people have seen previous programs on quality quietly sink into oblivion. Unfortunately, the skepticism is not vocalized. The result is that management fails to present a convincing case of (a) "proof of the need" for the quality effort and (b) their determination to make the effort a success.
4. *An assumption by management that the exhortation approach will work.* This approach involves a strong presentation on proof of the need (i.e., convincing everyone about the seriousness of the quality problem) coupled with relying principally on motivational techniques to inspire everyone to do better.
5. *Lack of accountability, reward, and recognition for completing assigned responsibility.* In conjunction with the preceding points, or by itself, this lack of oversight and review—which recognizes good as well as insufficient contributions during and after expected results—will limit motivation and enthusiasm both currently and for future cycles.
6. *Failure to "start small" and learn from pilot activities.* Sometimes, in haste to achieve sizable results rapidly, the small pilot phase is omitted; instead, massive training takes place with the expectation that the troops can then simultaneously advance on all fronts. This approach doesn't work. A much better alternative uses a small number of pilot projects, with the scope of each project carefully defined so that completion is likely within six months. Perhaps the most common error in quality projects is failure to limit their scope to a digestible bite. People quickly grow tired of projects that seem to take forever.
7. *Reliance on specific techniques as the primary means of achieving quality goals.* Examples of such techniques are statistical process control, quality cost, quality circles, quality function deployment, etc. These techniques are valuable and often necessary, but they address only specific parts of the problem.
8. *Underestimating the time and resources required.* About 10% of the time of upper and middle management and professional specialists is required to achieve breakthroughs in quality. Typically, this time must be found without adding personnel. Thus priorities must be changed, i.e., other activities must be delayed or eliminated.

Organizations embarking on a significant new effort on quality are well advised to seek out the advice of other organizations and particularly to learn about their reasons for failure.

SUMMARY

- Strategic quality management is the process of establishing long-range customer-focused goals and defining the approach to meeting those goals.
- Upper management develops and implements strategic quality management.
- Approaches to strategic quality management have some common elements: a focus on customer needs, upper management leadership, integrating quality strategies into business plans, and implementation by line departments.
- A policy is a broad guide to action.

- A goal is a statement of the desired result to be achieved within a specified time.
- Competitive benchmarking identifies key quality characteristics and uses other leading organizations as reference points to develop goals and strategies.
- We "deploy" quality goals by dividing them into specific deeds, allocating responsibility, and providing resources.
- Implementing a quality strategy involves five phases: decide, prepare, start, expand, and integrate.

PROBLEMS

2.1. Select one improvement program that you have observed in an organization. The program of improvement may focus on quality, safety, absenteeism, costs, or other matters. With respect to implementation of the program, what were its strengths? What were the weaknesses in implementation? What recommendations would you make to implement an improvement program in the future?

2.2. Propose a list of five quality policies for a specific organization, e.g., your company, a supermarket.

2.3. Propose a list of five quality goals for a specific organization, e.g., your company, a supermarket.

2.4. Many organizations have a general statement called a "quality policy." This statement is often vague, e.g., the organization will supply the customer with "high quality." Obtain an example of such a statement and explain how it could be made more specific.

2.5. Refer to the list of 13 quality policies under "Quality or Values" in Section 2.7. Comment on the extent to which each policy applies to a specific organization.

2.6. A bank requires its branches to develop an annual quality plan that deploys corporate strategy into quality goals for five subject areas. One area is human resource utilization. Propose four goals to support this area.

2.7. Four months ago, an organization decided that it needed a new approach to quality. One of the steps taken was the establishment of quality goals. As part of a "50% in 5" focus, each department now has a goal of reducing the cost of poor quality by 50% in 5 years. Comment on this approach.

2.8. For an organization with which you are familiar, define one strategic quality goal. For that goal, propose one specific quality project (improvement, planning, or control). For that project, prepare a brief mission statement and a list of the functions that should be included on the project team.

REFERENCES

Allen, R. L. and C. V. Bailes (1988). "Managing the Startup of a Corporate Quality Improvement Effort—Translating Corporate Strategies into Field Operations," *Impro Conference Proceedings*, Juran Institute, Inc., Wilton, CT, pp. 6A-13 to 6A-18.

Camp, R. C. (1994). *Business Process Benchmarking: Finding and Implementing Best Practices,* Quality Press, ASQ, Milwaukee; Quality Resources, White Plains, NY.

Camp, R. C., ed. (1998). *Global Cases in Benchmarking: Best Practices from Organizations around the World,* ASQ Quality Press, Milwaukee.

DeFeo, J. A. (1995). *Juran's Quality Handbook, 5th Edition,* McGraw-Hill, New York.

Juran Institute, Inc. (1992). *Strategic Quality Planning,* Wilton, CT.

Kaplan, R. S. and D. P. Norton (1996). *The Balanced Scorecard,* Harvard Business School Press, Boston.

Leo, R. J. (1994). "A Corporate Business Excellence Process," *Impro Conference Proceedings,* Juran Institute, Inc., Wilton, CT, pp. 6A.1-1 to 6A.1-13.

Luther, D. B. (1993). "Advanced TQM: Measurements, Missteps, and Progress through Key Result Indicators at Corning," *National Productivity Review,* Winter, pp. 23–36.

McCain, C. (1995). "Successfully Solving the Quality Puzzle in a Service Company," *Impro Conference Proceedings,* Juran Institute, Inc., Wilton, CT, pp. 6A.1-1 to 6A.1-13.

Perry, A. C. (1989). "From Teams to Total Quality," *Impro Conference Proceedings,* Juran Institute, Inc., Wilton, CT, pp. 3A-7 to 3A-16.

Porter, M. E. (1996). "What Is Strategy?" *Harvard Business Review,* November–December, pp. 61–78.

Slater, R. (1999). *Jack Welch and the GE Way,* McGraw-Hill, New York.

Thompson, A. A. Jr. and A. J. Strickland III (1998). *Strategic Management,* 10th ed., McGraw-Hill, New York.

Utzig, L. (1980). "Quality Reputation—A Precious Asset," *ASQ Technical Conference Transactions,* Milwaukee, pp. 145–154.

CHAPTER 3

Organizationwide Assessment of Quality

→ 3.1 WHY ASSESS PERFORMANCE?

An early step to transform with a strategy of quality is to understand the baseline of your culture, system, processes, people, and costs. A key role of the quality office is to conduct these assessments to educate your leaders so they understand what needs to be changed or improved. System assessments provide a comprehensive, cost-effective review that provides an objective baseline to assess the performance of organizations. There are many organizational assessments that can be used. We will examine six areas are important to being competitive in quality:

1. Assessing quality risk
2. Estimating the costs of poor quality
3. Assessing performance and standing in the marketplace
4. Assessing using national performance standards and awards
5. Assessing to the international system standards
6. Competitive benchmarking best practices

→ 3.2 QUALITY RISK ASSESSMENT

From 1970 to 1986, the Juran Institute was one of the first to document and conduct comprehensive assessments for quality. These assessments were based on Dr. Juran's early writings and the need for improved competitiveness in many industries in the United States. The Institute developed a set of guidelines that, if met, were and still are critical for business success and leadership in quality.

    The Juran comprehensive assessment of quality existed prior to the ISO 9000 standards and the national quality awards that came later in the 1980s. Today, the

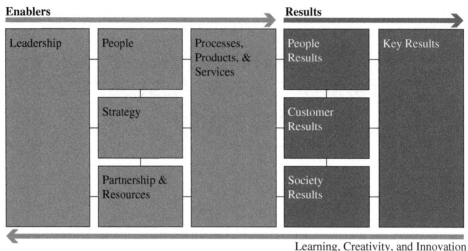

FIGURE 3.1
EFFQM excellence model.

Juran Quality Risk Assessment has been used by many organizations to better understand and create a plan to improve performance. For some organizations, the ISO Standards are used; others use regulatory requirements such as FDA standards or Sarbanes-Oxley; and still others use the U.S. Malcolm Baldrige Criteria for Performance Excellence, the EFQM Excellence Model, or their own country equivalent.

Quantifying the costs related to poorly performing quality (COP^3; DeFeo, 2003) is important to understand the total loss due to poor quality. Estimating the COP^3 is one of the best means to get management's attention. As you will see, the COP^3 for most organizations can exceed the value of their profits!

Assessing market standing and the acceptance of your quality from your customers is important. Knowing what the paying customers like and dislike can be beneficial in driving the organization to a new level of performance.

Each of these assessments brings with it a business model for integrating quality into the organization. One example is the EFQM model shown in Figure 3.1.

The Baldrige Model was named after a former Secretary of Commerce, Malcolm Baldrige. This model incorporates best practices from many companies since its inception in 1980s. It is used to measure and evaluate the total business system against a set of competencies and best practices (see Figure 3.2).

3.3 STRATEGIC ALIGNMENT, DEPLOYMENT, AND ASSESSMENT

We use the term *business system* to assess all parts of the organization rather than the quality system infrastructure. The quality system must be integrated into the business system. An effective assessment must analyze the organizational structure, roles and responsibilities, processes, and resources for implementing excellent performance throughout the organization.

FIGURE 3.2
2013 Baldrige model.
From Baldrige Performance Excellence Program, *2013, 2013–2014 Criteria for Performance Excellence* (Gaithersburg, MD: U.S. Department of Commerce, National Institute of Standards and Technology, http://www.nist.gov/baldrige/publications/business_nonprofit_criteria.cfm).

In order to properly assess any organization, start at the top and identify the organization's strategic plan and how it is associated with day-to-day activities. It is crucial that top-level management be involved when assessing these areas. Many organizations are negatively affected by not having aligned their strategic objectives and goals to specific daily activities and projects. The leadership of the organization needs to identify if their mission, vision, values, and policies are all aligned and communicated effectively, keeping the end-customer in mind (see Figure 3.3).

Selected definitions	
Mission	What business we are in
Vision	Desired future state of organization
Values	Principles to be observed to meet vision or Principle to be served by meeting vision
Policy	How we will operate and our commitment to customers and society

FIGURE 3.3
Organizational vision and mission.
Source: Juran Institute Inc., Southbury, CT.

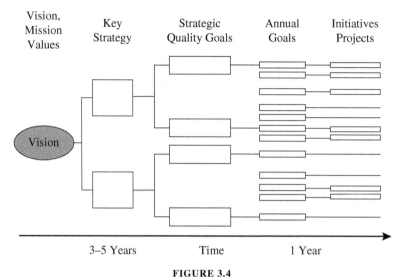

FIGURE 3.4
Deploying the vision.
Source: Juran Institute Inc., Southbury, CT.

From these objectives, strategies and annual goals are developed. *Strategies* are a means of achieving the vision of the company. Strategies are typically long term in nature and associated with key success factors such as increase revenue, reduce costs, or develop new products. In order to achieve these long-term goals, the leadership teams create *annual goals* that are short term in nature. As reassurance, the leaders develop policies that help manage the goals on day-to-day activities.

Figure 3.4 shows how the vision breaks down into key strategies, strategic goals, annual goals, and initiatives and projects. Typically, there many different levels of management from the executive group; when deploying the strategic plan, a lot of information and direction gets lost in translation. This leads to the strategic plan not being aligned with individual projects, which could lead to projects of low importance and very important resources being wasted. The organizational assessment will identify those areas that are disjointed to realign them.

The assessment approach described in this chapter provides a few options that can be used to stimulate upper management to take action on quality (that's why assessment is presented early in this book). This factual information not only inspires immediate action on quality, but also provides a solid basis for developing long-range strategies for quality.

We begin with a comprehensive assessment based on the *Juran Organization Health Check*. We use the term *quality risk assessment* to describe an organization-wide review of the status of quality. The quality risk assessment has proven to be efficient and effective. Other assessments, such as the ISO 9000 standards, are at times seen as a paper chase and do not look at business results. The National Award Criteria are comprehensive and do include business results, but their main use is to win the award. Therefore, organizations should choose what is most important when developing their own system and the assessment of that system.

An effective quality risk assessment should include three areas or three criteria for success:

1. **Approach:** What are the organizational approach, intent, and design, used within the organization?
2. **Deployment:** How broadly has the approach been deployed or executed in the organization? In other words, how many employees are knowledgeable and have been trained, understand, and can implement the approach?
3. **Results:** What are the measurable outcomes that demonstrate the approach and deployment are valid?

An effective assessment is conducted by observation, data collection, and interviews of key functions within the organization. Specific strengths and weaknesses are documented at the detailed dimension level. Alignment and themes are identified across each category and are then scored. The Juran Organization Health Check categories and points are shown in Table 3.1.

The results of the quality risk assessment must provide an organization with a review of the organization's performance, in terms of strengths, weaknesses, opportunities, and threats (SWOT) and become a formal part of the SWOT analysis for overall operations of an organization. Other elements may be added as circumstances require. An annual or biannual assessment is usually warranted. This assessment can be conducted for an entire organization or for a division, a plant, a department, or a process.

We begin to discuss how crucial an assessment is to ensure that the strategic plan is aligned and properly deployed. Later we discuss the different assessments that are used to drill down deeper into specific categories such as cost of poor quality, organization culture, and market standing.

3.4 PLAN THE ASSESSMENT

The assessment is an evaluation of current element activities in the organization. Such an evaluation could cover a wide range of both scope (global to a few activities) and examination (cursory to detailed). Our discussion concentrates on an organizationwide examination in sufficient depth to meet the needs of the overall quality assessment.

Often, an organization starts by having an outsider conduct a three- to five-day assessment of the current baseline of performance. The assessment is structured by initiating the assessment, evaluating the organization's elements, and preparing a detailed report. Proper planning will ensure that the assessment is efficient and completed quickly to allow leadership to make timely decisions. Planning requires that the organization:

1. Define the team.
2. Define objectives.
3. Define scope and criteria.
4. Establish a communication plan between the team and leadership.
5. Complete the pre-assessment (generalized version of OHC).

TABLE 3.1
Juran's organization quality and risk check scoring criteria

Categories and items (point values)

1. Strategic alignment and deployment (100)
    1.1 Strategic planning (30)
    1.2 Strategic deployment (45)
    1.3 Stakeholder (executive point of view) (15)
    1.4 Key support systems (10)
2. Quality management system (100)
    2.1 Quality system (40)
    2.2 Quality control (10)
    2.3 Continuous improvement (50)
3. Measurement and analysis capabilities (75)
    3.1 Measurement systems (25)
    3.2 Customer relationship and requirements (25)
    3.3 Competitors (5)
    3.4 Product and service quality (5)
    3.5 Support processes (5)
4. Effectiveness and efficiency of business processes (100)
    4.1 Core business processes (30)
    4.2 Voice of the customer (20)
    4.3 Product and service creation (20)
    4.4 Service and order fulfillment (10)
    4.5 Key support processes (10)
    4.6 Process efficiency and cost (10)
5. Employee engagement and culture (100)
    5.1 Human resource structure (30)
    5.2 Talent development (40)
    5.3 Culture (30)
6. Supplier management (75)
    6.1 Supply chain policy (25)
    6.2 Supplier procedures (50)
7. Scorecard and results (450)
    7.1 Customer (80)
    7.2 Product and service (80)
    7.3 Culture (80)
    7.4 Supplier performance (80)
    7.5 Financial results (130)

Total points = 1,000

Once the planning is complete, the assessment can begin. To collect the right data, there may be one-on-one discussions with selected employees typically from all levels of the organization within the scope of the assessment. These employees will validate and demonstrate the existence and effectiveness of the total business system and subsequent critical business processes.

Examples of items targeted during task observations, focus groups, ad-hoc interviews, and review of operational results are as follows:

- Is your organization's quality management system responsible for creating a focus on the customer and a quality foundation?
- Is the quality management system aligned to the organization's strategies, goals, and actions for creating a quality culture?

- Are improvement methodologies used to effectively identify projects and diagnose root causes of problems within the organization?
- Are there measurement systems in place to improve the overall performance of your organization?
- Does your organization have an effective process for measuring customer satisfaction and loyalty?
- Does your organization perform cost of poor quality? Have its standing in the marketplace and the quality culture been assessed?
- Are your key business processes and support process identified, evaluated, measured, and continuously improved?
- Does your organization incorporate customer requirements into the key business processes?
- Are the organization's workforce policies known and implemented?
- Is there an effective evaluation process in place to measure employees' capabilities?
- Does the workforce have control over the work processes?
- Are your suppliers certified and continuously verified to maintain performance?
- Are the trends in key measures for your product and service quality performance trending positively over the past five years?
- Is the financial impact of the results communicated, and does the impact trigger improvements where necessary?
- Are your results from improvement activities sustained over time?

As the assessment is completed, the observations, data collected, and information received from interviews are documented and measured against the quality risk assessment scoring criteria described in Table 3.1 and the level of approach, deployment, and results.

For example, if an organization scores 20% of the total points, there may be no systematic approach and an undocumented deployment, and there may be not data reported or poor results are reported. Additionally, if the organization scores 95% of the total points, the organization will probably have a strong, integrated prevention and fact-based system in place in all areas. There are continuous evaluation and improvement cycles with breakthrough refinements. The approach is fully deployed with no gaps, and the results are being sustained. Juran considers performance at this level "Best in Class" or "World-Class Quality." Typically, no organization can ever score a 100% because it will always be continuously improving the seven elements.

✧ 3.5 COST OF POOR QUALITY

During the 1950s, the concept of "quality costs" emerged. Different people assigned different meanings to the term. Some people equated quality costs with the costs of *attaining* quality; some people equated the term with the extra costs incurred because of poor quality. This book focuses on the cost of poor quality because this component of assessment is important in reducing costs and customer dissatisfaction.

The *cost of poor quality* is the annual monetary loss of products and processes that are not achieving their quality objectives. In DeFeo and Barnard (2003), the cost of poor quality (COPQ) is appropriately renamed the cost of poorly performing processes (or COP³, pronounced, C.O.P. cubed). This is to emphasize the fact that the cost of poor quality is not limited to quality, but is essentially the cost of waste associated with poor performance of processes.

Companies estimate the cost of poor quality for several reasons:

1. Quantifying the size of the quality problem in the language of money improves communication between middle managers and upper managers. In some companies, the need to improve communication on quality-related matters has become a major objective for embarking on a study of the costs of poor quality. Some managers say, "We don't need to spend time to translate the defects into dollars. We realize that quality is important, and we already know what the major problems are." But typically, when a study is done, these managers are surprised by two results. First, the quality costs turn out to be much higher than the managers had thought—in many industries in excess of 20% of sales. Second, though the distribution of the quality costs confirms some of the known problem areas, it also reveals other problem areas that were not previously recognized.
2. Major opportunities for cost reduction can be identified. Costs of poor quality do not exist as a homogeneous mass. Instead, they are the total of specific segments, each traceable to some specific cause. These segments are unequal in size, and relatively few of the segments account for the bulk of the costs. A major by-product of this evaluation is the identification of these vital few segments.
3. Opportunities for reducing customer dissatisfaction and associated threats to product salability can be identified. Some costs of poor quality are the result of product failures after a sale. In part, these costs are paid by the manufacturer in the form of warranty charges, claims, etc. But whether or not the costs are paid by the manufacturer, the failures add to customers' costs because of downtime and other forms of disturbance. Analysis of the manufacturer's costs, supplemented by market research into customers' costs of poor quality, can identify the vital few areas of high costs. These areas then lead to problem identification.
4. Measuring this cost provides a means of evaluating the progress of quality improvement activities and spotlighting obstacles to improvements.
5. Knowing the cost of poor quality (and the three other assessment elements) leads to the development of a strategic quality plan that is consistent with overall organization of goals.

The main components of the cost of poor quality (which apply to both manufacturing and service organizations) are shown in Figure 3.5. Note that this framework

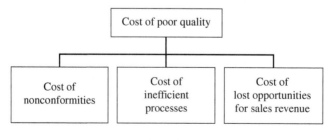

FIGURE 3.5
Cost of poor quality.

reflects not only the cost of nonconformities (sometimes called "quality costs") but also process inefficiencies and the impact of quality on sales revenue. These three categories are incorporated in Section 3.6, in the categories of quality costs. Each organization must decide which cost elements to include in its cost of poor quality.

⇥ 3.6 CATEGORIES OF QUALITY COSTS

Many organizations summarize the costs associated with quality in four categories: internal failures, external failures, appraisal, and prevention. These categories are discussed later. A useful reference on definitions, categories, and many other aspects is Campanella (1999). For an exhaustive listing of elements within the four categories, see Atkinson, Hamburg, and Ittner (1994).

Collectively, the four categories are often called the "cost of quality." The cost of poor quality includes the internal and external failure categories, whereas the appraisal and prevention categories are viewed as investments to achieve quality objectives.

Internal Failure Costs

Internal failure costs are the cost of deficiencies discovered before delivery that are associated with the failure to meet explicit requirements or implicit needs of customers. Also included are avoidable process losses and inefficiencies that occur even when requirements and needs are met. These costs would disappear if no deficiencies existed. Internal failure costs consist of (1) the cost of failure to meet customer requirements and needs and (2) the cost of inefficient processes.

Failure to meet customer requirements and needs

Examples of subcategories are costs associated with the following:

Scrap. The labor, material, and (usually) overhead on defective product that cannot be repaired economically. The titles are numerous—scrap, spoilage, defectives, etc.

Rework. Correcting defectives in physical products or errors in service products.

Lost or missing information. Retrieving information that should have been supplied.

Failure analysis. Analyzing nonconforming goods or services to determine causes.

Scrap and rework—supplier. Scrap and rework because of nonconforming product received from suppliers. This area also includes the costs to the buyer of resolving supplier quality problems.

One hundred percent sorting inspection. Finding defective units in product lots that contain unacceptably high levels of defectives.

Reinspection, retest. Reinspection and retest of products that have undergone rework or other revision.

Changing processes. Modifying manufacturing or service processes to correct deficiencies.

Redesign of hardware. Changing designs of hardware to correct deficiencies.

Redesign of software. Changing designs of software to correct deficiencies.

Scrapping of obsolete product. Disposing of products that have been superseded.
Scrap in support operations. Defective items in indirect operations.
Rework in internal support operations. Correcting defective items in indirect operations.
Downgrading. The difference between the normal selling price and the reduced price because of poor quality.

Cost of inefficient processes

Examples of subcategories include the following:

Variability of product characteristics. Losses that occur even with conforming product (e.g., overfill of packages due to variability of filling and measuring equipment).
Unplanned downtime of equipment. Loss of capacity of equipment due to failures.
Inventory shrinkage. Loss due to the difference between actual and recorded inventory amounts.
Variation of process characteristics from "best practice." Losses due to cycle time and costs of process compared to best practices in providing the same output. The best-practice process may be internal or external to the organization.
Non-value-added activities. Redundant operations, sorting inspections, and other non-value-added activities. A value-added activity increases the usefulness of a product to the customer; a non-value-added activity does not. The concept is similar to the 1950s idea of value engineering and value analysis.

External Failure Costs

External failure costs are associated with deficiencies that are found after the customer receives the product. Also included are lost opportunities for sales revenue. These costs also disappear if there were no deficiencies.

Failure to meet customer requirements and needs

Examples of subcategories include the following:

Warranty charges. The costs involved in replacing or making repairs to products that are still within the warranty period.
Complaint adjustments. The costs of investigation and adjustment of justified complaints attributable to defective product or installation.
Returned material. The costs associated with receipt and replacement of defective product received from the field.
Allowances. The costs of concessions made to customers due to substandard products accepted by the customer as is or to conforming product that does not meet customer needs.
Penalties due to poor quality. This category applies to goods or services delivered or to internal processes such as late payment of an invoice resulting in a lost discount for paying on time.

Rework on support operations. Correcting errors on billing and other external processes.
Revenue losses in support operations. An example is the failure to collect receivables from some customers.

Lost opportunities for sales revenue

Examples include the following:

Customer defections. Profit on potential customers lost because of poor quality.
New customers lost because of lack of capability to meet customer needs. Profit on potential revenue lost because of inadequate processes to meet customer needs.

Appraisal Costs

Appraisal costs are incurred to determine the degree of conformance to quality requirements. Examples include the following:

Incoming inspection and test. Determining the quality of purchased product, whether by inspection on receipt, by inspection at the source, or by surveillance.
In-process inspection and test. In-process evaluation of conformance to requirements.
Final inspection and test. Evaluation of conformance to requirements for product acceptance.
Document review. Examination of paperwork to be sent to customer.
Balancing. Examination of various accounts to assure internal consistency.
Product quality audits. Performing quality audits on in-process or finished products.
Maintaining accuracy of test equipment. Keeping measuring instruments and equipment calibrated.
Inspection and test materials and services. Materials and supplies in inspection and test work (e.g., X-ray film) and services (e.g., electric power) where significant.
Evaluation of stocks. Testing products in field storage or in stock to evaluate degradation.

In collecting appraisal costs, the decisive factor is the kind of work done and not the department name (the work may be done by chemists in the laboratory, by sorters in the operations department, by testers in the inspection department, or by an external firm engaged for the purpose of testing). Also note that industries use various terms for "appraisal," e.g., *checking, balancing, reconciliation, review.*

Prevention Costs

Prevention costs are incurred to keep failure and appraisal costs to a minimum. Examples include the following:

Quality planning. This category includes the broad array of activities that collectively create the overall quality plan and the numerous specialized plans.

It includes also the preparation of procedures needed to communicate these plans to all concerned.

New-products review. Reliability engineering and other quality-related activities associated with the launching of a new design.

Process planning. Process capability studies, inspection planning, and other activities associated with the manufacturing and service processes.

Process control. In-process inspection and test to determine the status of the process (rather than for product acceptance).

Quality audits. Evaluating the execution of activities in the overall quality plan.

Supplier quality evaluation. Evaluating supplier quality activities prior to supplier selection, auditing the activities during the contract, and performing associated effort with suppliers.

Training. Preparing and conducting quality-related training programs. As in the case of appraisal costs, some of this work may be done by personnel who are not on the payroll of the quality department. The decisive criterion is again the type of work, not the name of the department performing the work.

Note that prevention costs are costs of special planning, review, and analysis activities for quality. Prevention costs do not include basic activities such as product design, process design, process maintenance, and customer service.

The compilation of prevention costs is initially important because it highlights the small investment currently made (usually) in prevention activities and suggests the potential for an increase in prevention costs to reduce failure costs. Upper management immediately grasps this point. Continuing measurement of prevention costs can usually be excluded, however, to focus on the major opportunity, i.e., failure costs.

One of the issues in calculating the cost of poor quality is how to handle overhead costs. Three approaches are common: include total overhead using direct labor or some other base, include variable overhead only (the usual approach), or do not include overhead at all. The allocation of overhead has an impact on the total cost of poor quality and on the distribution over various departments. Activity-based costing (ABC) can help to provide a realistic allocation of overhead costs. ABC is an accounting method that allocates overhead based on the activities that cause overhead cost elements to be incurred. For an explanation of the concept, see *JQH5*, page 8.14. Cokins (1999) discusses the impact of traditional accounting and ABC on quality.

An example for one plant of a tire manufacturer is shown in Table 3.2. This example resulted in some conclusions that are typical of these studies.

- The total of almost $900,000 per year is large.
- Most (79.1%) of the total is concentrated in failure costs, specifically in waste scrap and consumer adjustments.
- Failure costs are about five times the appraisal costs.
- A small amount (4.3%) is spent on prevention.
- Some consequences of poor quality cannot be conveniently quantified, e.g., customer ill will and customer policy adjustment. However, these factors are listed as a reminder of their existence.

TABLE 3.2
Annual quality cost—tire manufacturer

Cost of quality failures—losses		
Defective stock	$ 3,276	0.37%
Repairs to product	73,229	8.31
Collect scrap	2,288	0.26
Waste—scrap	187,428	21.26
Consumer adjustments	408,200	46.31
Downgrading products	22,838	2.59
Customer ill will	Not counted	
Customer policy adjustment	Not counted	
Total	$697,259	79.10%
Cost of appraisal		
Incoming inspection	$ 32,655	2.68
Inspection 1	32,582	3.70
Inspection 2	25,200	2.86
Spot-check inspection	65,910	7.37
Total	$147,347	16.61%
Cost of prevention		
Local plant quality Control engineering	7,848	0.89
Corporate quality Control engineering	30,000	3.40
Total	$ 37,848	4.29%
Grand total	$882,454	100.00%

As a result of this study, management decided to increase the budget for prevention activities. Three engineers were assigned to identify and pursue specific quality improvement projects.

Strictly defined, the cost of poor quality is the sum of the internal and external failure cost categories. This definition assumes that the elements of appraisal costs—e.g., 100% sorting inspection—necessitated by inadequate processes are classified under internal failures. Some practitioners use the term *cost of quality* for the four broad categories.

Although many organizations use the categories of internal failure, external failure, appraisal, and prevention, the structure may not apply in all cases. A different approach employed by a bank is shown in Table 3.3.

TABLE 3.3
Revenue lost through poor quality

$10,000,000	Annual customer service revenue
1,000	Number of customers
× 25%	Percent dissatisfied
250	Number of dissatisfied
× 75%	Percent of switchers (60–90% of dissatisfied)
188	Number of switchers
× $10,000	Average revenue per customer
$ 1,880,000	Revenue lost through poor quality

Source: The University of Tampa.

Here the cost of poor service is calculated based on customer satisfaction data and customer loyalty/retention data. In this case, the cost of poor quality is the cost of lost opportunities for sales revenue (see Figure 3.5).

In still another approach, some organizations define the cost of poor quality with a focus on the cost of key activities or processes, i.e., the difference between actual costs and the cost in an organization that has the best practice for that activity. This approach addresses the cost of inefficient processes (see Figure 3.5). For a discussion of the traditional quality cost approach (four categories), the process cost approach, and the "quality loss approach," see Schottmiller (1996).

Additional examples of cost-of-poor-quality studies in both manufacturing and service industries are provided by Campanella (1999) and Atkinson et al. (1994).

Hidden Quality Costs

The cost of poor quality may be understated because of costs that are difficult to estimate. "Hidden" costs occur in both manufacturing and service industries and include the following:

1. Potential lost sales.
2. Costs of redesign of products due to poor quality.
3. Costs of changing processes due to inability to meet quality requirements for products.
4. Costs of software changes due to quality reasons.
5. Costs of downtime of equipment and systems including computer information systems.
6. Costs included in standards because history shows that a certain level of defects is inevitable and allowances should be included in standards.
   a. *Extra material purchased.* The purchasing buyer orders 6% more than the production quantity needed.
   b. *Allowances for scrap and rework during production.* History shows that 3% is "normal," and accountants have built this allowance into the cost standards. One accountant said, "Our scrap cost is zero. The production departments are able to stay within the 3% that has been added in the standard cost and therefore the scrap cost is zero." Ah, for the make-believe numbers game.
   c. *Allowances in time standards for scrap and rework.* One manufacturer allows 9.6% in the time standard for certain operations to cover scrap and rework.
   d. *Extra process equipment capacity.* One manufacturer plans for 5% unscheduled downtime of equipment and provides extra equipment to cover the downtime. In such cases, the alarm signals ring only when the standard value is exceeded. Even when operating within those standards, however, the costs should be a part of the cost of poor quality. They represent opportunities for improvement.
7. Extra indirect costs due to defects and errors. Examples are space charges and inventory charges.
8. Scrap and errors not reported. Scrap may never be reported because employees fear reprisals, or scrap may be charged to a general ledger account without being identified as scrap.
9. Extra process costs due to excessive product variability (even though within specification limits). For example, a process for filling packages with a dry soap

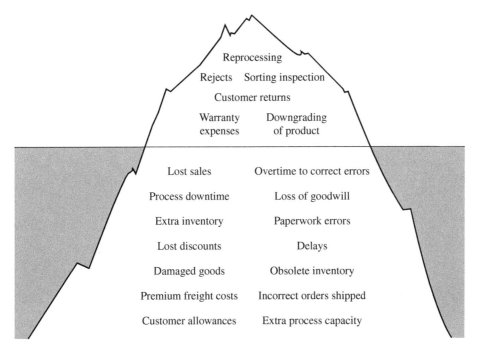

FIGURE 3.6
Hidden costs of poor quality.

mix meets requirements for label weight on the contents. The process aim, however, is set above label weight to account for variability in the filling process. See "Cost of Inefficient Processes" under "Internal Failure Costs."

10. Cost of errors in support operations, e.g., order filling, shipping, customer service, billing.
11. Cost of poor quality within a supplier's company. Such costs are included in the purchase price.

Hidden costs can accumulate to a large amount, three or four times the reported failure cost. Where agreement can be reached to include some of these costs and where credible data or estimates are available, then they should be included in the study. Otherwise, they should be left for future exploration.

Progress has been made in quantifying certain hidden costs, and therefore some of them have been included in the four categories discussed. Obvious costs of poor quality are the tip of the iceberg (Figure 3.6).

→ 3.7 RELATING THE COST OF POOR QUALITY TO BUSINESS MEASURES

Interpretation of the cost of poor quality is aided by relating this cost to other figures with which managers are familiar. Table 3.4 shows actual examples of the annual cost of poor quality related to various business measures.

TABLE 3.4
Languages of management

Money (annual cost of poor quality)
24% of sales revenue
15% of manufacturing cost
13 cents per share of common stock
$7.5 million per year for scrap and rework compared to a profit of $1.5 million per year
$176 million per year
40% of the operating cost of a department

Other languages
The equivalent of one plant in the company making 100% defective work all year
32% of engineering resources spent in finding and correcting design weaknesses
25% of manufacturing capacity devoted to correcting quality problems
13% of sales orders canceled
70% of inventory carried attributed to poor quality levels
25% of manufacturing personnel assigned to correcting quality problems

Reducing the cost of poor quality has a dramatic impact on company financial performance, as illustrated by the Dupont Financial Model (Werner and Stoner, 1995). This model states,

$$\text{Return on assets} = \text{Profit margin} \times \text{Asset turnover}$$

Suppose that the cost of poor quality (COPQ) was 10% of sales revenue, the profit margin was 7%, and the asset turnover was 3.0. The return on assets is then 7.0×3.0 or 21%. Further suppose that a quality improvement effort reduced the COPQ from 10 to 6% and the asset turnover remained at 3.0. The profit margin would then be $7.0 + 4.0$ or 11%, and the return on assets would be 11.0×3.0 or 33%. Note the impact of asset turnover.

How do we convince upper management to institute a quality improvement program to reduce the cost of poor quality? The steps are presented in Section 4.4.

3.8 OPTIMUM COST OF QUALITY

When cost summaries on quality are first presented to managers, one typical question is, What are the right costs? The managers are looking for a standard ("par") against which to compare their actual costs so that they can judge whether there is a need for action.

Unfortunately, few credible data are available because (1) companies almost never publish such data and (2) the definition of cost of poor quality varies by company. But we can cite some numbers. For manufacturing organizations the annual cost of poor quality is about 15% of sales income, varying from about 5 to 35% depending on product complexity. For service organizations the average is about 30% of operating expenses, varying from 25 to 40% depending on service complexity. In one study of healthcare costs, Mortimer, DeFeo, and Stepnick (2003) estimate that as much as 30% of all direct healthcare outlays today are the result of poor quality care, consisting

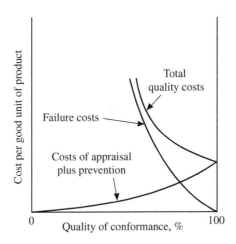

FIGURE 3.7
Model for optimum costs. (*From JQH4, p. 4.19.*)

primarily of overuse, misuse, and waste. With U.S. national health expenditures of roughly $1.4 trillion in 2001, the 30% figure translates into $420 billion spent each year as a direct result of poor quality. In addition, the indirect costs of poor quality (e.g., reduced productivity due to absenteeism) add an estimated 25 to 50%—$105 to $210 billion—to the national bill. Private purchasers absorb about one-third of these costs. In that study, it is estimated that poor quality healthcare costs the typical employer between $1,900 and $2,250 per covered employee each year. But three conclusions on cost data do stand out: Total costs are highest for complex industries, failure costs are the largest percentage of the total, and prevention costs are a small percentage of the total.

The study of the distribution of quality costs over the major categories can be further explored by using the model shown in Figure 3.7. The model shows three curves:

1. *The failure costs.* These costs equal zero when the product is 100% good and rise to infinity when the product is 100% defective. (Note that the vertical scale is cost per good unit of product. At 100% defective, the number of good units is zero, and hence the cost per good unit is infinity.)
2. *The costs of appraisal plus prevention.* These costs are zero at 100% defective and rise as perfection is approached.
3. *The sum of curves 1 and 2.* The third curve, marked "total quality costs," represents the total cost of quality per good unit of product.

Figure 3.7 suggests that the minimum level of total quality costs occurs when the quality of conformance is 100%, i.e., perfection. This result has not always been the case. During most of the 20th century, the predominant role of (fallible) human beings limited the efforts to attain perfection to finite costs. Also, the inability to quantify the impact of quality failures on sales revenue resulted in underestimating the failure costs. The result was to view the optimum value of quality of conformance as less than 100%.

Although perfection is obviously the goal for the long run, perfection is not necessarily the most economic goal for the short run or for every situation. Industries, however, are facing increasing pressure to reach perfection. The prospect is that the trend to 100% conformance will extend to more and more goods and services of greater and greater complexity.

To evaluate whether quality improvement has reached the economic limit, we need to compare the benefits possible from specific projects with the costs involved in achieving these benefits. When no justifiable projects can be found, the optimum has been reached.

Let's identify a few key points about the cost of poor quality. This cost is usually high (sometimes larger than the annual profit), quantifying this cost can be the key to gaining approval from management to assign resources to quality improvement, and the main uses of the cost-of-poor-quality study are to identify opportunities for improvement projects and to provide supporting data to assist in improvement.

For elaboration on finalizing the definitions of cost categories, steps in making the initial study, and data collection methods, see *JQH5,* Chapter 5.

Next, we move to another element of assessment—standing in the marketplace.

→ 3.9 ASSESSMENT AND STANDING IN THE MARKETPLACE

Estimating the cost of poor quality is an essential part of the assessment. But it is not enough. We also need to understand where the company stands on quality in the marketplace, relative to the competition. This component of assessment will prove important in increasing sales income.

Similar to the assessment of the cost of poor quality, the market study (1) gives a snapshot of standing relative to competition and (2) identifies opportunities and threats.

The approach must be based on a market research study. Such studies should be planned not by any one department but by a team involving members from marketing, product development, quality, operations, and other areas as needed. This team must agree beforehand on what questions need to be answered by the field study. Three types of questions should be considered: (1) What is the relative importance of various product qualities as seen by the user? (2) For each of the key qualities, how does our product compare with competitors' products, as seen by users? (3) How likely is the customer to purchase from us again or recommend us to others?

Chapter 11 includes other questions to help in developing new or modified products.

Answers to such questions must be based on input from current customers, lost customers, and noncustomers. Opinions of company personnel, no matter how extensive the experience base, cannot and should not substitute for the voice of the customer.

Examples of Field Studies

The first example comes from a manufacturer of health products. In a multiattribute study, customers were asked to consider several product attributes and indicate both

TABLE 3.5
Multiattribute study

Attribute	Relative importance, %	Company X Rating	Company X Weighted rating	Competitor A Rating	Competitor A Weighted rating	Competitor B Rating	Competitor B Weighted rating
Safety	28	6	168	5	140	4.5	126
Performance	20	6	120	7	140	6.5	130
Quality	20	6	120	7	140	4	80
Field service	12	4	48	8	96	5	60
Ease of use	8	4	32	6	48	5	40
Company image	8	8	64	4	32	4	32
Plant service	4	7.5	30	7.5	30	5	20
Total			582		626		488

their relative importance and a competitive rating. The results for one product are shown in Table 3.5. Note that an overall score is obtained for each manufacturer by multiplying the relative importance by the score for that attribute and then adding these results.

In another example, a manufacturer of equipment was experiencing a decline in market share. Complaint's about quality led to a proposal to "beef up the inspection." Discussions within the company revealed uncertainty about the nature of the complaints, so management decided to conduct a field study to learn more about customer views. A team was formed to plan and conduct the study. About 50 customers were visited.

In one part of the study, six attributes were identified, and customers were asked to rate the company as superior, competitive, or inferior to the competition on each attribute (see Table 3.6). The results were a surprise. The equipment problems were confirmed, but the study revealed the presence of both design and manufacturing causes. Also, documentation and field repair service were identified as weak areas; these were surprises to the company involved. The company took dramatic action. This manufacturer created a broad approach to quality, starting with the initial design and continuing throughout the spiral of all activities affecting fitness for use. This outcome was in stark contrast to the original proposal of adding inspectors. The study

TABLE 3.6
Heavy equipment case

Attribute	Comparison to competition, % Superior	Comparison to competition, % Competitive	Comparison to competition, % Inferior
Analysis of customer needs			
Preparation of quality requirements and purchase order			
Preparation of specifications and technical documentation			
Quality of equipment			
Quality and availability of spare parts			
Quality of field repair service			

Source: Private communication.

TABLE 3.7
Market research at a bank

Satisfaction with ...	Very satisfied, %	Sample size	Important and low in satisfaction, %	Important and high in satisfaction, %
1. Greeting you with a smile				
8. Processing transactions without error				
14. Easy-to-read and understand bank statements				
20. Prompt follow-up to questions and problems				

required about seven worker-months of effort, including planning, customer visits, analysis of results, and preparation of a report—a small price to pay to develop a proper strategy.

Many organizations in service industries have extensive experience in market research. For example, one bank periodically conducts market research as part of a quality system. This research probes 20 attributes of banking service by asking consumers about the relative importance of attributes and consumer degree of satisfaction. Table 3.7 shows the format of the summarized results.

In an example from the public sector, the United States Postal Service has used a Customer Satisfaction Index survey that asks customers for the relative importance and satisfaction rating on 10 attributes (e.g., responsiveness, carrier services, complaint handling). A separate question asks about "customer willingness to switch to another mail service."

Graphing the market research results can be helpful. Rust et al. (1994) show an example of how the mapping of satisfaction and importance ratings can relate customer views and potential action (Figure 3.8). In this approach, attributes in which importance is high and satisfaction is poor represent the greatest potential for gain.

In Figure 3.8, the four quadrants are roughly defined by the averages on the two axes. Interpretation of the quadrants is typically as follows:

> *Upper left (satisfaction strong, importance low).* Maintain the status quo.
> *Upper right (satisfaction strong, importance high).* Leverage this competitive strength through advertising and personal selling.

TABLE 3.8
Benchmark data for local telephone service

Attribute	Company score	Competitor scores
Overall quality	38.6	36.6–46.3
Local dial quality	40.6	36.7–47.9
Billing quality	34.5	28.7–37.2
Installation quality	41.2	43.2–53.3
Long distance quality	47.5	40.9–55.3
Operator quality	41.5	35.0–47.1

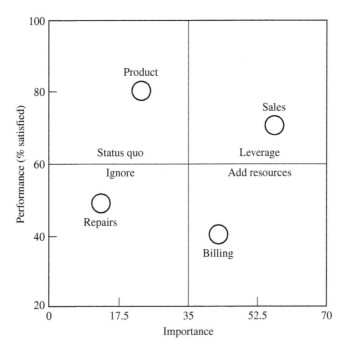

FIGURE 3.8
Performance versus importance in driving satisfaction: quadrant map. (*Rust et al., 1994.*)

Lower left (satisfaction weak, importance low). Assign little or no priority to action.
Lower right (satisfaction weak, importance high). Add resources to achieve an improvement.

Collecting data on customer loyalty and retention goes beyond customer satisfaction research. Lauter (1997) reports on customer retention results at a bank. The actual customer retention percentage was calculated for each quarter in 1996—the percentage retained was typically more than 95%. Further analysis of the customers lost revealed a surprising result (Figure 3.9). For example, in the first quarter, 120,000 households (HHs) were lost, and this loss translated into $26 million of profit. Although the information has been modified because of its proprietary nature, a small percentage of customers lost from a high volume of customers can result in a significant absolute dollar amount of revenue.

↦ 3.10 ASSESSING USING NATIONAL PERFORMANCE STANDARDS AND AWARDS

The national quality and performance excellence criteria and awards have become a major influence on attaining superior organization results. In fact, it could be argued that had it not been for the recognition Motorola received as one of the first recipients of the

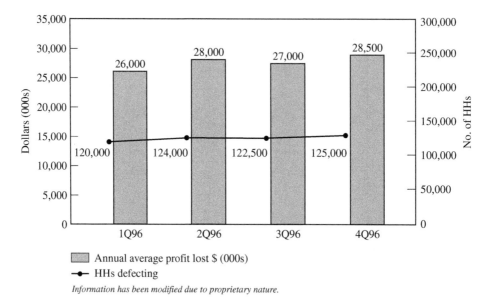

FIGURE 3.9
Average "opportunity cost of attrition." (*Lauter, 1997.*)

Malcolm Baldrige National Quality Award, it is possible that the practice of six sigma methodology may have never gained the level of exposure and practice that it has today.

National quality awards in particular have greatly influenced the way many organizations manage their systems for quality and performance excellence. The prestige of winning a national quality award has provided increased recognition to the award recipients' role-model approaches and created additional incentives for many organizations to apply for the award and stick to their approaches. However, the real benefit stated by many is not the award itself, but the organizational improvement that can occur from adopting the award models and criteria as a means for pursuing excellence.

Malcolm Baldrige National Quality Award Criteria

The Baldrige performance excellence criteria are a framework that any organization can use to improve overall performance.

The seven categories that make up the award criteria are connected and integrated as depicted earlier in the framework of Figure 3.2. The following is a list of the Baldrige Award categories:

1. **Leadership:** Examines how senior executives guide the organization and how the organization addresses its responsibilities to the public and practices good governance and citizenship.
2. **Strategic planning:** Examines how the organization sets strategic directions and how it develops key action plans.

3. **Customer and market focus:** Examines how the organization engages its customers for long-term marketplace success, builds a customer-focused culture, and listens to the voice of its customers and uses this information to improve and identify opportunities for innovation.
4. **Measurement, analysis, and knowledge management:** Examines the management, effective use, analysis, and improvement of data and information to support key organization processes and the organization's performance management system.
5. **Workforce focus:** Examines how the organization engages, manages, and develops its workforce to develop its full potential and how the workforce is aligned with the organization's objectives.
6. **Process management:** Examines aspects of how work systems are designed and how key work processes are designed, managed, and improved.
7. **Results:** Examines the organization's performance and improvement in its key business areas: product and service, customer-focused, financial and marketplace, workforce-focused, process effectiveness, and leadership. The category also examines how the organization performs relative to competitors and other organizations with similar product offerings.

Within each category of the Baldrige Criteria are the requirements and questions that are used as part of an assessment or an awards application. There are 18 criteria (see Figure 3.10). In addition, there are 36 areas to address contained within the 18 criteria. The numerous requirements are expressed as individual criteria questions.

Malcolm Baldrige National Quality Award Core Values

There are 11 core values and concepts embodied in the award criteria (see Figure 3.11):

1. **Visionary leadership:** An organization's senior leaders should set directions and create a customer focus, clear and visible values, and high expectations.
2. **Customer-driven excellence:** Performance and quality are judged by an organization's customers. Thus, an organization must take into account all product features and characteristics and all modes of customer access and support that contribute value to its customers.
3. **Organizational and personal learning:** Achieving the highest levels of organizational performance requires a well-executed approach to organizational and personal learning that includes sharing knowledge via systematic processes.
4. **Valuing workforce members and partners:** An organization's success depends increasingly on an engaged workforce that benefits from meaningful work, clear organizational direction, and performance accountability and that has a safe, trusting, and cooperative environment. Additionally, the successful organization capitalizes on the diverse backgrounds, knowledge, skills, creativity, and motivation of its workforce and partners.
5. **Agility:** Success in today's ever-changing, globally competitive environment demands agility—a capacity for rapid change and flexibility. Organizations face shorter cycles for the introduction of new/improved products, and nonprofit and government organizations are increasingly being asked to respond rapidly to new

Criteria for Performance Excellence—Item Listing

P **Preface: Organizational Profile**

P.1 Organizational Description
P.2 Organizational Situation

Categories and Items	Point Values
1 Leadership	**120**
1.1 Senior Leadership	70
1.2 Governance and Societal Responsibilities	50
2 Strategic Planning	**85**
2.1 Strategy Development	40
2.2 Strategy Deployment	45
3 Customer Focus	**85**
3.1 Customer Engagement	40
3.2 Voice of the Customer	45
4 Measurement, Analysis, and Knowledge Management	**90**
4.1 Measurement, Analysis, and Improvement of Organizational Performance	45
4.2 Management of Information, Knowledge, and Information Technology	45
5 Workforce Focus	**85**
5.1 Workforce Engagement	45
5.2 Workforce Environment	40
6 Process Management	**85**
6.1 Work Systems	35
6.2 Work Processes	50
7 Results	**450**
7.1 Product Outcomes	100
7.2 Customer-Focused Outcomes	70
7.3 Financial and Market Outcomes	70
7.4 Workforce-Focused Outcomes	70
7.5 Process Effectiveness Outcomes	70
7.6 Leadership Outcomes	70
TOTAL POINTS	**1,000**

Item listing from the Baldrige Criteria for performance excellence, including the total point values per item and category. *(Baldrige National Quality Program.)*

FIGURE 3.10
Criteria for performance excellence—item listing.
(Baldrige National Quality Program.)

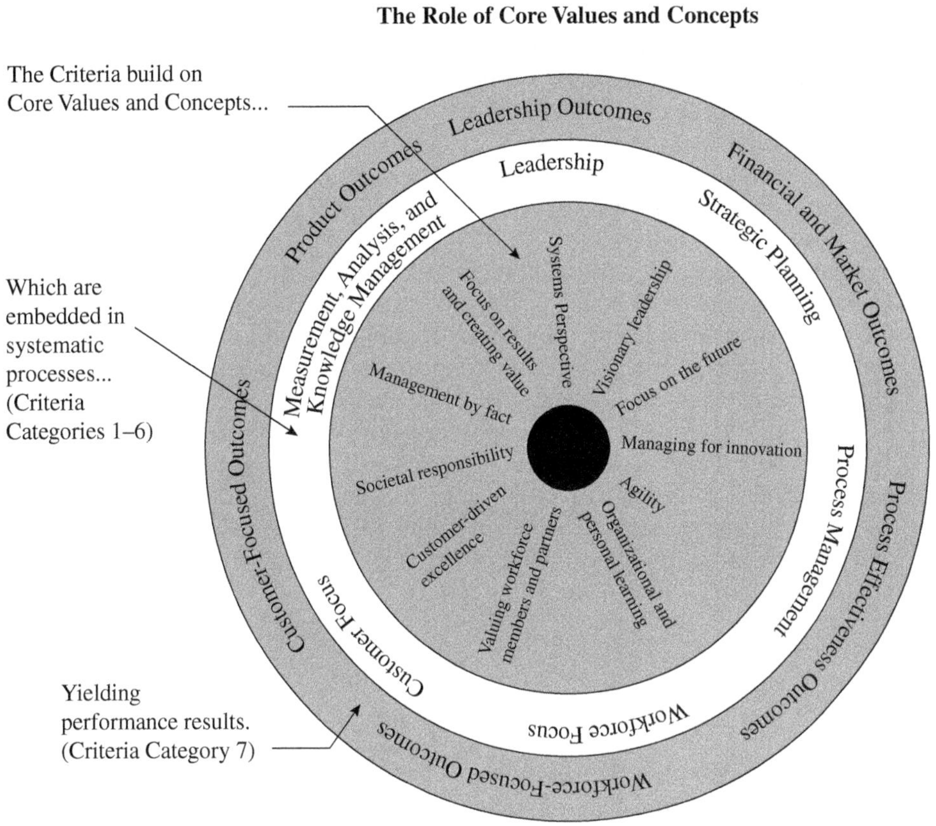

The Role of Core Values and Concepts

The Criteria build on Core Values and Concepts...

Which are embedded in systematic processes... (Criteria Categories 1–6)

Yielding performance results. (Criteria Category 7)

FIGURE 3.11
The role of core values and concepts.
(Baldrige National Quality Program.)

or emerging social issues. Major improvements in response times often require new work systems, simplification of work units and processes, or the ability for rapid changeover from one process to another.

6. **Focus on the future:** Creating a sustainable organization requires understanding the short- and long-term factors that affect an organization and marketplace.

7. **Managing for innovation:** Innovation means making meaningful change to improve an organization's products, services, programs, processes, operations, and business model to create new value for the organization's stakeholders. Innovation should lead an organization to new dimensions of performance.

8. **Management by fact:** Organizations depend on the measurement and analysis of performance. Such measurements should derive from business needs and strategy, and they should provide critical data and information about key processes, outputs, and results. Many types of data and information are needed for performance management.

9. **Societal responsibility:** An organization's leaders should stress responsibilities to the public, ethical behavior, and the need to consider societal well-being

and benefit. Leaders should be role models for the organization in focusing on ethics and the protection of public health, safety, and the environment.
10. **Focus on results and creating value:** An organization's performance measurements need to focus on key results. Results should be used to create and balance value for key stakeholders—customers, workforce, stockholders, suppliers, and partners; the public; and the community. By creating value for key stakeholders, an organization builds loyalty, contributes to growing the economy, and contributes to society.
11. **Systems Perspective:** The Baldrige Criteria provide a systems perspective for managing an organization and its key processes to achieve results—and to strive for performance excellence. The seven Baldrige Criteria Categories, the Core Values, and the Scoring Guidelines form the building blocks and the integrating mechanism for the system. However, successful management of overall performance requires organization-specific synthesis, alignment, and integration. Synthesis means looking at the organization as a whole and building on key business attributes, including core competencies, strategic objectives, action plans, and work systems. Alignment means using the key linkages among requirements given in the Baldrige Criteria Categories to ensure consistency of plans, processes, measures, and actions.

→ 3.11 BALDRIGE SCORING SYSTEM

The Baldrige scoring system is based on a 0–1,000 point scale. The available points are distributed among the categories and items as shown in Figure 3.10. There is a heavy focus on results with Category 7, Business Results, making up 450 of the total 1,000 points available. During an assessment or award application review, examiners assign scores based upon scoring guidelines. There are two evaluation dimensions with scoring guidelines for each: (1) process and (2) results.

Process refers to the work methods an organization uses and improves to address the item requirements in categories 1 through 6. Four factors used to evaluate process are approach, deployment, learning, and integration (ADLI):

1. **Approach:** Refers to the methods used to accomplish the process; the appropriateness of the methods to the item requirements and the organization's operating environment; the effectiveness of the use of the methods; the degree to which the approach is repeatable and based on reliable data and information (i.e., systematic).
2. **Deployment:** Refers to the extent to which the approach is applied in addressing item requirements relevant and important to the organization; the approach is consistently applied; the approach is used (executed) by all appropriate work units.
3. **Learning:** Refers to refining the approach through cycles of evaluation and improvement; encouraging breakthrough change to the approach through innovation; sharing refinements and innovations with other relevant work units and processes in the organization.
4. **Integration:** Refers to the extent to which the approach is aligned with the organizational needs identified in the organizational profile and other process items;

the measures, information, and improvement systems are complementary across processes and work units; the plans, processes, results, analyses, learning, and actions are harmonized across processes and work units to support organizationwide goals.

Results refers to an organization's outputs and outcomes in achieving the requirements in items 7.1–7.6 (Category 7). The four factors used to evaluate results are levels, trends, comparisons, and integration (LeTCI):

1. **Levels:** Refers to the current level of performance.
2. **Trends:** Refers to the rate of performance improvements or the sustainability of good performance (i.e., the slope of trend data); the breadth (i.e., the extent of deployment) of the performance results.
3. **Comparisons:** Refers to performance relative to appropriate comparisons, such as similar competitors or organizations; the performance relative to benchmarks or industry leaders.
4. **Integration:** Refers to the extent to which the results measure (often through segmentation) and address important customer, product, market, process, and action plan performance requirements identified in the organizational profile and in process items; results include valid indicators of future performance; results are harmonized across processes and work units to support organizationwide goals.

→ 3.12 USING QUALITY AWARDS AS A SYSTEM ASSESSMENT TOOL

Many organizations use the criteria from national quality awards for the benefit of getting a thorough organizational assessment, often with no intent of even applying for award recognition. There are various methods that can be used to complete an assessment based upon many factors, including the size and geography of the organization, the number of facilities, and the availability of internal expertise to conduct an assessment.

Written Responses

Without completing a formal application, organizations can use a question-and-answer approach to respond to the criteria questions from an awards program. There could be multiple responses for each question if more than one input is desired for the response. Some more-advanced formats for written responses could seek more probing information for each question. For example, a Baldrige-based written response questionnaire may be formatted to seek a specific response for approach, deployment, learning, and integration for each process-related question. This provides richer information for the purposes of scoring the application using the awards process scoring guidelines.

Survey

This approach can be used to gather assessment input from a large number of people. Questions are designed to gather the collective input on the performance of the organization as it relates to the awards criteria. The Baldrige National Quality Program provides a free survey for this purpose called, "Are We Making Progress" (2008). This is a 40-question survey that can be completed in about 10 minutes. A survey can be used as the primary information gathering method, or it can be used as supplemental or additional information as part of more comprehensive assessment.

Application

A formal application can be prepared that simulates an actual awards process application. This requires more effort than simple written responses but can also be more revealing because of the thought process of the writer in attempting to use consistent language, identify linkages, and identify the most important strengths of the organization. Organizations that are just starting out may decide to answer only the higher-level questions rather than the more specific multiple requirements in awards criteria. For example, if assessing the Category 3, Customer Focus, section of the Baldrige Criteria, the self-assessment team may only respond to the item-level questions:

> 3.1 Customer engagement: How do you engage customers to serve their needs and build relationships?

> 3.2 Voice of the customer: How do you obtain and use information from your customers?

This is a much simpler process, whereas the multiple requirements from Category 3, Customer Focus, include 35 separate questions compared to only 2 questions at the item level.

Interview

In this approach, the assessment team would schedule interviews with key individuals to respond to criteria questions. During the interviews, the assessment team will record notes about the approaches used and results attained for further evaluation after all of the information has been collected. This approach requires less preparation time than written responses. The disadvantage is that the respondents may not provide complete information during the interview setting, depending on their understanding of the question and their recall of all the relevant information in a single interview setting. There is also less learning on the part of the respondents when compared to written responses because the level of engagement is less.

Focus Group

This is similar to the interview approach except multiple people are involved in the sessions for different sections of the criteria review. For example, an assessment for item 1.1, Senior Leadership, of the Baldrige Criteria may involve several members of the senior leadership team in a focus group setting. Like the interview approach, the assessment team will record notes about the approaches used and results attained for further evaluation after all of the information has been collected. This approach has an advantage over individual interviews because of the shared knowledge of the group and the ability to build on the responses of others to provide complete assessment information.

Collaborative Assessment

This is a variation on a focus group approach. Instead of gathering specific answers to the approach questions, the respondents provide their opinions about strengths and opportunities for improvement for each section of the assessment criteria. Responses are captured in real time for review by the focus group. At the end of each focus group session, each individual can provide a rating of the importance of each opportunity for improvement. This collaborative approach to assessment has the advantage of involving many people in gathering the information and building consensus on both the strengths and opportunities for improvement within the organization (Hoyt and Ralston, 2001).

→ 3.13 ASSESSING TO THE INTERNATIONAL STANDARDS

Standards exist principally to facilitate international trade and to avoid harming customers and society. In the pre-standardization era (pre-1980), there were various national and multinational standards. Standards for electrical, mechanical, and chemical process compatibility have been around for decades. Other standards, such as military standards, were developed for military and others for the nuclear power industry, and, to a lesser extent, for commercial and industrial use. These standards have commonalities and historical linkages. However, they often were not consistent in terminology or content for widespread use in international trade. As a result, organizations were left to re-create their own standards or adapt the existing ones. This only led to even less commonality. In the 1980s, as most of the first world industrial organizations began to improve quality and safety at record paces, there became a need to fill a void. That void was a common quality management system that would be a nonbinding "contract" between the customer and the supplier. This void was filled by the ISO 176 Technical Committee in the form of the ISO 9000 set of standards. This was later followed by filling a similar void for environmental standards with ISO 14000. Many organizations globally began using these standards as a "certified" standard for performance. While their intent was important, the standards became more of an opportunity to get a certificate of compliance that could be used to "wow" customers, rather than a set of requirements that, when met, ensure customer needs are met.

Certain industry/economic sectors then began developing industrywide quality system standards, based on the verbatim adoption of ISO 9000, together with industrywide supplemental requirements. The automotive industry (QS 9000), the pharmaceutical and medical devices industry (cGMPs), government regulatory agencies, and military procurement agencies (AS 9100 and MAP) are adopting this approach in many places worldwide. Even software development uses the CMMi standard of software quality created in the early 1990s at Carnegie Mellon University to ensure a common approach to managing software quality. The standards play an important—but not always understood—role in the management of quality.

ISO 9000 Quality Management System Standard

The ISO 9000 standards have had great impact on international trade and quality systems implementation by organizations worldwide. These international standards have been adopted as national standards by more than 100 countries. They have been applied in a wide range of industry/economic sectors and government regulatory areas. The ISO 9000 standards deal with the management systems used by organizations to ensure quality in design, production, delivery, and support products. The standards apply to all generic product categories: hardware, software, processed materials, and services. The complete family of ISO 9000 standards provide quality management guidance, quality assurance requirements, and supporting technology for an organization's quality management system. The standards provide guidelines or requirements on what features are to be present in the management system of an organization but do not prescribe how the features are to be implemented. This nonprescriptive character gives the standards their wide applicability for various products and situations. Upon implementing ISO 9000, an organization can be registered as a Certified Quality Management System.

The standards in the ISO 9000 family were created and are produced and maintained by Technical Committee 176 of the International Organization for Standardization (ISO). The first meeting of ISO/TC176 was held in 1980. ISO 8402, the vocabulary standard, was first published in 1986. The initial ISO 9000 series was published in 1987, consisting of:

- The fundamental concepts and road map guideline standard ISO 9000
- Three alternative requirements standards for quality assurance (ISO 9001, ISO 9002, or ISO 9003)
- The quality management guideline standard ISO 9004

Since 1987, additional standards have been published. The ISO 9000 family now contains a variety of standards supplementary to the original series. In particular, revisions of the basic ISO 9000 series—ISO 9000 through ISO 9004—were published in 1994, 2000, and most recently in 2008 under the name ISO 9000:2008. This section is written in relation to the 2008 revisions after an initial introduction to the original standard. Table 3.9 displays the ISO 9000:2008 list of requirements.

ISO 9000 has been adopted and implemented worldwide for quality assurance purposes in both two-party contractual situations and third-party certification/registration situations. Their use grew in the 1990s and early 2000s but has since slowed. It will grow again with the newest update in eight years: the release of ISO 9000:2008. The infrastructure of certification and registration bodies, accreditation bodies, course providers, consultants, and auditors trained and certified for auditing to

TABLE 3.9
The clauses of ISO 9001 and their typical structure

Clause titles		
1		Scope
2		Normative reference
3		Definitions
4		Quality system requirements
	4.1	Management responsibility
	4.2	Quality system
	4.3	Contract review
	4.4	Design control
	4.5	Document and data control
	4.6	Purchasing
	4.7	Control of customer-supplied product
	4.8	Product identification and traceability
	4.9	Process control
	4.10	Inspection and testing
	4.11	Control of inspection, measuring, and test equipment
	4.12	Inspection and test status
	4.13	Control of nonconforming product
	4.14	Corrective and preventive action
	4.15	Handling, storage, packaging, preservation, and delivery
	4.16	Control of quality records
	4.17	Internal quality audits
	4.18	Training
	4.19	Servicing
	4.20	Statistical techniques

these standards followed a similar pattern. Mutual recognition arrangements between and among nations continue to develop, with the likelihood of ISO-sponsored quality system accreditation recognition in the near future. The periodic surveillance audits that are part of the third-party certification/registration arrangements worldwide provide continuing motivation for supplier organizations to maintain their quality systems in complete conformance and to improve the systems to continually meet their objectives for quality.

The market for quality management and quality assurance standards itself grew rapidly, partly in response to trade agreements such as the European Union (EU), General Agreement on Tariffs and Trade (GATT), and North American Free Trade Association (NAFTA). These agreements all are dependent on standards that implement the reduction of nontariff trade barriers. The ISO 9000 family occupies a key role in the implementation of such agreements.

The globalization of business is a reality even for many small- and medium-size organizations. These smaller organizations, as well as large organizations, now find that some of their prime competitors are likely to be based in another country. Fewer and fewer businesses are able to survive by considering only the competition within the local community. This affects the strategic approach and the product planning of organizations of all sizes.

An important question that is often asked is: In this world of rapid change, how can a single family of standards, ISO 9000, apply to all industry and economic sectors, all products, and all sizes of organizations? The ISO 9000 standards are founded on the concept that the assurance of consistent product quality is best achieved by simultaneous application of two kinds of standards:

- Product standards (technical specifications)
- Quality system (management system) standards

Product standards provide the technical specifications that apply to the characteristics of the product and, often, the characteristics of the process by which the product is produced. Product standards are specific to the particular product: both its intended functionality and the end-use situations the product may encounter.

The management system is the domain of the ISO 9000 standards. It is by means of the distinction between product specifications and management system features that the ISO 9000 standards apply to all industry/economic sectors, all products, and all sizes of organizations.

The standards in the ISO 9000 family, both guidance and requirements, are written in terms of what features are to be present in the management system of an organization but do not prescribe how the features are to be implemented. The technology selected by an organization determines how the relevant features will be incorporated in its own management system. Likewise, an organization is free to determine its own management structure.

→ 3.14 QUALITY SYSTEM CERTIFICATION/REGISTRATION

The earliest users of quality assurance requirements standards were large customer organizations such as electric power providers and military organizations. These customers often purchase complex products to specific functional design. In such situations, the quality assurance requirements are called up in a two-party contract, where the providing organization (i.e., the supplier) is referred to as the "first party" and the customer organization is referred to as the "second party." Such quality assurance requirements typically include provisions for the providing organization to have internal audits sponsored by its management to verify that its quality system meets the contract requirements. These are first-party audits. Such contracts typically also include provisions to have external audits sponsored by the management of the customer organization to verify that the supplier organization's quality system meets the contract requirements. These are second-party audits. Within a contractual arrangement between two such parties, it is possible to tailor the requirements, as appropriate, and to maintain an ongoing dialogue between customer and supplier.

When such assurance arrangements become a widespread practice throughout the economy, the two-party, individual-contract approach becomes burdensome. There develops a situation where each organization in the supply chain is subject to periodic management system audits by many customers and is itself subjecting many of its sub-suppliers to such audits. There is a lot of redundant effort throughout the supply chain because each organization is audited multiple times for essentially the same requirements. The conduct of audits becomes a significant cost element for both the auditor organizations and audited organizations.

Certification/Registration-Level Activities

The development of quality system certification/registration is a means to reduce the redundant, non-value-adding effort of these multiple audits. A third-party organization, which is called a "certification body" in some countries, or a "registrar" in other countries (including the United States), conducts a formal audit of a supplier organization to assess conformance to the appropriate quality system standard, say, ISO 9001 or ISO 9002. When the supplier organization is judged to be in complete conformance, the third party issues a certificate to the supplying organization and registers the organization's quality system in a publicly available register. Thus, the terms *certification* and *registration* carry the same marketplace meaning because they are two successive steps signifying successful completion of the same process.

To maintain its registered status, the supplier organization must pass periodic surveillance audits by the registrar. Surveillance audits are often conducted semiannually. They may be less comprehensive than the full audit. If so, a full audit is performed every few years.

In the world today, there are hundreds of certification bodies/registrars. Most of them are private, for-profit organizations. Their services are valued by the supplier organizations they register—and by the customer organizations of the supplier organizations—because the registration service adds value in the supply chain. It is critical that the registrars do their work competently and objectively and that all registrars meet standard requirements for their business activities. They are, in fact, supplier organizations that provide a needed service product in the economy.

→ 3.15 INDUSTRY-SPECIFIC ADOPTIONS AND EXTENSIONS OF ISO 9000 STANDARDS

In some sectors of the global economy, there are industry-specific adoptions and extensions of the ISO 9000 standards. These situations are a classic example of a problem opportunity. As problems, such adaptations and extensions strain the goal of nonproliferation. As opportunities, they have been found effective in a very few industries where there are special circumstances and where appropriate ground rules can be developed and implemented consistently. These special circumstances have been characterized by:

1. Industries where the product impact on the health, safety, or environmental aspects is potentially severe; as a consequence most nations have regulatory requirements regarding the quality management system of a supplier.
2. Industries that have had well-established, internationally deployed industry-specific or supplier specific quality system requirements documents prior to publication of the ISO 9000 standards.

Fortunately, in the very few instances so far, the operational nonproliferation criteria of the ISO/IEC directives have been followed.

Medical Device Industry

Circumstance 1 relates to the medical device manufacturing industry. For example, in the United States, the Food and Drug Administration (FDA) developed and promulgated the Good Manufacturing Practice (GMP) regulations. The GMPs operate under the legal imprimatur of the FDA regulations, which predate the ISO 9000 standards. The FDA regularly inspects medical device manufacturers for compliance with the GMP requirements. Many of these requirements are quality management system requirements that parallel the subsequently published ISO 9002:1987 requirements. Other GMP regulatory requirements relate more specifically to health, safety, or environmental aspects. Many other nations have similar regulatory requirements for such products.

In the United States, the FDA has created revised GMPs that parallel closely the ISO 9000 standard, plus specific regulatory requirements related to health, safety, or environment. The expansion of scope to include quality system requirements related to product design reflects the recognition of the importance of product design and the greater maturity of quality management practices in the medical device industry worldwide. Similar trends are taking place in other nations, many of which are adopting ISO 9001 verbatim for their equivalent of the GMP regulations.

Current Good Manufacturing Practices (cGMPs) for human pharmaceuticals affect every American. Consumers expect that each batch of medicines they take will meet quality standards so that they will be safe and effective. Most people, however, are not aware of cGMPs, or how the FDA ensures that drug manufacturing processes meet these basic objectives. Recently, the FDA announced a number of regulatory actions taken against drug manufacturers based on the lack of cGMPs.

What Are cGMPs?

Current Good Manufacturing Practice regulations—or cGMPs—are enforced by the U.S. Food and Drug Administration (FDA). cGMPs provide for systems that ensure proper design, monitoring, and control of manufacturing processes and facilities. Adherence to the cGMP regulations ensures the identity, strength, quality, and purity of drug products by requiring that manufacturers of medications adequately control manufacturing operations. This includes establishing strong quality management systems, obtaining appropriate quality raw materials, establishing robust operating procedures, detecting and investigating product quality deviations, and maintaining reliable testing laboratories. This formal system of controls at a pharmaceutical organization, if adequately put into practice, helps to prevent instances of contamination, mix-ups, deviations, failures, and errors and ensures that drug products meet their quality standards.

The cGMP requirements were established to be flexible in order to allow each manufacturer to decide individually how to best implement the necessary controls by using scientifically sound design, processing methods, and testing procedures. The flexibility in these regulations allows companies to use modern technologies and innovative approaches to achieve higher quality through continual improvement.

Accordingly, the "c" in cGMP stands for "current," requiring companies to use technologies and systems that are up-to-date in order to comply with the regulations. Systems and equipment that may have been "top-of-the-line" to prevent contamination, mix-ups, and errors 10 or 20 years ago may be less than adequate by today's standards.

It is important to note that cGMPs are minimum requirements. Many pharmaceutical manufacturers are already implementing comprehensive, modern quality systems and risk management approaches that exceed these minimum standards.

Why Are cGMPs Important to Software Development?

A consumer usually cannot detect (through smell, touch, or sight) that a drug product is safe or if it will work. While cGMPs require testing, testing alone is not adequate to ensure quality. In most instances, testing is done on a small sample of a batch (e.g., a drug manufacturer may test 100 tablets from a batch that contains 2 million tablets), so that most of the batch can be used for patients rather than destroyed by testing. Therefore, it is important that drugs are manufactured under conditions and practices required by the cGMP regulations to ensure that quality is built into the design and manufacturing process at every step. Facilities that are in good condition, equipment that is properly maintained and calibrated, employees who are qualified and fully trained, and processes that are reliable and reproducible are a few examples of how cGMP requirements help to ensure the safety and efficacy of drug products.

How Does the FDA Determine if an Organization Is Complying with cGMP Regulations?

The FDA inspects pharmaceutical manufacturing facilities worldwide using scientifically and cGMP-trained individuals whose job it is to evaluate whether the organization is following the cGMP regulations. The FDA also relies on reports of potentially defective drug products from the public and the industry. The FDA will often use these reports to identify sites for which an inspection or investigation is needed. Most companies that are inspected are found to be fully compliant with the cGMP regulations.

In August 2002, the FDA announced the Pharmaceutical CGMPs for the 21st Century Initiative. In that announcement, the FDA explained the agency's intent to integrate quality systems and risk management approaches into its existing programs with the goal of encouraging industry to adopt modern and innovative manufacturing technologies. The CGMP initiative was spurred by the fact that since 1978, when the last major revision of the CGMP regulations was published, there have been many advances in manufacturing science and in our understanding of quality systems. In addition, many pharmaceutical manufacturers are already implementing comprehensive, modern quality systems and risk management approaches. This guidance is intended to help manufacturers implementing modern quality systems and risk management approaches to meet the requirements of the agency's CGMP regulations. The agency also saw a need to harmonize the CGMPs with other non-U.S. pharmaceutical regulatory systems and with the FDA's own medical device quality systems regulations. This guidance supports these goals. It also supports the objectives of the Critical Path Initiative, which intends to make the development of innovative medical products more efficient so that safe and effective therapies can reach patients sooner.

The CGMPs for the 21st Century Initiative steering committee created a Quality System Guidance Development working group (QS working group) to compare the current CGMP regulations, which call for some specific quality management elements, to other existing quality management systems. The QS working group mapped the relationship between CGMP regulations (parts 210 and 211 and the 1978 Preamble to the CGMP regulations) and various quality system models, such as the Drug Manufacturing Inspections Program (i.e., systems-based inspectional program), the Environmental Protection Agency's Guidance for Developing Quality Systems for Environmental Programs, ISO Quality Standards, other quality publications, and experience from regulatory cases. The QS working group determined that, although the CGMP regulations do provide great flexibility, they do not incorporate explicitly all of the elements that today constitute most quality management systems.

The CGMP regulations and other quality management systems differ somewhat in organization and in certain constituent elements; however, they are very similar and share underlying principles. For example, the CGMP regulations stress quality control. More recently developed quality systems stress quality management, quality assurance, and the use of risk management tools in addition to quality control. The QS working group decided that it would be very useful to examine exactly how the CGMP regulations and the elements of a modern, comprehensive quality system fit together in today's manufacturing world. This guidance is the result of that examination.

In ISO, a new technical committee, ISO/TC210, has been formed specifically for medical device systems. TC210 has developed standards that provide supplements to ISO 9001 clauses. These supplements primarily reflect the health, safety, and environment aspects of medical devices and tend to parallel the regulatory requirements in various nations. These standards are in late stages of development and international approval at this time.

Software Quality

The global economy has become permeated with electronic information technology (IT). The IT industry now plays a major role in shaping and driving the global economy. As in past major technological advances, the world seems fundamentally very different and, paradoxically, fundamentally the same. Computer software development occupies a central position in this paradox.

First, it should be noted that computer software development is not so much an industry as it is a discipline.

Second, many IT practitioners emphasize that computer software issues are complicated by the multiplicity of ways that computer software quality may be critical in a supplier organization's business. For example:

- The supplier's product may be complex software whose functional design requirements are specified by the customer.

- The supplier may actually write most of its software product or may integrate off-the-shelf packaged software from sub-suppliers.
- The supplier may incorporate computer software/firmware into its product, which may be primarily hardware and/or services.
- The supplier may develop and/or purchase from sub-suppliers software that will be used in the supplier's own design and/or production processes of its product.

However, it is important to acknowledge that hardware, processed materials, and services are often involved in a supplier organization's business in these same multiple ways, too.

What, then, are the issues in applying ISO 9001 to computer software development? There is general consensus worldwide that:

- The generic quality management system activities and associated requirements in ISO 9001 are relevant to computer software, just as they are relevant in other generic product categories (hardware, other forms of software, processed materials, and services).
- There are some things that are different in applying ISO 9001 to computer software.

There is, at this time, no worldwide consensus as to which things, if any, are different enough to make a difference and what to do about any things that are different enough to make a difference.

ISO/TC176 developed and published ISO 9000-3:1991 as a means of dealing with this important, paradoxical issue. ISO 9000-3 provides guidelines for applying ISO 9001 to the development, supply, and maintenance of (computer) software. ISO 9000-3 has been useful and is widely used. ISO 9000-3 offers guidance that goes beyond the requirements of ISO 9001, and it makes some assumptions about the life-cycle model for software development, supply, and maintenance. In the United Kingdom, a separate certification scheme (TickIT) for software development has operated for several years, using a combination of ISO 9001 and ISO 9003. The scheme has received both praise and criticism from various constituencies worldwide. Those who praise the scheme claim that it:

- Addresses an important need in the economy to provide assurance for customer organizations that the requirements for quality in software they purchase (as a separate product, or incorporated in a hardware product) will be satisfied.
- Includes explicit provisions beyond those for conventional certification to ISO 9001 to ensure competency of software auditors, their training, and audit program administration by the certification body.
- Provides a separate certification scheme and logo to exhibit this status publicly.

Those who criticize the scheme claim that it:

- Is inflexible and attempts to prescribe a particular life-cycle approach to computer software development that is out of tune with current best practices for developing many types of computer software.
- Includes unrealistically stringent auditor qualifications in the technology aspects of software development and qualifications whose technical depth is not necessary for effective auditing of management systems for software development.

- Is almost totally redundant with conventional third-party certification to ISO 9001, under which the certification body/registrar already is responsible for competency of auditors and accreditation bodies verify the competency as part of accreditation procedures.
- Adds substantial cost beyond conventional certification to ISO 9001 and provides little added value to the supply chain.

In the United States a proposal to adopt a TickIT-like software scheme was presented to the ANSI/RAB accreditation program. The proposal was rejected, primarily on the basis that there was not consensus and support in the IT industry and the IT-user community.

CMMi: Software and Systems Development

Another standard that gained popularity is the Capability Maturity Model (CMM), which is a service mark owned by Carnegie-Mellon University (CMU) and refers to a development model elicited from actual data. The data was collected from organizations that contracted with the U.S. Department of Defense, which funded the research and became the foundation from which CMU created the Software Engineering Institute (SEI). Like any model, it is an abstraction of an existing system. Unlike many that are derived in academia, this model is based on observation rather than on theory.

When applied to an existing organization's software development processes, CMM allowed an effective approach toward improving them. Eventually, it became clear that the model could be applied to other processes, which gave rise to a more general concept that is applied to business processes and to developing people.

The Capability Maturity Model was originally developed as a tool for objectively assessing the ability of government contractors' processes to perform a contracted software project. CMM is based on the process maturity framework first described in the 1989 book *Managing the Software Process* by Watts Humphrey. It was later published in a report in 1993 (Technical Report CMU/SEI-93-TR-024 ESC-TR-93-177 February 1993, Capability Maturity Model SM for Software, Version 1.1) and as a book by the same authors in 1995.

Though the CMM comes from the field of software development, it is used as a general model to aid in improving organizational business processes in diverse areas; e.g., in software engineering, system engineering, project management, software maintenance, risk management, system acquisition, information technology (IT), services, business processes generally, and human capital management. The CMM has been used extensively worldwide in government, commerce, industry, and software development organizations.

An organization may be assessed by an SEI-authorized lead appraiser and will then be able to claim that it has been assessed as CMM level X, where X is from 1 to 5 (maturity levels). Maturity Level 1 = Initial; Maturity Level 2 = Managed; Maturity Level 3 = Defined; Maturity Level 4 = Quantitatively Managed; Maturity Level 5 = Optimizing (read further for more explanation of the levels). Although sometimes called CMM certification, the SEI doesn't use this term due to certain legal implications.

In the 1970s, the use of computers became more widespread, flexible, and less expensive. Organizations began to adopt computerized information systems, and the demand for software development grew significantly. The processes for software development were in their infancy, with few standard or "best practice" approaches defined.

As a result, the growth was accompanied by growing pains: Project failure was common, the field of computer science was still in its infancy, and the ambitions for project scale and complexity exceeded the market capability to deliver.

In the 1980s, several U.S. military projects involving software subcontractors ran over budget and were completed much later than planned, if they were completed at all. In an effort to determine why this was occurring, the U.S. Air Force funded a study at the SEI.

The Standard CMMI Appraisal Method for Process Improvement (SCAMPI) is the official SEI method to provide benchmark-quality ratings relative to CMMI models. The CMMI model is used as a "ruler" to measure an organization's process definition as the model is a collection of process best practices assimilated into process areas. The SCAMPI appraisal methodology is utilized to measure how well an organization has institutionalized the process definition into its everyday way of doing business. SCAMPI appraisals are used to identify strengths and weaknesses of current processes, reveal development/acquisition risks, and determine capability and maturity level ratings. They are mostly used either as part of a process improvement program or for rating prospective suppliers. The method defines the appraisal process as consisting of preparation; on-site activities; preliminary observations, findings, and ratings; final reporting; and follow-on activities.

Active development of the model by the U.S. Department of Defense Software Engineering Institute (SEI) began in 1986 when Humphrey joined the Software Engineering Institute located at Carnegie Mellon University in Pittsburgh, Pennsylvania, after retiring from IBM. At the request of the U.S. Air Force, he began formalizing his Process Maturity Framework to aid the U.S. Department of Defense in evaluating the capability of software contractors as part of awarding contracts.

➔ 3.16 BENCHMARKING BEST PRACTICES: MOVING TOWARD SUSTAINABILITY

There are several organizations that ask, "We have been implementing improvement activities and have received amazing results. We continue to reduce costs, improve efficiency, and add Belts to the workforce. If we have done all of this, what is next to continually manage quality and ensure that results are sustained?" The answer to this question is to perform internal benchmarks of the organization or perform external benchmarks. This may seem like a methodology that has been done before, but with the technology capabilities many organizations have, benchmarking can be used to facilitate communication among plants, regions, and corporate headquarters. The type of communication that may occur is how to solve a quality problem faster and before it happens. For example, there are two plants that manufacture the same car, one in the United States and one in South Korea. The plant in South Korea may encounter a problem; through benchmarking, it could reach out to the plant in the United States to see if it had a similar problem. Most likely, the plants will have similar issues. The U.S. plant may have found a solution and can provide that solution to the South Korean plant and quickly fix the issue, rather than going through the process of diagnosing the root cause. This type of thinking and communication is how organizations can continually evolve, adapt, and create more efficiencies by itself. A person may consider this as an organization self-actualizing its maximum potential.

Our experience and research has led us to the development of the 7-Step Benchmarking Process©. The key to the 7-Step Benchmarking Process is the importance of taking benchmarking beyond its conventional level of analysis to facilitate the understanding of best practices and their implementation. In this connection, the performance measurement part of our benchmarking process, positioning analysis, is augmented with the proprietary performance measurement methodology that is expressed in the form of a single index, the Juran Quality Index©. Furthermore, the learning phase of the benchmarking process puts in place mechanisms that encourage the best practitioners to share their knowledge and experience with the rest of the organization. We have developed a unique and effective method for data normalization based upon the technical complexity of assets and processes. The Juran Complexity Factor© enables the direct comparison of differing assets and processes across an entire organization. Figure 3.12 is the model of the 7-Step process.

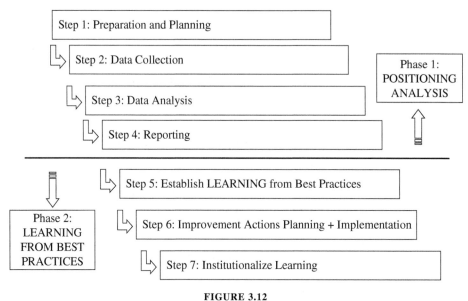

FIGURE 3.12
The Juran 7-Step Benchmarking Process.
(Juran Institute, Inc. Copyright 1994. Used by permission.)

Step 1: Preparation and Planning

Thorough advance planning is critical to success of benchmarking. As a result of the preparation and planning, an organization will:

- Determine the areas to be included.
- Review metric models, key performance indicators (KPIs), and KPI definitions.
- Create scorecards and structure to report findings.
- Determine logistics and project roll-out of the plan.
- Communicate key issues.
- Agree on roles and responsibilities.

It is very important to clearly and unambiguously define metrics and KPIs in Step 1 to ensure consistency in data collection.

Step 2: Data Collection and Validation

Valid data are the key to success of any benchmarking program. Incorrect or inaccurate data can easily result in misguided conclusions and inappropriate actions and can lead to the failure of any improvement program.

Step 3: Data Analysis

Effective data normalization is essential to meaningful benchmarking because it enables "apples to apples" comparisons in greatly differing operational situations. In this program, the Juran Complexity Factor is used to normalize data, and the Juran Quality Index is used for the comparative analysis between sites.

The Juran Complexity Factor (CF) is an overall measurement of the complexity of the routine operation and maintenance of any type of site. The CF is calculated by breaking down the processes and assets at each site into modules of standard equipment. It is based upon an empirical system based on 25 years of worldwide experience in operation maintenance. The system has subsequently been refined and modified by Juran. The CF relies on a weighting factor that is calculated depending on the types of process and products and the inputs and outputs to that process. The identification of items is systematic and consistent in order to avoid omissions and duplications. In order to maintain an objective and unbiased assessment of each of the processes or sites, the calculation of the CF will be carried out by a Juran consultant.

The Juran Quality Index is a cumulative indicator of overall performance based on both efficiency and effectiveness. It is a numerical value indicating the percentage deviation of a site's performance, relative to the corresponding group of sites. Thus, it provides a balanced viewpoint, accounting for all aspects of sites' performance and comparing these to the average. It has the advantage of providing a single index figure that immediately gives a high-level indication of relative performance, but it is also possible to further break down the index into its main constituent elements. The following example of a

"Management Dashboard" shows how the index may be broken down into its constituent elements and represented graphically. Juran will supply a JQI for each site.

Step 4: Reporting—Benchmark Report

All data will be analyzed and conclusions drawn from all sites. The report will include all normalized data sets/charts, all KPIs, key findings per system (gap analysis with prioritization, strengths and weaknesses, and improvement suggestions). The benchmarking report will be made up of seven components:

1. High-level overview
2. Key findings and site comparisons
3. Gap analysis
4. Complexity factor assessments for all benchmarked processes and assets
5. KPI/metric definitions
6. Process and asset descriptions
7. Next steps and sustainability

This step is similar to the one identified in the Organization Health Check but continually expands and more information is extracted and shared amongst the organization.

Step 5: Establish Learning from Best Practices

After the benchmark report has been delivered to the leadership, the organization should conduct workshops and have representatives from each area included in the benchmark. The objective of this workshop is the identification of a master class of opportunities and the transfer of knowledge from the best sites to other sites. The site that has the best practices will present to the other sites the "why's and the how's" of their best practices. The intention is that the audience learns from these presentations to help them subsequently formulate their own improvement programs. The best practice workshops are recommended to be conducted on an annual basis to ensure sustainability of desired goals.

Step 6: Improvement Action Planning and Implementation

This road map shows how the organization will continue to sustain the best practices and continually improve and meet the organization's strategic goals in a timely manner.

Step 7: Institutionalized Learning

As part of the action planning, there may be requirements to teach or reteach to achieve the best practices identified. This may seem like a repetitious step, but it is essential for an organization to evolve and adapt on its own.

→ 3.17 BENCHMARKING: WHAT IT IS AND WHAT IT IS NOT

Benchmarking has been in existence for a many years. The concept of one individual observing how another performs a given task and then applying any learning from that to adapt and improve how the task is executed is one of the fundamental ways in which human beings learn and develop. In the context of business, learning from one's competitors has also been in existence for as long as business has. However, the application of learning from best practices to the business environment in a structured, methodical, and indeed legal and ethical way is relatively new. The Xerox Corp. is most commonly credited with developing the modern form of benchmarking, and it is fair to say that the majority of today's benchmarking practices are built upon the approach developed by Xerox in the 1970s.

Although its story has previously been well told in a multitude of management texts, it is still worthy of brief comment here to set the scene. A combination of poor product quality, high overheads, and increasing competition from a growing number of Japanese organizations had left Xerox in a precarious position in the late 1970s. A visit to Japan provided the wakeup call that change was essential if the company were to survive (Kearns and Nadler, 1993). Xerox put in place a series of benchmarking activities aimed at identifying the best performing organizations in various aspects of its business and determining what it was that these organizations were doing that enabled a superior performance. Most famous is the benchmarking of logistics operations that they undertook with L.L. Bean (Camp, 1989). With this, modern benchmarking was born.

Benchmarking has evolved to become an essential element of the business performance improvement toolkit and is now frequently used by many organizations in a wide range of different industries. Despite this, it remains one of the most widely misunderstood improvement tools because it means many different things to many different people and because, all too frequently, benchmarking projects fail to deliver on their promise of improvement or real results.

However, executed correctly, benchmarking can provide a powerful focus for organizations, driving home the facts and convincing the organization of the need to embark upon improvement strategies. Benchmarking is a tool that enables the identification and, ultimately, the achievement of excellence, based on the realities of the business environment rather than on internal standards and historical trends.

Benchmarking is not what we would term "industrial tourism," in which superficial industrial visits are undertaken in the absence of any point of reference and do not assist in the improvement process. It is impossible to acquire detailed knowledge after only a quick glance or one short visit, and it is rare for such visits to result in an action plan that will lead to improvement. In the absence of prior benchmarking, it is also difficult to identify which organizations should be visited, so there is a real risk that visits are made to organizations that are perceived as being the best or at least better, when the reality may be very different. However, there is a valuable role to be played by this type of site visit when it is conducted following a structured benchmarking analysis and the organization being visited has been identified as a best performer.

Benchmarking should not be considered a personal performance appraisal tool. The focus should be on the organization and the individuals within it. Failure to adopt this philosophy will only lead to resistance and will undoubtedly add roadblocks to a successful benchmarking journey. Nor should benchmarking be a momentary glimpse, but rather should be considered a continuous process. Organizations must change performance rapidly to remain competitive in business environments today. This fast-paced tempo is further accelerated in sectors where benchmarking is commonplace and where businesses rapidly and continuously learn from one another. A prime example comes from the oil and gas industry, where organizations have to respond to ever-increasing business, technological, and regulatory demands. The majority of the key players in this industry are participating in focused benchmarking consortia on an annual basis. It is also much more than a competitive analysis. Benchmarking goes further than examining the pricing and features of competitor's products or services. It considers not only the output, but also the process by which the output was obtained. Benchmarking is also much more than market research, considering the business practices in place that are enabling the satisfaction of customer needs and thus realizing superior business performance. It provides evidence-based input offering a powerful focus for management, driving home the facts and convincing the organization of the need to embark on improvement activities.

Participating in benchmarking should also not be viewed as a standalone activity. To succeed, it must be part of a continuous improvement strategy, it must be conducted regularly, and it should be enveloped in the continuous improvement culture of an organization. Like any other project, it has to have the full support of senior management, the resources necessary to fulfill the objectives, and a robust project plan that is adhered to.

Finally, benchmarking should not be viewed as the answer in itself. It is a means to an end. An organization will not improve performance by benchmarking alone. It must act upon the findings of the benchmarking in order to improve. The output of benchmarking should provide input to decision making or improvement action planning. This requires detailed consideration of the benchmarking analysis, formulation of learning points, and development of action plans in order to implement change and realize improvements.

So how can we define benchmarking? A scan of the literature quickly reveals myriad definitions (see Anand and Kodali, 2008), each offering a slight variance on a common theme. Rather than repeat these here we prefer to offer our own definition:

> Benchmarking is a systematic and continuous process that facilitates the measurement and comparison of performance and the identification of best practices that enable superior performance.

This definition is deliberately generic such that it can encompass all types of benchmarking. In this context, measurement and comparison may be between organizations, business units, business functions, business processes, products, or services. The benchmarking may be internal or external, between competitors, within the same industry, or cross-industry. Irrespective of the category of benchmarking, this definition still applies.

Objectives of Benchmarking

The objectives of benchmarking can be summarized as follows:

1. Determine superior performance levels.
2. Quantify any performance gaps.
3. Identify best practices.
4. Evaluate reasons for superior performance.

5. Understand for performance gaps in key business areas.
6. Share knowledge of working practices that enable superior performance.
7. Enable learning to build foundations for performance improvement.

When talking of superior performance, ultimately of course the aim should be for world-class performance. However, in reality it is often difficult to be able to ensure that the world's leading performers are participating in a given benchmarking exercise. Instead, benchmarking partners should be selected carefully to ensure the output will provide the required added value.

Having determined superior performance, the gap between this and the performance level of the benchmarker is quantified. The working practices enabling superior performance are identified and the enablers evaluated. This knowledge is then shared between benchmarkers to enable the learning to be taken away and implemented as part of a performance improvement program.

Thus, benchmarking can be viewed as a two-phase process, where phase 1 is a positioning analysis aimed at identifying gaps in performance and phase 2 is focused on learning from those best practices that enable superior performance.

Why Benchmark?

There are two good reasons for organizations to benchmark themselves. First, it will help them stay in business by offering opportunities to become better than other similar organizations, competitors or not. Second, it ensures that an organization is continually striving to improve its performance through learning. Benchmarking opens minds to new ideas from other sources—whether they are within the same industry or from many other unrelated industries—identifying how those who have demonstrated performance leadership work.

On the other hand, many organizations benchmark simply to be able to demonstrate to stakeholders—be they customers, shareholders, lenders, regulators, etc.—that the organization is performing at an acceptable level. Of course, this is a perfectly legitimate reason for benchmarking, although the real potential value of the technique is missed by confining the focus in this way.

Benchmarking also provides a very effective input to an organization's strategic planning processes by establishing credible goals and realistic targets based on external references.

To really grasp the intent of benchmarking, an organization should be benchmarking not only to demonstrate good performance, but to identify ways in which it can change its practice to significantly improve its performance. Those organizations with a strong performance improvement culture will be benchmarking continuously because this provides them with objective evidence of where to focus improvement activities, how much they should be improving, and what changes to their working practices they might consider in order to realize improvements.

Classifying Benchmarking

There are many different ways to classify benchmarking, and the literature is full of different classifications (see Anand and Kodali, 2008), which make it very confusing for someone new to the topic to really understand what benchmarking is and which approach is best for that person or company. The fact of the matter is that there is an underlying process that can be considered generic to almost all types of benchmarking. However, in order to provide some clarity on the differences in classification, we have considered benchmarking in terms of what it is that is to be benchmarked, who the benchmarking is going to involve, and how the benchmarking is to be conducted:

- Subject matter and scope (what)
- Internal and external, competitive and noncompetitive benchmarking (who)
- Data and information sources (how)
- Subject matter and scope (what)

Benchmarking is often categorized according to what it is that is being benchmarked. Typical categories include:

- Functional benchmarking
- Process benchmarking
- Business unit or site (location) benchmarking
- Projects benchmarking
- Generic benchmarking
- Business excellence models

Functional Benchmarking

Functional benchmarking describes the process whereby a specific business function forms the focus for the benchmarking. In the context of the organization, this may involve benchmarking several different business units or site locations. Typical examples of functional benchmarking include the analysis of the procurement, finance, IT, safety, operations, or maintenance functions. The analysis focuses on all aspects of the function rather than the processes involved and the specific activities conducted.

Process Benchmarking

Process benchmarking is where the focus of the study is on a specific business process or a part thereof. Examples include product development, invoicing, order fulfillment, contractor management, customer satisfaction management, etc. Process benchmarking will often involve several functional groups and may also involve many different site locations. There is often a great deal of overlap between what is termed *functional benchmarking* and *process benchmarking* (e.g., a benchmarking of the procurement process may look very similar to a benchmarking of the procurement function). Many business processes are not specific to any one industry and so can benefit from broadening participation in the analysis to organizations from a multitude of industries.

Business Unit or Site (Location) Benchmarking

Benchmarking individual business units or site locations against each other is often (but not always) seen in internal benchmarking studies within a single organization. The performance of each unit is analyzed and compared to that of the other units. This analysis may incorporate all activities of each unit in their entirety or may be confined to selected functional groups or business processes. For example, Juran manages an annual benchmarking consortium comparing the performance of many of the world's oil- and gas-processing facilities. Each of the key business processes is included in the analysis, and participants come from a wide range of different organizations.

Projects Benchmarking

This type of benchmarking focuses on projects undertaken by organizations. Because projects vary widely in their nature, these studies are normally tailored for specific project types. For example, oil pipeline construction projects, software implementation projects, facility decommissioning projects, etc., may all be benchmarked. Normally included are all the business processes pertaining to the project being analyzed, although the scope may often be limited to a subset of processes. For example, a construction project benchmarking may focus specifically on contractor selection, procurement, commissioning, etc.

Generic Benchmarking

Generic benchmarking considers all business processes required to achieve a certain level of performance in a given area. The focus is upon the result and what is required to achieve it. For example, a hospital may undertake a generic benchmarking exercise to identify ways in which it can reduce treatment waiting times. In so doing, the hospital may benchmark across a number of different industries where customer waiting times are of paramount importance—e.g., insurance claims processing, vessels clearance procedures for major waterways (e.g., Suez Canal), and calamity response times for the different emergencies services (police, fire, ambulance). Inevitably, there is a great deal of overlap between process benchmarking and generic benchmarking, although in the latter there is often less emphasis on gap analysis and more upon a detailed consideration of working practices.

→ 3.18 INTERNAL AND EXTERNAL COMPETITIVE AND NONCOMPETITIVE BENCHMARKING (WHO)

Benchmarking studies are also frequently classified by type of participant. Depending on the type of benchmarking being undertaken, it is not always possible to have any control over participant selection. But, where this is possible, the selection of whom to benchmark with is one of the first and often the most difficult tasks at the outset of any benchmarking study. Potential participants will be identified according to a range of criteria, the main one being the perceived performance level (where superior/world-class performance is the aim).

The four main types of benchmarking are:

1. Internal benchmarking
2. External benchmarking
3. Competitive benchmarking
4. Noncompetitive benchmarking

Each of these has specific benefits and drawbacks that need to be considered when selecting the most suitable benchmarking approach. These include:

- The similarity between the participants in terms of the subjects to be benchmarked
- The level of control over the benchmarking process
- The cost and time input required to conduct the benchmarking
- The degree of openness that is possible and the level of confidentiality necessary
- The potential for learning and therefore performance improvement

Internal Benchmarking

Internal benchmarking is the comparison of performance and practices of similar operations within the same organization. Depending on the size of the organization and the nature of its business, this may or may not be feasible for the organization to have duplicate groups conducting the same activities. Should this be the case, internal benchmarking is often a popular first step because it allows organizations to prepare themselves for broader benchmarking activities within the safety of their own environment where they have full control over the process. This is likely also to be the least costly and time-consuming way to benchmark. However, the potential for finding performance leaders is much smaller, and the opportunity for learning is usually more limited.

External Benchmarking

External benchmarking involves participants from different organizations. The opportunity for learning is normally greater than that achievable by internal benchmarking, but there is obviously a requirement to share information outside of the organization. This brings with it some potential restraints. There will almost certainly be limits on the data organizations are willing to share, especially if the other participants are competitors, and there will of course be a need for stricter confidentiality. External benchmarking is further categorized according to nature of the participants, who can either be competitors (competitive benchmarking) or not (noncompetitive benchmarking).

Competitive Benchmarking

Competitive benchmarking is a form of external benchmarking where the participants are all in competition with each other. By definition, the participants in a competitive benchmarking program are from the same industry, and the focus is normally on industry-specific processes. For example, Juran has studied patient safety performance among different hospitals. This normally brings with it a high degree of sensitivity that needs to be carefully managed for a successful outcome to be realized, but when conducted properly, the results can be very valuable. Conversely, topics that are not directly related to the core business of the competing organizations are usually less sensitive to benchmarking between competitors, but ironically, these more generic topics are not normally those that an organization wishes to benchmark with its competitors because greater value is more frequently gained from benchmarking such topics cross-industry. There are also likely to be some subject areas that most organizations will not be willing to benchmark with competitors, such as proprietary processes or products and innovations that provide competitive advantage.

Noncompetitive Benchmarking

Noncompetitive benchmarking is a form of external benchmarking where the participants are not in direct competition with each other. They may be from within the same industry or cross-industry. For example, an organization operating a container port in the United States may benchmark with another in Europe. While they are in the same industry, they are unlikely to be competitors because they are operating in different markets. Juran manages an annual global benchmarking consortium for gas pipelines. The participants are all in the same industry and primarily interested in benchmarking those processes specific to their industry. But they are not in direct competition with each other because they operate in totally different marketplaces, delineated by geographical region. This means that they are very willing to share knowledge and practices openly for mutual benefit without fear of giving anything away that may affect their competitiveness.

In cross-industry noncompetitive benchmarking, it is those subject areas that are not industry-specific that are most commonly analyzed, and it is in this classification that most generic benchmarking studies sit. These tend to be support processes such as administration, human resources, research and development, finance, procurement, IT, and health, safety, and environmental (HSE). Cross-industry external benchmarking potentially offers the greatest opportunities for learning and performance improvement for a number of reasons. First, the pool of potential participants is much bigger. Second, participants thought to be superior performers in the subject area being benchmarked can be identified and invited to participate. Third, the willingness to share knowledge will be greatest where there is no fear of competitive sensitivity.

→ 3.19 DATA AND INFORMATION SOURCES (HOW)

Benchmarking can also be classified according to the source of the data used in the comparative analysis. Such classification can be made in many ways, but the following list addresses the main categories found:

- Database benchmarking
- Survey benchmarking
- Self-assessment benchmarking
- One-to-one benchmarking
- Consortium benchmarking

Database Benchmarking

In this type of benchmarking, data from a participant are compared to an existing database containing performance data. An analysis is performed, and the results are provided to the participant. Benchmarking in this way normally requires a third party to administer the database and produce the analyses. The development of the Internet led to the growth in this type of benchmarking because it can be easily administered online. The participating organization can submit its data via an online questionnaire and receive a report of the analyses online, usually in a very short time period. A quick search of the Internet reveals the large number and wide range of online benchmarking databases available.

This type of benchmarking is also sometimes offered by consultancy organizations that have accumulated performance data pertaining to specific activities. For example, Juran has been benchmarking in the oil and gas industry since 1995 and, during this time, has developed a comprehensive database of performance figures relating to this industry. And because the data have been well defined and thoroughly validated during the collection process, the data are extremely reliable. It is, therefore, an excellent data source against which oil and gas organizations can be benchmarked.

Many organizations start out on their benchmarking journey by purchasing data from proprietors of such databases. While this type of benchmarking can be very useful in providing fast feedback on performance, it can have drawbacks. The participant has no control over the content of the analysis and has to accept the metrics that are used to determine performance. Often the source of the data is not disclosed; thus, it may be difficult for the benchmarker to assure itself of its relevance. The metrics used may not be clearly defined and may not be validated effectively, resulting in poor data quality and flawed analyses. Care should therefore be taken when entering into this type of benchmarking, and the participant should realize the potential shortcomings. For best results, only a bone fide consultant with a strong reputation and a good track record should be sought.

Survey Benchmarking

This term is used to describe benchmarking exercises conducted via the completion of a survey or a review process. Typically, a survey document is sent to participating organizations to be completed and returned. Sometimes the survey documents are sent to organizations without their prior agreement to participate in the hope that they will complete the survey and return it. Of course, this approach is nearly always less successful, with a relatively poor return rate.

The survey may be organized by a third-party consultant or by one of the participating organizations themselves, although in the latter case there will be greater restrictions on what data can be shared directly between participants in order to ensure compliance with antitrust legislation. Sometimes, there may be a fee involved for organizations to participate, and sometimes a single organization or even a consultancy may sponsor the entire exercise, in which case the output for the other participants is often less sophisticated.

The potential drawbacks of this approach are similar to those encountered with database benchmarking in that each organization has minimal control over the benchmarking process, the metrics may not be defined adequately, and the validation of submitted data may be limited. Nonetheless, this type of approach can provide a useful albeit limited comparative analysis with limited effort required on the part of each participating organization.

The survey process can be extended to include a review element, whereby the benchmarking coordinator (normally a third-party consultant) will visit each of the participating organizations as part of the survey process. This allows the consultant to delve more deeply into specific areas to gain richer data (often qualitative data), which can better inform the learning process of the participants, particularly in the area of assessing working practices that underpin superior performance.

Self-Assessment Benchmarking

As previously discussed, self-assessment is an integral part of many performance excellence models. These self-assessments can be used for generic benchmarking between and among organizations across all industries. These models provide an excellent framework for comparative benchmarking, whereby all elements required for excellence are considered. The analysis will often focus not just on quantitative data analysis, but also on a qualitative view of working practices. However, there is an inherent weakness in the process associated with the subjective nature of self-assessment. Sometimes, third parties (consultants) are employed to oversee the process in order to introduce some level of objectivity or even to conduct the assessments.

One-to-One Benchmarking

This type of benchmarking is probably the most common reported in the literature, but as we pointed out, benchmarking is not industrial tourism, whereby relatively superficial site visits are conducted between organizations to explore performance. Such exchanges rarely bring fruitful insights, and there is always the uncertainty of whether the organization being visited is really a superior performer or not.

However, if, having conducted a benchmarking study, an organization is identified as delivering superior performance levels in the subject area of interest, then a one-to-one benchmarking can investigate specific areas in much greater depth and deliver rich information pertaining to the drivers of superior performance. This approach is common in consortium benchmarking where, having received the benchmarking analysis, two organizations will agree to benchmark further one-to-one in order to obtain more detailed understanding of specific working practices. A good example of this comes from a cross-industry procurement benchmarking consortium managed by Juran. Following participation in the study, two of the participants agreed to undertake a one-to-one benchmarking. One organization was particularly strong in contract tendering and contractor selection, while the other demonstrated superiority in strategic procurement. Cooperating in this way led to a greater understanding of leading working practices in each of these areas and improved performance for both parties.

Consortium Benchmarking

Without doubt, this form of benchmarking has the greatest potential to deliver improved performance for its participants. A consortium is formed between participants, usually (but not always) supported by a third-party facilitator. They agree on the participants to be invited; the subjects to be benchmarked; the methodology to be followed; the metrics (and their definitions) to be used; the validation criteria; the nature of the analysis, reporting, and deliverables; and the timescales to be adhered to. Thus, the participants have a very high level of control over the entire process, and the outcomes from the process are normally reliable data, thorough analysis, and valuable results. This approach does require a great effort on behalf of each of the participants in order to achieve the desired outcome. It is, therefore, more time-consuming and often more costly to undertake, but the added value is normally far in excess of that achieved through other benchmarking approaches.

We have demonstrated that benchmarking can be classified in many different ways according to the subject matter of the analysis, the nature of the benchmarking participants, data sources, and the methodologies employed. However, differentiating in this way is largely academic, and while these different approaches have their inherent pros and cons and some are clearly more effective than others, they all should have the same ultimate objective: to provide learning on how to improve business performance.

↷ 3.20 DATA NORMALIZATION

The single biggest problem in any benchmarking exercise is how to compare benchmarked subjects on a like-for-like basis (i.e., how to compare apples with pears). In some circumstances, the benchmarkers will be similar enough to enable direct comparisons of performance between them. However, more typically, the subjects being benchmarked will all be different from each other—be they organizations as a whole, business units, different sites, or different functional groups, business processes, or products. No two subjects will be identical, although the extent of difference between them will vary considerably depending on what and who is being benchmarked. Thus, in order to be able to compare differences in performance levels, some intervention must be exercised. Some form of data normalization is usually required to enable like comparisons to be made between what may be very different subjects. Without it, direct comparisons of performance are normally impossible and may lead to misinformed conclusions. Normalization can be made on the basis of a wide range of factors including scope, scale, contractual arrangements, regulatory requirements, and geographical and political differences.

One solution is to organize benchmarkers into categories or peer groups with other benchmarkers or datasets with similar characteristics. The key is to be able to identify the factors that are driving the performance and then develop a method by which these drivers can be considered when comparing performance metrics. In its simplest form, this may involve stratifying the data according to underlying criteria. For example, if a health authority wishes to compare death rates of people in different regions, it may stratify these according to gender or age. Another example comes from the chemical industry. A series of chemical organizations may decide to benchmark their performance in the field of managing for the environment, and in so doing, they may wish to compare emission levels for a variety of polluting gases (e.g., oxides of nitrogen and sulfur, carbon dioxide and methane). These data could be stratified according to their harmful impact to the environment through the use of a standardized measurement such as the environmental impact unit (EIU). A further example might be those organizations comparing the efficiency of their R&D activities using a KPI that measures the percentage of sales attributable to new products or services (e.g., products that have been on the market for less than two years).

But, of course, even within these groups there may be differences between the benchmarking subjects. In order to be able to conduct a valid comparison of performance, these differences in characteristics need to be taken into consideration in the analysis. The most effective way of doing this is through normalization of the performance data.

Normalization is essentially the process of converting metrics into a form that enables their comparison on a like-for-like basis, accounting for all (or as many as possible) of the variation between the benchmarking subjects. What is essential is that the normalizing factor used is truly a driver for the performance being benchmarked. For example, if benchmarking the operating costs of the invoicing function of an organization, perhaps a suitable normalizing factor is the number of invoices raised. For example, the costs could be compared on a per-invoice-raised basis. However, it may be that some invoices are more complicated to produce than others (e.g., they may contain more line items or be for a higher total value, which requires more checks before the invoice is raised), so this way of normalizing may not after all be appropriate. The most common way of normalizing is by looking at performance per unit or per hour. For example, if we are measuring the cost of manufacturing a motor car, we might compare the cost per vehicle produced, or if we are looking at the time taken to treat a hospital patient with a given ailment, we might consider the number of patients examined per hour.

In some cases, a simple measurement per unit is not sufficient to accommodate for the variation observed between benchmarking subjects, and a more sophisticated approach has to be developed. In such cases, the use of weighting factors that represent the variation of the different benchmarking subject is often a very effective way of normalization. Weighting factors may be developed in relation to cost, time, and efficacy. An example of a highly effective weighting factor is the Juran Complexity Factor (JCF). The JCF was developed to enable like-for-like comparisons to be made between oil and gas production facilities of very different size and design. The normalizing factor takes into consideration the equipment present in the facility and the time it takes to operate and maintain this equipment under normal conditions. The JCF is then used to normalize all cost performance between facilities in the benchmarking. This enables organizations to directly benchmark their facilities with those of other organizations, even though they may be very different is design and size.

The efficacy of any normalization method should be fully tested before it is implemented. As mentioned, for a normalizing factor to be effective it must be representative of the driving force for the performance subjects being benchmarked. This means that there must be a good relationship between the performance metric and the normalizing factor. A good way of testing this is to examine the correlation between the normalizing factor and the performance metric being normalized. There should be a strong direct relationship between the two. For example, an increase in the normalizing factor should lead to an increase in the metric being benchmarked (and vice versa), although this relationship may or may not be linear.

→ 3.21 ANALYSIS AND IDENTIFICATION OF BEST PRACTICES

The aim of the analysis is to determine the findings from the data collected in the benchmarking in conjunction, where appropriate, with other pertinent data and information from a number of different sources, including the public domain, the participants themselves, and from any previous editions of the benchmarking study. The level of analysis will depend on the scope and objectives agreed at the commencement of the benchmarking.

It is essential that the analysis be impartial and totally objective. It must also be aligned to the benchmarking objectives, and to be of value, it must indicate the benchmarker's strengths and weaknesses and determine and (where possible) quantify the gaps to the best performers, and identify (as far as is possible) the reasons for these gaps. It is important that the metrics are considered collectively and not in isolation because the results from one metric may help to explain those of another. The strategies and working practices of each of the participants should also be explored and used to determine how they may influence their performance.

The performance data and any normalization data streams are analyzed to compare participant performance and determine performance gaps. It is also important to consider a level of statistical testing of the data to ensure that the comparisons being made are statistically significant and the conclusions drawn thereafter are valid.

Quantitative analyses are typically made in relation to the top quartile (i.e., the boundary to the 25th percentile), the best in class (i.e., the single best performer), or the average (mean) of the benchmarking population. There are pros and cons to comparisons to each of these criteria. Analysis of the gap to the best in class is probably the most common and, on the face of it, seems the most obvious—after all the objective is to close the gap to the best performer. However, making comparisons to the data of a single benchmarker at a single point in time will always carry a risk that there is error in the data value (although the validation process should minimize this error) or that the performance level reported is not sustainable in the longer term and therefore not realistic. In contrast, comparison to the top quartile and, in particular, to the average is more stable and reliable because these comprise data from more than one participant.

Reasons for apparent differences in performance should be considered during the analyses. With multinational or globally benchmarking studies, it is important to consider the impact that may be attributed to differences in geographical location. For example, when analyzing costs, it is clear that cost levels (e.g., salaries) in the West cannot be easily compared to Asia, the Middle East, Russia, Africa, or Latin America. In addition, fluctuations in exchange rates between currencies can have a dramatic effect. Likewise different tax regimes, regulatory requirements, political policies, and cultural difference can all significantly influence performance.

Report Development

Once the analysis is complete, it must be reported to the benchmarking participants. The content of the report and the medium used for reporting will have been agreed upon at the outset of the benchmarking exercise and will, in part, be determined by the type of benchmarking being undertaken.

Reports may be delivered online or electronically or in paper, hard-copy format. Whatever the medium selected, the report must present the benchmarking findings in a clear, concise, and easily understood form. Optimum use of color, diagrams, pictures, and charts should be made in order to optimize communication of the findings. Charts and tables should be annotated to provide guidance to the reader. The analysis should be reported in full, together with recommendations for the focus of performance improvement efforts required in order to close gaps.

One very important point that must be addressed is the level of data anonymization that will be employed in the report. There is always a trade-off between confidentiality and learning opportunity. The higher the level of confidentiality, the lower the potential for learning. If the identity of the superior performers cannot be revealed, then the opportunity to learn from them is minimized. However, compliance with antitrust legislation is always a primary requirement. Thus, the report must always be in line with any confidentiality agreement made between the benchmarkers and must meet legal requirements. But, in order to maximize the learning potential, the degree of openness should also be maximized.

Unfortunately, many benchmarking exercises will stop at this point. But to maximize the value gained from benchmarking, organizations must go further to try to understand the practices that enable the leaders to attain their superior performance levels. This is the purpose of phase 2 of Juran's 7-Step Benchmarking Process.

SUMMARY

- All organizations periodically need a companywide assessment of quality.
- Quality assessment comprises four elements:

   Cost of poor quality
   Standing in the marketplace
   Quality culture
   Operation of the quality system

- The Malcolm Baldrige National Quality Award provides criteria to identify organizations that have achieved the highest levels of quality. This award recognizes superior achievement.
- The ISO 9000 standards provide minimum criteria for a quality system. These documents provide some assurance to potential customers that an organization certified as meeting the standard has an adequate quality system.

PROBLEMS

3.1. The Federated Screw Company manufactures a wide variety of made-to-order screws for industrial companies. The designs are usually supplied by customers. Total manufacturing payroll is 260 people with sales of about $28 million. Operations are relatively simple but geared to high-volume production. Wire in rolls is fed at high speeds to heading machines, where the contour of the screw is formed. Pointers and slotters perform secondary operations. The thread-rolling operation completes the screw configuration. Heat treatment, plating, and sometimes baking are the final steps and are performed by an outside contractor located nearby.

You have been asked to prepare a quality cost summary for the company and have made the following notes:

- The quality control department is primarily a final inspection department (eight inspectors), which also inspects the incoming wire. Patrol inspection (one inspector) is performed in the Heading Room by checking the first and last pieces of each run. The quality control department also checks and sets all gages used by that department and by production personnel. An inspector's salary is approximately $24,000 a year.
- Quality during manufacture is the responsibility of the operator setup teams assigned to batteries of about four machines each. It is difficult to estimate how much of their time is spent checking setups or checking the running of the machines, so you have not tried to do this as yet. Production has two sorting inspectors, each earning $18,000, who sort lots rejected by final inspection.
- The engineering department prepares quotations, designs tools, plans the routing of jobs, and establishes quality requirements, working from customers' prints. The engineers also do troubleshooting, at a cost of about $20,000 a year. Another $16,000 is spent in previewing customers' prints to identify critical dimensions, trying to get such items changed by the customer, and interpreting customers' quality requirements into specifications for use by Federated inspectors and manufacturing personnel.
- Records of scrap, rework, and customer returns are meager, but you have been able to piece together a certain amount of information from records and estimates:

    Scrap from final inspection rejections and customer returns amounted to 438,000 and 667,000 pieces, respectively, for the last two months.
    Customer returns requiring rework average about 1 million pieces per month.
    Scrap generated during production is believed to be about half of the total floor scrap (the rest is not quality related) of 30,000 pounds per month.
    Final inspection rejects an average of 400,000 reworkable pieces per month. These items can then be flat rolled or rerolled.

- Rough cost figures have been obtained from the accountants, who say that scrap items can be figured at $12.00 per thousand pieces, floor scrap at $800 per thousand pounds, reworking of customer returns at $4.00 per thousand pieces, and flat rolling or rerolling at $1.20 per thousand pieces. These figures are supposed to include factory overhead.

    Prepare a quality cost summary on an annual basis. (This example was adapted from one originally prepared by L. A. Seder.)

3.2. Review the explanation of the market research study depicted in Table 3.6. What additional question would have been useful to ask during the study?

3.3. Information on the present quality culture in an organization is an important input to assessment. Asking employees their opinions about the quality culture incurs some risks. State two such risks.

3.4. In a survey used to learn about the quality culture in an organization, employees are asked questions that relate to three levels—upper management, their own managers, and the people in their work group. For each of these levels, state two questions that would identify employee perceptions about the quality culture.

3.5. When consumers eat at a fine restaurant, they expect excellent food and service. Research has identified seven features of restaurant service. What are these service features?

REFERENCES

Atkinson, H., J. Hamburg, and C. Ittner (1994). *Linking Quality to Profits,* ASQ Quality Press, Milwaukee, and Institute of Management Accountants, Montvale, NJ.

Campanella, J. ed. (1999). *Principles of Quality Costs,* 3rd ed., ASQ, Milwaukee.

Cokins, G. (1999). "Why Is Traditional Accounting Failing Quality Managers? Activity Based Costing Is the Solution," *Annual Quality Congress Proceedings,* ASQ, Milwaukee.

DeFeo, J. A. and W. Barnard (2003). *Juran's Six Sigma Breakthrough and Beyond,* McGraw-Hill, New York.

Lauter, B. E. (1997). "Determining the State of Your Customers," Sterling Quality Conference, Orlando, FL.

Mortimer, J., J. A. DeFeo, and L. Stepnick (2003). "Reducing the Costs of Poor Quality Health Care" a report published by the Midwest Business Group on Health in collaboration with the Juran Institute, Inc. and The Severyn Group, Inc.

Rust, R. T., A. J. Zahorik, and T. L. Keiningham (1994). *Return on Quality,* Probus, Chicago.

Schottmiller, J. C. (1996). "ISO 9000 and Quality Costs," *Annual Quality Congress Proceedings,* ASQ, Milwaukee, pp. 194–199.

Werner, F. M. and J. A. F. Stoner (1995). *Modern Financial Managing,* Harper Collins College Publishers, New York, pp. 143–144.

CHAPTER 4

Improving Quality While Decreasing Cost

→ 4.1 LEAN SIX SIGMA AS A BREAKTHROUGH IMPROVEMENT MODEL

All quality management systems must include a quality improvement program focused on creating breakthroughs in current performance. Quality improvement methods play a dominant role in reducing the costs of deficiencies, waste, and driving out defects as seen by the customer.

There are numerous methods of quality improvement, such as Lean, Kaizen, Six Sigma PDCA, PDSA, and the theory of constraints. All have merit and use at the right time for the right problem. Each method comes with a set of ingredients that make it work. Before we explore these methods, it is important to note that there are different types of problems businesses face, and thus different methods may be required to deal with them. No one set fits all when it comes to continuous improvement.

The costs associated with waste (cost, quality, time related) are due to both *sporadic* and *chronic* quality problems (see Figure 4.1). A sporadic problem is a sudden, adverse change in the status quo, which requires remedy by *restoring* the status quo (e.g., changing a depleted reagent chemical). A chronic problem is a long-standing adverse situation, which requires remedy by *changing* the status quo (e.g., revising an unrealistic specification).

"Continuous improvement" (called *kaizen* by the Japanese) has acquired a broad meaning, i.e., enduring efforts to act upon both chronic and sporadic problems and to make refinements to processes. For chronic problems, it means achieving better and better levels of performance each year; for sporadic problems, it means taking corrective action on periodic problems; for process refinements, it means taking such action as reducing variation around a target value.

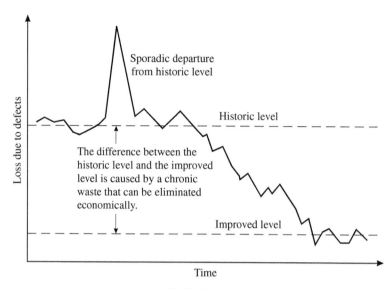

FIGURE 4.1
Sporadic and chronic quality problems.

The distinction between chronic and sporadic problems is important for two reasons:

1. The approach to solving sporadic problems differs from that to solving chronic problems. Sporadic problems are attacked by the control process defined and developed in Chapter 6. Chronic problems use the improvement process discussed in this chapter.
2. Sporadic problems are dramatic (e.g., an irate customer reacting to a shipment of bad parts) and must receive immediate attention. Chronic problems are not dramatic because they occur for a long time (e.g., 2% scrap has been typical for the past five years), are often difficult to solve, and are accepted as inevitable. The danger is that the firefighting on sporadic problems may take continuing priority over efforts to achieve the larger savings that are possible, i.e., on chronic problems.
3. A key reason for the chronic waste present in organizations is the lack of a structured approach to identify and reduce the waste. This chapter provides that structure.

We need to explain the sequence of this book's quality chapters. Logic would have it that first we plan (Chapter 5), then we measure and control (Chapter 6), and then we improve (Chapter 4). In the real world, logic does not always prevail. We will first cover the improvement process because most organizations have serious quality problems that practical people want to correct. People are more motivated to correct current problems and less motivated to do better planning. (Later, you will see that the cause of many current problems is poor planning, but this relationship needs to be personally discovered before there is a burning desire to do good planning.)

A structured approach to improvement applies not only to quality but also to other parameters, e.g., productivity, cycle time, and safety. Addressing chronic problems achieves breakthrough to an improved level (Figure 4.1). Acting on chronic problems is best achieved by the "project-by-project" approach.

✦ 4.2 BREAKTHROUGH IMPROVEMENT: PROJECT-BY-PROJECT APPROACH

The most effective approach to breakthrough improvement must be project by project. Here, a project is a chronic problem that has been chosen for a solution. An improvement project addresses the deficiencies dimension of quality. In practice, all organizations benefit most by implementing a systematic improvement program and starting with quality improvement projects.

Preparing to launch a quality improvement program, comprises three main steps:

1. Proving the need (describing the business case)
2. Selecting projects
3. Organizing and launching project teams

Carrying out an improvement project involves these tasks:

1. Defining the project scope and charter
2. Diagnosing the causes by obtaining information
3. Providing a remedy or solution to eliminate the cause
4. Dealing with resistance to change
5. Instituting controls to hold the gains

This approach was originally proposed by Juran in 1964 as the "breakthrough sequence" for improvement (Juran, 1964). Other approaches to improvement include plan, do, study, act; reengineering, theory of constraints, six sigma and lean six sigma. Each approach brings fresh ideas, and organizations have learned to integrate them continuously with the older successful methods.

Today, six sigma is the most practical method for creating breakthroughs. The phases of the six sigma approach are define, measure, analyze, improve, and control. This chapter presents an integrated view of Juran's breakthrough sequence and the six sigma approach.

To place individual projects in perspective, we first provide a summarized example. Next we discuss how to prove the need for a major improvement initiative consisting of many projects. Experiences with the project approach follow. Then we are ready to describe the steps for setting up and carrying out each project.

✦ 4.3 SIX SIGMA IMPROVEMENT PROJECT EXAMPLE

The problem concerns the soldering process used in the manufacture of printed circuit boards (PCBs). Any solder connection can cause testing problems or performance and reliability problems for the customer. In the following discussion, we trace the steps of the breakthrough sequence for improvement (see "Project-by-Project Approach"). Each step in the breakthrough sequence is related to the corresponding step in the six sigma approach.

Verify the Project Need and Mission (the Define Step in Six Sigma)

The process was statistically out of control and numerous solder connections required touch-up. The project team's mission: reduce the number of defective solder connections.

Diagnose the Causes (the Measure and Analyze Steps in Six Sigma)

A team of people, not from one department but from several cross-functional departments, was set up to guide the project and do the diagnosis. Figure 4.2 (a Pareto diagram; for elaboration, see Section 4.7) depicts the distribution of symptoms by type of solder defect. Data on the defects were analyzed and theories were offered on the causes of the defects. Figure 4.3 is a cause-and-effect diagram summarizing the theories. These theories were grouped into three categories, thereby allowing a checklist to be developed that supervisors and the control inspector used to evaluate the theories. After additional data collection and analysis, low solder temperature was found to be the main cause of defects. Figure 4.4 shows part of the analysis.

Provide a Remedy and Prove Its Effectiveness (the Improve Step in Six Sigma)

Data and further analysis revealed that, for ideal soldering conditions, either the temperature of the solder should be raised or the conveyor speed of the wave soldering machine should be reduced. These were remedies to remove the cause. A trial was conducted using a higher temperature. This resulted in an improvement in solder defects without any adverse effects.

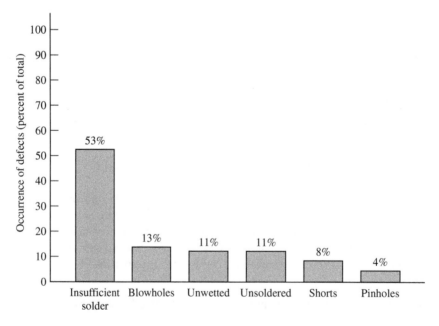

FIGURE 4.2
Solder defect types, Pareto analysis. (*From Betker, 1983.*)

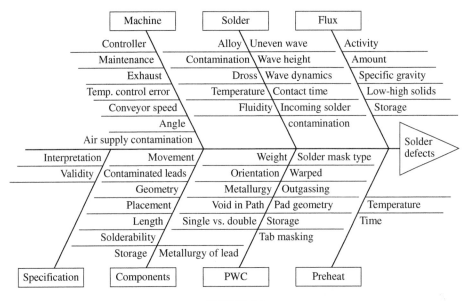

FIGURE 4.3
Ishikawa cause-and-effect diagram. (*From Betker, 1983.*)

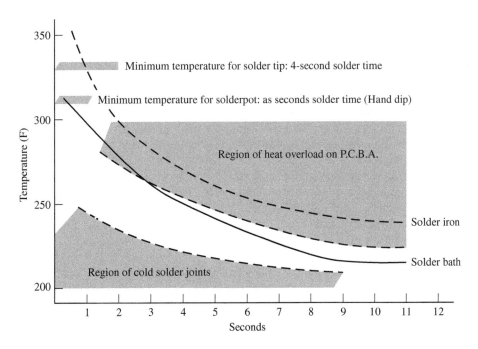

FIGURE 4.4
Temperature/time relationship when flow or dip soldering is applied. (*From Betker, 1983.*)

Deal with Resistance to Change (the Improve Step in Six Sigma)

From the start of the project, a manufacturing engineer on the team argued that the cause was outside the control of the machine. The diagnosis explained previously convinced him otherwise. But he felt that raising the temperature would result in "reflow of tin under the solder mask, thereby causing solder shorts and peeling of the solder mask." This belief, based on a trial conducted 10 years earlier on other equipment, had been expressed so often that it was no longer questioned. The trial of the remedy overcame this resistance.

Institute Controls to Hold the Gains (the Control Step in Six Sigma)

The defect level was reduced by 62%, and the out-of-control points on statistical control charts were eliminated. To ensure that the improved level was maintained, the process was monitored. Not only was the improved level maintained but the elimination of the dominant cause (low temperature) unmasked other causes. Performance has been improved to the point that the hand solder touch-up operation may be eliminated.

This example provides an overview of the breakthrough sequence and the six sigma approach as applied to individual quality projects. Carrying out quality improvement projects on chronic problems takes resources, and so we next address how to prove the need for such resources.

→ 4.4 THE BUSINESS CASE FOR QUALITY IMPROVEMENT

To gain management approval for a major quality initiative requires several steps:

1. *Estimate the size of the chronic waste or other quality-related losses.* The key data for this step comes from studies on the cost of poor quality and on competitive standing in the marketplace, as described in Chapter 3, "Organizationwide Assessment of Quality."

    Usually, managers are stunned by the size of the cost of waste and/or lost sales income due to poor quality. To establish proof of the need, different "languages" may be required for different levels of management. For upper management, the language of money works best; for lower levels, other languages are effective. Table 3.4 shows examples of statements in money and other languages.

2. *Estimate the savings and other benefits:*
    a. If the organization has never before undertaken a program to reduce quality-related costs, then a reasonable goal is to cut these costs in half, within five years.
    b. Don't imply that the quality costs can be reduced to zero.
    c. For any benefits that cannot be quantified as part of the return on quality, present these benefits as intangible factors to help justify the improvement program. Some benefits can be related to problems of high priority to upper

management, such as meeting delivery schedules, controlling capital expenditures, or reducing a delivery backlog. In a chemical company, a key factor in justifying an improvement program was the ability to reduce significantly a major capital expenditure to expand plant capacity. A large part of the cost of poor quality was due to having to rework 40% of the batches every year. The improvement effort was expected to reduce the rework from 40 to 10%, thus making available production capacity that was no longer needed for rework. What convinced management was not the savings in rework costs, but the reduction of capital expenditures.

3. *Calculate the return on investment resulting from improvement.* This return should reflect savings in the traditional cost of poor quality, savings in process capability improvement, and increases in sales revenue due to a reduction in customer defections and an increase in new customers.
4. *Use a successful case history (a bellwether project) in the organization to justify a broader program.* One or more of these pilot projects not only shows the tangible benefits that can result but also demonstrates that a successful step-by-step approach has been developed in the organization. These short-term wins must be genuine, visible, and related to the overall improvement effort. The following scenario illustrates how the ABC electronics company used this approach.

The estimated cost of poor quality was $200 million per year, and a notorious quality problem was scrap for a certain major electronic component. This scrap ran to about $9 million per year. The principal defect type was defect X, and it was costing about $3 million per year ("proof of the need" for eliminating defect X).

The company took on a project to reduce the incidence of defect X. The project was a stunning success. The cost of defect X was cut from $3 million to $1 million—an annual profit improvement of $2 million. An investment of about $250,000 was needed.

Then followed an exciting extrapolation and contrast. Engineers estimated that extending the improvement to the entire $200-million cost of poor quality could cut the total in half—thus creating a profit improvement of $100 million annually.

The defect X project proved to ABC managers that they could get a big return on investment by improving quality. This in-house project was more convincing to results-oriented managers than any number of lectures, books, or success stories about other companies.

Next we summarize experiences with the project-by-project-approach and then examine the specific steps in more detail.

➔ 4.5 LESSONS LEARNED WITH THE PROJECT-BY-PROJECT APPROACH

Experiences in both manufacturing and service industries have led to encouraging conclusions:

- Large cost reductions and improved quality to customers have been achieved. For each dollar invested in improvement activity, the return is between $5 and $10.

- Investment required for improvement has been modest and *not* capital intensive. Most of the investment is in the time of people diagnosing the projects.
- Most projects can be completed in six months if the scope in the mission statement is carefully defined.
- The key chronic quality-related problems cut across departments and thereby require cross-functional project teams.
- Some organizations consist of multiple autonomous units that have similar activities, e.g., manufacturing plants, hospitals, hotels. An improvement project that is successful in one unit may apply to another unit without the need for duplicating the diagnostic work. For elaboration on this "cloning," see *JQH5,* pages 5.29–5.30.
- The improvement approach should include suppliers—both by encouraging or requiring suppliers to have their own improvement program and by executing joint quality improvement projects with suppliers (see Chapter 15 for more details).
- Like any major activity (e.g., new product development), improvement must include goals in the annual business plan that are aligned with other business goals and strategies, organizational infrastructure to execute improvement (process for selecting projects and the formation, training, and support of improvement teams), training, review of progress by upper management, and reward and recognition. All of these matters should be integrated into a road map for implementation.

Of particular concern is the need for sufficient time and resources for people to carry out their improvement responsibilities. In many companies, employees believe that they are asked to work on improvement as an "add on" to their regular responsibilities and are not given sufficient time to do the improvement work.

An increasing number of companies have reported the completion of more than 1000 projects in about four years. Using the six-sigma approach, General Electric reports an increase in projects from 200 in 1995 to 47,000 in 1999 (Slater, 1999). Today's competitive business conditions dictate a revolutionary rate of improvement to replace the evolutionary rate of the past. Such improvement means replication of projects using the project-by-project approach.

4.6 LEAN AND SIX SIGMA IMPROVEMENT

The six sigma approach is a collection of managerial and statistical concepts and techniques that focus on reducing variation in processes and preventing deficiencies in product.

Variation in a process is denoted by sigma—the standard deviation of measurements around the process mean (see Chapter 17 under "The Concept of Variation"). In a process that has achieved six sigma capability, the variation is small compared to the range of the specification limits; i.e., there are six standard deviations between the process mean and either specification limit. (In the earlier days of the quality movement, a process was considered adequate if there were three standard deviations between the process mean and either specification limit.) Even if the process mean shifts (by 1.5 sigma), no more that 3.4 units per million fall outside the specification limits.

("Units" may be parts, lines of code, transactions, or other forms of output.) Thus the higher the number of sigmas the better. Most processes are at about 3 to 4 sigma. For elaboration on the statistical concept, see Chapter 20 under the "Six Sigma Concept of Process Capability."

A key focus is the relationship between the input variables and the outputs of a process, expressed as

$$Y = f(X_1 \ldots X_n)$$

Conceptually, product results (Y) are a function (f) of many process variables $X_1 \ldots X_n$. Thus Y is an output, an effect, a dependent variable; X are inputs, causes, dependent variables. The six sigma approach identifies the process variables that cause variation in product results. Some of these process variables are critical to quality, are set at a certain value, and are maintained within a specified range (i.e., "controllable variables"). Other variables cannot be easily maintained around a certain value and are considered uncontrollable or "noise." For elaboration, see Sanders, Ross, and Coleman (1999).

Six sigma uses five phases:

1. *Define.* This step identifies potential projects, selects and defines a project, and sets up the project team.
2. *Measure.* This step documents the process and measures the current process capability.
3. *Analyze.* This step collects and analyzes data to determine the critical process variables.
4. *Improve.* This step conducts formal experiments, if necessary, to focus on the most important process variables and determine the process settings to optimize product results.
5. *Control.* This step measures the new process capability, documents the improved process, and institutes controls to maintain the gains.

The six sigma approach and the breakthrough sequence are identical in objective and similar in the steps to achieve the objective. Both have an objective of finding the causes of deficiencies and developing remedies to prevent future deficiencies. Both employ the same tools and techniques, most of which were developed in the 1950s. Many of the quantitative tools were under used in the past because of their relative complexity (for some users), but recent advances in information technology have simplified data collection and analysis, thereby facilitating the use of these tools.

Chua (2001), Harry and Schroeder (2000), Hahn et al. (2000), Chua and Janssen (2000), and Henson (1999) present overviews of six sigma, including examples from General Electric and other organizations. Bajaria (1999) identifies certain elements ("points") of six sigma but then raises some issues ("counterpoints") to provide perspective. Yun and Chua (2002) showcase Samsung Electronic's six sigma success story. DeFeo and Barnard (2003) present six sigma in the larger context of achieving breakthrough improvement. Bottome and Chua (2005) showcase Genetech's successful Error-Proofing efforts using six sigma and lean.

The remainder of this chapter examines the five phases of six sigma.

114   Quality Management and Analysis

→ 4.7   DEFINE PHASE

This phase identifies potential projects, selects and defines a project, and sets up the project team. The steps are

- Identify potential projects
- Evaluate projects
- Select project
- Prepare problem and mission statement for project
- Select and launch project team

Identify Potential Projects

Project identification consists of nominating, screening, and selecting projects. *The focus must be on the vital few opportunities that will increase customer satisfaction and reduce the cost of poor quality.* All projects should be aligned with the organization's mission and quality goals. The six sigma approach particularly uses financial measurements to select and measure the success of projects.

Nomination for projects

Nominations come from several sources:

- Analysis of data on the cost of poor quality, quality standing in the marketplace, employee satisfaction studies, or other forms of assessment.
- Improvement goals needed to fulfill annual goals and achieve strategic goals.
- Analysis of other field intelligence, e.g., input from sales, customer service, and other personnel to retain current customers and attract new customers.
- Gaps between goals and actual performance.
- Inputs from all levels of management and the workforce.
- Benchmarking studies of other organizations.
- Developments arising from the impact of product quality on society, e.g., government regulation, growth of product liability lawsuits.

Project nominations address both improvement of effectiveness (meeting customer needs) and efficiency (meeting those needs at minimum cost). Note that the six sigma and the breakthrough sequence approaches are strategic because projects should be based on gaps between actual performance and goals.

The Pareto principle is a data analysis tool for generating project nominations.

Juran's Pareto principle

As applied to the cost of poor quality, the Pareto principle (named by J. M. Juran) states that a few contributors to the cost are responsible for the bulk of the cost. These vital few contributors need to be identified so that quality improvement resources can be concentrated in those areas.

A cost of poor quality (COPQ) study conducted at a manufacturer of orthopedic implants found that, in the previous year, internal failure costs amounted to

$115 million (see Table 4.1a). The top five categories (excess inventory reserves, safety stock, immediate stock inventory and obsolete inventory reserves and production scrap) accounted for 83% of the total internal costs of poor quality.

With reference to Table 4.1a: **Excess inventory reserves** are reserves for on-hand, slow moving inventory that exceeds the 24 month sales forecast. **Safety stock** are excess inventory maintained as an in-house bank due to the inability to produce reliable yields, forecast consistently, or receive procured products on a timely basis. **Intermediate stock inventory (net)** are work-in-process material used as a buffer for inadequate foundry yields, inaccurate sales forecasts and launch plans, design changes which necessitate rapid response, and to support the complexities of instrument assembles. (Net inventory is utilized to not duplicate excess and obsolete reserve costs.) **Obsolete inventory reserves** are reserves for inventory of inactive product (removed from the price list). **Production scrap** is material discarded in and reported by cell and product family.

Excess inventory reserves alone accounted for $36 million last year, or 31% of the total internal failure costs. When excess inventory reserves is analyzed using Juran's Pareto principle, knees, trauma fixation, and hips were the top three categories (see Table 4.1b). Studying and reducing the cost of these vital few categories results in attacking 86% of the excess inventory reserves problem. No major reduction in excess inventory costs can be achieved without attacking the excess inventory reserves for knees, trauma fixation, and hip implants. Such analysis would be helpful in nominating and selecting projects for cost reduction.

TABLE 4.1a
Pareto analysis of Internal Failure Costs at a manufacturer of orthopedic implants

INTERNAL COSTS	DOLLARS	INDIV %	CUM %
Excess inventory reserves	$36,253,810	31.38%	31.4%
Safety stock	$16,213,000	22.32%	53.7%
Intermediate stock inventory	$25,785,999	14.03%	67.7%
Obsolete inventory reserves	$11,552,776	10.00%	77.7%
Production scrap	$6,469,000	5.60%	83.3%
Excess inventory reserves carrying costs	$5,075,533	4.39%	87.7%
Safety stock carrying costs	$3,610,040	3.12%	90.8%
Dispositions (Unreserved outermediate scrap)	$2,473,000	2.14%	95.1%
Production rework	$2,470,000	2.14%	95.1%
Intermediate stock carrying costs	$2,269,540	1.96%	97.1%
Obsolete inventory reserves carrying costs	$1,617,389	1.40%	98.5%
QC reinspection indirect costs	$642,114	0.56%	99.0%
Investigation of failures-Mfg. Eng.	$445,536	0.39%	99.4%
Design changes	$333,000	0.29%	99.7%
Downtime	$212,834	0.18%	99.9%
Vendor rework charges	$115,000	0.10%	100.0%
Back-log late payments lost interest		0.00%	100.0%
Accounts receivables		0.00%	100.0%
TOTAL	**$115,538,571**	**100.00%**	

TABLE 4.1b
Pareto analysis of Excess Inventory Reserves

EXCESSIVE RESERVE EFFECT	DOLLARS	INDIV %	CUM %
Knees	$15,153,840	41.80%	41.8%
Trauma internal and external fixation	$8,682,144	23.95%	65.7%
Hips	$7,368,144	20.32%	86.1%
Custom	$4,301,665	11.87%	97.9%
Endos	$552,631	1.52%	99.5%
Other Orthopedics	$109,447	0.30%	99.8%
Other accessories	$80,283	0.22%	100.0%
Miscellaneous	$5,000	0.01%	100.0%
TOTAL	**$36,253,154**	**99.99%**	

Evaluate potential projects

Project nominations are usually reviewed by middle management, and recommendations are then made to upper management for final approval.

The review varies from an analysis of the project scope and potential benefit to a formal examination of several factors to help set priorities. For example, an insurance company screens potential projects by asking six questions: Can we impact? Can we analyze? Are data available? Are they measurable? What areas are affected? What is the level of control?

Hartman (1983) describes an approach at AT&T that uses a Pareto priority index (PPI) to evaluate each project.

$$\text{PPI} = \frac{\text{Savings} \times \text{probability of success}}{\text{Cost} \times \text{time to completion (years)}}$$

Table 4.2 shows the application of this index to five potential projects. High PPI values suggest high priority. Note how the ranking of projects A and C is affected when the criterion is changed from cost savings alone to the index covering the four factors.

The result of the review by middle management is a recommended list of projects. Typically, one responsibility of an upper management quality council is reviewing the recommendations or creating the organizational machinery for review and final approval.

Pareto analysis is an example of *data mining,* i.e., a process for analyzing data to extract information that is not offered by the raw data alone. Koch (1998) presents a

TABLE 4.2
Ranking by use of Pareto Priority Index (PPI)

Project	Savings, $ thousands	Probability	Cost, $ thousands	Time, years	PPI
A	100	0.7	10.0	2.0	3.5
B	50	0.7	2.0	1.0	17.5
C	30	0.8	1.6	0.25	60.0
D	10	0.9	0.5	0.50	36.0
E	1.5	0.6	1.0	0.10	9.0

wide-ranging discussion of the Pareto principle that covers applications to organizations and individuals for both business and personal effectiveness. The Pareto concept is called "key-factor analysis" in disciplines such as sociology, animal behavior, ecology, and toxicology.

The Pareto principle identifies the "vital few" projects for improvement. These projects are the major contributors to quality leadership in terms of sales revenue and lower costs. Such projects usually go across functions and require cross-functional quality improvement teams. This chapter addresses the vital few projects. Beyond the vital few projects are the "useful many" projects. Typically, the useful many projects are not cross functional and provide an opportunity for employee participation through workforce teams within a department; see Chapter 8. The total improvement effort must include both the vital few and the useful many projects.

Selection of Initial Projects

"The first project should be a winner." A successful project is a form of evidence to the project team members that the improvement process does lead to useful results. Ideally,

- The project should deal with a chronic problem—one that has been awaiting solution for a long time.
- The project should be feasible, i.e., have a good likelihood of being successful within about six months.
- The project should be significant. The results should be sufficiently useful to merit attention and recognition.
- The results should be measurable in money as well as in technological terms.
- The project should serve as a learning experience for the process of problem solving.

Problem and Mission Statement for Project

A problem statement identifies a visible deficiency in a planned outcome. For example, "During the past year, 7% of invoices sent to customers included errors." Note that this statement is specific and manageable (names a specific and limited process) and is observable and measurable (describes the size of the problem). A problem statement should never imply a cause or a solution.

A mission statement is based on the problem statement but provides direction to the project team. A goal or other measure of project completion and a target date should be defined. For example, the team is asked to reduce the error rate in invoices to 2% or less within the next three months. A mission statement should not imply cause or solution—that is the role of later steps based on data.

Select and Launch the Project Team

A project team usually consists of about six to eight persons drawn from multiple departments and assigned to address the chronic problem selected. Suppliers and customers may also be part of the team.

The team meets periodically, and members usually serve part time in addition to performing their regular functional responsibilities. When the project is finished, the team disbands.

A project team typically has a sponsor ("champion"), a leader, a recorder, team members, and a facilitator. Their roles on the project team are critical. Other improvement teams include "blitz teams" (for accelerated improvement) and "virtual teams" (involving different geographical locations). These matters are discussed in Chapter 8.

To help launch a team, some organizations develop a charter that defines *what* the team will do (e.g., mission) and *how* the team will function (e.g., principles used to make decisions). Van Aken and Sink (1996) describe 14 elements of a charter. In addition, their research concluded that teams having a charter ("team reviewed" or "team developed") had both higher team performance and higher team satisfaction than teams with no charter.

Early and Godfrey (1995) summarize a multi-industry study of 20 improvement projects that were identified as taking too long. Of the wasted time, 39% related to management not providing preparation and support (e.g., vague mission, limited time dedicated to the project, management delay in confronting resistance), and 24% related to the teams not using best practices (e.g., not focusing on the vital few causes, too much flow diagramming too soon, prematurely jumping to a remedy).

In practice, conducting effective team meetings requires certain skills that, frankly, many people do not possess. These skills include planning for team meetings and conducting the meetings. Planning involves matters such as logistics (time, location, use of electronic meeting software), setting meeting objectives, and preparing and distributing an agenda and other documentation. Conducting the meeting requires skills in developing participation, listening, and trust; handling problems (overly talkative members, quiet members, side conversations, absent members); resolving disagreements; and guiding the team to decisions. Some organizations have found it necessary to provide special training for team leaders.

4.8 MEASURE PHASE

This phase identifies key product parameters and process characteristics and measures the current process capability. The steps are

- Verify the project need, Y [in $Y = f(X)$].
- Document the process.
- Plan for data collection.
- Validate the measurement system.
- Measure the baseline performance of Y.
- Measure the process capability.

Verify the Project Need

Presumably, a project has been selected because it is important. It is useful, however, to verify the size of the problem *in numbers*. This process serves two purposes:

(1) ensure that the time to be spent by the project team is justified and (2) helps to overcome resistance to accepting and implementing a remedy. Verifying the need for an individual project uses the same type of information discussed earlier under "Prove the Need." In addition, the *scope* of the project must be reviewed after the team has met once or twice, to be sure that the mission assigned to the team can be accomplished within, say, about six months. Otherwise, the project should be divided into several projects. Failure is likely if a project stretches out like a freight train.

Document the Process

This step records the activities under study and information relating to actual or potential problems. A useful tool is the process flow diagram (or "process map"). An example of the process of admitting patients to a hospital emergency room is given in Figure 4.5 (Lenhardt, 1993). The diagram shows the sequence of steps and their relationships in the process. Study of the flow diagram, in conjunction with other information (see following discussion), enables us to develop a list of product characteristics [or key process output variables (KPOVs)] and key process parameters [or key process input variables (KPIVs)]. The KPOVs and KPIVs provide the detail for the conceptual framework of $Y = f(X)$, i.e., relating process variables to process results.

Other information is useful in documenting the process, but first we need to define some terms:

A *defect* (or disconnect) is any nonfulfillment of intended use requirements, e.g., oversize, low mean time between failures, illegible invoice. A defect can also go by other names, e.g., error, discrepancy, nonconformance.

A *symptom* is an observable phenomenon arising from and accompanying a defect. Sometimes, but not always, the same word is used both as a defect description and as a symptom description, e.g., open circuit. More usually, a defect will have multiple symptoms; e.g., "insufficient torque" may include the symptoms of vibration, overheating, erratic function, etc.

A *theory* is an unproved assertion of reasons for the existence of defects and symptoms. Usually, several theories are advanced to explain the presence of the observed phenomena.

A *cause* is a proven reason for the existence of the defect. Multiple causes are common, in which case they follow the Pareto principle; i.e., the vital few causes will dominate the rest.

A *remedy* is a change that can successfully eliminate or neutralize a cause of defects.

Two journeys are required for quality improvement: the diagnostic journey from symptom to cause and the remedial journey from cause to remedy. This distinction is critical. To illustrate, three supervisors were faced with a problem of burrs on screws at the final assembly of kitchen stoves. In their haste to act, they skipped the diagnostic journey and concluded that better screws were needed (a remedy). Fortunately, a diagnostician interceded. He pointed out that three separate assembly lines were feeding product into one inspection station and suggested that the data be segregated by assembly line. The data revealed that the burrs occurred only on line 3.

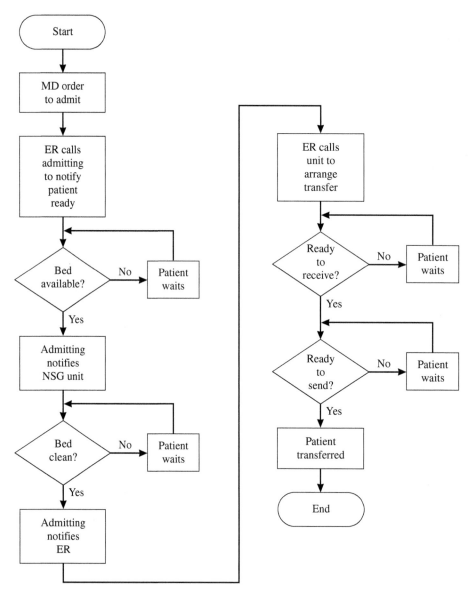

FIGURE 4.5
High-level flow diagram to admit ER patient.

Further diagnosis based on data led to agreement that the true cause was an improperly trained assembler. Then the remedy came easily.

The diagnostic journey has three steps:

- Study the symptoms surrounding the defects as a basis for theorizing about causes.
- Theorize on the causes of these symptoms.
- Collect and analyze data to test the theories and thereby determine the causes.

Many analytical techniques are available to assist in these three steps. Some are illustrated in Section 4.9.

Plan for Data Collection

Chronic problems are usually not easy to solve and require careful planning and collection of data to confirm and analyze the input and output variables. This "management by facts" concept is basic to all problem-solving approaches. The time and effort involved may be considerable but is a necessary investment for analysis and improvement. Klenz (1999) discusses the role of a "quality data warehouse" in quality improvement.

Planning for data collection involves matters such as where in the process data will be collected, who will provide the data and how often, data collection forms, data accuracy, separation of data into categories ("stratification"), and whether the data are sufficient in content and quantity for the data analysis tools. Sometimes an apparently simple matter of describing the symptoms of a problem requires careful planning.

Description of symptoms

Understanding symptoms is often hindered because some key word or phrase has multiple meanings.

In one example, a Pareto analysis of inspection data in a wire mill indicated a high percentage of defects due to "contamination." Various remedies were tried to prevent the contamination. All were unsuccessful. The desperate investigators finally spoke with the inspectors to learn more about the contamination. The inspectors explained that the inspection form contained 12 defect categories. If the observed defect did not fit any of the categories, the inspectors would report the defect as "contamination."

Imprecise wording also occurs because of generic terminology. For example, a software problem is described in a discrepancy report as a "coding error." Such a description is useless for analysis because there are many types of coding errors, e.g., undefined variables, violation of language rules, violation of programming standards.

A way out of such semantic tangles is to think through the meanings of the words used, reach an agreement, and record the agreement in the form of a glossary. The published glossary simplifies subsequent analysis.

Quantification of symptoms

The frequency and intensity of symptoms are of great significance in pointing to directions for analysis. The Pareto principle, when applied to records of past performance, can help to quantify the symptom pattern. Figure 4.6 displays a Pareto diagram for errors in sales order forms sent from a field sales office to the home office for order processing. The four vital few categories account for about 86% of the total errors. Note that the diagram includes three elements: (1) the contributors to the total effect, ranked by the magnitude of the contribution, (2) the magnitude of the contribution expressed numerically, and (3) the cumulative percentage of total effect of the ranked contributors.

The Pareto principle applies to several levels of diagnosis: finding the vital few defects, finding the vital few symptoms of a defect, and finding the vital few causes of one symptom.

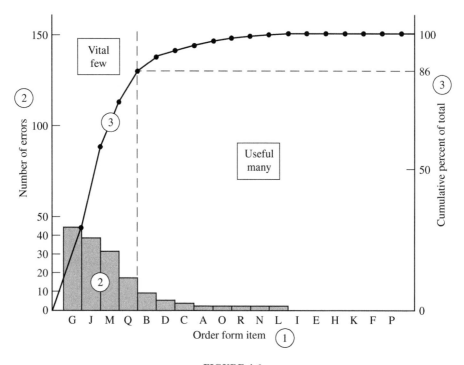

FIGURE 4.6
Pareto diagram of errors on order forms.

Formulation of theories

This process has three steps: generation of theories, arrangement of theories, and choice of theories to be tested.

Generation of theories. The best sources of theories are the line managers, the technologists, the line supervisors, and the workforce. A systematic way to generate theories is the brainstorming technique. Potential contributors are assembled for the purpose of generating theories. Creative thinking is encouraged by asking each person, in turn, to propose a theory. No criticism or discussion of ideas is allowed, and all ideas are recorded. Following the brainstorming session, the resulting list of theories is critically reviewed.

A useful supplement to the brainstorming technique is "storyboarding." Each theory proposed is recorded on an index card or adhesive note. The cards are arranged on a board to form a visual display of the theories. Storyboarding provides a visual system for organizing theories and planning subsequent evaluation of these theories.

Arrangement of theories. Normally, the list of theories should be extensive, 20 or more. As a list grows, it is essential to create an orderly arrangement. Such order helps us to understand the interrelationships among theories and to plan for testing the theories. Table 4.3 shows a tabular arrangement of theories contributing to low yield of a process making fine-powder chemicals. The theories consist of major variables and contributing subvariables. A second method, which is highly effective, is a graphical

TABLE 4.3
Orderly arrangement of theories

Raw material	Moisture content
Shortage of weight	Charging speed of wet powder
Method of discharge	Dryer, rpm
	Temperature
Catalyzer	Steam pressure
Types	Steam flow
Quantity	
Quality	Overweight of package
	Type of balance
Reaction	Accuracy of balance
Solution and concentration	Maintenance of balance
B solution temperature	Method of weighing
Solution and pouring speed	Operator
pH	
Stirrer, rpm	Transportation
Time	Road
	Cover
Crystallization	Spill
Temperature	Container
Time	
Concentration	
Mother crystal	
Weight	
Size	

arrangement called the Ishikawa cause-and-effect (or "fishbone") diagram. Figure 4.7 shows such a diagram, which presents the same information as listed in Table 4.3.

Writing and arranging theories on adhesive notes (storyboarding) can be a useful tool. Forsha (1995) describes storyboarding.

Sanders, Ross, and Coleman (1999) describe how combining the flow diagram and the cause-effect diagram can help to classify process parameters as controllable or noise.

Thus the cause-and-effect diagram provides information for identifying the input and output variables. Additional tools include the quality function deployment matrix and the failure mode, effects, and criticality analysis. Thoroughly documenting the current ("as is") process is an essential step in developing an improved ("should") process.

Choice of theories to be tested. After the theories are arranged in an orderly fashion, priorities must be established for testing them. In practice, the improvement team reaches a consensus on the most likely theory for testing. Whether to test one theory at a time, one group of interrelated theories at a time, or all theories simultaneously requires judgment based on the experience and creativity of the team.

Validate the Measurement System

The variation in observed measurements from a process is from the variation of the process itself and variation of the measurement system. Often the variation of the

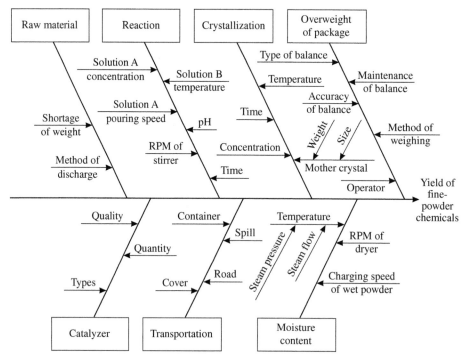

FIGURE 4.7
Ishikawa cause-and-effect diagram.

measurement system is assumed to be zero or at least small compared to process variation. Although this assumption simplifies matters, it is typically made without any data. Thus we often do not know how close the observed measurements are to their true (or "master") values.

With the current emphasis on defining quality in units of parts per million rather than percentage, the capability of the measurement system must be recognized as important and evaluated before measuring the capability of the process. Measurement capability involves both the ability of the people making the measurements and the capability of the measuring instruments. When necessary, a complete measurement capability study can involve matters such as reproducibility, repeatability, accuracy, stability, and linearity. Fortunately, software tools such as Minitab® are of great help in analyzing the measurement study data. A measurement study ensures that the measurement system data can be trusted. With such assurance, then data collection for product and process analysis and improvement can start.

Measure the Process Capability

Process capability refers to the inherent ability of a process to meet the specification limits for a product. In the measure phase, the initial process capability is established

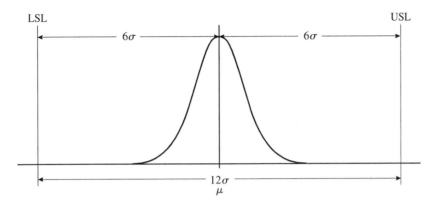

FIGURE 4.8
Process capability and six sigma.

by obtaining measurements and observing how the process variability compares to the specification limits. For a static process to be capable at the six-sigma level, the specification limits must be at least six sigma above and below the process mean (see Figure 4.8). In practice, the capability study must recognize both short-term variation, in which the process average and spread are stable, and long-term variation, in which the process average may shift up or down. Knowing the initial process capability helps to define the work to be done in the analysis and improve phases to achieve capability at the six sigma level. The methodology for measuring process capability is discussed in Chapter 20.

❖ 4.9 ANALYZE PHASE

This phase analyzes past and current performance data to identify the causes of variation and process performance. The steps are

- Plan for data collection.
- Collect and analyze data.
- Test theories (hypotheses) on sources of variation and cause–effect relationships (i.e., identify the determinants of process performance).

In this phase we diagnose—study symptoms of a problem and determine their cause(s). The beginning of diagnosis is collecting data on the symptoms; the end is agreement on the causes.

Planning for Data Collection

Teams are seeking the answers to questions, "How often does the problem occur?" or "What is causing the problem?" In other words, they are seeking information. However, though good information is always based on data (the facts), simply collecting some data does not necessarily ensure that a team will have useful information. The data may not be relevant or specific enough to answer the question at hand.

The key issue, then, is not "How do we collect data?" Rather, the key issue is "How do we generate useful information?" Although most organizations have vast stores of data about their operations, when teams begin working on a project, they often find that the information they need does not exist.

Information generation begins and ends with questions. To generate information, we need to

- Formulate precisely the question we are trying to answer.
- Collect data relating to that question.
- Analyze the data to determine the factual answer to the question.
- Present the data in a way that clearly communicates the answer to the question.

Learning to "ask the right questions" is the key skill in effective data collection. Accurate, precise data, collected through an elaborately designed statistical sampling plan, is useless if it does not clearly address a question that someone cares about.

Notice in Figure 4.9 how this planning process "works backward" through the model. We start by defining the question. Then, rather than diving into the details of data collection, we consider how we might communicate the answer to the question and what types of analysis we will need to perform. This helps us define our data needs and clarifies which characteristics are most important in the data. With this understanding as a foundation, we can deal more coherently with the where, who, how, and what else issues of data collection.

To generate useful information, planning for good data collection proceeds along the following lines.

1. Establish data collection objectives. Formulate the question or theory:
    - What is your goal for collecting data?
    - What process or product will you monitor to collect the data?

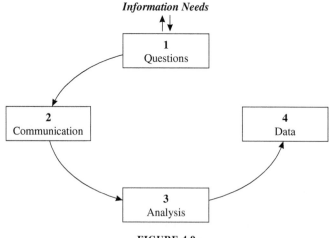

FIGURE 4.9
Planning for data collection.

- What is the theory you are trying to prove?
- What is the question you are attempting to answer?
2. Decide what to measure. How will data be communicated and analyzed?
   - What data do you need?
   - What type of measure is it?
   - What type of data is it?
   - What is the operational definition of each measure?
   - How will the data be communicated and analyzed?
3. Decide how to measure a population or sample:
   - What measurement tool will you use?
   - What is your sampling strategy?
   - How much data will be collected?
   - What is the measurement method?
4. Collect data with a minimum of bias:
   4a. Define comprehensive data-collection points.
       - Where in the process can we get these data?
   4b. Select unbiased collectors. Understand data collectors and their environment. Train data collectors.
       - Who in the process can give us these data?
       - How can we collect these data from these people with minimum effort and least chance of error?
   4c. Design, prepare, and then test the forms and instructions.
       - What additional information do we need to capture for future analysis, reference, and traceability?
   4d. Audit the collection process and validate the results.

The example in Figure 4.10 is based on a data collection plan spreadsheet developed by Chua (1999) for planning tests for the Analyze Phase in six sigma projects.

Ref.	Theories To Be Tested (Selected From The C-E Diagram, FMECA, and/or FDM)	List Of Questions That Must Be Answered To Test Each Selected Theory	Where Applicable, State The Null and Alternative Hypotheses		Tools To Be Used	Description/ Data Type	Sample Size, Number of Samples	Where To Collect Data	Who Will Collect Data	How Will Data Be Recorded	Remarks
			H_O	H_A							
1	The shortage problem is systematic throughout the entire loading/shipping process	Is there a difference in shortage rate by type of product?	Shortage rate is independent of type of product	Shortage rate is not independent of type of product	Chi square	Y = Discrete X = Discrete	Selected customers by plant for approximately 1500 units shipped per plant	Completed load sheets	Six sigma team members	Shipping shortage spreadsheet	
2		Is there a difference in shortage rate by Supervisor of the cell that shipped the product?	Shortage rate is independent of supervisor	Shortage rate is not independent of supervisor	Chi square	Y = Discrete X = Discrete	Selected customers by plant for approximately 2000–3000 units shipped per plant	Completed load sheets	Six sigma team members	Shipping shortage spreadsheet	

FIGURE 4.10
Example: data collection plan for shipping shortages.
(Courtesy of Merillat Industries.)

For each theory (or group of theories), list all data required, each tool to be used, the results that will support the theory, and results that will rule out the theory.

Collect and Analyze Data

Many managers harbor deep-seated beliefs that most defects are caused during operations and specifically are due to worker errors, i.e., that defects are mainly worker-controllable. The facts seldom bear this out, but the belief persists. To deal with such deep-seated beliefs, it can be useful to conduct studies to separate defects into broad categories of responsibility. For example:

A study to determine whether defects are primarily management-controllable or worker-controllable ("management" here includes not only people in supervisory positions but also others who influence quality, e.g., design engineers, process engineers, buyers, etc.). In general, defects are more than 80% management-controllable and less than 20% worker-controllable. Some authors use the term "system-controllable" for "management-controllable."

Such broad studies provide important direction for improvement, but individual projects require their own unique data collection and analysis as discussed later.

We proceed next to the test of theories—first management-controllable, then worker-controllable.

Test Theories of Management-Controllable Problems

Basic to the concept of diagnosis is the factual approach—the use of facts, rather than opinions, to reach conclusions about the causes of a quality problem. Obvious? In practice, the urgency to take action on a dramatic problem often results in premature (and incorrect) decisions. Some problems become chronic because the true causes have not been determined, even though a flurry of action takes place. The factual approach not only determines the true cause but also helps to *gain agreement* on the true cause by all of the parties involved. The eloquence of facts carries the day.

Numerous diagnostic methods have been created to test theories. These "incisions" into product and process data can be fascinating and some of the basic methods are discussed later.

Flow diagram

Preparing a flow diagram (see Section 4.8) helps us understand the progression of steps in a process. The flow diagram provides an important tool for initiating process capability and process dissection.

Process-capability analysis

One of the theories most widely encountered is "the process can't meet the specifications." To test this theory, measurements from the process must be taken and analyzed to determine the amount of variability in the process. This variability is then compared to the specification limits. Figure 4.11a shows an example of the weight of

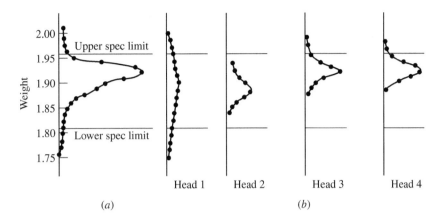

FIGURE 4.11
Distribution of glass bead weights: *(a)* sum of four heads, *(b)* weight distribution on each of the four heads. (*From QCH4, p. 22.42.*)

glass rings from a glass-tube-cutting process. Note that the process is currently not meeting the specifications.

The weight of the rings was the critical element in determining the properties of the finished glass product. A sample of data (Figure 4.11a) apparently confirmed a theory that the machine was not functioning correctly. However, the machine contained four heads. Data collected separately from each head (stream) revealed that nothing was wrong with heads 2, 3, or 4—except for a need to recenter heads 3 and 4 (Figure 4.11b). However, something was wrong with head 1. Ultimately, the remedy was proper maintenance of the machine rather than a redesign of the machine, as had originally been contemplated based on Figure 4.11a.

Initially, operations people believed that the process was not capable, but the capability study revealed that, with proper maintenance, the process was capable of meeting the specifications.

The concept of process capability applies not only to attributes of physical products but also to attributes of service activities, e.g., the time taken to respond to a customer request at a call center or the time taken to conduct an automatic teller machine (ATM) transaction at a bank. When data show that a process is not meeting a standard, then we must use the tools of product and process dissection to determine the cause.

Product and process dissection

Some products are produced by a "procession" type of process, i.e., a series of sequential operations. At the end of the series, the product is found to be defective, but which operation did the damage is not known. In some of these cases, it is feasible to dissect the process, i.e., make measurements at intermediate steps in the process to discover at which step the defect appears. Such a discovery can drastically reduce the subsequent effort in testing theories. Dissection of a process is aided by constructing a flow diagram showing the various steps in the process (see Section 4.8). Several other forms of dissection are discussed later.

Test at intermediate stages. When defects or errors are found at the end of a procession of steps, we must determine which operational step did the damage. Measuring or testing the product at intermediate steps to discover where the defect first appears can greatly reduce the effort of testing theories.

Stream-to-stream analysis

To meet production volume requirements, several sources of production ("streams") are often necessary. Streams take the form of different machines, machine operators, call center operator shifts, suppliers, etc. Although the streams may seem identical, the resulting products may not be. Stream-to-stream analysis consists of recording and examining data separately for each stream.

In the glass-ring example, the four heads on the machine were really four streams. Another example involves invoice errors (Juran Institute, Inc., 1989). A quality improvement team studied 60 incorrect invoices. The data showed the day of the week the invoice was issued, the week of the month, and which of four employees prepared the invoice. First, the data were separated into five "streams" representing Monday through Friday to test the theory that day of the week was the cause. No real difference was found. Next, to test the theory that employees were the cause, the data were separated into streams by employee. Again, no difference. Finally, the data were separated into streams by week of the month (see Figure 4.12). It was clear that at the end of the month, the number and percentage of errors were about double what they were during the first three weeks. Further study revealed that during the last week of the month, the workload increased dramatically, resulting in a higher error rate. At the start of the study, researchers believed that some employees were the most likely cause of errors. But the study concluded that the excessive workload of the fourth week was the primary factor in errors.

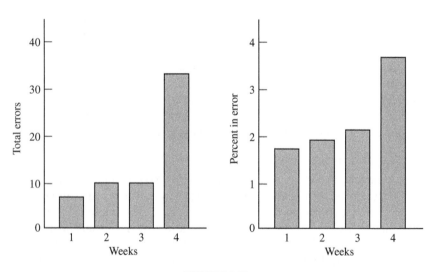

FIGURE 4.12
Stratification of invoice errors by week.

Stream-to-stream analysis uses the concept of "stratification." Stratification separates data into categories and helps to identify which categories (strata) are the main contributors to the problem.

Time-to-time analysis

Time-to-time analyses include (1) a simple plot of data on a timescale; (2) analysis of the time between abnormalities or problems; (3) analysis of the rate of change, or "drift," of a characteristic; and (4) cumulative data techniques with respect to time. Examples are given here.

In one example, field failures of oil coolers were assumed to be due to manufacturing. A parade of remedies (skipping the journey from symptom to cause) resulted in zero improvement. An engineer decided to plot the frequency of failures by month of the year, which led to an important discovery. Of 70 failures during a nine-month period, 44 occurred during January, February, and March. These facts shifted the search to other causes such as winter climatic conditions. Subsequent diagnosis revealed that the cause was in design rather than manufacturing.

In analyzing time-to-time variations, the length of time between abnormalities can be a major clue to the cause. A textile-carding operation experienced a cyclic rise and fall in yarn weights; the cycle was about 12 minutes. The reaction of the production superintendent was immediate: "The only thing we do every 12 minutes or so is to stuff that feed box."

Within many streams, there is a time-to-time "drift," e.g., the processing solution gradually becomes more dilute, the tools gradually wear, the worker becomes fatigued. Such drifts can often be quantified to determine the magnitude of the effect.

Cumulative data plots can help to discover differences that are hidden when the data are in noncumulative form. Figure 4.13 compares histograms (noncumulative)

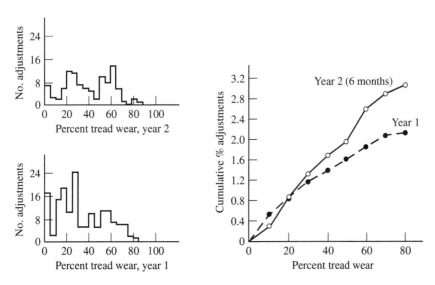

FIGURE 4.13
Comparison of histograms and cumulative plots. (*From QCH4, p. 22.44.*)

and cumulative plots for data from two years. A difference in adjustments for year 1 versus year 2 is apparent from the cumulative plot but is hidden in the histogram.

Control charts are a powerful diagnostic tool. Data are plotted chronologically, and the chart then shows whether the variability from sample to sample is due to chance or assignable causes of variation. Detection of assignable causes of variation can be the link to discovering the cause of a problem. Chapter 20, "Tools to Maintain Control of Quality," explains the concept.

Other helpful statistical tools include box plots and probability paper (see Chapter 17).

Simultaneous dissection

Some products exhibit several types of variation, e.g., piece to piece, within piece, and time to time. The multivari chart is a clever tool for analyzing such variation. In this chart, a vertical line depicts the range of variation within a single piece of product. Figure 4.14 depicts three examples of the relationship of product variation to tolerance limits. In case 1, the within-piece variation alone is too great in relation to the tolerance. Hence no solution is possible unless within-piece variation is reduced. In case 2, within-piece variation is comfortable, occupying only about 20% of the tolerance. The problem, then, is piece-to-piece variation. The problem in case 3 is excess time-to-time variability. For elaboration on the multivari chart, see Bhote (1991), Chapter 7.

Defect-concentration analysis

A different form of piece-to-piece variation is the defect-concentration study used for attribute types of defects. The purpose is to discover whether defects are located in the same physical area. The technique has long been used by shop personnel when

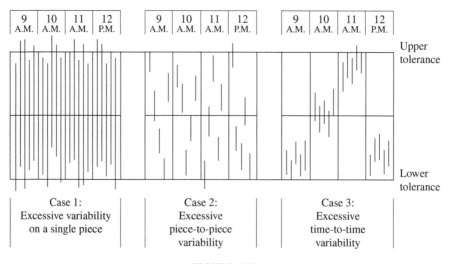

FIGURE 4.14
Multivari chart.

they observe that all pieces are defective and in precisely the same way. However, when the defects are intermittent or become evident only in later departments, the analysis can be beyond the unaided memory of the shop personnel.

For example, a problem of pitted castings was analyzed by dividing the castings into 12 zones and tallying up the number of pits in each zone over many units of product. The concentration at the gates (through which the metal flows) became evident, as did areas that were free of pits.

A quality improvement team at the Aid Association for Lutherans used the defect-concentration concept to analyze errors in changing beneficiaries on insurance policies. The team recorded data on the type of error, the frequency, the form on which the error occurred, and the location of the error on the form.

Association searches

Sometimes diagnosis can be advanced by analyzing data relating symptoms to some theory of causation, pinpointing the process, piece of equipment, employee(s), or other factor(s). Possible relationships can be examined by using tools such as correlation and ranking.

Correlation analysis. This approach plots data that relate the incidence of symptoms of the problem to values of a potential causal variable.

In one case (Juran Institute, Inc., 1991), a previously successful drug was losing its reputation for effectiveness, and the medical community was complaining to the manufacturer. One theory suggested that the active ingredient in the compound could be breaking down more quickly than expected while the medication sat on the shelf. Researchers decided to investigate this shelf-life theory.

Samples of the drug were pulled from shelves and returned to the lab to determine the milligrams of active ingredient (the minimum for the drug to be effective was 120 milligrams). Data on time since production and the weight of the ingredient were examined in several ways. A key analysis tool was the scatter diagram shown in Figure 4.15. Note that the data were stratified by two suppliers who provided a certain "filler" ingredient. Lots produced with filler from supplier A showed slow degradation of the active ingredient; lots associated with supplier B showed rapid degradation. Further investigation revealed a substance in B's formulation that led to accelerated breakdown of the active ingredient.

Ranking. In this approach, past or current data are collected on two or more variables of a problem and summarized in a table to see whether any pattern exists. In a study involving 23 types of torque tubes, the symptom was dynamic unbalance. One theory was that a swaging operation was a dominant cause. Table 4.4 tabulates the percentage of defective pieces (dynamic unbalance) and shows whether swaging was part of the process. The result was dramatic—the worst seven types of tubes were all swaged; the best seven were all unswaged. This pattern partially confirmed that swaging was a dominant cause. Later analysis revealed an inadequate specification on an important coaxial dimension.

Later in this chapter, a matrix-ranking technique for analyzing problems in insurance contracts is illustrated.

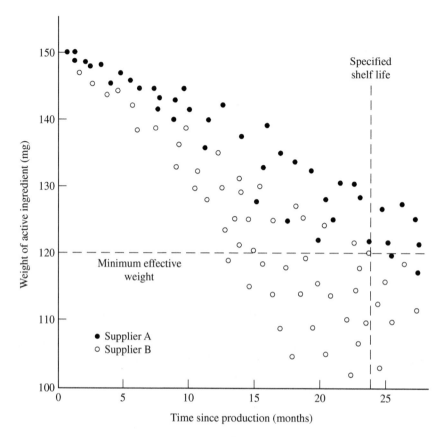

FIGURE 4.15
Stratified scatter diagram of shelf-life data.

**TABLE 4.4
Test of theories by ranking**

Type	Defective, %	Swaged (marked X)	Type	Defective, %	Swaged (marked X)
A	52.3	X	M	19.2	X
B	36.7	X	N	18.0	X
C	30.8	X	O	17.3	
D	29.9	X	P	16.9	X
E	25.3	X	Q	15.8	
F	23.3	X	R	15.3	
G	23.1	X	S	14.9	
H	22.5		T	14.7	
I	21.8	X	U	14.2	
J	21.7	X	V	13.5	
K	20.7	X	W	12.3	
L	20.3				

Test of theories by collection of new data

In some cases, discovery of causes requires careful examination of additional stages in the process. This "cutting of new windows" can take several forms:

1. *Measurement at intermediate stages of a single operation.* An example concerned a defect known as "voids" in welded joints of pressure vessels. Initial diagnosis established six sources of variation: operator, time to time, joint to joint, layer to layer, within layers, and within one welding "bead." Available past data permitted analysis of the first two sources as possible causes of voids. The remaining three could not be analyzed because the critical X-ray test was performed only when a joint was completely finished. The answer was to "cut a new window" by making an X ray after each of the several beads needed to make a joint. The data established that the main variable was within-bead variation and that the problem was concentrated at the start of the bead.

    An example from a human resources process concerns the time required for hiring new engineers. Measurements taken at six steps in the hiring process formed the basis for a diagnosis of excessive time taken to hire engineers.
2. *Measurement following noncontrolled operations.* This type of diagnosis includes the collection of additional information at individual steps in a process. One example involved the diagnosis of excessive shutdown time for removing a hot mold from a molding machine. The process of shutdown was divided into 11 steps, and measurements were taken to estimate the time required for each step. Two of the 11 steps accounted for 62% of the shutdown time. This Pareto effect was important in further diagnosis.
3. *Measurement of additional or related properties of the product or process.* Diagnosis sometimes requires measuring of characteristics other than those for which the specification is not being met. In the manufacture of disks, the symptom was a high percentage of disks with surface defects. Automatic timers controlled the processing cycle. Diagnosis revealed that the times for various steps should not be fixed, but should be determined based on additional measurements taken periodically. The remedy was to monitor pressure, temperature, viscosity, and other factors. An online computer evaluates these data for each disk and decides on the optimum molding conditions. Only then, and no sooner, does the process create the product.
4. *Study of worker methods.* In some situations, there are consistent differences in the defect levels coming from various workers. Month after month, some workers produce more "good" product than others do. This consistent difference in observed performance must have a cause. Diagnosis of problems related to human performance is discussed later in this section.

Sometimes formally designed experiments are needed to test theories (see Section 4.10 in this chapter).

Most quality problems are management/systems controllable—although many managers find this statement impossible to believe. Consider the situation at a bank. A middle manager faced the problem of an employee whose work output was poor. The manager sincerely believed that most problems were due to the employee. When the situation was examined more closely (as part of a classroom assignment), the

TABLE 4.5
Matrix of errors by insurance policy writers

	Policy writer						
Entry	A	B	C	D	E	F	Total
1	0	0	1	0	2	1	4
2	1	0	0	0	1	0	2
3	0	(16)	1	0	2	0	(19)
4	0	0	0	0	1	0	1
5	2	1	3	1	4	2	(13)
6	0	0	0	0	3	0	3
.							
.							
.							
27							
28							
29							
Total	6	(20)	8	3	(36)	7	80

employee and her boss concluded that the process was much of the problem. In her words, "If you put a good guy up against a bad process, the bad process will win every time." Exactly. With that reminder, we next discuss human errors.

Test of theories involving human error

Diagnosis of human errors reveals "multiple species" of errors. To illustrate these species, Table 4.5 shows the distribution of 80 errors made by six office workers engaged in preparing insurance policy contracts.

The 29 entries follow the Pareto principle. Notice the errors for entry 3. There were 19 of these, and worker B made 16 of the 19. The table also shows the rest of the work done by worker B. Except for entry 3, B made few errors. There is nothing basically wrong with the job specification or the method because the other five workers had little or no trouble with entry 3. There is nothing basically wrong with worker B except for entry 3. It follows that worker B and no one else is misinterpreting some instruction, resulting in that cluster of 16 errors of entry 3.

Entry 5 is of a different species. There is a cluster of 13 of these and all the workers made this error, more or less uniformly. This pattern suggests some difference in approach between all the workers, on the one hand, and the inspector, on the other. Such a difference is usually of management-controllable origin, but the reality can soon be established by interviews with the respective employees.

Notice also the column of numbers associated with worker E. The total is 36 errors, the largest cluster in the table. Worker E made nearly half the errors for the entire team, and that worker made them in virtually all entries. Why did worker E make so many errors? It might be any of a variety of reasons, e.g., inadequate training, lack of capacity to do such exacting work. Further study is needed, but it might be easier to go from symptom directly to remedy—find a less demanding job for that worker.

TABLE 4.6
Interrelationship among error pattern, likely subspecies of worker error, and likely solution

Pattern disclosed by analysis of worker error	Likely subspecies of error causing this pattern	Likely solution
On certain defects, no one is error-prone; defect pattern is random.	Errors are due to inadvertence.	Error-proof the process.
On certain defects, some workers are consistently error-prone, and others are consistently "good."	Errors are due to lack of technique (ability, know-how, etc.). Lack of technique may take the form of secret ignorance. Technique may consist of known knack or of secret knowledge.	Discovery and propagation of knack. Discovery and elimination of secret ignorance.
Some workers are consistently error-prone over a wide range of defects.	There are several potential causes: Conscious failure to comply with standards; Inherent incapacity to perform this task; Lack of training	Solution follows the cause: Increase motivation; Transfer worker; Supply training
On certain defects all workers are error-prone.	Errors are management-controllable.	Meet the criteria for self-control.

This one table shows the presence of multiple species of worker errors. The remedy is not as simplistic as "motivate the worker." Understanding these species through diagnosis is important in identifying causes. The great majority of worker errors fall into one of four categories: inadvertent, technique, conscious, and communication. Table 4.6 shows the interrelationship among the error pattern, the likely subcategory, and the likely remedies. The four categories are examined here.

Inadvertent errors. Workers are unable to avoid inadvertent errors because of human inability to maintain attention. Centuries of experience have demonstrated that human beings are simply unable to maintain constant attention.

The usual examples involve a component omitted from an assembly or a process adjustment that is set incorrectly.

Diagnosis to identify errors as inadvertent is aided by understanding their distinguishing features. They are

- *Unintentional.* The worker does not want to make errors.
- *Unwitting.* At the time of making an error, the worker is unaware of having made it.
- *Unpredictable.* There is nothing systematic as to when an error will be made, what type of error will be made, or which worker will make the error. As a consequence of this unpredictability, the error pattern exhibits randomness. *A set of data that shows a random pattern of worker error suggests that the errors are inadvertent.* The randomness of the data may apply to the types of errors, to the persons who make errors, and to the times when the errors are made.

Remedies for inadvertent errors involve two approaches:

1. *Reducing the extent of dependence on human attention.* All the tools used here are of the error-proofing type: fail-safe designs, validation of processes, countdowns, redundant verifications, cutoffs, interlocks, alarm signals, automation, robots. Large reductions in errors can result from the use of bar codes to help identify items.
2. *Helping workers remain attentive.* Examples of remedies are reorganization of work to reduce fatigue and monotony; job rotation; and use of sense multipliers, templates, masks, and overlays.

Technique errors. These errors arise because the worker lacks some essential technique, skill, or knowledge needed to prevent the error from happening. Diagnosis to identify errors due to technique is aided by understanding their features. They are

- *Unintentional.* The worker does not want to make errors.
- *Specific.* Technique errors are unique to certain defect types—those types for which the missing technique is essential.
- *Consistent.* Workers who lack the essential technique consistently make more errors than workers who possess the technique. This consistency is readily evident from data on worker errors.
- *Unavoidable.* Inferior workers are unable to match the performance of the superior workers because the former do not know "what to do differently."

THE KNACK. The study of work methods showed that superior performers used a file to cut down one of the dimensions on a complex component; the inferior performers did not file that component. This filing constituted a "knack"—a small difference in method that accounts for a large difference in results. (Until the diagnosis was made, the superior assemblers had not realized that filing greatly reduced the incidence of defects.)

Usually the difference in worker performance is traceable to some superior knack used by successful performers to benefit the product. In the case of the gun assemblers, the knack consisted of filing one component. In some cases, however, the difference in worker performance is due to unwitting *damage* done to the product by inferior performers.

A useful rule for predicting whether the difference in worker performance is due to a beneficial knack or to a negative knack considers which workers are in the minority. If the superior performers are in the minority, the difference is probably due to the beneficial knack. If the inferior performers are in the minority, the difference in performance is probably due to a negative knack.

SUMMARY OF TECHNIQUE ERRORS. The sequence of events to identify, analyze, and remedy technique errors follows.

1. For the defect types under study, create and collect data that can disclose any significant worker-to-worker differences.
2. Analyze the data on a time-to-time basis to discover whether consistency is present.
3. Identify the consistently best and consistently worst performers.
4. Study the work methods used by the best and worst performers to identify their differences in technique.

5. Study these differences further so as to discover the beneficial knack that produces superior results or the negative knack that is damaging the product.
6. Bring everyone up to the level of the best through appropriate remedial action such as
   a. Training inferior performers in use of the knack or in avoidance of damage.
   b. Changing the technology so that the process embodies the knack.
   c. Error-proofing the process in ways that either require use of the knack or prohibit the technique that is damaging to the product.

As illustrated by the preceding examples, causes can sometimes be determined by a simple analysis of available data (current and past) from a process. Surprising agreement on causes can occur when people are confronted with data—particularly if data had never been collected before.

In other cases, however, the data analyses require the use of statistical techniques, i.e., setting up quantitative hypotheses and using statistical tests to accept or reject the hypotheses. As explained in Chapter 17, a failure to understand the need for a statistical analysis of data can easily lead to incorrect conclusions.

Many techniques are available to assist in the analysis phase. The Japanese concluded that most quality-related problems could be solved with seven basic tools: the cause–effect diagram, stratification analysis, check sheet, histogram, scatter diagram, Pareto analysis, and control charts. Later, seven new tools were recommended (Table 4.7): affinity diagram, tree diagram (systematic diagram), process decision program chart, matrix diagram, interrelationship diagraph (relations diagram), prioritization matrix (matrix data analysis), and activities network diagram (arrow diagram).

These tools are drawn from the industrial engineering, operations research, statistics, and management disciplines and focus on understanding relationships among quality-related activities in a broad system. In contrast, the seven basic tools focus on problem solving for a specific product or process.

TABLE 4.7
Seven new Japanese tools

Tool	Concept
Affinity diagram	Organize facts, opinions, and issues into natural groupings as an aid to diagnosis
Tree diagram (systematic diagram)	Dissect a problem into subproblems and causes
Process decision program chart	Assess alternative processes to help select the best process
Matrix diagram	Show the presence or absence of relationships among collected pairs of elements in a problem situation
Interrelationship diagraph (relations diagram)	Help identify the key factors for inclusion in a tree diagram
Prioritization matrix (matrix data analysis)	Evaluate options through a systematic approach of identifying, weighing, and applying criteria to the options
Activities network diagram (arrow diagram)	Analyze the sequence of tasks necessary to complete a project and determine the critical tasks to monitor to execute the project efficiently

140   Quality Management and Analysis

Another broad approach to improvement, the theory of constraints (TOC), deserves mention. A *constraint* is the weakest link in a system (process) and therefore should be the focus for improvement. Constraints are mostly policy (procedures, past practice) but may also be physical (machines, people, other resources). TOC identifies the constraint(s) and analyzes the system to ensure that all parts are aligned and adjusted to support the maximum effectiveness of the constraint.

Some recent and classic books that provide extensive discussion on analysis include Wadsworth, Stephens, and Godfrey (2001); Smith (1998); Kolarik (1995); Mears (1995); Wilson, Dell, and Anderson (1993); and Juran Institute, Inc. (1989). Plsek (1997) applies the tools of creative thinking and innovation to various aspects of quality management, including process design.

4.10   IMPROVE PHASE

This phase designs a remedy, proves its effectiveness, and prepares an implementation plan. The steps are

- Evaluate alternative remedies.
- If necessary, design formal experiments to optimize process performance.
- Design a remedy.
- Prove the effectiveness of the remedy.
- Deal with resistance to change.
- Transfer the remedy to operations.

Evaluate Alternative Remedies

Remedial action responds to the findings of diagnosis. Usually several alternative remedies (solutions) are proposed to remove the cause found in diagnosis. The remedy selected should significantly improve on the original problem and should optimize both company costs and customer costs. In evaluating remedies, a useful tool for the improvement team is the remedy selection matrix (Table 4.8).

The team first discusses and agrees on the criteria it will use. If desired, weights can be assigned to the criteria. Then the team evaluates each remedy for each criterion. The rating can be quantitative or stated simply as high, medium, or low desirability.

Design of Experiments

Experiments in the laboratory or outside world may be necessary to determine and analyze the dominant causes of a quality problem and to design a remedy. Five types of experiments are summarized in Table 4.9.

Exploratory (screening) experiments are often conducted to verify the vital few causes and dominant variables, and laboratory and production experiments are conducted to generate a mathematical model of the process and optimize process performance.

TABLE 4.8
Remedy selection matrix

Criterion	Remedy 1	Remedy 2	Remedy 3
Remedy name			
Total cost			
Impact on the problem			
Benefit/cost relationship			
Cultural impact/resistance to change			
Implementation time			
Uncertainty about effectiveness			
Health and safety			
Environment			
Summary (1 for best, 2 for next, and so on.)			

Source: Juran Institute, Inc.

TABLE 4.9
Types of diagnostic experiments

Type of experiment	Purpose and approach
Evaluating suspected dominant variables	Evaluate changes in values of a variable by dividing a lot into several parts and processing each portion at some different value, e.g., temperature.
Exploratory experiments to determine dominant variables	Statistically plan an experiment in which a number of characteristics are carefully varied to yield data for quantifying each dominant variable and the interactions among variables.
Production experiments (evolutionary operation)	Make small changes in selected variables of a process, and evaluate the effect to find the optimum combination of variables.
Response to surface experiments	Make changes in selected variables of a process to generate an empirical map and mathematical model of the response.
Simulation	Use the computer to study the variability of several dependent variables that interact to yield a final result.

These experiments require definition of the objective, response (output) variables, independent (input) variables, test levels for the variables, and the selection of the design of the experiment. Chapter 18 discusses several types of experimental designs, some tools for sound experimentation, and the classical (one factor at a time) versus modern methods of experimentation. *JQH5,* Section 47, provides a comprehensive coverage of both the design and analysis of experiments as part of the learning process in the scientific method.

The application of the formal design of experiments (DOE) has been hindered by both the difficulties of selecting an experimental design and the relatively complex statistical calculations and data analysis required. Now the personal computer and associated software reduce these burdens and facilitate the use of powerful tools like DOE. Software such as MINITAB® can help to select designs, analyze data (including computer-constructed graphical displays of data), and help to create a mathematical model of process performance.

A well-organized exploratory experiment has a high probability of identifying the dominant causes of variability. However, there is a risk of overloading the experimental plan with too much detail. A check on overextension of the experiment is to require that the analyst prepare a written plan for review. This written plan must define

1. The characteristics of material, process, environment, and product to be observed.
2. The control of these characteristics during the experiment; a characteristic may be
    a. Allowed to vary as it will and measured as is.
    b. Held at a standard value.
    c. Deliberately randomized.
    d. Deliberately varied, in several classes or treatments.
3. The means of measurement to be used (if different from standard practice).

If the plan shows that the experiment may be overloaded, a dry run in the form of a small-scale experiment is in order. A review of the dry-run experiment can then help decide the final plan.

Production experiments

Experimentation is often regarded as an activity that can be performed only under laboratory conditions. To achieve maximum performance from some manufacturing processes, however, the effect of key process variables on process yield or product properties must be demonstrated under shop conditions. Laboratory experimentation to evaluate these variables does not always yield conclusions that are completely applicable to shop conditions. When justified, a "pilot plant" may be set up to evaluate process variables. However, the final determination of the effect of process variables must often be done during the regular production run by informally observing results and making any changes that are necessary. Thus informal experimentation *does* take place on the manufacturing floor.

To systematize informal experimentation and provide a methodical approach for process improvement, G. E. P. Box developed a technique known as "evolutionary operations" (EVOP). EVOP is based on the concept that every manufactured lot has information to contribute about the effects of process variables on a quality characteristic. Although such variables could be analyzed by an experimental design, EVOP introduces *small* changes into these variables according to a planned pattern of changes. These changes are small enough to avoid nonconformance but large enough to establish gradually (1) which variables are important and (2) the optimum process values for these variables. Although this approach is slower than a formal experimental design, results are achieved in a production environment without the additional costs of a special experiment.

The steps are

1. Select two or three independent process variables that are likely to influence quality. For example, time and temperature were selected as variables affecting the yield of a chemical process.

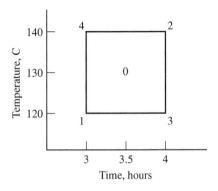

FIGURE 4.16
An EVOP plan. Numbers are in run
order. 0 is the reference run.

2. Change these steps according to a plan (see Figure 4.16). This diagram shows the *plan*, not any data. For example, a reference run was made with the production process set to run at 130°C for 3.5 hours. The next batch (point 1 in Figure 4.16) was run at 120°C for three hours. The first *cycle* contains five runs, one at each condition. Samples were taken from each batch, and analyses were made.
3. After the second repetition of the plan (cycle 2) and each succeeding cycle, calculate the effects (see *JQH5,* pp. 47.46–47.51).
4. When one or more of the effects is significant, change the midpoints of the variables and perhaps their ranges.
5. If no variable has been effective after eight cycles, change the ranges or select new variables.
6. Continue moving the midpoint of the EVOP plan and adjust the ranges as necessary.
7. When a maximum has been obtained or the rate of gain is too slow, drop the current variables from the plan and run a new plan with different variables.

EVOP is a highly structured form of production experimentation. An associated technique that is applied in research and development laboratories is response surface methodology (RSM). RSM provides empirical maps (contour diagrams) that show how factors under the experimenter's control influence the response. Factor settings are viewed as defining points in the factor space (may be multidimensional) at which the response will be recorded. Thus the maps illustrate the nature of the response surface.

Simulation experiments

From the field of operations research comes a technique called "simulation" that can be useful in analyzing quality problems. Simulation provides a method of studying the effect of a number of variables on a final quality characteristic—but all of this is done on paper without conducting experiments! A simulation study requires the following inputs:

1. Definition of the output variable(s)
2. Definition of the input variable(s)
3. Description of the complete system relating the input and output variables

4. Data on the distribution of each input variable; thus variability is accepted as inherent in the process

In simulation, a system model is developed and translated into a computer program. This program not only defines the relationship between input and output variables but also makes provision for storing the distribution of each input variable. The computer then selects values at random from each input distribution and combines these values, using the relationship defined, to generate a simulated value of the output variable. Each repetition of this process results in a simulated output result. These can then be formed into a frequency distribution. The payoff is to make *changes* in the input variables or the relationships, run another simulation, and observe the effect of the change. Thus the significance of variables can be evaluated on paper, providing one more way of evaluating theories on the causes of problems.

Simulation has been applied to many quality problems, including interacting tolerances, circuit design, and reliability.

Simulation has also been used in the service industry. Anderson, Abetti, and Savage (1995) present an application at Aetna Health Plans. The improvement project involved reducing the cycle time of patient cases that required additional medical review. The original average turnaround time (TAT) was 18.1 days and was a major source of customer complaints. The goals of the improvement project were to (1) reduce TAT to 50% of files resolved within one week, (2) improve overall quality, and (3) encourage a sense of ownership and accountability. The people involved included customer service representatives (CSRs), nurses, mailroom personnel, medical directors, and various supervisors. A simulation model for the computer was developed by using the process model (how work flows through the area), the resource model (who is available to do the work), and the input model (what kind of work is done). For example, detailed process flow diagrams, including times for each task, were developed for the simulation model. The model was then tested (and refined) by using the computer simulation to generate three months of TAT data and comparing the simulation results to three months of actual data. The result was a computer simulation model that could be used to evaluate proposed process changes.

Three sets of process changes were tested by using the simulation model. The simulation results identified the key drivers of TAT. The remedy in the form of process changes included assigning a dedicated CSR resource, scheduling meetings among resources to replace interoffice mail routing, using preprinted form letters, undertaking aggressive action to obtain outside information, and assigning work not requiring medical expertise to clerical resources. The improved process reduced TAT from 18.1 days to 6.6 days.

Aetna concluded that the simulation model provides an experimental test bed (in the computer) to try ideas, encourages the need for data, and expedites implementation of the improved process because "to delay implementation in the face of such compelling data would have been unthinkable." The model was a powerful force in obtaining upper management focus on the problem.

Batson and Williams (1998) present brief descriptions of seven case studies in quality simulation in both manufacturing and service industries. The *New York Times* (January 11, 2000) describes a fascinating application of simulation to study the movement of bubbles in a glass of Guiness stout. Using computational fluid dynamics software, researchers ran simulations to alter various parameters, including the temperature of the stout, the size and concentration of the bubbles, and even the shape of the glass. Of course, the computer simulation was supplemented by field trials in local pubs—sample size not reported.

For a complete discussion of the methodology and application of simulation, see Banks (1998). Also, an emerging development is the combining of simulation with "fuzzy logic" concepts. *Fuzzy logic* is the application of mathematics to represent and manipulate data that possess nonstatistical uncertainty and uses imprecise information, e.g., cool, warm, or hot, rather than specific temperatures (for a brief explanation, see *JQH5,* page 10.8).

Design a Remedy

The remedy must fulfill the original project mission, particularly with respect to meeting customer needs. This step identifies the customers, defines their needs, and proves the effectiveness of the remedy. In some cases, designing a remedy may require a major replanning of the product or process and may involve a structured approach to the replanning.

Prove Effectiveness of the Remedy

Before a remedy is finally adopted, it must be proven effective. Two steps are involved:

1. *Preliminary evaluation of the remedy under conditions that simulate the real world.* Such evaluation can use a "paper" reliability prediction, a dry run in a pilot plant, or testing of a prototype. But these preliminary evaluations have assumptions that are never fully met, e.g., the prototype is assumed to be made under typical manufacturing conditions, when actually it is made in the engineering model shop.
2. *Final evaluation under real-world conditions.* There is no substitute for testing remedies in the real world. If the remedy is a design change on a component, the final evaluation must be a test of the redesigned component operating in the complete system under field conditions; if the remedy is a change in a manufacturing procedure, the new procedure must be tried under typical (not ideal) factory conditions; if the remedy is a change in a maintenance procedure, the effectiveness must be demonstrated in the field environment by personnel with representative skill levels.

Finally, after a remedy is proven effective, an issue of communication remains. A remedy on one project may also apply to similar problems elsewhere in an organization. It is useful, therefore, to communicate the remedy (1) to others who may face similar problems and (2) to those responsible for planning future products and processes. In one approach, the remedy is entered into a database that can easily be examined by means of key words.

Deal with Resistance to Change

Various objections to the remedy may be voiced by different parties, e.g., through delaying tactics or outright rejection of the remedy by a manager, the workforce, or the union. "Resistance to change" is the usual name. Change consists of two parts: (1) a technological change and (2) a social consequence of the technological change.

People often voice objections to technological change, although the true reason for their objection is the social effect. Thus those proposing the change can be misled by the objections stated. For example, an industrial engineer once proposed a change in work method that involved moving the storage of finished parts from a specific machine to a central storage area. The engineer was confused by the affected worker's resistance to the new method. The method seemed to benefit all parties concerned, but the worker argued that it "would not work." The supervisor was perceptive enough to know the true reason for resistance—the worker's production was superb, and many people stopped at his machine to admire and compliment him. Who would want to give up that pleasure? To cite another example, some design engineers resist the use of computer-aided design (CAD), claiming that the technology is not as effective as design analysis by a human being. The real reason, for some older designers, may include the fear of having difficulty adapting to CAD. To achieve change, we must

- Be aware that we are dealing with a pattern of human habits, beliefs, and traditions (culture) that may differ from our own.
- Discover the exact social effects of the proposed technological changes.

Based on the scars of experience, some rules can be identified for introducing change.

Rules of the road for introducing change

Important among these are

- *Provide for participation.* The most important rule for introducing change is to provide for participation. Those who are likely to be affected by the change should be members of the project team and participate in both diagnosis and remedy. Lack of participation leads to resentment, which can harden into a rock of resistance.
- *Establish the need for the change.* The change should be explained in terms that are important to the people involved rather than on the basis of the logic of the change.
- *Provide enough time.* How long does it take the members of a culture to accept a change? They must take enough time to evaluate the impact of the change and find an accommodation with the advocates of the change. Providing enough time takes various forms:
  a. *Starting small.* Conducting a small-scale tryout before going "all out" reduces the risks for the advocates as well as for the members of the culture.
  b. *Avoiding surprises.* A major benefit of the cultural pattern is its predictability. A surprise is a shock to this predictability and disturber of the peace.
  c. *Choosing the right year.* There are right and wrong years—even decades—for a change.
- *Keep the proposals free of excess baggage.* Avoid cluttering the proposals with extraneous matters not closely concerned with getting results. The risk is that debate will get off the main subject and onto side issues.
- *Work with the recognized leadership of the culture.* The culture is best understood by its members. They have their own leadership, which is often informal. Convincing the leadership is a significant step in getting the change accepted.

- *Treat people with dignity.* The classic example is that of the relay assemblers in the "Hawthorne experiments." Their productivity kept rising, under good illumination or poor, because in the laboratory they were being treated with dignity.
- *Reverse the positions.* Ask the question: What position would I take if I were a member of the culture? It is even useful to get into role playing to stimulate understanding of the other person's position.
- *Deal directly with the resistance.* There are many ways of dealing directly with resistance to change:
  a. Trying a program of persuasion.
  b. Offering a quid pro quo—something for something.
  c. Changing the proposals to meet specific objections.
  d. Changing the social climate to make the change more acceptable.
  e. Forgetting it; sometimes the correct alternative is to drop the proposal.

Dealing with resistance to change is an art. However, some approaches provide a methodical way of (1) understanding the impact of change and (2) resolving differences among the parties involved. One approach to understanding the impact is to identify the restraining forces and the driving forces for change ("force field analysis"). Another approach to resolving differences focuses on having the parties clearly state their positions to identify the exact areas of disagreement (see *JQH5,* page 5.66, for elaboration).

Transfer the Remedy to Operations

Transfer to operations may include revisions in operating standards and procedures; changes in staffing and responsibilities; additional equipment, materials, and supplies; and extensive training on the why and how of the changes. A detailed schedule of tasks and dates can help to plan for the implementation. We move now to control, the final phase of six sigma.

4.11 CONTROL PHASE

In this phase, we design and implement certain activities to hold the gains of improvement. The steps are

- Design controls and document the improved process.
- Validate the measurement system.
- Determine the final process capability.
- Implement and monitor the process controls.

Design Controls and Document the Improved Process

In this step we provide a systematic means for holding the gains—the process of *control.* Control during operations is done through use of a feedback loop—a measurement of actual performance, comparison with the standard of performance, and action on the difference. Documentation of the improved process should include the steps taken during the define, measure, analyze, and improve phases.

Validate the Measurement System

The measurement system for the improved process must be evaluated and made capable. This step may involve new measurement devices, the collection of new data, and additional training for process personnel.

Determine the Final Process Capability

This step means providing the operating forces with a process capable of holding the gains under operating conditions. To the extent that is economically feasible, the process changes should be designed to be *irreversible*. For example, changing from hand insertion of components for printed circuit boards to automatic insertion by programmed tape rolls is an irreversible remedy. In wave soldering, a remedy that requires a different specific gravity for a flux could be reversible because contamination or other factors may result in a flux having a specific gravity that was previously unacceptable.

Implement and Monitor the Improved Process

In this step, the improved process is placed into operation, and the control steps described are used to monitor process conditions and product performance. The team should provide for measuring the cost of poor quality to confirm that the remedies have worked.

Implementing and monitoring the improved process is the final step in a quality improvement project. We conclude the chapter with a discussion of how to maintain a focus on continuous improvement.

→ 4.12 LEAN SIX SIGMA SUMMARY AND PROJECT EXAMPLE

The following is a summary of the deliverables and tools typically used in each phase of Six Sigma Improvement (DMAIC), reproduced with permission from the Juran Institute, Inc.

Phase-by-Phase Deliverables and Tools

Define Phase

Deliverables:
- Project charter:
  - Problem statement
  - Project mission (objectives)

- Cost of poor quality and other financial impact
  - Operational/strategic impact
  - Project team
- Project plan
- Operational definitions
- High-level process map (SIPOC)
- Critical-to-quality requirements (CTQs)

Tools:
- Strategic alignment and goal deployment
- Cost of poor quality analysis
- Project selection and chartering
- Project management
- SIPOC (supplier-input-process-output-customer) map
- Voice of customer (VOC) analysis

Measure Phase

Deliverables:
- Measurable definition of Y (or Y's) in $Y = f(X)$
- Detailed process map(s) of current state (as-is)
- Process failure mode and effects analysis (FMEA)
- Data collection plan
- Measurement system analysis (MSA)
- Baseline performance for Y (or Y's)
- Process capability and sigma level
- Pareto chart(s)
- Cause–effect diagram(s)
- List of possible X's (or theories) that impact Y

Tools:
- Process mapping
- Process FMEA
- Data collection planning
- MSA
- Graphs and charts
- Stratification
- Process capability analysis
- Sigma calculation
- Pareto chart(s)
- Brainstorming
- Cause–effect (or fishbone) diagrams

Analyze Phase

Deliverables:
- List of theories (X's) to be tested
- Data collection plan for analyze phase
- Each theory tested (proven or disproven) with data using hypothesis testing protocol
- List of proven X's or proven root causes, i.e., $Y = f(X_1, X_2, \ldots X_n)$

Tools:
- Data collection planning
- Power and sample size
- Confidence intervals
- Hypothesis testing protocol
- T-tests, ANOVA, test for normality, tests for equal variances
- Nonparametric tests
- Correlation, regression
- Chi-square contingency tables (test of independence)
- Proportions tests

Improve Phase

Deliverables:
- Improvement strategy
- Outputs of solution alternative generation
- If applicable, DOE plan and results
- Pugh matrix, criteria selection matrix, payoff matrix
- If pilot is applicable, plan and results of pilot implementation
- List of remedies (solutions) selected
- Revised process map (and photos if applicable)
- Updated FMEA
- Implementation plan showing start and end dates of solution activities and by whom

Tools:
- Design of experiments (DOE)
- Creative thinking
- Benchmarking
- Lean (or kaizen) event
- Mistake proofing
- Pugh matrix, criteria selection matrix, payoff matrix
- Process mapping
- Process FMEA
- Stakeholder analysis
- Change management
- Project management

Control Phase

Deliverables:
- Process control plan
- List of procedures (revised or created)
- Communication and training plans
- Control charts showing results and comparison of process performance (before vs. after improvements)
- Graphical and statistical analysis to prove significance of improvement
- Results (metrics and COPQ) of actual vs. target vs. baseline
- Signoff on results by champion and finance

- Updated project plan (timeline/Gantt chart)
- Lessons learned
- Final project report

Tools:
- Self-control analysis
- Process control plan
- Mistake proofing
- 5S
- Statistical process control
- Standard operating procedures
- Change management

Example of a Six Sigma DMAIC Project

Paint Line First Run Yield Improvement

(Adapted from the final report of a six sigma project led by Chris Arquette at a Juran Institute client; used with acknowledgment of thanks)

Executive Summary

Problem Statement

The paint line has a first run yield of 74%. This means that 26% of all frames need to be reworked at least once. Defects due to finish issues account for 15% and material (wood) issues account for 11%. This project looks only at finish defects because these are readily within our control. Any rework is essentially nonvalue added and contributes to wasted paint/primer, labor, utilities, w.i.p., capacity, and more hazardous waste. Our goal is to improve first run yield to 90% for finish defects only.

Chronicle of Effort

The paint line is very old and somewhat neglected. There is generally a lot of "low hanging fruit." Our first MSA results were expectedly poor and the appraisers were contributing to the defect rate by rejecting good frames. We improved this by continued training of the appraisers by Q.C. We did two more MSAs with acceptable results. This will need to be an ongoing test/train routine.

We discovered in the Measure Phase that blisters, contamination, and dirt in finish made up about 77% of all finish defects, so we decided to narrow our scope to just those three. We set up our end of line appraisers to mark defects on the hour and to count frames run per hour by the off-loader. This enabled us to follow the defect rate with variables as they change. We had cause and effect sessions with representatives from the paint supplier, plant, R & D, and corporate facilities. This really helped us to get very detailed cause and effects diagrams for our defects and was crucial in determining potential key factors. We performed Design of Experiments on our key factors for blisters and dirt in finish. The team built a "wind tunnel" around the primer flash conveyor to better control the flash environment. We wanted to use heated air

Selected Slides

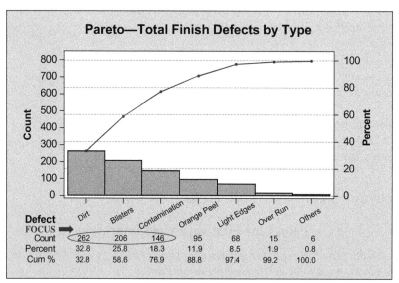

CONCLUSION: Dirt in finish, blisters, and contamination are the most frequent finish defects on all frames.

for this but were unable to come up with a viable heat source that was inexpensive to implement. This will be looked at in the future. We discovered that wax crayon residue from the component plant was the cause in 75% of the contamination we looked at. We worked with the component plant to resolve this issue. To try to reduce dirt in finish, we installed a new air knife and conveyor brush to loosen dirt and then blow it off with high velocity air.

Improvements and Results

We achieved a 7.5% increase in first run yield, from a baseline of 85% to 92% overall for finish defects. This was mainly due to reduction of blisters and contamination defects. Dirt in finish defects decreased slightly but was not statistically significant. Here are the changes that were made to the process:

1. Optimum gram weight settings for primer and top coat were verified. These will be monitored daily with X Bar-R control charts.
2. The wind tunnel for primer flash will stay as a permanent part of the process.
3. Optimum substrate temperature was verified and will be controlled through primer oven temperature. This will be monitored daily on an I-MR chart.
4. MSAs will be conducted monthly. Results will be monitored on an I-MR chart.
5. The component plant will add a standard procedure to inspect and clean machine surfaces of wax crayon residue prior to every shift. This will be marked off on a check sheet.
6. Defect rate for blisters, contamination, and dirt in finish will be monitored continuously on p-charts and updated daily.

Cause–Effect Diagram Showing Theories for Dirt in Finish

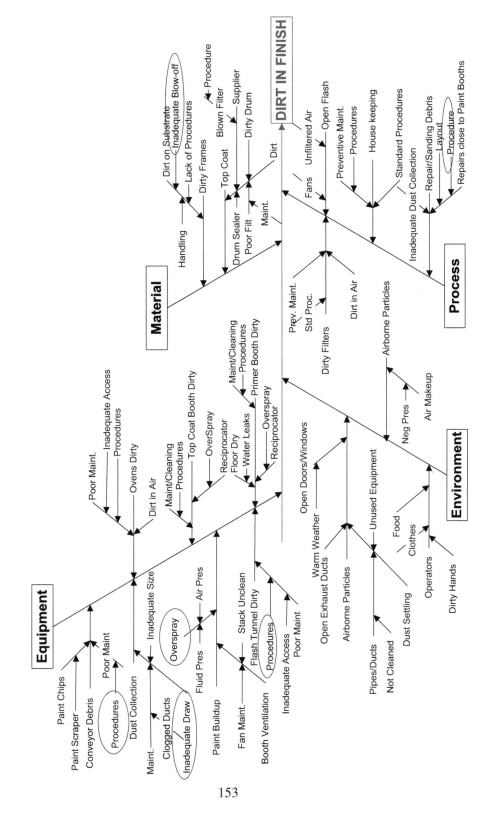

Cause–Effect Diagram Showing Theories for Blisters

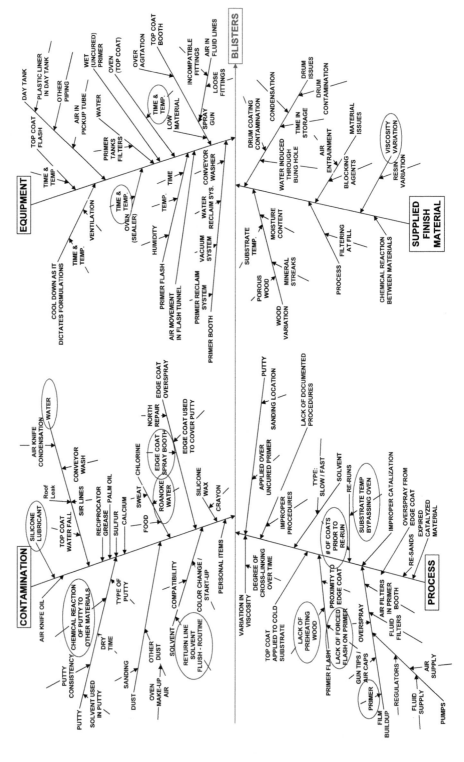

Results achieved: Increased first time yield for finish defects from 85 to 92.5%. Sigma level went from 2.6 to 2.9. Cost of poor quality went down substantially. Long-term capability went up from 0.35 to 0.48. We achieved substantial total cost savings. This includes the reduction in C.O.P.Q. and also savings from reducing gram weight, which reduces paint usage and associated waste. We potentially have additional cost savings through future Kaizen events identified from our Value Stream Map.

Lessons Learned

- MSA is a very important step prior to any data collection.
- Do not trust existing data.
- Be creative in coming up with improvement ideas.
- Take time to discuss the control plan with everyone involved.
- Monitor data collection activities for accuracy.

Theories Were Then Selected to Be Tested in the Analyze Phase

Blisters' Potential *X*'s—Theories to be tested

$X1$: Insufficient flash-off of primer increases occurrence of blisters.

$X2$: Inconsistent oven temperatures causes blisters.

$X3$: Variation in gram weight for film build increases occurrence of blisters.

$X4$: Blisters occur more frequently on reworked frames than first run.

$X5$: Variation in substrate temperature prior to topcoat increases occurrence of blisters.

$X6$: Blisters occur more frequently after solvent flush than during normal run.

Dirt's Potential *X*'s—Theories to be tested

$X7$: Line cleaning causes an increase in the ratio of frames with dirt.

$X8$: Dirt in finish occurs more frequently at the topcoat stage than before topcoat.

$X9$: Dirt in finish occurs more frequently on the small side vs. large side.

Contamination's Potential *X*'s—Theories to be tested

$X10$: Wax markings cause higher frequency of contamination than unmarked substrate.

$X11$: Lubrication from the topcoat reciprocator causes higher ratio of contaminated frames than other process equipment.

→ 4.13 MAINTAINING A FOCUS ON CONTINUOUS IMPROVEMENT

Competitive pressures require many improvement projects—at a revolutionary rate. We must replicate the results of current projects, nominate new projects, and take other actions for improvement.

Data Collection Plan for Analyze Phase
(data collection plan template developed by Dr. Richard Chua, Juran Institute, Inc., all rights reserved)

	Ref.	Theories To Be Tested (Selected From The C-E Diagram, FMECA, and/or FDM)	List Of Questions That Must Be Answered To Test Each Selected Theory	Where Applicable, State The Null and Alternative Hypotheses		Data To Be Collected					
				H_o	H_A	Tools To Be Used	Description/ Data Type	Sample Size, Number of Samples	Where To Collect Data	Who Will Collect Data	How Will Data Be Recorded
BLISTERS	1	Insufficient flash of primer causes blisters.	Does ambient humidity level affect blister ratio?	Median blister ratio at humidity level 1 = median ratio at humidity level 2	Median blister ratio at humidity level 1 does not = median ratio at humidity level 2	Mann-Whitney	Y = Continuous X = Discrete	1 week's production	End of line inspection	Inspectors	Data sheet
			Does ambient temperature affect blister ratio?	Median blister ratio at temp 1 = median ratio at temp 2	Median blister ratio at humidity level 1 does not = median ratio at humidity level 2	Mann-Whitney	Y = Continuous X = Discrete	1 week's production	End of line inspection	Inspectors	Data sheet
			Does primer viscosity affect blister ratio?	Median blister ratio at viscosity 1 = median ratio at viscosity 2	Median blister ratio at humidity level 1 does not = median ratio at humidity level 2	Mann-Whitney	Y = Continuous X = Discrete	1 week's production	End of line inspection	Inspectors	Data sheet
	2	Inconsistent oven temperatures causes blisters	Does topcoat oven temperature variation affect blister ratio?	There is no correlation between blister ratio and topcoat oven temp	There is a correlation between blister ratio and topcoat oven temp	Correlation/ regression	Y = Continuous X = Continuous	2 days production @ 2 hour intervals	In topcoat oven using data pak	Six-sigma team members	On data pak print out sheet
	3	Variation in gram weight for film build increases occurrences of blisters	Does gram weight variation affect blister ratio?	Median blister ratio @ grasm wt. level 1 is = to median blister ratio @ gram wt. 2	Median blister ratio @ grasm wt. level 1 is not = to median blister ratio @ gram wt. 2	Kruskal-Wallace	Y = Continuous X = Discrete	1 week's production	Topcoat booth and end of line inspection	Topcoat booth operator	On data sheet
	4	Blisters occur more frequently on reworked frames than on first run frames	Is there a difference in blister ratio between first run vs. reworked frames?	Blister occurence is independent of number of runs	Blister occurence is dependent on number of runs	Chi square	Y = Discrete X = Discrete	3 rework runs	End of line inspection	Inspectors	On defect data sheet
	5	Variation in substrate temperature prior to topcoat affects occurence of blisters	Does variation in substrate temperature affect blister ratio	Blister occurence is independent of substrate temperature	Blister occurence is dependent on substrate temperature	Chi square	Y = Discrete X = Discrete	4 hours of production	After primer oven prior to topcoat application	Six-sigma team members	On data sheet
	6	Blisters occur more frequently after solvent flush than during normal run	Is there a difference in blister ratio between frames run right after solvent flush vs. those run before?	Blister occurence is independent of solvent flush	Blister occurence is dependent on solvent flush	Chi square	Y = Discrete X = Discrete	2 hours prior and 2 hours following line flush	End of line inspection	Six-sigma team members	On data sheet
DIRT IN FINISH	7	Line cleaning causes an increase in the ratio of frames with dirt	Is there a difference in the ratio of frames with dirt directly after line cleaning than before cleaning?	Occurence of dirt in finish is independent of line cleaning	Occurence of dirt in finish is dependent on line cleaning	Chi square	Y = Discrete X = Discrete	Half a day of production prior to and after cleaning line	End of line inspection	Six-sigma team members	On defect data sheet
	8	Dirt in finish occurs more frequently at the topcoat stage vs. after sanding	Is there a difference in the ratio of frames with dirt at the topcoat stage vs the primer stage?	Occurence of dirt in finish is independent of process stage	Occurence of dirt in finish is dependent on process stage	DOE	Y = Continuous X = Discrete	1 day's production	After sanding and end of line inspection	Six-sigma team members	On data sheet
	9	Dirt in finish occurs more frequently on the small side vs. large side	Is there a difference in the ratio of frames with dirt between the small side vs. large side of line?	Occurence of dirt in finish is independent of process stream	Occurence of dirt in finish is dependent on process stream	Chi square	Y = Discrete X = Discrete	1 week's production	End of line inspection	Six-sigma team members	On defect data sheet
CONTAMINATION	10	Wax markings are a significant cause of contamination on frames	Are wax markings a significant cause of contamination on frames?	Contamination is independent of wax marking	Contamination is dependent on wax marking	Pareto	Y = Discrete X = Discrete	40 contaminated sample frames	End of line inspection	Inspectors	On defect data sheet
	11	Lubrication from the reciprocators causes more contamination than lubrication from other equipment	Is there a difference in the ratio of contaminated frames from the reciprocator vs. other equipment?	Mean contamination ratio after reciprocator is = to mean before reciprocator	Mean contamination ratio after reciprocator is not = to mean before reciprocator	DOE	Y = Continuous X = Discrete	2 hours per day over 3 days	End of line inspection		

Examples of Testing of Theories

Test of Theory X_3

Theory: Variation in gram weight for film build affects the occurrences of blisters.

H₀: Median blister ratio @ gram wt. level 1 is = to median blister ratio @ gram wt. 2

Analysis:

Normality Test:
 H₀: data are normal
 H_A: data are not normal
Results: P values: .000; Reject null, data are not normal.

Test for Equal Variance:
 H₀: $\sigma^2 = \sigma^2$
 H_A: $\sigma^2 > \sigma^2$
Results: $p = .018$; reject the null, variances are not equal.

Compare Medians–Mann-Whitney
 H₀: medians are equal
 H_A: Medians are not equal
Results: $p = .003$; Reject the null, medians are not equal.

Conclusion: Gram weight level does affect blister ratios.

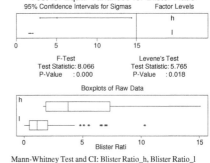

Mann-Whitney Test and CI: Blister Ratio_h, Blister Ratio_l
 Blister N = 6   Median = 3.750
 Blister N = 133 Median = 1.100
 Point estimate for ETA1-ETA2 is   2.400
 95.0 Percent CI for ETA1-ETA2 is (0.903,3.901)
               W = 704.0
Test of ETA1 = ETA2 vs ETA1 not = ETA2 is significant at 0.0033
The test is significant at 0.0032 (adjusted for ties)

Test of Theory X_6

Theory: Blisters occur more frequently right after solvent flush than during normal run.

H₀: The occurrence of blisters is independent of solvent flush.

Analysis:

Chi-square test: Before flush, after flush

Expected counts are printed below observed counts

```
        Before F   After Fl   Total
   1     14021       931      1495
         14015.23   936.77

   2       252        23       275
          257.77     17.23

Total    14273       954     15227

Chi-Sq = 0.002 + 0.036 +
        0.129 + 1.933 = 2.100
DF = 1, p-Value = .147
```

Conclusion: Fail to reject H₀, Blister occurrence is independent of production before solvent flush vs. right after solvent flush.

Blisters Before vs. After Improvements

Is the occurrence of blisters after improvements significantly better (lower) than blisters before improvements?

H_0: Blister occurrence is independent of improvements made.

Analysis:

```
Chi-Square Test: Blisters: Before, After Improvements
Expected counts are printed below observed counts
            Before      After      Total
    Bad     16133       348        16481
            14302.61    2178.39
    Good    225176      36405      261581
            2.27E+05    34574.61
    Total   241309      36753      278062

Chi-sq =234.245 +1.5E+03 +
        14.759 + 96.901 = 1883.883
DF = 1,  p-Value = 0.000
```

p is <.05, Reject H_0.

Conclusion: Occurrence of blisters is not independent of improvements made (after is lower).

Contamination Before vs. After Improvements

Is the occurrence of contamination after improvements significantly better (lower) than contamination before improvements?

H_0: Contamination occurrence is independent of improvements made.

Analysis:

```
Chi-Square Test: Contamination Before, After
Expected counts are printed below observed counts
            Before      After      Total
    Bad     6695        1219       7914
            6189.36     1724.64
    Good    398862      111788     510650
            3.99E+05    1.11E+05
    Total   405557      113007     518564
Chi-sq = 41.309 +148.248 +
         0.640 +  2.298 = 192.494
DF = 1,  p-Value = 0.000
```

p is <.05, Reject H_0.

Conclusion: Occurrence of contamination is not independent of improvements made (after is lower).

Project Results
Baseline Performance vs. Target vs. Achieved

BASE LINE	TARGET	ACHIEVED
COPQ: $ 2X per year	COPQ: $ X per year	COPQ: $ Saved more than target savings
First Run Yield: 85%	First Run Yield: 90%	First Run Yield: 92.5%
Capability: $P_{PK} = 0.35$	Capability: $P_{PK} = 0.44$	Capability: $P_{PK} = 0.48$
DPMO: 146247	DPMO: 91293	DPMO: 75368
Sigma Level: 2.6 (Finish Defects)	Sigma Level: 2.8 (Finish Defects)	Sigma Level: 2.9 (Finish Defects)

To replicate the results of a current project means to apply a proven remedy to similar problems in an organization. As part of completing a project, the improvement team documents all the steps taken in diagnosing causes and implementing remedies. A computer database of improvement projects (completed and in progress) should be maintained and made known to all managers. In addition, managers should be asked to identify problems in the organization that are similar to those solved by completed improvement projects. When a similar problem is identified, a "replication team" can be appointed to validate the problem in the new situation and determine whether the causes and remedy of the original problem apply to the new problem.

The foundation for maintaining a focus on improvement is the infrastructure for improvement. The infrastructure includes formalizing a process for nominating and selecting projects, forming improvement teams, and providing the training and support for the teams. Additional actions can help to maintain the focus:

1. Use information from assessment studies to identify improvement opportunities. These assessments include the cost of poor quality, market research on customer satisfaction and loyalty, quality culture, and the broad system. Note that these areas go far beyond nonconformance to specifications.
2. Address process improvement in terms of effectiveness, efficiency, adaptability, and cycle time.
3. Pursue radical forms of improvement.

4. Apply all the tools of improvement—technical and behavioral, simple and sophisticated. Increasingly, savings from improvement projects have "skimmed the cream off the top." The next round requires deeper analysis.
5. Build up a database of improvement information. We can start by simply recording completed and in-progress improvement projects, but other steps can also be taken. For example, a fast-food firm is creating an "intellectual network" of computer bulletin boards of information to include best-practices information. Rethmeier (1995) explains how an alliance of 300 hospitals uses a "learning center" to create and transfer knowledge for continuous improvement. The learning center uses lessons derived from the experience of prior events. Juran (*JQH5,* Section 5) calls this retrospective analysis a "Santayana review" in honor of George Santayana who observed, "Those who cannot remember the past are condemned to repeat it." Taking an example from early Native American tribes, organizations will require a "wisdomkeeper."

SUMMARY

- The quality improvement process addresses *chronic* quality problems.
- The six sigma sequence and the breakthrough sequence are strategic approaches to improvement—strategic because the projects selected are based on gaps between actual performance and goals. The two approaches are complementary in both objective and content.
- The six sigma steps are define, measure, analyze, improve, and control.
- The breakthrough steps are proving the need, identifying projects, organizing project teams, verifying the project need and mission, diagnosing the causes, providing a remedy and proving its effectiveness, dealing with resistance to change, and instituting control to hold the gains.

PROBLEMS

4.1. Selwitchka (1980) presents data on 10 types of errors using two measures—frequency and cost:

Type of error	Frequency	Cost ($, DM)
A	960	20,000
B	870	28,460
C	420	375,000
D	210	42,000
E	180	124,300
F	180	9,000
G	60	77,800
H	60	12,125
I	30	9,000
J	30	9,125

Note that the first two columns present a Pareto analysis based on frequency of occurrence. Prepare a second Pareto table based on cost. Comment on the ranking of errors using frequency versus ranking based on cost.

4.2. Unplanned shutdowns of reactors have been a chronic problem. After much discussion, the consensus—called the "wisdom"—identified "cylinder changes" and "human error" as the primary causes. A diagnostic approach based on facts was instituted. Here are data on the causes of previous shutdowns:

Cause	Frequency
Cylinder changes	21
Human error	16
Hot melt system	65
Initiator system	25
Interlock malfunction	19
Other	23
	169

(a) Convert the preceding data into a Pareto table having three columns: cause, frequency, and percentage of total frequency.
(b) Calculate the cumulative frequencies for the table in part (a).
(c) Calculate the percentage cumulative frequencies. Make a Pareto diagram by plotting percentage cumulative frequency versus causes.
(d) Comment about the wisdom versus the fact.

4.3. Draw an Ishikawa diagram for one of the following: (a) the quality of a specific activity at a university, bank, or automobile repair shop; (b) the quality of *one* important characteristic of a product at a local plant. Base the diagram on discussions with the organization involved.

4.4. The following data summarize the total number of defects for each worker at a company during the past six months:

Worker	Number of defects	Worker	Number of defects
A	46	H	9
B	22	I	130
C	64	J	10
D	5	K	125
E	65	L	39
F	79	M	26
G	188	N	94

A quality cost study indicates that the cost of these defects is excessive. There is much discussion about the type of quality improvement program. Analysis indicates that the manufacturing equipment is adequate, the specifications are clear, and workers are given periodic information about their quality record. What do you suggest as the next step?

4.5. An engineer in a research organization has twice proposed that the department be authorized to conduct a research project. The project involves redesigning a component to reduce the frequency of failures. The research approach has been meticulously defined by the engineer and verified as valid by an outside expert. Management has not authorized the project because "other projects seem more important." What further action should the engineer consider?

4.6. A company that manufactures small household appliances has experienced high scrap and rework for several years. The total cost of the scrap and rework was recently estimated on an annual basis. The number shocked top management. Discussions by the management team degenerated into arguments. Finally, top management proposed that all departments reduce scrap and rework costs by 20% in the coming year. Comment on this proposal.

4.7. A small steel manufacturing firm has had a chronic problem of scrap and rework in the wire mill. The costs involved have reached a level where they are a major factor in the profits of the division. All levels of personnel in the wire mill are aware of the problem, and there is agreement on the vital few product lines that account for most of the problem. However, no reduction in scrap and rework costs has been achieved. What do you propose as a next step?

4.8. The medical profession makes the journey from symptom to cause and cause to remedy for medical problems of human beings. Compare this process with the task of diagnosing quality problems of physical products. If possible, speak with a physician to learn the diagnostic approach used in medicine.

4.9. From your experience, recall a chronic quality-related problem that was acted upon by an organization. Critique the approach for handling the problem in terms of the use—or nonuse—of the steps in the breakthrough sequence.

4.10. Select one chronic quality-related problem in your organization.
(a) Write a brief problem statement.
(b) Write a mission statement for a quality improvement team.
(c) What data could be collected to prove the need to address the problem?
(d) What departments should be represented on the team?
(e) State one or more symptoms of the problem.
(f) State at least three theories for the cause(s).
(g) Select one theory. What data or other information is needed to test the theory?
(h) Assume the data show that your selected theory is the true cause. State a remedy to remove the cause.
(i) What forms of resistance to the proposed remedy are you likely to encounter, and how will you deal with this resistance to change?
(j) What methods should be instituted to hold the gains?

4.11. You are trying to convince the upper management of an insurance company to embark on a new approach to quality. Studies on the cost of poor quality, market standing, quality culture, and the quality system have shown a need for action. You have proposed a course of action, but management has not acted on your proposal. Which additional step would be most useful to convince management?

4.12. Your hospital has embarked on the project-by-project approach to quality improvement. Projects have been selected, a team formed for each project, and diagnostic and other forms of training provided to the teams. Many of the teams have been working on their project for more than one year, and some teams have become discouraged. What is the most likely reason for the lengthy project times?

4.13. Describe how the principles of quality improvement could help to improve your game of golf. Consider matters such as setting goals, defining measurements, collecting and analyzing data, identifying weaknesses, implementing improvement actions, and training. For the views of a quality professional and a golf professional, see Karlin and Hanewinckel (1998).

REFERENCES

Anderson, D., F. Abetti, and P. Savage (1995). "Process Improvement Utilizing Computer Simulation: Case Study," *Annual Quality Congress Proceedings,* ASQ, Milwaukee, pp. 713–724.

Banks, J., ed. (1998). *Handbook of Simulation,* Engineering and Management Press, Institute of Industrial Engineers, Norcross, GA.

Batson, R. G. and T. K. Williams (1998). "Process Simulation in Quality and BPR Teams," *Annual Quality Congress Proceedings,* ASQ, Milwaukee, pp. 368–374.

Betker, H. A. (1983). "Quality Improvement Program: Reducing Solder Defects on Printed Circuit Board Assemblies," *Juran Report Number Two,* Juran Institute, Inc., Wilton, CT, pp. 53–58.

Bhote, K. R. (1991). *World Class Quality,* AMACOM, New York.

Bottome, R., and R. C. H. Chua (2005). "Genentech Error Proofs Its Batch Records," *Quality Progress,* July pp. 25–34.

Chua, R. C. H. (1999). "Analyze Phase Data Collection Plan," Juran Institute's *Improvement Breakthrough Black Belt Workshop,* second ed., Juran Institute, Inc.

Chua, R. C. H. (2001). "What You Need to Know About Six Sigma," *Productivity Digest,* Singapore, December, pp. 37–44.

Chua, R. C. H., and A. Janssen (2000). "Six Sigma: A Pursuit of Bottom-Line Results," *European Quality,* Vol. 8, No. 3.

DeFeo, J. A. and W. Barnard (2003). *Six Sigma Breakthroughs and Beyond,* McGraw-Hill, New York.

Forsha, H. I. (1995). *Show Me: The Complete Guide to Storyboarding and Problem Solving,* Quality Press, ASQ, Milwaukee.

Hartman, B. (1983). "Implementing Quality Improvement," *Juran Report Number Two,* Juran Institute, Inc., pp. 124–131.

Juran Institute, Inc. (1989). *Quality Improvement Tools,* Wilton, CT.

Juran, J. M. (1995). *Managerial Breakthrough,* rev. ed., McGraw-Hill, New York.

Karlin, E. W. and E. Hanewinckel (1998). "A Personal Quality Improvement Program for Golfers," *Quality Progress,* July, pp. 71–78.

Klenz, B. W. (1999). "The Quality Data Warehouse: Serving the Analytical Needs of the Manufacturing Enterprise," *Annual Quality Congress Proceedings,* ASQ, Milwaukee, pp. 521–529.

Kolarik, W. J. (1995). *Creating Quality,* McGraw-Hill, New York.

Lenhardt, L. (1993). "Quality Improvement in the Emergency Department Admission Process," *Impro Proceedings,* Juran Institute, Inc., Wilton, CT, pp. 3A.3-1 to 3A.3-9.

Mears, P. (1995). *Quality Improvement Tools and Techniques,* McGraw-Hill, New York.

Plsek, P. E. (1997). *Creativity, Innovation, and Quality,* Quality Press, ASQ, Milwaukee.

Rethmeier, K. A. (1995). "Creating the Learning Organization: Toward Excellence in Knowledge Transfer," *Impro Proceedings,* Juran Institute, Inc., Wilton, CT, pp. 3C.1-1 to 3C.1-11.

Sanders, D., B. Ross, and J. Coleman (1999). "The Process Map," *Quality Engineering,* vol. 11, no. 4, pp. 555–561.

Selwitchka, R. (1980). "The Priority List on Measures for Reducing Quality Related Costs," *EOQC Quality,* vol. 24, no. 5, pp. 3–7.
Slater, R. (1999). *Jack Welch and the GE Way*, McGraw-Hill, New York.
Smith, G. F. (1998). *Quality Problem Solving*, Quality Press, ASQ, Milwaukee.
Wadsworth, H. M., K. S. Stephens, and A. B. Godfrey (2001). *Modern Methods for Quality Control and Improvement,* John Wiley and Sons, New York.
Wilson, P. F., L. D. Dell, and G. F. Anderson (1993). *Root Cause Analysis*, Quality Press, ASQ, Milwaukee.
Yun, J. Y. and R. C. H. Chua (2002). "Samsung Uses Six Sigma to Change Its Image," *Six Sigma Forum*, November 2002, ASQ pp. 13–16.

CHAPTER 5

Quality by Design to Increase Sales

→ 5.1 CONTRIBUTION OF QUALITY TO SALES INCOME

Planning for quality always takes time and effort. Unfortunately, people simply won't spend the time on planning unless they are first convinced that thorough quality planning is essential for business success. To cite an old cliche—they don't have time to do thorough quality planning, but they do have the time to correct the mistakes of poor planning. Fortunately, we know how to do good quality planning—it's one of the three processes in the trilogy. But before we discuss the steps in quality planning, we will first discuss why the time spent on quality planning is an investment that yields a strong measurable return.

In defining quality for goods and services, two dimensions have been identified: features and freedom from deficiencies (Figure 1.1). Although both components are essential to generating sales, in general the features component is more dominant. This chapter discusses the impact of quality on sales income and then provides an overview of an approach that provides the features necessary to meet sales goals. Chapter 11, "Marketing's Role in Quality," Chapter 12, "Designing for Quality," Chapter 13, "Quality in Operations—Manufacturing Sector," and Chapter 14, "Managing Quality in Operations—Healthcare Sector," provide further elaboration. Also, note that this chapter discusses quality planning for products (both goods and services) and processes (operational quality planning) in contrast to strategic quality planning, which is discussed in Chapter 2, "Integrating Quality into the Enterprise Strategic Plan."

For profit-making organizations, the contribution of quality to sales income occurs by several means:

- Increasing market share
- Securing premium prices
- Achieving economies of scale through increased production
- Achieving unique competitive advantages that cement brand loyalties

For both profit and nonprofit organizations, quality is synonymous with customer satisfaction, both internal and external. For most organizations, satisfaction must be viewed relative to the competition and thus goes far beyond the document called a "specification."

5.2 QUALITY AND FINANCIAL PERFORMANCE

Understanding the effect of quality on sales and financial performance can be aided by looking at some valuable research efforts. This research is based on a database known as Profit Impact of Market Strategies (PIMS). The PIMS program aims at determining how key dimensions of strategy affect profitability and growth. More than 450 companies from numerous manufacturing and service industries have contributed data.

Analysis of the PIMS data has led to important conclusions that center on quality relative to the competition. Assessment of relative quality involves the use of multi-attribute comparisons (see Section 3.6, "Assessing Performance and Standing in the Marketplace"). A key conclusion is this: In the long run, the most important factor affecting business performance is quality relative to the competition. Market share and profitability measure business performance. Table 5.1 shows return on investment (ROI) data. Note that businesses having both a larger market share and better quality earn much higher returns than businesses with a smaller market share and inferior quality. Further, although quality and market share are correlated, each has a strong separate relationship to profitability. Figure 5.1 shows the relationship of quality to profitability [return on sales (ROS) or ROI].

Of course, the greater profitability could be due either to higher prices or lower costs. But analysis of the PIMS data sheds some additional light. Quality affects relative price, but separate from quality, market share has little effect on price. The quality/price relationship is shown in Table 5.2. Superiority in quality commands a premium price.

According to other PIMS data, relative quality has little effect on cost. Because relative quality in this study is the customer's perception of quality, the customer may see two products as being of approximate equal quality even though one of them has "inspected quality in" at high cost and another has "built quality in" at lower cost. These and other similar outcomes may obscure the quality–cost relationship from the producer's viewpoint.

TABLE 5.1
Quality and market share both drive profitability (ROI)

Relative quality percentile	Relative market share		
	Below 25%	25–59%	60% and above
Below 33%	7	14	21
33–66%	13	20	27
67% and above	20	29	38 (ROI)

Source: Buzzell and Gale (1987).

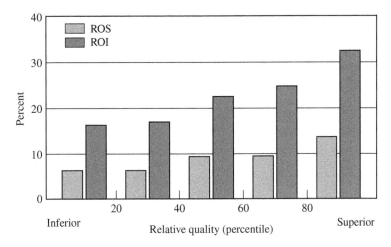

FIGURE 5.1
Relative quality boosts rates of return. (*From Buzzell and Gale, 1987.*)

TABLE 5.2
Quality and price

Relative quality	Relative price
Lowest	100
Commodity	102
Average	104
Better	106
Best	108

Source: Gale and Klavans (1985).

➔ 5.3 ACHIEVING MARKETPLACE SUPERIORITY

Marketplace superiority for a good or service usually arises from a higher level of quality performance than the competition. This advantage will come from noticeably better results for one or more parameters of quality such as performance features, long life, ease of use, freedom from deficiencies, personal service, or fast service.

In some cases superior quality may be obvious, e.g., a hotel room has outstanding features or an insurance claim adjuster arrives within two hours after a car accident is reported.

In other cases the superiority can be translated into users' economics, e.g., an automobile achieves higher mileage per gallon than a competitor's. Sometimes the superiority is minor but can be demonstrated, e.g., one chemical has slightly less variation in a property than a competitor's chemical, even though both meet the same specification. In still other cases, clever marketing can present convincing illustrations of superior performance that are accepted by buyers. Many organizations try to establish superiority in value (the combination of quality and price). Value is difficult to quantify, but because of the importance to buyers, value is often chosen as the focus for establishing superiority and quality leadership. Chapters 11 and 12 provide some examples of using value to achieve leadership. Also, Rust et al. (1994) elaborates on the concepts of value, quality, and utility. Gale (1994) discusses how customer value analyses can help to plan improvement actions.

Market leadership requires demonstrable superiority on one or more product features. Achieving that superiority requires that we know exactly where we stand with respect to the competition and what changes are required to achieve competitively better performance. Competitive benchmarking is a tool that helps companies both measure their current standing and lay out a plan to close any gap and move into the lead.

Competitive Benchmarking

A *benchmark* is a point of reference by which performance is judged or measured. For quality, possible benchmarks go from the traditional to the unusual:

- The specification
- Customer desires
- Competition (see Chapter 3 under "Assessing Performance and Standing in the Marketplace")
- Best in our industry
- Best in any industry

For survival in the marketplace, the traditional benchmark (the product specification) must be complemented by measuring quality relative to competition; for quality leadership, the benchmark must be the "best." Xerox was one of the earliest systematic practitioners of benchmarking. It defined competitive benchmarking as "the continuous process of measuring our products, services and practices against our toughest competitors or those companies renowned as leaders." Thus a benchmark for a digital printer would be the best competitor in the digital printer industry; a benchmark for the Xerox system of packing warehouse orders might be the performance of a company in any industry, e.g., a mail-order firm selling consumer products. The initial benchmarking steps are

1. Determining the characteristics to be benchmarked
2. Determining the organizations from which data will be collected
3. Collecting and analyzing the data
4. Determining the "best in class"

Plans are then prepared to develop or adapt the "best practices." Such plans are, of course, directed at retaining present customers and generating new ones.

Some additional detail on competitive benchmarking as an approach to achieving quality leadership is a topic in Chapter 2. Also, more technical considerations on competitive as well as other kinds of benchmarking will be found in Chapter 15, "Benchmarking: Defining Best Practices for Market Leadership" (*JQH6*, pp 439–466).

TABLE 5.3
Customer satisfaction versus customer loyalty

Customer satisfaction	Customer loyalty
What customers say—opinions about a product	What customers do—buying decisions
Customer may intend to buy from several suppliers in the future	Customer expects to buy primarily from one or two suppliers in the future
Company aims to satisfy a broad spectrum of customers	Company identifies key customers and "delights" them
Company measures satisfaction primarily with the product for a broad spectrum of customers	Company measures satisfaction with all aspects of interaction with key customers and also their intention to repurchase
Company measures satisfaction primarily for current customers	Company also analyzes and quantifies the reasons for lost customers (defections)
Company measures customer opinions and attitudes	Company measures and evaluates customer buying behavior

↦ 5.4 CUSTOMER SATISFACTION VERSUS CUSTOMER LOYALTY

It is useful to distinguish between customer satisfaction and customer loyalty (see Table 5.3).

In brief, a satisfied customer will buy from our company but also from our competitors; a loyal customer will buy primarily (or exclusively) from our company. A dissatisfied customer is unlikely to be loyal, but surprisingly, a satisfied customer is not necessarily loyal (see Section 5.6).

The sales revenue lost due to customers with problems can be calculated—see Section 5.5.

↦ 5.5 CUSTOMER LOYALTY AND RETENTION

Customer satisfaction and customer loyalty are distinct, although related, concepts. In brief, a satisfied customer will buy from our company but also from our competitors; loyal customers will buy primarily (or exclusively) from our company (see Section 5.4). Also, customer satisfaction concerns what customers *say*, their opinions about a product; customer loyalty relates to what customers *do*, their buying decisions.

Loyal customers not only provide continuing sales revenue but also contribute other benefits:

- Adding new sales by referring other potential customers
- Paying (often) a price premium
- Buying other products from the company
- Cooperating in the development of new products
- Reducing company internal costs such as selling costs

Thus the benefits of having a high percentage of loyal (not just satisfied) customers warrants specific steps to achieve high customer loyalty and minimize defections. To provide an overall perspective, the following nine actions are collectively a road map for achieving high customer loyalty.

1. Continually Assess Customer Needs and Translate These Needs into Better Products

Product development must be based on a thorough understanding of customer needs. But those needs are continually changing. Thus market research to define customer needs must be ongoing (see Chapter 11). The results of the research are then translated into new or modified products (see this chapter and Chapter 12).

2. Periodically Assess Market Standing Relative to Competition

This process typically means conducting multiattribute market research studies on quality. These studies not only provide status in the marketplace but also identify differences in satisfaction that are likely to result in customer defections. This research should incorporate one or more questions on the likelihood that the customer will repurchase or recommend the product.

3. Track Retention and Loyalty Information

As a simple customer retention measure, an insurance company measures the percentage of customers who do not allow a policy to lapse due to nonpayment of the annual premium. But as a loyalty measure, the company estimates what percentage of insurance products purchased by its customers are purchased from the company versus competitor insurance companies, i.e., the share of spending a firm earns from its customers.

Fidelity Investments (Nash, 1998) has determined that highly loyal customers have almost twice the assets and share of wallet with Fidelity as low loyal customers. Reichheld (1996, Chapter 8) discusses the share-of-wallet concept and other measures of customer loyalty. Higher profitability from loyal customers grows with each additional year. Often, the acquisition costs for a new customer are so high that customers who defect after one are two years may actually be money-losing propositions for the company. See Figure 5.2 for a typical example of the composition of the margin generated by a customer over five years.

Sun Microsystems, now a part of Oracle, calculated a customer loyalty index based on four components of loyalty questions: customer satisfaction, likelihood to repurchase, likelihood to recommend, and customer delight (Lynch, 1998). In addition, it backs it up with tracking customer purchase behavior.

It is useful to set goals for customer retention and customer loyalty. For example, Lexus set an owner retention goal of 75% (Waltz, 1996).

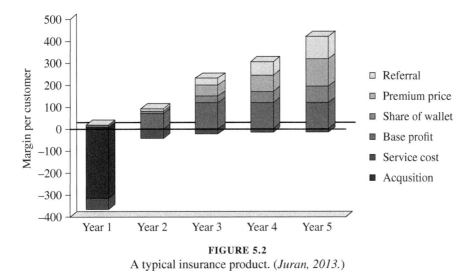

FIGURE 5.2
A typical insurance product. (*Juran, 2013.*)

4. Determine the Drivers of Customer Loyalty

These drivers are the specific elements that have a significant effect on customer loyalty.

Fidelity Investments (Nash, 1998) analyzed data obtained from a 12-page questionnaire mailed to a random sample of customers. Loyalty data were matched against marketing data to determine actual behavior. Regression analysis was used to determine 10 drivers of loyalty. For example, the top driver (satisfaction with customer service) had a 17% impact on loyalty; a driver called satisfaction with investment performance had a 14% impact on loyalty.

Sun Microsystems (Lynch, 1998) identified 20 loyalty drivers and classified them by relative importance and Sun performance relative to competition.

5. Determine the Impact on Profit of Reducing Customer Defections

Customer problems must be converted to cost and revenue implication. The sales revenue from a loyal customer measured over the period of potential repeat purchases can be dramatic. Research concluded that a five percentage point decrease in the defection rate can increase profit by 35 to 95%, depending on the service industry involved (see Section 5.6).

6. Understand the Impact of Handling Complaints on the Likelihood of Repurchasing

Customer satisfaction with the handling of complaints has a significant impact on repurchasing and intention to recommend purchase to others. The numbers can be dramatic (see Section 5.7) and should be determined and made known throughout the organization.

7. Analyze Complaints

Complaints are an early indicator of potential customer defections. The frequency and nature of complaints must be analyzed.

8. Determine the Reasons for Customer Defections

This step means conducting research to discover the reasons for defections by asking customers why they left. But the reasons stated by customers are often not the real reasons, e.g., price is often mentioned as a key reason, but probing usually reveals other reasons. Reichheld (1996, Chapter 7) discusses how a "failure analysis" (a *failure* is a customer defection) can help to discover the causes of defections. The analysis includes applying the classical improvement technique of five whys—i.e., asking why an event occurred at least five times to get to the *root* cause of a defection.

The financial impact of the defections can be displayed in a disloyalty map similar to the one in Figure 5.3. This kind of detailed quantitative result requires a combination of systematic analysis of in-house data coupled with skilled third-party probing of the disaffected customer. Without the five-why and interviewing skill, the answer may typically be: "It's the price." But only rarely is that the true root cause. The results in this example are not unusual. Some of the biggest drivers are related to the service and responsiveness of the company, not just the physical product.

9. Present the Results of Loyalty Analyses for Action

Market research and loyalty analysis results must be acted upon if they are to contribute to customer retention and loyalty. To assure action, one bank employs a guide to help each branch manager interpret and act on the research results (for elaboration, see JQH5, page 18.19).

A related issue is designing customer surveys to collect the right information in sufficient detail so that line managers can act on the results. This activity highlights the importance of obtaining input from the managers in designing customer surveys.

Graphing the research results can be an effective stimulus for action. For example, mapping customer satisfaction and importance ratings to relate customer views and potential action. For examples of maps for customer retention modeling, see Lowenstein (1995, Chapter 9).

Finally, it is useful to relate satisfaction research results to operational processes or to specific activities. Figure 5.4 is one example of this.

The nine actions discussed earlier involve many units in an organization. Therefore, some mechanism is necessary to plan, coordinate, and ensure that the activities are assigned sufficient resources and are executed effectively. A quality department could perform this role, perhaps by using the project management concept.

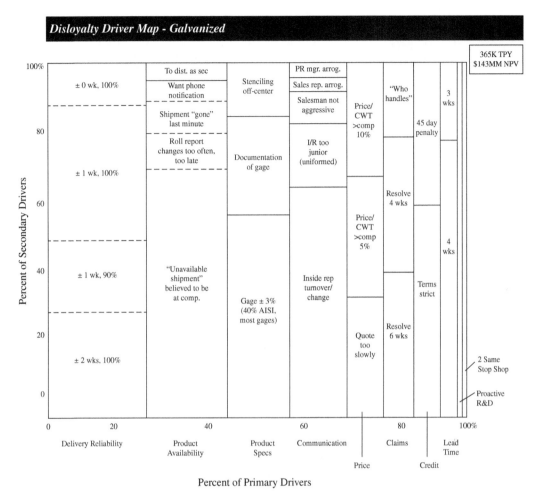

FIGURE 5.3
Disloyalty drive map for a gauge manufacturer.

→ 5.6 ECONOMIC WORTH OF A LOYAL CUSTOMER

The sales revenue from a loyal customer measured over the period of repeat purchases can be dramatic, e.g., $5000 for a pizza eater; $25,000 for a business person staying at a favorite hotel chain; $300,000 for a loyalist to a brand of automobile. The economic worth is calculated as the net present value (NPV) of the net cash flow (profits) over the expected lifetime of repeat purchases. The NPV is the value in today's dollars of the profits over time. Profits generally rise as the customer defection rate (customers who do not repurchase) decreases, i.e., the retention rate increases. Reichheld (1996) shows, for a variety of service industries, the increase in NPV of the customer profit stream in the future from a decrease of five percentage points in the defection rate

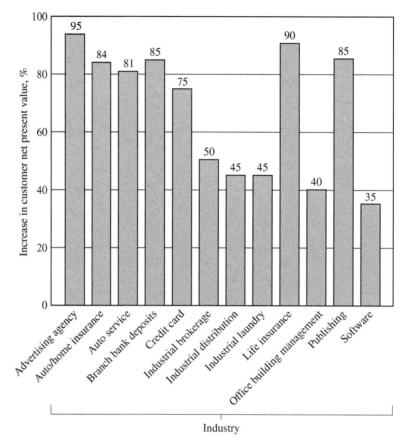

FIGURE 5.4
Profit impact of 5% increase in retention rate. From *The Loyalty Effect,* by F. F. Reichheld. Boston, MA, 1996. Copyright © 1996 by the Harvard Business School Publishing Corporation; all rights reserved. Reprinted by permission of Harvard Business School Press.

(Figure 5.2). He concludes that a five percentage point decrease in the defection rate can increase profits by 35 to 95%.

→ 5.7 IMPACT OF POOR QUALITY ON LOST SALES

Sometimes the evidence of the importance of quality in retaining present customers is dramatic. Manufacturer B of household appliances had a leadership position for two of the four models of a product (see Table 5.4). During a four-year period, leadership was lost, even though the company was competitive in product features, price, and delivery dates. It was not competitive, however, in field failures and warranty costs. The president personally had to lead the development and execution of a strategy to reduce the failures and warranty costs. He was successful.

TABLE 5.4
Change in suppliers of product

Model	1979	1980	1981	1982
High price	A	C	C	C
Middle price	B	B	C	C
Low price	C	C	C	C
Special model	B	B	B	C

Note: A, B, and C indicate suppliers.

Sometimes the sales lost because of poor quality can be quantified. Two manufacturers of washing machines were compared. One measure was the percentage of present customers who would not purchase the same brand again. For brand A, only 1.1% said they would not purchase it again; for brand B, 10.5% declined to purchase it again. For brand B, the overwhelming reason was "poor quality." When the 10.5% was translated to the total present customer population for brand B, researchers concluded that additional sales income of $5 million would be needed to make up the profit on the lost replacement sales. In a case involving an industrial product, a survey was made of present customers who were buying the product made by one manufacturer. Some of those customers had purchased a different brand. One-third of these former customers said the primary reason was poor quality. These lost sales due to poor quality amounted to $1.3 billion in sales income (and 2000 jobs).

For elaboration on quantifying the impact of quality on sales, see Rust et al. (1994).

➔ 5.8 LEVEL OF SATISFACTION TO RETAIN PRESENT CUSTOMERS

Sometimes acceptable levels of customer satisfaction with the product still result in a significant loss of new sales. Table 5.5 presents two examples from the service industry. Note that even when the customer view of quality is good, a quarter or more of the present customers may not return. A similar story exists in the manufacturing sector (Burns and Smith, 1991). At the Harris Corporation, satisfaction with specific attributes of the product (goods and service) is measured on a scale of 1 to 10; loyalty is measured as the percentage of customers who will continue to purchase from Harris. The data showed that to achieve 90% loyalty required a satisfaction level of

TABLE 5.5
Satisfaction and sales

Customer view of quality	Willing to recommend supplier, % (GTE)	Very willing to repurchase, % (AT&T)
Excellent	96	92
Good	76	63
Fair	35	18
Poor	3	0

Source: Bultmann (1989); Scanlan (1989).

TABLE 5.6
Repurchase intent

Type of customer	Percentage who intend to repurchase	Percentage who will recommend purchase to others
Typical satisfied customer with no problem	51	65
Customer with problem; unsatisfied with recovery	30	46
Customer with problem; very satisfied with recovery	79	88

at least 8.5. In some industries, more than 90% of customers report that they are satisfied or very satisfied with the product—but repurchase rates are only about 35%.

Another dimension of this phenomenon is the level of customer satisfaction with the handling of complaints. Table 5.6 shows the percentage of customers at a financial services firm who intend to purchase a product or service again, based on their level of satisfaction with the resolution of their complaints (*JQH5,* page 33.22).

Customers who have a problem but are unsatisfied with the resolution ("recovery") are unlikely to repurchase (30%). Customers who are very satisfied with the handling of complaints have much higher intention to repurchase (79%) and recommend purchase to others (88%). Some companies seize upon a complaint as a special opportunity to generate additional sales revenue by providing dramatic and memorable recovery action. Finally, note that some satisfied customers with no problem will not repurchase.

✧ 5.9 LIFE-CYCLE COSTS

For simple consumable products such as food or a bus ride, the purchase price is also the cost of using the product. As products grow in complexity and as the length of use increases, the total cost of using the product is more than just the purchase price. Total cost also includes operational, maintenance, and other special costs. For some products, the after-sale costs can easily exceed the original purchase price.

A *life-cycle cost* can be defined as the "total cost to the user of purchasing, using, and maintaining a product over its life." This is also sometimes called the *total cost of ownership.* A study of all of the cost elements can lead to redesign of a product that could result in a significantly lower life-cycle cost, perhaps at the expense of a small increase in the original price. This presents a marketing opportunity to provide a product that will result in savings over the life of the product to potential customers. These customers are urged to make purchasing decisions by comparing life-cycle costs for competing products. But the initial purchase price may be higher, and marketing people may find it more difficult to sell the product to potential customers whose first priority is initial price. Table 5.7 shows the ratio of life-cycle costs to original price for various consumer products.

An associated concept, user failure cost, calculates the cost to the user of failures during the product life. Table 5.8 shows an example of annual failure-related costs (Gryna, 1977).

TABLE 5.7
Life-cycle costs: consumer products

Product	Ratio, life-cycle cost to original price
Room air conditioner	3.3
Dishwasher	2.5
Freezer	4.8
Range, electric	4.4
Range, gas	1.9
Refrigerator	3.5
TV (black and white)	2.5
TV (color)	1.9
Washing machine	3.6

The life-cycle cost concept is logically sound, but implementation has been slow. Two reasons for the slow pace of adoption predominate. First, estimating the future costs of operation and maintenance is difficult. A greater obstacle, however, is the cultural resistance of purchasing managers, marketing people, and product designers. The skills, habits, and practices of these people have long been built around the concept that the original purchase price has primary importance (see *JQH5,* pages 7.22–7.27, for further elaboration).

→ 5.10 SPECTRUM OF CUSTOMERS

For both consumer and industrial products and for both physical goods and services, the variety of customers forms a wide spectrum. Some companies choose to address one part of the spectrum, whereas other companies pursue several types of customers. For purposes of planning for quality, we will identify three types of customers:

1. Those who emphasize initial purchase price as equal to or more important than quality
2. Those who evaluate alternative products on both initial price and quality simultaneously
3. Those who emphasize obtaining "the best"

TABLE 5.8
Annual quality cost of foundry equipment

	11 shell core makers	12 induction furnaces
Repairs	$16,000	$ 70,000
Effectiveness loss	25,000	90,000
Lost income	8,000	50,000
Extra capacity	2,500	8,000
Preventive maintenance	8,000	20,000
Total	$59,500	$238,000
Initial investment	$70,000	$700,000

TABLE 5.9
Categories of customer emphasis related to quality

Emphasis	Features	Freedom from deficiencies
Initial economy	Willing to forego some features Like do-it-yourself features and add options later	Will tolerate some product deficiencies at delivery and during use
	Will tolerate a relatively short product life	Will tolerate some deficiencies in service before and after purchase
Value	Willing to make trade-offs between quality and price	Warranty provisions can be important
	Features must be justified by benefits and related price	Concerned about operating and repair costs
The "best"	Desire many convenience features	Greatly annoyed at deficiencies and associated inconveniences
	Emphasis on luxury, esthetics, brand image	Demand complete and timely response to all problems
	Desire high level of performance from product and from all personnel	

Table 5.9 links these three categories with a translation into desires for product features and freedom from deficiencies.

All three categories need to be satisfied in the marketplace. In consumer products particularly, some customers change categories over a lifetime, e.g., some young couples with growing children change from "economy minded" to "value minded."

→ 5.11 THE QUALITY BY DESIGN ROAD MAP

The quality by design road map (see Table 1.4) presents a framework for planning (or replanning) new products (or product revisions). This road map applies to both the manufacturing and service sectors and to products for both external and internal customers. Developing a remedial solution for a quality improvement project (see Chapter 4) may require one or more steps of this quality by design process as well. Early and Colletti (1999) and Juran (1988) provide extensive discussions of the steps, in slightly different language. These quality planning steps must be incorporated with the technological tools for the product being developed. Designing an automobile requires automotive engineering disciplines; designing a path for treating diabetes requires medical disciplines. But both need the tools of quality planning to ensure that customer needs are met.

The road map is presented in more detail in Figure 5.5, and the steps are discussed in later chapters. It is useful, however, to present an overview now to explain briefly the steps (see Early and Colletti, 1999).

One of the earliest rigorous modern applications of these methods was in the early 1980s when Ford began initial planning for a new front-wheel-drive midsize car.

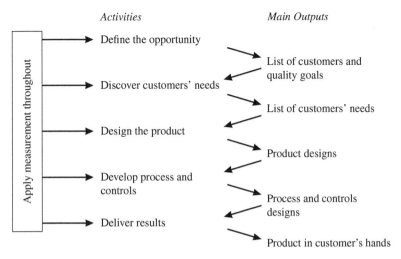

FIGURE 5.5
Quality planning road map. (*From Juran Institute, 2013.*)

The business environment included some ominous elements: strong foreign competition, decreasing market share, and a projection of greatly increased fuel prices. Ford concluded that a new approach to designing the model was essential. Basic to the new approach was "customer satisfaction" with the objective that the Taurus would be the best car in its class. This "best in class" focus spawned some unusual approaches to planning. One of the breaks with tradition was the organization of the planning for the Taurus. Historically, new-car design used the traditional organizational structure (Figure 5.6a). With such a structure, the main activities were executed *sequentially*, e.g., the planning department studied customer desires and then presented the results to design; design performed its tasks and handed the results to engineering; engineering created the detailed specifications; and the results were then given to manufacturing. Unfortunately, the sequential approach results in a minimum of communication between the departments as the planning proceeds—each department hands its output "over the wall" to the next department. This lack of communication often leads to problems for the next internal customer department. Under Taurus, activities were organized as a team (Figure 5.6b) from the beginning of the project. Thus, for example, manufacturing worked *simultaneously* with design and engineering before the detailed specifications were finalized. This approach allowed the team to address predictability issues during the preparation of the specifications.

While subsequent experience in many organizations has provided improvements and expansions to the methods, the Ford organizational practice continues to be a best practice. We will now refer to the steps in Figure 5.5 to explain the high-level elements of effective quality by design.

1. *Define the opportunity.* This step has three substeps:
   a. Identify the opportunity. Strategic plans and market research identify unserved, underserved, or vulnerable market segments that need new or renewed products.

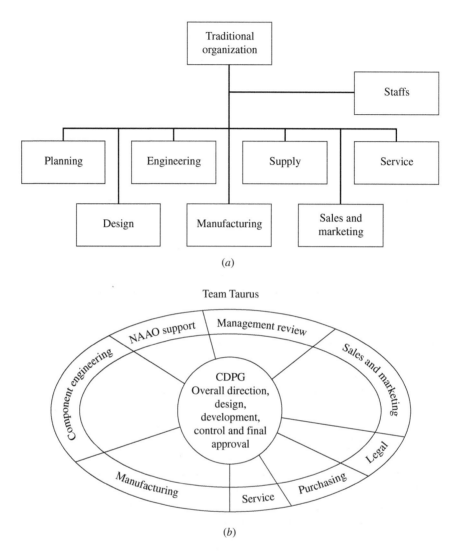

FIGURE 5.6
(a) Traditional organization. (b) Organization for Taurus. (*From Juran, 1990.*)

    b. Define the target customers. Which customers do we want to buy and use this product?
    c. Set goals with respect to market, customer, quality, financials, and performance.
    d. Establish the product team to lead and complete the project.
    e. Construct high-level flow of product production and use.
    f. Identify all customers, internal and external. Include users, processors, distributors, sales, regulators, and others. The customer is anyone who is affected by the product—not just the target purchasers or users who were designated in the Define step. This step includes identifying both the external and internal

customers (see Chapter 1 under "Quality Defined"). Examples of tools employed are flowcharts, Pareto analysis, and spreadsheets.
g. Detailed responsibilities, schedules, resources, and follow-up.

Some customers are obvious; others are not. Thus, for an automobile:

Company function	Customer
Sales	Consumer
Legal	U.S. Dept. of Transportation
Design engineering	Manufacturing
Parts manufacturing	Assembly plant
Advertising	Media

(Note that the customers are both external and internal.)

Some tools used to establish the project are competitive analysis, benchmarking, and deployment of goals.

For the original Ford Taurus (in the early 1980s), the quality goal was to be best in class. For the 1994 Ford Mustang, four numerical quality goals were set, e.g., "things gone wrong/1000."

The Mustang team was cross-functional and included representatives from various sections of design engineering, manufacturing and assembly, sales and marketing, purchasing, and finance. This team functioned as a quality planning project team. In addition, the design and development process benefited from the strong involvement of inputs from customers (external and internal) and suppliers—before the release of the design to production. Chapter 8 explains several types of teams and the roles of a team leader, team facilitator, and team members.

2. *Discover customers' needs.* This step has five substeps. This step is discussed further in Chapter 11, "Understanding the Market's Role in Quality."
   a. Plan how to discover customer needs—e.g., market research, customer complaints, dealer input, and competitive evaluations.
   b. Collect information on customer needs in their language of benefit.
   c. Analyze and prioritize customer needs.

   Some tools for discovering customers' needs are multiattribute studies, focus groups, questionnaires, site visits, and customer need spreadsheets (see Chapter 11).

   We develop this step in more detail in Chapter 11.
   d. Translate needs into CTQs. CTQs are the critical-to-quality elements. These are the customers' needs translated into the language of the producer and expressed in measurable terms.

3. *Design the product.* This step has four substeps:
   a. Select high-level product features and goals (functional design).
   b. Develop detailed product features and goals.
   c. Optimize product features and goals.
   d. Finalize the product design.

   Some tools used to develop the product are competitive analysis, reliability, safety and value analysis, prototype tests, designed experiments, and design spreadsheets (see Chapter 12, "New Product Development's Role in Designing for Quality").

   The Taurus team used market research to provide product development with detailed guidelines for 429 product features that were important for high

product salability. These guidelines then became the basis of specific design projects. Two examples of these features were the amount of effort required to raise the hood of the car and the level of wind noise. For each feature, a means of measurement was defined, competitive data were obtained, and a numerical goal was set. For example, the effort required to raise the hood was measured in pounds by a spring scale. Competitive data showed that the best competitor had a design requiring 9 pounds of effort. For Taurus, a goal of 8 pounds was set; the final design exceeded the goal by requiring only 7 pounds. Taurus achieved best in class for 80% of the product features. Certain other product features were also incorporated, even though they were not directly related to product salability, e.g., the high mounting of the brake light, a feature that was desired by the Department of Transportation.

This step is discussed further in Chapter 12.

4. *Develop the process and controls.* Developing the process portion of this step has seven substeps:
   a. Collect known baselines.
   b. Select general process design.
   c. Identify high-level process features and goals.
   d. Develop detailed process features and goals.
   e. Optimize process features and goals.
   f. Establish initial process capability.
   g. Finalize the process design.

   Developing the controls portion of this step has five substeps:
   (1) Identify control subjects.
   (2) Develop feedback loops.
   (3) Develop control plans.
   (4) Establish audit.
   (5) Demonstrate process capability and controllability.

   Some tools used to develop process features are flowcharts, process capability studies, pilot runs, and spreadsheets (see Chapter 13, "Operations—Manufacturing Sector").

   Because the team for developing the new automobile was organized at the beginning of the project, manufacturing worked simultaneously with design and engineering before the detailed specifications were finalized. This approach enabled the team to address production issues during the preparation of the specifications; e.g., the assembly plan identified specific manufacturing issues to be addressed during design and manufacturing planning. The assembly plant listed 1400 issues, ranging from a desire to automate body-side assembly to having an annual assembly-plant shutdown for vacation. Much effort to achieve process capability ("process qualification") and to optimize the processes was also part of the planning. As a result, a set of process plans was ready at the start of production. This step is elaborated in Chapter 13.

5. *Deliver results.* This step has three substeps:
   a. Plan for transfer to operations.
   b. Implement plan.
   c. Validate transfer.

As these plans were put into production at Ford, the coordination among all functions continued and resulted in final refinements to the product and process design.

For elaboration, see Chapter 13, "Managing Quality in Operations—Manufacturing," and Chapter 20, "Tools to Maintain the Control of Quality."

An Example of Quality by Design for Services

In an example from the service sector, the quality by design process was applied to replanning the process of acquiring corporate and commercial credit customers for a major affiliate of a large banking corporation. Here is a summary of the steps in the quality planning process.

1. *Define opportunity.* A goal of $43 million of sales revenue from credit customers was set for the year.
2. *Discover customers' needs.* Internal customers had 27 needs; external customers had 34 needs.
3. *Design the product.* The product had nine product features to meet customers' needs.
4. *Develop the process and controls.* To produce the product features, 13 processes were developed, along with appropriate controls for each.
5. *Deliver results.* The plans were placed in operation.

The revised process achieved the goal on revenue. Also, the cost of acquiring the customers was only one-quarter of the average of other affiliates in the bank.

Quality by design generates a large amount of information that must be organized and analyzed systematically. The alignment and linkages of this information are essential for effective quality planning for a product. A useful tool is the quality by design spreadsheet or matrix (basically, a table). Figure 5.7a illustrates the four major components of a typical design spreadsheet and illustrates it with a portion of a customer-needs spreadsheet for a college student credit card. The high-medium-low designations can be replaced with symbols or numeric ratings. The use of numeric ratings increases the power of the analysis, but we have used words here to show the concept.

Figure 5.7b shows five spreadsheets corresponding to steps in the quality by design process. Note how the spreadsheets interact and build on one another; they cover both quality planning for the product and quality planning for the process that creates the product. The approach is sometimes called "quality function deployment" (QFD). Sometimes QFD is used to mean a particular form of the planning spreadsheet that predates the common availability of computers. The use of spreadsheets in the quality by design process unfolds in later chapters.

These six quality planning steps apply to a new or modified product (goods or service) or process in any industry. In the service sector the "product" could be a credit card approval, a mortgage approval, a response system for call centers, or hospital care. Also, the product may be a service provided to internal customers. Endres (2000) describes the application of the six quality planning steps at the Aid Association for Lutherans insurance company and the Stanford University Hospital.

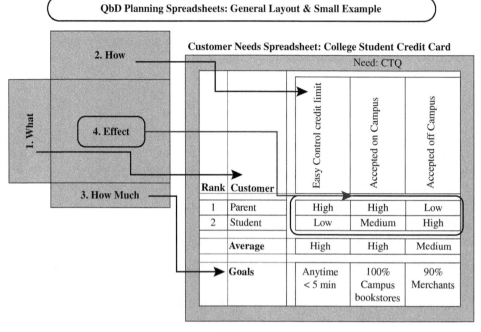

FIGURE 5.7a
Contruction of spreadsheets in quality by design.
(*Juran Institute, Inc. Copyright 2013. Used with permission.*)

5.12 QUALITY BY DESIGN (QbD) AND DESIGN FOR SIX SIGMA (DFSS)

Section 4.6 describes the six sigma approach in terms of five phases: define, measure, analyze, improve, and control (DMAIC). The DMAIC process aimed mainly at reducing defects and errors in existing products, services, and processes. The importance of design has led to the incorporation of six sigma principles into design projects as well. Unlike the widespread adoption of the DMAIC steps in six sigma–based improvement, DFSS has had a number of different formulations. In addition to the original define-discover-design-develop-deliver discussed earlier, there have been formulations as define-measure-analyze-design-verify and define-measure-explore-develop-implement.

Design for Six Sigma (DFSS)

Design for six sigma (DFSS) is focused on creating new or modified designs that are capable of significantly higher levels of performance (approaching six sigma). The discover-design-develop-deliver sequence is a design methodology applicable to developing new or revised products, services, and processes.

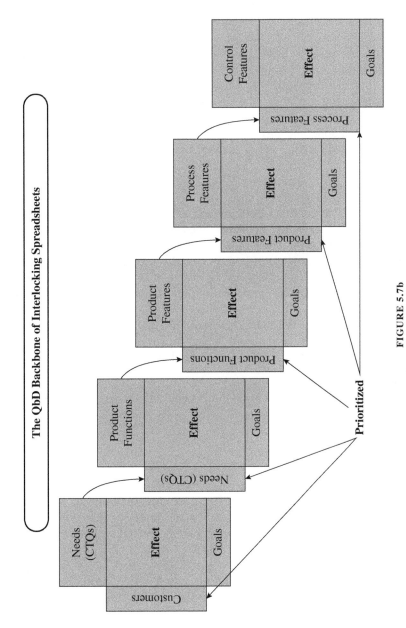

FIGURE 5.7b
Interlocking quality by design spreadsheets.
(Juran Institute, Inc. Copyright 2013. Used with permission.)

Define Step

The Define Step sets the tone for the entire design project; it establishes its goals, charter, and infrastructure. During this phase, activities are shared between the management team and the chartered project design team. Management has the ultimate responsibility to define the design problem: what is to be modified, redesigned, or newly created.

Projects are nominated consistent with the overall business strategy and selected based upon their optimal contribution to that strategy.

A key task in the Define Step is to create the initial business case that validates the selection rationale and establishes the business justification through reduced product cost, increased sales, or entirely new market opportunities. The initial business casework is conducted under the auspices of the management team and then validated and updated continuously by the design team through the subsequent phases of the design project.

The management team nominates a Black Belt to lead the design project. The Champion, who is the management sponsor with vested interest in the success of the design, in conjunction with the Black Belt, is responsible for selecting a cross-functional team that will conduct all activities to complete the design and carry it into production.

With the nomination of the cross-functional team, the project is launched. Full responsibility for design success is transferred to the Black Belt and his/her team. Management participation continues throughout the design effort through the Champion's advisory and monitoring role that includes periodic updates from the Black Belt and team.

The design team under the Black Belt's leadership establishes the project plan that includes resource allocation, task lists, and project timelines.

In summary, the key deliverables that are required to complete the Define Step are:

- Established Design Project
- Project Charter, including specific measurable goals and the target customers
- Project Plan
- Initial Business Case

Discover Step

The Discover Step is concerned mainly with identifying the critical needs of key customers, and what are the measurable critical-to-quality requirements (CTQs) necessary for a successfully designed product, service, or process.

An initial assessment of markets and customer segmentation by various factors is required to identify the key customers. This assessment is normally completed by the marketing organization and is then reviewed and verified by the design team. However, it is the design team's responsibility to complete the customer needs analysis and compile its results into a prioritized tabulation of customer needs. Methods to determine customer needs include focus groups, interviews, and surveys of key customer groups.

In discovering needs, design teams should focus on the benefit that the customer is seeking, not a specification of the product. The better a team understands the benefit for the customer, the more creative and specific will be its design, and the more delighted the customer will be with the outcome. If we settle for a customer specifying the dimensions of a personal electronic device (phone, music player, GPS, etc.) and don't discover the benefit of that size, we might meet the size requirement but miss an opportunity to delight. For example, if the size is to enable carrying with you while exercising, the design might also therefore need to accommodate moisture and external shocks that were not otherwise identified.

The design team transforms the critical customer needs into measurable terms from a design perspective. These translated needs become the critical-to-quality requirements (CTQs) that must be satisfied by the design solution. Competitive benchmarking and creative internal development are two additional sources for CTQs. These areas probe into design requirements that are not generally addressed or possibly not even known by the customer. The result is a set of consistently defined and documented design CTQs.

In summary, the key deliverables that are required to complete the Discover Step are:

- Prioritized list of customer needs
- Prioritized list of CTQs
- Customer needs spreadsheet
  - Production process capability, supported by process flow diagram
  - Product and process risk assessment (using a design FMEA and a process FMEA)
- Initial design scorecard

Design Step

The main purpose of the Design Step is to select a high-level design from several design alternatives and develop the detailed design requirements against which a detailed design will then be optimized.

The starting point in the high-level design is to perform a functional analysis of the CTQs established in the Discover Step that results in a high-level functional design. This should satisfy the CTQs in an economically viable manner.

The design team develops several high-level design alternatives that represent different functional solutions to the stated functional design requirement. Benchmarking and innovative concept creations are encouraged in developing alternative concepts. These alternatives are analyzed against a set of evaluative criteria and one of them, or a combination of alternatives, is selected to carry forward as the preferred "high-level design." The development of alternatives and the selection of the preferred alternative are both iterative processes: as progressively more design information is developed, the iterative nature inherent in design requires several passes to ensure that the most capable high-level design is carried forward.

Several tools are used by the team to establish baseline performance and predict the performance of the high-level design against the CTQs. Process capability studies, product functionality and capability analysis, risk analysis, fault tree analysis, trade-off analysis, value analysis, carry-over analysis, and financial analysis are the analytical instruments used to predict performance.

After the evaluation is completed and the best high-level design is validated, the design team performs an optimization analysis that results in a "best fit design."

Using the vital few design parameters determined in the previous Design Step, designed experiments (DOEs), trade-off analysis, criticality analysis, and other analyses are conducted to optimize the detail design around key design parameters. Ideally, they result in an optimum detailed parametric design represented by a mathematical prediction equation. Most of the experimentation assumes linear relationships between the design parameters and overall performance. However, tools such as response surface methods are employed to handle nonlinear models.

Finally, the team develops and tracks a design scorecard to track the quality performance of each component of the product and its impact on the overall quality performance.

In summary, the key deliverables that are required to complete the Design Step are:

- Design alternative concepts
- Selected High-Level Design concept
- Results of High-Level Design capability/Risk Analysis
- High-level (functional) design spreadsheet
- Best-fit design
- Detailed design requirements
- Design scorecard

Develop Step

The Develop Step builds on the detailed design requirements to deliver an optimum detailed process and controls that ensures reliable and economical delivery of the product as designed.

Designed experiments provide the inputs for the tolerance design that concentrates on the selective reduction of tolerances to reduce variation and quality loss.

Using appropriate design methods and tools such as design for manufacturability (DFM), design for assembly (DFA), and reliability and serviceability analysis, the design team examines the capabilities of the current production process and related systems against the new design. A final risk analysis using traditional tools is conducted. Based upon these analyses, the final process and controls are developed to match the projected operational (manufacturing and service) capabilities.

DFSS Scorecards											
Final Product (Output) Scorecard											
Output Component	CTQ	Target	LSL	USL	Mean	Standard Deviation	Short- or Long-term	DPMO	# Times applied	DPU	Short-term Sigma
Connector	Pitch (mm)	0.30	0.29	0.031	0.30	0.002	LT	858	2	0.0017	4.64
Validation rules	Errors < 3. 4 ppm						LT	4	700	0.0028	5.97
Code Update	Every 30 days	30		45			LT	1000	12	0.012	4.59
Web access	Mbps	100	50		90	10	LT	32	1	0.000032	5.50
Total Product								23	715	0.0165	5.58

FIGURE 5.8
Design Scorecard

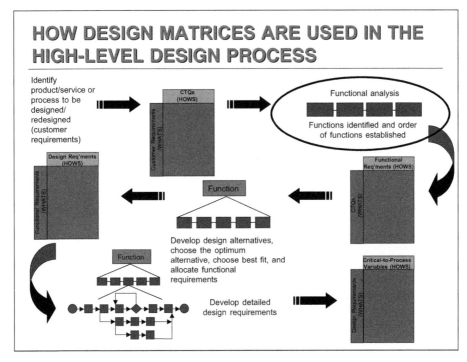

FIGURE 5.9
Reprinted with permission from Juran Institute, Inc. All rights reserved.

The Develop Step is concluded when the design team conducts a design verification test (DVT) that validates the detail design by using such tools as simulation, prototyping, and pilot testing. The results of the DVT are summarized and presented in a formal design review.

The design scorecard is updated again with the final design information and the latest DVT results. (See Figure 5.8 for an example of a design scorecard.)

In summary, the key deliverables that are required to complete the Design Phase are

- Optimal tolerances and design settings
- Detailed process design
- Reliability/lifetime analysis results
- DVT results
- Updated design scorecard
  - Control plan
  - Control audit

Deliver Results Step

The purpose of the Deliver Results Step is to ensure that the new design is in fact delivered as designed and that the financial, market, customer, and quality goals of the project are met.

The design team should ensure appropriate testing in a service and field support environment to uncover potential lifetime or serviceability issues. These tests will vary greatly depending on the product and industry.

A key task is to record all design documents and process control plans (including guidelines for self-control) into a robust set of standard operating procedures (see Figure 5.9). These formal documents are handed off to downstream process owners (e.g., manufacturing, logistics, and service). They should outline the required controls and tolerance limits that should be adhered to and maintained by manufacturing and service.

A final design scorecard should be completed and all key findings recorded and archived for future reference. The team should complete a final report and celebrate the successful completion of a six sigma design.

The Deliver Step is not complete until pilot tests are successful and scale-up plans are in place and validated.

In summary, the key deliverables that are required to complete the Verify Phase are

- Implementation and transfer plans
- Control plans (including plans for self-control and mistake proofing)
- Pilot test results
- Scale-up plans and validation
- Final design scorecard
- Final project report, including established audit plan and initial results against goals

SUMMARY

- Businesses having larger market share and better quality earn much higher returns than their competitors. Quality and market share have strong separate relationships to profitability.
- Competitive benchmarking is the continuous process of measuring products, services, and practices against those of the toughest competitors or leading companies.
- Quality can be the decisive factor in lost sales, and sometimes its impact can be quantified.
- Customer complaints resolved with less than complete customer satisfaction will result in significant lost sales.
- Planning for product quality must be based on meeting customer needs, not just meeting product specifications.
- In-depth market research can identify suddenly arising customer needs.
- Planning for quality must recognize a spectrum of customers with different needs.
- For some products, we need to plan for perfection; for other products, we need to plan for value.
- Life-cycle cost is the total cost to the user of purchasing, using, and maintaining a product over its life.
- Quality superiority can be translated into a higher market share or a premium price.

- Quality planning for a new or modified product follows these steps: establish the project, identify the customers, discover customers' needs, develop the product, develop process features, develop process controls, and transfer the plans to operations. The measurement process must be applied during all steps.

PROBLEMS

5.1. In the explanation of the Taurus case, examples were given to illustrate each step in the planning road map, e.g., five customers were listed.

For each step in the planning road map, give two new examples. The examples may apply to any product or service.

5.2. For a consumer or industrial physical product, make a quality comparison of three brands at the low, medium, and high price levels. The comparison should list differences in product features and freedom from deficiencies.

5.3. For a consumer or business service, make a quality comparison, as described in problem 5.2.

5.4. Select two physical products for which *Consumer Reports* magazine has conducted an analysis, provided ratings on quality, and reported the purchase price. For each product, summarize the relation between high quality and price, i.e., to what extent is high quality associated with a higher price.

5.5. Form a discussion group of five people. Select a physical product that you are likely to purchase and discuss how you might define quality for this product. How does the definition relate to customer satisfaction and loyalty, product features, and freedom from deficiencies?

5.6. Form a discussion group of five people. For a consumer or business service, conduct a discussion as described in problem 5.5.

REFERENCES

Bultmann, C. (1989). "How to Define Customer Needs and Expectations: An Overview," *Customer Satisfaction Measurement Conference Notes,* American Marketing Association and ASQ, February 26, 28. The written paper was "New Ways of Understanding Customers' Service Needs" by T. F. Gillett.
Burns, R. K. and W. Smith (1991). "Customer Satisfaction—Assessing Its Economic Value," *Annual Quality Congress Proceedings,* ASQ, Milwaukee, pp. 316–321.
Buzzell, R. D. and B. T. Gale (1987). *The PIMS Principles: Linking Strategy to Performance,* The Free Press, Macmillan, New York, reprinted with permission.
Early, J. F. and O. J. Coletti (1999). "The Quality Planning Process," in *Juran's Quality Handbook,* 5th ed., McGraw-Hill, New York, Section 3.
Endres, A. (2000). *Implementing Juran's Roadmap for Quality Leadership,* John Wiley & Sons, New York.

Gale, B. T. (1994). *Managing Customer Value,* The Free Press, New York.

Gale, B. T. and R. Klavans (1985). "Formulating a Quality Improvement Strategy," *Journal of Business Strategy,* Winter, pp. 21–32, Warren, Gorham, and Lamont, used with permission.

Gryna, F. M. (1977). "Quality Costs: User vs Manufacturer," *Quality Progress,* June, pp. 10–15.

Juran, J. M. (1988). *Juran on Planning for Quality,* The Free Press, New York.

Juran Institute (1990). Designs for World Class Quality; Juran Institute, Wilton, CT.

Juran Institute (2012). Quality by Design: An Action Workshop for Six Sigma Black Belts, Juran Institute, Southbury, CT.

Lynch, J. (1998). "Measuring the Pursuit of Customer Loyalty at Sun," *Proceedings of the 10th Annual Customer Satisfaction and Quality Measurement Conference,* ASQ and American Marketing Association, Atlanta.

Nash, M. (1998). "The Business Impact of Customer Loyalty," *Proceedings of the 10th Annual Customer Satisfaction and Quality Measurement Conference,* ASQ and American Marketing Association, Atlanta.

Reichheld, F. F. (1996). *The Loyalty Effect,* Harvard Business School Press, Boston.

Rust, R. T., A. J. Zahorik, and T. L. Keiningham (1994). *Return on Quality,* Probus, Chicago.

Scanlan, P. M. (1989). "Integrating Quality and Customer Satisfaction Measurement," *Customer Satisfaction Measurement Conference Notes,* American Management Association and ASQC, February, pp. 26–28.

Waltz, B. (1996). "Managing Customer Loyalty," *Proceedings of the Impro Conference,* Juran Institute, Inc., Wilton, CT, p. 7A-9.

CHAPTER 6

Control of Quality to Maintain Superior Performance

→ ## 6.1 COMPLIANCE AND CONTROL SYSTEMS

As used in this book, *control* refers to the process employed to meet standards consistently and to hold the gains. The control process involves observing actual performance, comparing it with some standard, and then taking action if the observed performance is significantly different from the standard.

Control happens as a proactive approach to establishing control points from the design process. It is also the outcome of a project—to hold the gains. In either case, control is best performed as close to the operation as possible. It requires establishing an effective feedback loop so all employees can maintain self-control of processes.

The control process requires a feedback loop (Figure 6.1). Control involves a universal sequence of steps as follows:

1. Choose the control subject, i.e., choose what we intend to regulate.
2. Establish measurement.
3. Establish standards of performance: product goals and process goals.
4. Measure actual performance.
5. Compare actual measured performance to standards.
6. Take action on the difference.

This universal sequence applies to individuals at all levels from the chief executive officer to members of the workforce. The sequence can be applied as a framework for helping supervisors and work teams to understand and run everyday work processes. Such a framework becomes increasingly important as the team concept—particularly self-directed teams—emerges as an important form of business life.

194 Quality Management and Analysis

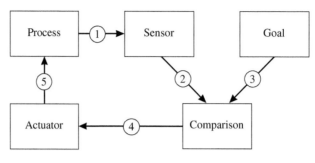

FIGURE 6.1
The feedback loop.

When the natural work team in a department puts the control process into practice, three purposes are served:

- Maintain the gains from improvement projects.
- Promote analysis of process variation, based on data, to identify improvement opportunities.
- Allow team members to clarify their responsibilities and work to achieve a state of self-control.

The first three steps in the control process (choose the subject, establish measurement, and establish standards) require the *participation* of the department work team. The last three steps (measure, compare to standards, and take action) can be the *responsibility* of the department work team.

Control, one of the trilogy of quality processes, is largely directed at meeting goals and preventing adverse change, i.e., holding the status quo. In contrast, improvement focuses on creating change, i.e., changing the status quo. The control process addresses sporadic quality problems; the improvement process addresses chronic problems.

✧ 6.2 THE IMPORTANCE OF INFORMATION AND MEASUREMENT

Quality measurement is central to the process of quality control: "What gets measured, gets done." Measurement is basic for all three operational quality processes and for strategic management: For quality control, measurement provides feedback and early warnings of problems; for operational quality planning, measurement quantifies customer needs and product and process capabilities; for quality improvement, measurement can motivate people, prioritize improvement opportunities, and help in diagnosing causes; and for strategic quality management, measurement provides input for setting goals and later supplies the data for performance review.

Figure 6.2 shows the far-reaching impact of measurement in quality management. Note how measurement provides both alignment and linkages at several levels from daily work to strategic quality planning. These elements, in turn, become

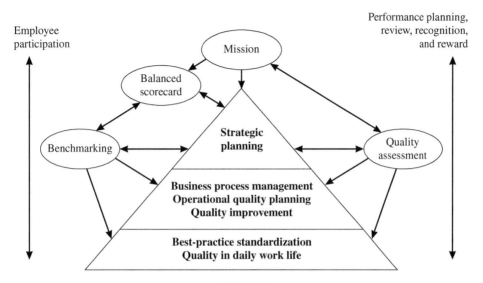

FIGURE 6.2
Measurement drivers. From *Implementing Juran's Road Map for Quality Leadership: Benchmarks and Results* by Al Endres, 2000. Reprinted with permission of John Wiley & Sons, Inc.

drivers to encourage the use of measurements for quality. This chapter presents concepts underlying measurement; later chapters present examples of quality measurement at both the operational and strategic levels.

The following principles can help to develop effective measurements for quality:

1. Define the purpose and use that will be made of the measurement. An example of particular importance is the application of measurements in quality improvement. Final measurements must be supplemented with intermediate measurements needed for diagnosis.
2. Emphasize customer-related measurements; be sure to include both external and internal customers.
3. Focus on measurements that are useful—not just easy to collect. When quantification is difficult, surrogate measures can at least provide a partial understanding of an output.
4. Provide for participation from all levels in both the planning and implementation of measurements. Measurements that are not used will eventually be ignored.
5. Provide for making measurements as close as possible to the activities they impact. This timing facilitates diagnosis and decision making.
6. Provide not only concurrent indicators but also leading and lagging indicators. Current and historical measurements are necessary, but leading indicators help to look into the future.
7. Define, in advance, plans for data collection and storage, analysis, and presentation of measurements. Plans are incomplete unless the expected use of the measurements is carefully examined.

8. Seek simplicity in data recording, analysis, and presentation. Simple check sheets, coding of data, and automatic gaging are useful. Graphical presentations can be especially effective.
9. Provide for periodic evaluations of the accuracy, integrity, and usefulness of measurements. Usefulness includes relevance, comprehensiveness, level of detail, readability, and interpretability.
10. Realize that measurements alone cannot improve in products and processes. Measurements must be supplemented with the resources and training to enable people to achieve improvement. For elaboration on these and other principles of measurement systems, see *JQH5,* Section 9, and Zairi (1994).

This chapter presents concepts underlying measurement, but measurement is spread throughout the book. For example, Chapter 3 discusses measurements for broad quality assessment; Chapter 2 addresses strategic measurement, including the balanced scoreboard; Chapters 12, 13, 14, and 15 present examples of functional measurements. Thus measurements are for both product process control and management control.

✣ 6.3 THE ULTIMATE INSPECTOR: SELF-MANAGED PROCESSES

Ideally, quality planning for any task should put the employee into a state of self-control. When work is organized so that a person has full mastery over the attainment of planned results, that person is said to be in a state of self-control and can therefore be held responsible for the results. Self-control is a universal concept, applicable to a general manager responsible for running a company division at a profit, a plant manager responsible for meeting the various goals set for that plant, a technician running a chemical reactor, or a bank teller serving customers.

To be in a state of self-control, people must be provided with

1. Knowledge of what they are supposed to do, e.g., the budgeted profit, the schedule, and the specification.
2. Knowledge of their performance, e.g., the actual profit, the delivery rate, the extent of conformance to specification (this is quality measurement).
3. Means of regulating performance if they fail to meet the goals. These means must always include both the authority to regulate and the ability to regulate by varying either (a) the process under the person's authority or (b) the person's own conduct.

If all the foregoing parameters have been met, the person is said to be in a state of self-control and can properly be held responsible for any deficiencies in performance. If any parameter has not been met, the person is not in a state of self-control and, to the extent of the deficiency, cannot properly be held responsible.

In practice, these three criteria are not fully met. For example, some specifications may be vague or disregarded (the first criterion); feedback of data may be insufficient, often vague, or too late (the second criterion); people may not be provided with the knowledge and process adjustment mechanisms to correct a process (the third criterion). Thus if we have a quality problem and we fail to meet any of the three criteria, the problem is "management controllable" (or "system controllable"); if we have a quality

TABLE 6.1
Classical control and self-control

Classical control	Self-control
Standard or goal	Knowledge of what people are supposed to do
Measurement	Knowledge of performance
Action on the difference	Means of regulating a process
Primary emphasis during execution	Primary emphasis before execution

problem and if all three criteria are fully met, the problem is "worker controllable." Chapters 13 and 14 apply the concept of self-control to manufacturing and service industries.

Classical control and self-control are complementary (Table 6.1). An important difference, however, involves timing. Classical control takes place *during* the execution of a task; self-control provides useful criteria for evaluating plans *before* a task is executed.

Kondo and Kano (1999, p. 41.3) submit that there is a relationship among the control process; the "plan, do, check, act" cycle; and the concept of self-control. Figure 6.3a depicts the plan, do, check, act cycle, which corresponds to the main elements of the feedback loop (Figure 6.1) of the control process. They observe that individual worker performance during the "do" step comprises a plan, do, check, act cycle (Figure 6.3b). The extent to which the task of the worker is adequately planned reflects the degree to which the worker is placed in a state of self-control. The plan, do, check, act cycle is often called the "Deming cycle."

Some authors refer to the cycle as plan, do, study, act. Gitlow et al. (1995) emphasize that the cycle repeats over and over and provides a means of never-ending improvement.

For both self-control and the Deming cycle, the concept of standardization of work practices is important. Here employees apply a standardize, do, study, act (SDSA)

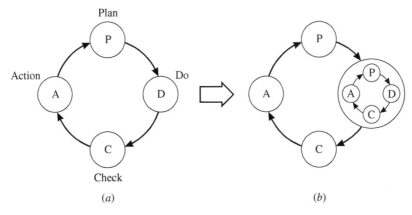

FIGURE 6.3
Deming's cycle. (*From JQH5, p. 41.4.*)

cycle. Employees analyze the process to develop best-practice methods, use the best-practice methods on a trial basis, evaluate the effectiveness of the best practices, and document the standardized process. This standard process helps to stabilize the process and reduce variation. For elaboration, see Gitlow et al. (1995) and Imai (1986). One tool of standardization is the "5S method" for achieving an organized workplace. This method is discussed in Chapter 13.

Schonberger (1999) describes the concept of a self-adjusting system where front-line personnel employ simple, direct methods continuously. He proposes four elements: (1) process capacity management to minimize queues ("kanban"), (2) operator plotting of process data ("statistical process control"), (3) prevention of errors ("fail-safing"), and (4) quality checks before passing work output to the next worker ("source inspection"). As with self-control, Schonberger's concept aims to provide personnel with all that is needed for them to control their work output directly.

Be aware that another concept, self-inspection, is *not* the same as self-control. Self-inspection addresses the examination of the product; self-control addresses the process of accomplishing a task. Self-inspection is discussed in Chapter 13.

We now proceed with an examination of the steps in the control sequence.

➔ 6.4 UNDERSTANDING WHAT IS CRITICAL TO CONTROL

Control subjects for quality are the critical parameters. At the technological level each division of a product—components, units, subsystems, and systems—has quality characteristics. Processing conditions (e.g., time to pay an insurance claim, oven temperature) and processing facilities also have quality characteristics. In addition, input materials and services have quality characteristics. Still more quality control subjects are imposed by external forces: clients, government regulations, and standardization bodies.

Beyond technological quality control subjects are managerial quality control subjects. These are mainly the performance goals for organization units and the associated managers. Managerial goals extend to nontechnological matters such as customer relations, financial trends (e.g., progress in reducing the cost of poor quality), employee relations, and community relations.

To identify and choose quality control subjects, several principles apply:

1. Quality control subjects should be aligned and linked with customer parameters, that is, the subjects should directly measure customer needs, satisfaction, and loyalty or measure product and process features that correlate with these customer parameters. External customers who affect sales income are paramount; equally important are internal customers, who affect internal costs such as the cost of poor quality. But let's face reality. Sometimes our control subjects are incomplete. For example, although advances have been made in measuring the quality of medical care, it is difficult to measure whether a physician detects a medical problem as early as possible.

    Table 6.2 shows examples of quality control subjects from different organizations. Later in this chapter, we specify these categories further by defining the units of measure.

TABLE 6.2
Control subjects

Electronics manufacturer	A bank
Document quality	Operations—timeliness
Software quality	Retail banking—accuracy
Hardware quality	Commercial banking—loan payment posting
Process quality	Credit card and ATM cards—transactions
System quality	Financial and investments—transactions
	Human resources—personnel requisitions
	Information services—system downtime
	Administrative—work order status

2. Defining quality control subjects for work processes starts with defining work processes in terms of objectives, process steps, process customers, and customer needs.
3. Quality control subjects should recognize both components of the definition of quality, i.e., freedom from deficiencies and product features. The number of errors per thousand lines of computer code (KLOC) is important, but even perfect code does not mean that a customer will be satisfied with the software.
4. Potential quality control subjects can be identified by obtaining ideas from both customers and employees. Customers can be asked, "How do you evaluate the product or service that you receive from me?" A focus group of customers can provide valuable responses. Again, we are addressing both external and internal customers. All employees are sources of ideas, but employees who have direct contact with external customers can be a fertile source of imaginative ideas on quality control subjects.
5. Quality control subjects must be viewed by those who will be measured as valid, appropriate, and easy to understand when translated into numbers. These are nice notions, surely. But in the real world they can be pretty elusive.

Next we must establish the measurement process for these control subjects.

➔ 6.5 ESTABLISH MEASUREMENT AND ESTABLISH STANDARDS OF PERFORMANCE

To quantify, we must create a system of measurement consisting of

- A *unit of measure:* the unit used to report the value of a control subject, e.g., pounds, seconds, dollars.
- A *sensor:* a method or instrument that can carry out the evaluation and state the findings in terms of the unit of measure.

Units of measure for product and process performance are usually expressed in technological terms, for example, fuel efficiency is measured in terms of distance traveled per volume of fuel; timeliness of service is expressed in minutes (hours, days, etc.) required to provide service.

Units of measure for product deficiencies usually take the form of a fraction:

$$\frac{\text{Number of occurrences}}{\text{Opportunity for occurrence}}$$

The numerator may be in such terms as defects per million, number of field failures, or cost of warranty charges. The denominator may be in such terms as number of units produced, dollar volume of sales, number of units in service, or length of time in service.

Units of measure for product features are more difficult to create. The number and variety of these features may be large. Sometimes inventing a new unit of measure is a fascinating technical challenge. In one example, a manufacturer of a newly developed polystyrene product had to invent a unit of measure and a sensor to evaluate an important product feature. It was then possible to measure that feature of both the product and of competitors' products before releasing the product for manufacture. In another case, the process of harvesting peas in the field required a unit of measure for tenderness and the invention of a "tenderometer" gauge. A numerical scale was created, and measurements were taken in the field to determine when the peas were ready for harvesting.

Table 6.3 shows examples of units of measure for a manufacturing organization and for a service organization. It should be noted that for many service industries, the *time* taken to deliver a service to an external customer is the decisive control subject for measurement.

Often a number of important product features exist. To develop an overall unit of measure, we can identify the important product features and then define the relative importance of each feature. In subsequent measurement, each feature receives a score. The overall measure is calculated as the weighted average of the scores for all features. In using such an approach for periodic or continuous measurement, some cautions should be cited (Early, 1989). First, the relative importance of each feature is not precise and may change greatly over time. Second, improvement in certain features can result in an improved overall measure but can hide deterioration in one feature that has great importance.

Measurement scales are part of a system of measurement. The most useful scale is the *ratio scale* in which we record the actual amounts of a parameter such as weight. An *interval scale* records ordered numbers but lacks an arithmetic origin such as zero—clock time is an example. An *ordinal scale* records information in ranked categories—an example is customer preference for the flavor of various soft drinks. An unusual example of a measurement scale is the Wong-Baker FACES pain rating scale used widely in hospitals for children to communicate the intensity of pain felt to nurses (Wong and Baker, 1998). The scale shows six faces to which a child can point, ranging from a very happy face (to indicate no hurt) to a very sad face (hurts most). Finally, the *nominal scale* classifies objects into categories without an ordering or origin point—an example is the count of population in each state. The type of measurement scale determines the statistical analysis that can be applied to the data. In this regard, the ratio scale is the most powerful scale. For elaboration, see Emory and Cooper (1991).

TABLE 6.3
Units of measure—examples

Electronics manufacturer	A bank

Document quality
  Defects per thousand formatted
  output pages
	Operations
	$\dfrac{\text{Number of statements mailed late}}{\text{Total number of statements processed}}$

Software quality
  Defects corrected per thousand
  noncomment source statements
	Retail banking
	$\dfrac{\text{Number of teller entry errors}}{\text{Total number of teller entries}}$

Hardware quality Field removal rate	Commercial banking $\dfrac{\text{Loan payments posted incorrectly}}{\text{Total loan payments}}$
Process quality Functional yields	Credit card and ATM cards $\dfrac{\text{Number of mispostings}}{\text{Total number of transactions}}$
System quality Total outages	Financial/investments $\dfrac{\text{Number of trading corrections}}{\text{Number of trades made}}$
	Human resources $\dfrac{\text{Requisitions not filled in 30 days}}{\text{Total number of requisitions}}$
	Information services $\dfrac{\text{Customer information system (CIS) downtime}}{\text{Total CIS time}}$
	Administrative $\dfrac{\text{Number of work orders not completed within 10 days}}{\text{Number of work orders completed}}$

The Sensor

The sensor is the means used to make the actual measurement. Most sensors are designed to provide information in terms of units of measure. For operational control subjects, the sensors are usually technological instruments or human beings employed as instruments (e.g., inspectors, auditors); for managerial subjects, the sensors are data systems. Choosing the sensor includes defining how the measurements will be made—how, when, and who will make the measurements—and the criteria for taking action. This information can be conveniently summarized in a control spreadsheet (see Figure 6.4a and b).

There has been a continuing trend toward providing sensors with additional functions of the feedback loop: data recording, data analysis, comparison of performance with standards, and initiating corrective action. A useful tool for operationalizing self-control and the feedback loop is the control plan, also called a process control plan. It can be used both as a blueprint to plan for control and as a work procedure to implement self-control and the feedback loop. An example of a control plan for the

Expedite Order Entry Control Plan

Process Name: Processing of Expedite Orders Date: 4/14/2003 Revision Level: Approved By:

Control subject	Subject goals	Unit of measure	Sensor	Frequency of measurement	Sample size	Recording of measurement/ tool used	Measured by whom	Criteria for action (when to take)	What actions to take	Who decides	Who acts	Record of actions taken	Comments
Processing and entry of expedite orders	90% on-time entry of expedite orders	% expedite orders not entered before 12:30p.m.	Sorter	Calculate daily	All expedite orders	P-Chart	Sorter/ supervisor	Performance level drops below 90%	Investigate and determine cause. Adjust process as necessary	Management	Supervisor	Process SOPs, Standard work	
Adequate fax machine paper level to print faxes	No late orders due to fax machine out of paper	N/A	Sorter	Four times per day	Population (two fax machines)	N/A	Sorter	Add paper at 8:00a.m. 11:00a.m. 12:00p.m. 3:00p.m.	Re-fill paper at every check.	Sorter	Sorter	N/A	See Sorter SOP
Adequate fax machine toner level to print faxes	No late orders due to fax machine out of toner	N/A	Sorter	Four times per day	Population (two fax machines)	N/A	Sorter	Low toner indicator illuminated	Replace toner cartridge	Sorter	Sorter	N/A	See Sorter SOP

Process	Specification	Measurement	Method	Sample Size/Freq	Sample Size	Recording	Who Measures	Decision Rule	Corrective Action	Who Acts	Who Decides	SPC Chart	Reference
Distribution of faxed expedite orders to CCC's	Sort and distribute all submitted expedites every 30 min before 10:30a.m. Every 15 min. 10:30 to 12:00	Elapsed time	Clock watch	One delivery per 30 min. One delivery per 15 min.	All faxed orders on two fax machines	Sorter stamps date/time of distribution on orders	CCCs	CCC does not receive any orders for more than 30 min. Undistributed orders remain at 12:00p.m.	Investigate status of sorter. Refer to sorting of expedites SOP	CCC or supervisor	CCC/supervisor	N/A	See sorter SOP
Usage of message board when CCC is not available	Return time of CCC indicated on message board for every absence >15 min.	Message board used per absence	Sorter	Check every fax distribution cycle	All fax distribution cycles	N/A	Sorter	Message board is blank and CCC is not available	Hand off order to another CCC. Refer CCC to message board SOP	Sorter/supervisor	Supervisor	N/A	See message board SOP
Elapsed time to enter order	CCC/keyer enters within 30 minutes of receipt (delivery by sorter)	Elapsed time	Stamped time and JDE entry time	Whenever performance level drops below 90% on a given day	30 expedite orders	Data collection form	Supervisor	When the median time for entering orders is greater than 60 minutes	Investigate and determine cause. Adjust process as necessary	Supervisor	Supervisor	N/A	Expedites process SOP

FIGURE 6.4a

Control Plan for Distribution and Order Entry of Faxed Expedite Orders, courtesy of Merillat Industries, Inc.

Process control features / Control subject	Unit of measure	Type of sensor	Goal	Frequency of measurement	Sample size	Criteria for decision making	Responsibility for decision making	
Wave solder conditions Solder temperature	Degree F (F)	Thermo-couple	505 F	Continuous	N/A	510°F reduce heat 500°F increase heat	Operator	...
Conveyor speed	Feet per minute (ft/min)	Timer	4.5 ft/min	1/hour	N/A	5 ft/min reduce speed; 4 ft/min increase speed	Operator	...
Alloy purity	% total contaminates	Lab chemical analysis	1.5% max	1/month	15 grams	At 1.5%, drain bath, replace solder	Process engineer	...
	...	...	...	...	...	...	...	

FIGURE 6.4b
Spreadsheet for who does what.
(*Making Quality Happen, Juran Institute, Inc., senior executive workshop, p. F-8, Wilton, CT.*)

timely distribution and order entry of faxed expedite orders at a customer care center (CCC) is shown in Figure 6.4a.

Despite the large number of control subjects, relatively few human beings are needed to carry out the control process. Imagine a pyramid of control subjects: A few vital controls are carried out by supervisors and managers; another segment is carried out by the workforce; the remaining majority of control subjects is handled by nonhuman means (stable processes, automated processes, servomechanisms).

Clearly, sensors must be economical and easy to use. In addition, because sensors provide data that can lead to critical decisions on products and processes, sensors must be both accurate and precise. The meaning, measurement, and impact of accuracy and precision are discussed in Chapter 10.

Standards of Performance

Each control subject must have a quality goal. Table 6.4 shows examples of control subjects and associated goals for a variety of control subjects ranging from those for products, processes, and departments to that of an entire organization.

This chapter concentrates on goals at operational levels.

To set operational goals, certain criteria must be met. The goals should be

- *Legitimate:* have official status.
- *Customer focused:* external and internal.
- *Measurable:* numbers.
- *Understandable:* clear to all.
- *In alignment:* integrated with higher levels.
- *Equitable:* fair for all individuals.

Goals for product features and process features are based on technological analysis. To encourage continuous improvement, goals should be based on high levels achieved by others (see Chapter 2 under "Benchmarking"). The deployment and alignment of company quality goals to operational goals is discussed in Chapter 2 under "Deployment of Goals."

TABLE 6.4
Control subjects and goals

Control subject	Goals
Mean time between failures	Minimum of 5000 hours
Solder temperature of soldering process	500° F
Overnight delivery	99.5% delivered prior to 10:30 a.m. next morning
Relative quality ranking	At least equal in quality to competitors A and B
Customer retention	95% of key customers from year to year

→ 6.6 MEASURE ACTUAL PERFORMANCE

In organizing for control, a useful technique is to establish a limited number of control stations for measurement. Each such control station is then given the responsibility for carrying out the steps of the feedback loop for a selected list of control subjects. A review of numerous control stations discloses that they are usually located at one of several principal junctures:

- At changes of jurisdiction, e.g., where products are moved between companies or between major departments.
- Before embarking on an irreversible path, e.g., setup approval before production.
- After creation of a critical quality.
- At dominant process variables, e.g., the vital few.
- At natural windows, for economical control.

The choice of control stations is aided by preparation of a flow diagram that shows the progression of events through which the product is produced.

It is essential to measure both the quality of the output going to the external customer ("final yield") and the quality at earlier points in the process, including the "first-time yield."

In Figure 6.5, 100 units of input enter a process. After operations A, B, and C, an inspection is conducted; 87 acceptable units continue on to operation D, 8 units are reprocessed at previous operations, and 5 units are discarded. The first-time yield is thus 87%. After operations D and E, a second inspection is conducted; 82 acceptable units (of the 87) are available for delivery, 2 units are reprocessed, and 3 units are discarded. Assuming that all reprocessed units are acceptable, the final yield is 92 (82 + 8 + 2), or 92% of the original input. Note how the measurement of yield at several places highlights several opportunities for improvement. This concept applies to both manufacturing and nonmanufacturing processes. Don't let different

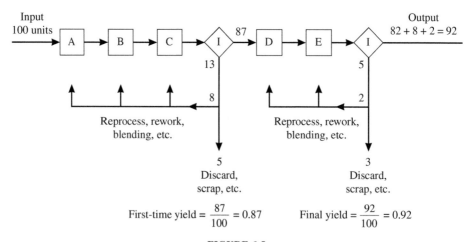

FIGURE 6.5

First-time yield and final yield (A, B, C, D, E = operations or tasks; I = inspections, checks, reviews).

terminology (e.g., inspection versus checking) obscure the concept. For example, in a software development organization, the average number of software errors was about two errors per thousand lines of code, just before delivery to the customer. The average level of errors, however, when measured earlier in the development process, was 50 errors per thousand lines of code. Huge resources were needed to screen out these errors. Ironically, the head of the organization was unaware of this first-time yield until it was revealed by a consultant.

For each control station, it is necessary to define the work to be done: which control subjects are to be measured; goals and standards to be met; procedures, instruments to be used; data to be recorded; and decisions to be made, including the criteria and responsibility for making each decision.

See the example of a control spreadsheet in Section 6.5, "Establish Measurement and Establish Standards of Performance." Keep in mind that control subjects include measurements on both product parameters and process variables. With all of this information, the feedback loop can function well.

The "flag diagram" (Figure 6.6) is an innovative illustration of how measurement can be combined with control subjects for tracking improvement. This diagram uses measurement data in combination with the Pareto concept and the cause-and-effect diagram (both discussed in Chapter 4).

The overall control subject (reduction of machining time) is divided into five major subjects, e.g., improving machining procedure. Each major subject is then further divided into secondary subjects, e.g., improving operation. Goals for each subject are shown as dotted lines on the charts and then performance is plotted on the same charts. The diagrams become a basis for review by the responsible manager and for action if there is a significant deviation from a goal.

❖ 6.7 COMPARE TO STANDARDS

This phase of the control process consists of comparing the measurement to the goal and deciding if any difference is significant enough to justify action. The criteria for taking action (or not taking action) should be numerically defined before measurements are taken, and training should be provided to ensure that the criteria are properly applied. Often the criteria can be simply stated: If a solder temperature exceeds 510°F, decrease the heat; if the temperature is between 500°F and 510°F, then take no action on temperature. Other cases present a need to distinguish between real and apparent differences in measurements on a product or process. This task can be done by using the concept of statistical significance.

Statistical Significance

An observed difference between performance and a goal can be the result of (1) a real difference due to some cause or (2) an apparent difference arising from random variation. Further, differences between a measurement and a goal should not be viewed

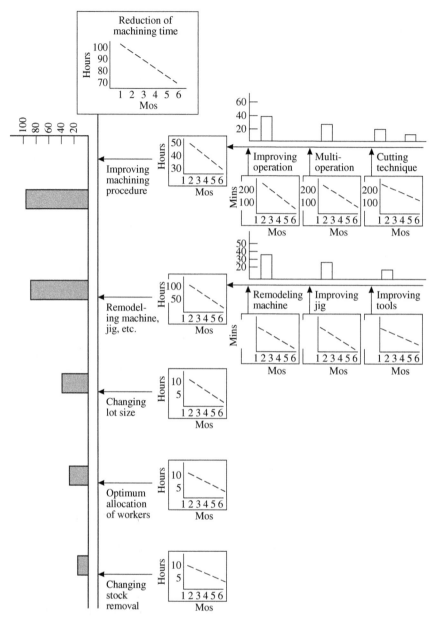

FIGURE 6.6
Example of a "flag diagram." (*Adapted from Kondo and Kano, 1999, p. 41.17.*)

individually. Knowing the pattern of differences over time is essential to drawing correct conclusions. In Figure 6.7, the measurements at A and B and the trend at C represent real ("statistically significant") differences from the goal; the other measurements are due to random variation. Figure 6.7 is a statistical control chart—one of the elegant statistical tools used to evaluate statistical significance.

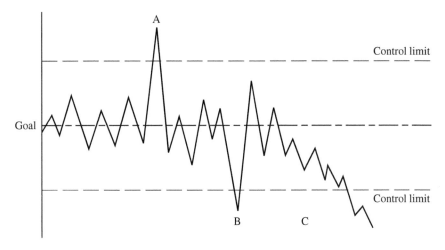

FIGURE 6.7
Control chart.

A control chart is a graphic comparison of process performance data to computed "control limits" drawn as limit lines on the chart. The process performance data usually consist of groups of measurements ("rational subgroups") selected in regular sequence of production.

A prime use of the control chart is to detect assignable causes of variation in the process. The term *assignable causes* has a special meaning, and understanding this meaning is a prerequisite to understanding the control chart concept (see Table 6.5).

Process variations are traceable to two kinds of causes: (1) random, i.e., due solely to chance; and (2) assignable, i.e., due to specific "special" causes. Ideally, only random (also called "common") causes should be present in a process. A process that is operating without assignable causes of variation is said to be "in a state of statistical control," which is usually abbreviated to "in control."

The control chart distinguishes between random and assignable causes of variation through its choice of control limits. These are calculated from the laws of probability so that highly improbable random variations are presumed to be due not to random causes, but to assignable causes. When the actual variation exceeds the control limits, it is a signal that assignable causes entered the process and the process should be investigated. Variation within the control limits means that only random causes are present.

The important advantages of statistical control and the methodology of constructing and interpreting control charts are given in Chapter 20, "Tools to Maintain the Control of Quality."

The control chart not only evaluates statistical significance but also provides an early warning of problems that could have major economic significance.

Random causes are usually chronic, associated with many minor variables, and thus difficult to diagnose and fix; assignable causes are typically sporadic and often originate in single variables, making diagnosis easier. A problem that exists when only random causes are present requires a basic analysis using quality improvement concepts

TABLE 6.5
Distinction between random and assignable causes of variation

Random (common) causes	Assignable (special) causes
Description	
Consists of many individual causes.	Consists of one or just a few individual causes.
Any one random cause results in a minute amount of variation (but many random causes act together to yield a substantial total).	Any one assignable cause can result in a large amount of variation.
Examples are human variation in setting control dials; slight vibration in machines; slight variation in raw material.	Examples are operator blunder, a faulty setup, or a batch of defective raw material.
Interpretation	
Random variation cannot be eliminated from a process economically.	Assignable variation can be detected; action to eliminate the causes is usually economically justified.
An observation within the control limits of random variation means that the process should not be adjusted.	An observation beyond control limits means that the process should be investigated and corrected.
With only random variation, the process is sufficiently stable to use sampling procedures to predict the quality of total production or do process optimization studies.	With assignable variation present, the process is not sufficiently stable to use sampling procedures for prediction.

or quality planning concepts. For example, a process may be in statistical control but does not have the inherent process capability (i.e., small variation) to meet a customer specification. A study is needed to improve the process capability. If a problem exists when assignable causes are present, then the quality control concepts in this section are appropriate. For example, a sudden increase in errors in processing insurance claims may be traced to one untrained person. We elaborate on these ideas in Chapter 20 under "Advantages of Decreasing Process Variability" and in Chapter 10 under "Conformance to Specification" and "Fitness for Use."

Economic Significance

Tools such as the statistical control chart serve several purposes—e.g., they document process performance and identify special situations such as assignable causes or trends. This type of tool provides an early warning of impending problems in the product. But identifying a statistically significant difference between a measurement and a goal does not always lead to corrective action. The presence of assignable causes does mean that the process is unstable, but sometimes assignable causes are so numerous that it is necessary to establish priorities for action based on economic significance and related parameters. When product problems are serious and/or frequent, then setting up a formal quality improvement project or taking other action is warranted.

→ 6.8 TAKE ACTION ON THE DIFFERENCE

In the closing step of the feedback loop, action is taken to restore the process to a state of meeting the goal. Action may be needed for three types of conditions:

1. *Elimination of chronic sources of deficiency.* The feedback loop is not suitable for dealing with such chronic problems. Instead, the quality improvement process described in Chapter 4 or the operational quality planning process described in Chapter 5 should be employed.
2. *Elimination of sporadic sources of deficiency.* The feedback loop is well designed for this purpose. In these cases, the cardinal issue is determining which changes caused the sporadic difference. Discovery of those changes, plus action to restore control, can usually be carried out by local operating supervisors using troubleshooting procedures (discussed later).
3. *Continuous process regulation to minimize variation.* This situation requires linking each product characteristic to one or more process variables, providing a means for convenient adjustment of the setting for the process variables, and determining the relationship between the change in the setting of a process variable and the resulting effect on the product characteristic. These matters are discussed in Section 13.2 under "Correlation of Process Variables with Product Results" and in Chapter 20, "Statistical Process Control."

Section 13.10 provides guidance to operations on when to take action in the form of troubleshooting (quality control), quality improvement, or operational quality planning.

Troubleshooting

Troubleshooting (also called "firefighting") is the process of dealing with sporadic problems and restoring quality to the original level. For organizations that do not have a formal effort to reduce chronic and sporadic problems, operations managers often spend 30% of their time on troubleshooting; for the supervisors reporting to these managers, the time consumed frequently exceeds 60%. In a moment of jest, an executive once said: "Managers who are good at putting out fires can become heroes. I think some of our managers may be arsonists."

Troubleshooting is diagnostic and remedial action applied to sporadic problems and involves three steps:

1. *Identify the problem.* Identification means pinpointing the problem in terms of a single process indicator, the time of occurrence, and its effect. For example, the billing process at a hospital requires an average of 5.2 working days from patient discharge to mailing a final bill. For one week, the average time was 6.7 days, exceeding an upper control limit of 5.9 days on a control chart (Juran Institute, 1995).
2. *Diagnose the problem.* Diagnosis means investigating, developing, and testing theories for the cause of the problem. Analysis of the bills for the particular week revealed one specific set of bills that was delayed. The bills for that week

were then classified by hospital department, payer, clerk preparing the bill, and discharging nursing unit. A Pareto diagram plotted percentage of all bills over seven days versus payer organization. This diagram showed that two-thirds of the delayed bills were for services to be paid for by a particular managed care plan. Further investigation revealed that the plan had just made significant revisions in procedures for submitting bills. These changes resulted in difficulty in the billing department and turned out to be the primary cause of the delayed bills for that week.
3. *Take remedial action.* Remediation requires taking steps to remove the cause identified in step 2. In this case, immediate action was taken to modify the new procedures, change certain software, and identify a single point of contact with the insurance plan until the problem was resolved.

Note that the diagnostic and remedial journeys for troubleshooting are similar to those for quality improvement (see Chapter 4, "Improving Quality While Decreasing Cost"). The approach to troubleshooting is usually less complex because the problem is localized to a specific sporadic time; in contrast, chronic problems are present for a sustained period of time.

Troubleshooting can be made more effective by anticipating problems and planning in advance for troubleshooting. A contingency planning matrix (see Figure 6.8) can be useful in this regard. Note how the planning tries to prevent problems and provide for action, when necessary, in a billing process. In manufacturing operations, each product characteristic (quality control subject) must be linked to one or more process variables so that employees have a contingency plan to adjust the process when necessary. For elaboration, see Chapter 14 under "Review of Process Design."

Example: billing department				
Process indicator: average days to bill				
Condition: weekly average exceeds 5.9 days				
Who	What	Where	When	How
Supervisor	If weekly bill volume is more than 800, add hours to part-time billing clerks	Weekly volume report	By 8:30 a.m. Monday	Inform both clerks and personnel
	Otherwise, convene troubleshooting team	Supervisor's office	By 11:00 am. Monday	By telephone
Troubleshooting team—supervisor, system-support technician, discharge planner	Troubleshoot the problem	—	Begin by 11:00 a.m. Monday	Standard methods

FIGURE 6.8
Contingency planning matrix.

→ 6.9 DEVELOPING PROCESS CONTROL

The GE Capital Mortgage Insurance Corporation provides us with an example of a process control system in the service industry. The company offers mortgage insurance to major lenders of mortgage funds for individual home buyers. Four key processes are involved: underwriting, billing, claims, and sales. The process control system employs both customer information and internal measurements. The elements of the process control system are illustrated in the six-step feedback loop discussed in this chapter.

1. *Choose the control subjects.* Figure 6.9 calls these subjects "measurement categories," e.g., a measurement category (control subject) for the underwriting process is turnaround time. A flow diagram documents the process and helps to identify process measures and outcome indications (process and product features).
2. *Establish measurement.* Nine units of measure (metrics) are employed, e.g., average turnaround time for underwriting.

Quality score card 1st quarter 1997

Measurement category	Customer specifications	Actual performed	Actual σ	Evaluation Excellent ←——→ Poor
Underwriting				
Turnaround time	4 hours	99.9%	4.6 σ	(5) 4 3 2 1
Accessibility	100%	99.5%	4.2 σ	5 (4) 3 2 1
Knowledgeable	Consistent application of guidelines	95.5%	3.2 σ	5 (4) 3 2 1
Billing				
Timeliness	3rd – 5th of month	99.9%	4.6 σ	(5) 4 3 2 1
Completeness	100%	98.9%	3.8 σ	5 (4) 3 2 1
Claims				
Timely payments	30 days	84%	2.5 σ	5 (4) 3 2 1
Work out cycle time	To guidelines 100%	95%	3.1 σ	5 (4) 3 2 1
Sales				
Meeting frequency	Monthly/ quarterly	100%	6+ σ	(5) 4 3 2 1
Knowledge	Answer questions when asked	86%	2.6 σ	5 (4) 3 2 1

FIGURE 6.9
Quality scorecard. From "Using Dashboards and Scorecards in a Service Industry" from *Annual Quality Congress Proceedings,* by S. J. Pautz, 1998. Reprinted with permission from *Annual Quality Congress Proceedings,* © 1998 American Society for Quality.

3. *Establish standards of performance.* For each metric, customers provide input to establish a numerical specification, e.g., turnaround time for underwriting is four hours.
4. *Measure actual performance.* Data collection includes the percentage of transactions meeting the specification, the actual sigma, and a customer evaluation (on a scale of 5 to 1 with 5 meaning excellent). For turnaround time on underwriting, actual performance is 99.9%, which is at the 4.6 sigma level, with a customer evaluation rating of 5. At the 4.6 sigma level, the average opportunity for defects is about 970 defects per million opportunities; at the 6.0 sigma level, the average opportunity is only 3.4 defects.
5. *Compare to standards.* Control charts monitor the processes and provide the linkage between top-level measurements and lower level process indicators. These charts along with the process flow diagram are displayed in the business area. A scorecard with data trends is presented to the customer (the major lenders).
6. *Take action on the difference.* Data are constantly reviewed to achieve process improvements aimed at a 6.0 sigma level by the year 2000. Periodic meetings with customers are held to review numerical performance and identify any changing customer needs.

For elaboration of this system, see Pautz (1998).

The elements of the feedback loop discussed in this chapter are universal. The concepts apply not only to manufacturing and service industries but also to executive and operational activities within all industries.

SUMMARY

- Control is the process we employ to meet standards.
- Control involves a universal sequence of steps: choosing the control subject, choosing a unit of measure, setting a goal, creating a sensor, measuring performance, interpreting the difference between actual performance and the goal, and taking action on the difference. Measurement is a quiet source of action.
- Self-control involves three elements: People must have knowledge of what they are supposed to do, knowledge of their performance, and means of regulating their performance.
- Troubleshooting is diagnostic and remedial action applied to sporadic troubles.

PROBLEMS

6.1. Select a specific task that you have regularly performed for an organization. Evaluate the degree to which this task meets the three criteria for self-control.

6.2. Interview someone who regularly performs a specific task for an organization. Explain the three criteria of self-control to that person and document the degree to which this task meets the criteria, as viewed by the person performing the task.

6.3. Place yourself in the role of a customer for any product—a good or a service. Identify at least four quality control subjects that are important to you as a customer and that the supplier should measure. For each quality control subject, propose a unit of measure.

6.4. Place yourself in the role of upper management for any organization producing goods or service. Identify at least four quality control subjects that are important to internal organizational performance and that the organization should measure. For each quality control subject, propose a unit of measure.

6.5. Select one process consisting of a series of tasks within an organization. Identify the location and the data that are collected on quality-related control subjects throughout the process.

6.6. Interview someone who regularly performs a manufacturing task that includes taking periodic measurements on a product or process characteristic and comparing the result to a specification. Determine how the person makes the following decisions:
   (a) How large a deviation of a measurement from a specification is permitted before the person takes action to adjust the process?
   (b) If a process adjustment is needed, what *amount* of adjustment is made?

6.7. You are designing process controls for the replacement of lost or stolen credit cards. Customers are concerned about responsibility for charges and quick replacement of cards. Identify two potential control subjects that will help manage the process while meeting customer needs. Then define a unit of measure and a sensor for each control subject.

6.8. A major hotel chain is the site for many business conferences. A key customer need is for comfortable meeting rooms that have adequate lighting, temperature control, and visual aid equipment. Identify two control subjects and a unit of measure and sensor for each subject.

REFERENCES

Early, J. F. (1989). "Strategies for Measuring Service Quality," *ASQC Quality Congress Transactions,* Milwaukee, pp. 2–9.

Emory, C. W. and D. R. Cooper (1991). *Business Research Methods,* Irwin, Homewood, IL.

Endres, A. C. (2000). *Implementing Juran's Roadmap for Quality Leadership,* John Wiley & Sons, New York.

Gitlow, H., A. Oppenheim, and R. Oppenheim (1995). *Quality Management—Tools and Methods for Improvement,* Irwin, Burr Ridge, IL.

Imai, M. (1986). *Kaizen,* McGraw-Hill, New York.

Juran Institute, Inc. (1995). *Work Team Excellence,* Wilton, CT.

Kondo, Y. and N. Kano (1999). "Quality in Japan," *JQH5,* Section 41.

Pautz, S. J. (1998). "Using Dashboards and Scorecards in a Service Industry," *ASQ Annual Quality Congress Proceedings,* Milwaukee, pp. 324–330.

Schonberger, R. J. (1999). "Economy of Control," *Quality Management Journal,* vol. 6, no. 1, pp. 10–18.

Wong, D. and C. Baker (1988). "Pain in Children: Comparison of Assessment Scales," *Pediatric Nursing,* 14(1):9–17.

Zairi, M. (1994). *Measuring Performance for Business Results,* Chapman and Hall, London.

CHAPTER 7

Business Process Management

➜ 7.1 FUNCTIONAL VERSUS BUSINESS PROCESS MANAGEMENT

In the future, traditional organizational structure and concepts may see remarkable—indeed, extraordinary—change. Historically, most enterprises have organized around functional departments, e.g., design engineering, manufacturing (operations), marketing, service, and support (the columns in Figure 7.1). Management direction, objectives, and review come from the top and proceed downward through a vertical hierarchy. The functional departments ("silos") focus on achieving their functional (departmental) objectives. Thus a development function has an objective to develop a number of new products; a sales function has an objective in terms of sales quotas. The management of each function emphasizes priorities that will help meet the functional objectives. An examination of the overall objectives of the enterprise, however, reveals that success depends on certain critical results that are the outcome of cross-functional processes (the rows in Figure 7.1). The traditional emphasis on functional objectives can be a serious obstacle to achieving company business objectives that require cross-functional processes.

A *process* is a collection of activities that converts inputs into outputs or results. Thus a process may simply be several steps in a manufacturing or service area. But experience suggests that achieving business goals depends mostly on large, complex processes that go across functional departments. Examples of such cross-functional processes are billing, product development, and distribution. Other examples are materials procurement, hospital patient care, and insurance claims servicing.

We define a *primary process* as a collection of cross-functional activities that are essential for external customer satisfaction and achieve the mission of the organization. These activities integrate people, materials, energy, equipment, and information.

218 Quality Management and Analysis

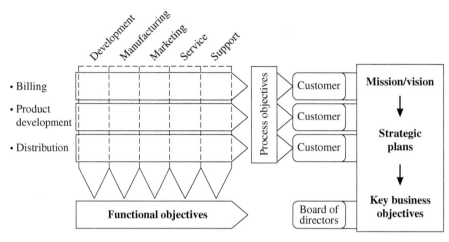

FIGURE 7.1
Work flow in a functional organization. (*From Juran Institute, Inc.*)

The products and services furnished to external customers are produced mostly by cross-functional primary processes. The managers of functional departments are responsible for functional pieces of the process, but no one is accountable for the whole process. Problems often arise because these managers focus on meeting functional objectives rather than process objectives. Problems frequently arise at the functional interfaces between departments (the "white space")—see Rummler and Brache (1995).

The greatest opportunities for improvement exist at the cross-functional process level, which leads us to the concept of business process management.

→ 7.2 BUSINESS PROCESS MANAGEMENT

In this book, business process management is an advanced approach for planning, controlling, and improving the key business processes in an organization by using permanently installed process owners and teams. The distinguishing features of business process management are:

- Emphasis on customer needs rather than functional needs.
- Focus on a few key cross-functional processes.
- Process owners who are responsible for all aspects of the process.
- Permanent cross-functional process teams responsible for operating the process (permanent for the life of the process).
- Application at the process level of the trilogy of quality processes—quality planning, quality control, and quality improvement (see Chapters 4, 5, and 6).

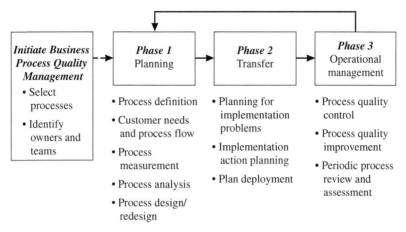

FIGURE 7.2
Road map for business process management. (*From Juran Institute, Inc.*)

Business process management replaces the hierarchical vertical organization with a horizontal view of an organization. (See the final section of this chapter for a discussion of the impact on functional departments that still exist under business process management.) Under business process management, functions have extensive interactions with each other, which leads to a healthy understanding of the interdependencies among functions, i.e., a "systems viewpoint."

One of the early titles of this approach was business process quality management. Other names include process quality management and process management. Part of the evolution includes an emphasis on both the effectiveness and efficiency of performance (discussed later) rather than a narrower conformance to the requirements of quality.

Business process management is commanding much attention and increased use (as contrasted to functional management), but full business process management as described here is still a minority (perhaps 30%) form of management.

A road map for business process management is shown in Figure 7.2. Process management starts when upper management selects key processes, appoints process owners and teams, and provides process mission statements and goals. We next consider, then, process selection.

7.3 SELECTION OF PROCESSES

Organizations have many important cross-functional processes. From these, upper management should select a few primary processes for the business process management approach. The processes selected should, of course, be aligned with the organization mission, strategic plans, and key business objectives (see Figure 7.1).

	Critical Success Factors (CSF)								
Key business processes	Product quality	Supplier quality	Empowered, skilled employees	Customer satisfaction	Lowest cost of poor quality	Lowest delivered cost	Count	Current performance	Total
	1	2	3	4	5	6			
A									
B									
C									
D									
E									
...									

FIGURE 7.3
Critical success factors.

The selection of the processes is based on the critical success factors of the organization, i.e., the few events that must occur for the organization to be successful. Examples of critical success factors are reduction of product development times, increased perception of value, and higher yields. Candidate processes can then be ranked by assessing the importance of the process with regard to the critical factors and also the current process performance. This step is illustrated in Figure 7.3. The relevance of each factor is entered in the body of the matrix (e.g., 5 is high relevance of the process to the critical success factor, 3 is moderate, and 1 is low relevance). The *count* is simply the total of the relevance ratings. The *current performance* of a process is a rating of how the process is performing now (e.g., 1 is good performance, 3 is fair, and 5 is poor). The count on relevance is then multiplied by the current performance rating to obtain a total score for the candidate process. Processes with the highest total would be likely selections for the formal business process management approach. Hardaker and Ward (1987) provide a classic paper describing the use of critical success factors in process selection. *JQH5,* Section 6, describes other approaches for selecting processes for business process management.

When a process is selected, the upper management quality council should prepare a statement of the process mission and process goals. For the accounts payable process, the statement might be something like this:

Process mission
 The mission of the accounts payable process is to pay suppliers in a timely and accurate manner.

Process goals
 The process owner and business process management team will be responsible for increasing payment accuracy by 50% and for reducing the total process cost by 25% within two years.

This statement should be presented to the team for review and to provide direction.

7.4 ORGANIZE THE PROCESS TEAM

After selecting the processes, the quality council (see Section 8.3) appoints a process owner who is responsible for all aspects of process performance. Specifically,

- Be responsible for making the process effective, efficient, and adaptable (discussed later).
- Schedule, set agendas, and conduct process team meetings.
- Establish cooperative working relationships among all functions contributing to the process.
- Guide the process team in analyzing the current process and achieving improvement.
- Make assignments to team members.
- Resolve or escalate issues that may hinder improvement.
- Assure that team members receive training in business process management.
- Manage the implementation of process changes.
- Schedule process reviews.
- Report progress of the process team to the quality council.

For critical cross-functional processes, the responsibility is heavy because the owner does not have line responsibility and authority for all of the component activities of the process. But the owner is responsible to upper management for the overall performance of the process. In practice, the owner focuses on establishing working relationships through a process team, installing quality concepts, resolving or escalating cross-functional issues, and encouraging continuous progress. The typical owner is from a high level of management and is often either the manager with the most resources in the process or the one who is affected the most when problems occur. For example, the purchasing manager is a good choice for the purchasing process.

Some processes have an "executive owner" serving as a champion and a "working owner" responsible for day-to-day activities. This structure provides the involvement and support of upper management and continuous management of the process details. The process owner (with these one or two tiers) is a permanent position.

The process team includes a manager or supervisor from each major function with work activities in the process. In contrast to the ad hoc teams described in Chapter 4 that solve specific quality improvement problems, the process team is permanent. Typically the team has a maximum of eight members and a facilitator (see Chapter 8 under "Quality Project Team").

Figure 7.4 shows a multifunctional organization and one of its major processes. The shaded portions show the executive owner, the working owner, the business process management team, and the functional managers who have work activities.

7.5 EXAMPLE OF BUSINESS PROCESS MANAGEMENT

An example of business process management is provided by a process that involves preparing price quotations and delivery schedules for large orders that go beyond the quantities in standard price discount tables (Juran Institute, Inc., 1990). The process,

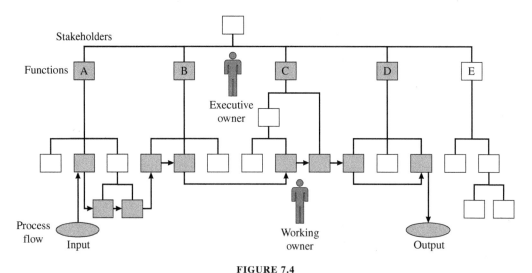

FIGURE 7.4
Organization infrastructure for business process management in multifunctional organizations. (*From Juran Institute, Inc.*)

called "contract management," required a review of the sales order, an analysis, and preparation of the final quotation. A process owner was selected, and a process team appointed. The previously used process, a manual one, is depicted in Figure 7.5. Note that the customer's request for a quotation was received at a branch office, traveled through various offices at different locations, and required 28 sequential approvals before the decision was given to the branch office, which notified the customer. Typically, customers waited 14 weeks for a quotation, and only 20% of these quotations resulted in firm sales orders.

Analysis of the process revealed that much of the delay was due to nonconformance to *internal* customer requirements on the format and content of the proposal for the *external* customer. A redesigned electronic process eliminated many reasons for internal delays. Also, separate procedures were created to handle the "vital few" and the "useful many" customer requests, with authority delegated to regional offices to make final decisions on the useful many. Finally, two review boards were set up to break the bottleneck of the 28 sequential reviews. The cycle time required to respond to the customer was reduced to an average of 17 days (later refinements achieved further savings). This shortened response time was instrumental in raising the yield of firm sales orders from 20% to a solid 60%.

For elaboration of this example, see *JQH5,* Section 6.

➢ 7.6 THE PLANNING PHASE OF BUSINESS PROCESS MANAGEMENT

The planning phase consists of five steps: (1) define the current process, (2) discover customer needs and flowchart the process, (3) establish process measurements, (4) analyze process data, and (5) design (or redesign) the process. Note the similarity

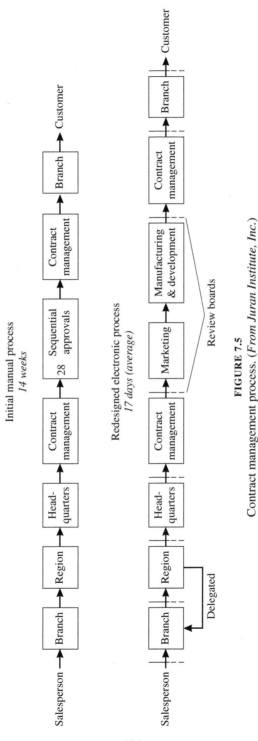

FIGURE 7.5
Contract management process. (*From Juran Institute, Inc.*)

of the planning phase with the six sigma approach and the trilogy of quality planning, quality control, and quality improvement.

Define the Current Process

This process definition step establishes the process mission, goals, scope, and major subprocesses of the current (or "as is") process.

To start, the process team thoroughly reviews the process mission and goals statement provided by the quality council, along with other information such as the strengths and weaknesses and performance history of the process. An example of a statement on mission and goals was given earlier for the accounts payable process.

The major subprocesses are described in a high-level or macroflow diagram, e.g., Figure 7.5. Another example is provided in Figure 7.6 for the billing and collection process at the AAL insurance organization (Hooyman and Harshbarger, 1994). This step in process definition should "bound" the process in terms of where the process starts, which activities are included (and excluded), and where the process ends.

Discover Customer Needs and Flowchart the Process

In this step, the team identifies the customers (i.e., all external and internal parties who are affected by the process), determines customer needs, and prioritizes those needs. The specifics of this process are described in Chapter 5, "Quality by Design to Increase Sales," and Chapter 11, "Understanding Customer Needs."

A more detailed flowchart (a "process map") is prepared showing the major activities, key customers and suppliers, and their roles in the process. This flowchart creates an understanding between the process owner and team members of how the process works. The use of the flowchart to gain this understanding is important because initial discussions usually reveal disagreements on how the process really works. Creating the flowchart clarifies—often after much discussion—how the process works. It is useful to schedule a work session of several hours in which the team discusses the process and prepares the flowchart. A facilitator describes how the work session will proceed, asks for inputs on the sequence of activities, and uses devices such as sticky notes to create the flowchart physically on a large board prepared for the purpose. The result is a starting point for analysis and improvement. An example of such a flowchart is given in Figure 7.7 (the measurement concept, M, is explained next).

Van Aken and Hacker (1997) describe in seven steps how the work session and follow-up are conducted. After an initial flowchart is created by hand, software (such as Visio® or Optima!®) should be used to generate the flowchart.

Establish Process Measurements

Measurements from a process are initially needed to describe how well the process is doing and to set the stage for process analysis and improvement. Later, measurements

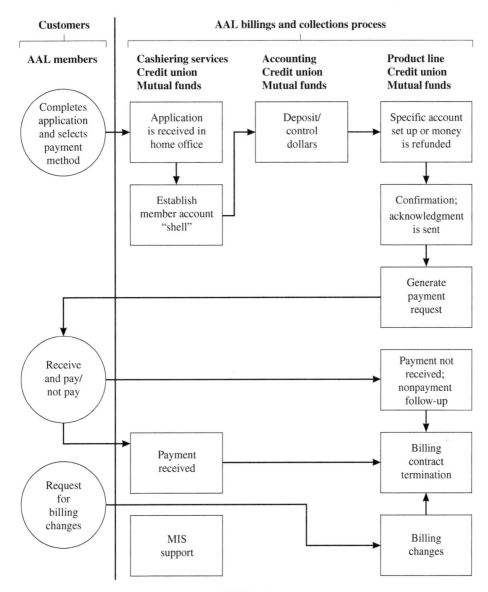

FIGURE 7.6
Macro-level process flow diagram for the billing and collections process.

are employed to help control process performance and periodically determine process capability.

In deciding which measurements to collect from a process, the emphasis should be on the process mission statement, goals, and customer needs discussed previously. In the framework shown in Figure 7.8, effectiveness focuses on meeting customer needs, efficiency addresses meeting the needs at least cost, and adaptability reflects

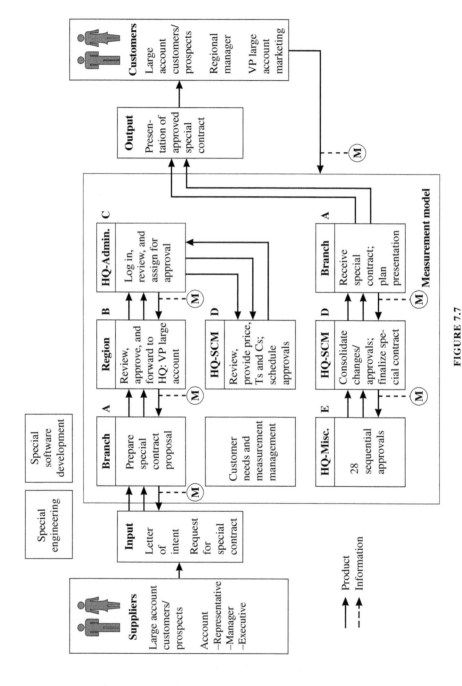

FIGURE 7.7 Flowchart of the special-contract management process, including process control measurement (M) points. (*From Juran Institute, Inc.*)

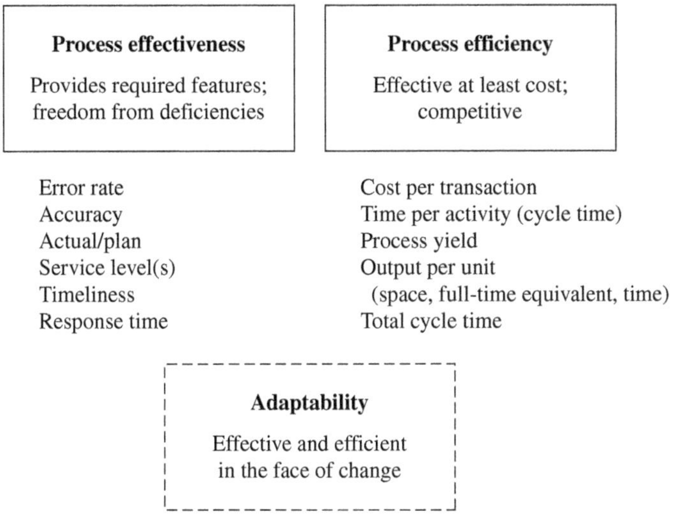

Process effectiveness

Provides required features; freedom from deficiencies

Process efficiency

Effective at least cost; competitive

Error rate
Accuracy
Actual/plan
Service level(s)
Timeliness
Response time

Cost per transaction
Time per activity (cycle time)
Process yield
Output per unit
   (space, full-time equivalent, time)
Total cycle time

Adaptability

Effective and efficient in the face of change

FIGURE 7.8
Business process measurements.

the ability of the process to react positively when process or external conditions change. Although effectiveness and efficiency can be quantified easily, quantifying adaptability is more difficult. Snee (1993) discusses methods for making processes adaptable to changing conditions. The basic steps in quantifying parameters are discussed in Chapter 6, "Control of Quality to Maintain Superior Performance." Note, in Figure 7.7, how the measurement control points are indicated on the flowchart.

Process measurements should be linked to traditional business indicators. Some examples are shown in Figure 7.9.

Analyze Process Data

In this step, we evaluate the process performance data, identify opportunities for improvement, and determine the causes of process problems. The specific approach and tools are described in Chapter 4, "Improving Quality While Decreasing Cost," Chapter 5, "Quality by Design to Increase Sales," and Chapter 6, "Control of Quality to Maintain Superior Performance," but with special emphasis on the total process and subprocesses.

Performance data are evaluated for both process effectiveness and process efficiency (see previous page), and problems are identified using Pareto analysis, flow diagrams, and other means discussed in Chapter 4 under "Identify Potential Projects."

The high-level and detailed flow diagrams are key tools at this point. For example, a team at a telecommunications service company required four hours to construct a high-level flow diagram for a new service provided to corporate customers. The time consumed was a classic case of each member having a different view of how the process operated. The discussion and resulting high-level diagram exposed missing

	Traditional business view		**Process view**	
Business objective	Business indicator		Key process	Process measure
Higher revenue	Percentage of sales quota achieved		Contract management	Contract close rate
			Product development	Development cycle time
	Percentage of revenue plan achieved		Account management	Backlog management and system assurance timeliness
	Value of orders canceled after shipment			
				Billing quality index
	Receivable days outstanding		Manufacturing	Manufacturing cycle time
Reduce costs	Inventory turns			

FIGURE 7.9
Linkages among business objectives, traditional business indicators, and process measures generated by the process-management approach—a few examples.
(*From Juran Institute, Inc.*)

links, bottlenecks, unnecessary steps, and redundancies in the process. For the first time, all team members had a common understanding of the process.

The team then divided the high-level diagram into four segments and constructed detailed flow diagrams for each segment. A portion of one detailed diagram is shown in Figure 7.10. Examination of the four detached flow diagrams revealed 30 rework loops (and Pareto analysis showed that 6 of the 30 loops accounted for 82% of the total rework time). In Figure 7.10, two of the rework loops (26 and 27) are identified by a triangle.

Flow diagrams often use four symbols for events: a diamond for a decision-making event, a small square for an activity, a group of consecutive squares for a rework loop, and a paper symbol for a document or database. Figure 7.11 provides a checklist of questions for each of these symbols to help discover opportunities for improvement. The construction and analysis of flowcharts along with other simple and complex process analysis techniques was originated and refined by the industrial engineering profession. A useful reference is Salvendy (1992).

An example of a powerful (but more complex) technique is computer simulation of a process. Here a computer model is developed based on the logic sequence of process activities, along with data on the activities. The computer model then generates simulated results of process outputs. The time spent in developing the computer model is an investment to help reveal bottlenecks, underused process activities, and key causes of problems and to understand the process. Later, the model can also help to evaluate potential solutions to problems. Batson and Williams (1998) describe seven cases, from both manufacturing and service industries, of using simulation in process improvement.

An excellent source of ideas to design or redesign a process is other organizations with similar processes. Some processes are common across many types of manufacturing and service industries (e.g., the hiring process), and the experience of other organizations can provide ideas that have been tested in practice. This approach is

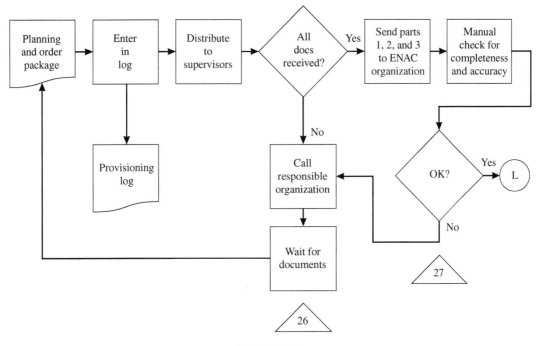

FIGURE 7.10
Flow diagram with rework loops (identified with triangles).
(From Juran Institute, Inc.)

really an application of benchmarking (see Chapter 2 under "Competitive Benchmarking" and also Camp, 1995). If it is possible to obtain a copy of process flow charts from another organization, so much the better.

Dow Corning now uses software to manage its process knowledge. This software creates interlinked maps of key processes, and the maps are stored on the company intranet. The "process repository" organizes information by process types and process parts. The Massachusetts Institute of Technology is licensing companies to use the process repository and software. A general process repository will be placed on the World Wide Web, giving managers access to a wealth of knowledge on process design (Carr, 1999).

At the end of the analyze phase, we have a clear understanding of the current process, have identified actual problems and causes, and have initial thoughts on types of improvement action necessary. This information should be reviewed by the executive process owner and other managers before proceeding to design/redesign.

Design (Redesign) the Process

This final step may involve radical change, incremental change, or both. The approach and tools of quality improvement, quality planning, and quality control are described

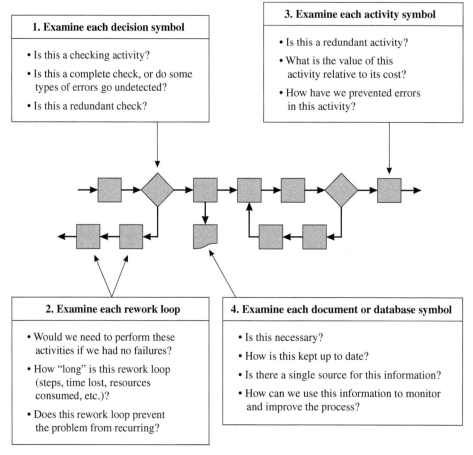

FIGURE 7.11
Analyzing a flow diagram. (*From Juran Institute, Inc.*)

in Chapters 4, 5, and 6, but these are applied with special emphasis on the process and subprocesses. We start with the flowchart analysis of the current (as is) process and then redesign the process to create a flowchart for the revised (should) process. Design changes involve workflow, information and other technology, people, physical locations, and policies and regulations.

Radical redesign of a process is associated with the term *reengineering*. Hammer and Champy (1993) define *reengineering* as "the fundamental rethinking and radical redesign of business processes to achieve dramatic improvements in critical contemporary measures of performance, such as cost, quality, service, and speed." Clearly, the four key words are *fundamental, radical, dramatic,* and *processes.*

In reality, reengineering is a blend of the old and the new. Many concepts and techniques of industrial engineering, quality management, and other disciplines have long been an integral part of process improvement and redesign. Some key aspects of reengineering are

- Apply state-of-the-art information technology (discussed later).
- Analyze process activities for improvement, i.e., eliminate non-value-added work; simplify, combine, resequence activities; and minimize transfer of material and information especially across departments. This aspect of reengineering includes combining process steps so that an employee produces the final output rather than transferring the work for subsequent steps to other specialists.
- Benchmark against other organizations.
- Empower employees to make decisions to minimize time required for approvals.
- Remove causes of errors in processes to reduce rework and minimize checking and controls.
- Consider transfer of some activities upstream to suppliers or downstream to customers.
- Establish a single point of contact for customers so that one employee handles an entire service, rather than transferring the customer to other employees.
- Use creative thinking techniques, including brainstorming.

This list presents highlights and is not all-inclusive.

A major contribution of reengineering is the emphasis on the application of the ever-increasing benefits of information technology on work. Hammer and Champy (1993) provide some illustrations:

- Information can appear simultaneously in as many places as it is needed.
- A generalist can do the work of an expert. For example, supported by integrated systems of information, a call center representative can handle all steps to service a customer.
- Businesses can simultaneously reap the benefits of centralization and decentralization. Decentralized divisions, at distant locations, can exchange information instantaneously with corporate headquarters and other divisions.
- Decision making is part of everyone's job. Database technology now provides many people with information that was previously available only to management.
- Field personnel can send and receive information wherever they are.
- The best contact with a potential buyer is effective contact (with or without human intervention). For example, banks, real estate brokers, and others can now provide customers with interactive video that provides a wealth of carefully prepared information.
- Things tell you where they are. For example, the locations of trucks and railroad cars can now easily be determined in real time.
- Plans get revised instantaneously. Instead of periodically revising plans and updating status, all work can now be done instantaneously and made available to all parties immediately.

The message is clear—information technology now has a major impact on how we organize and execute work.

In practice, design changes are both radical and incremental. Radical changes are dramatic, but incremental changes are collectively also essential.

Before the new design is placed into operation, conduct a design review and trial implementation. Typically, the process owner assembles a group of experts (from outside the process) to evaluate the design alternatives and the chosen design. Finally, the selected design should be tested under operating conditions using trial runs with regular operating personnel to predict both process effectiveness and process efficiency.

AT&T's similar approach to business process management is summarized in Table 7.1. Note how many of the tools of quality improvement, quality planning, and quality control play an integral role in business process management.

7.7 TRANSFERRING AND MANAGING THE NEW PROCESS

Transferring a new or revised process to operations involves carefully planning for the many aspects of the change. Managing the new process includes setting up appropriate process controls and planning for continuous improvement and periodic assessment. For elaboration, see Chapter 5, "Quality by Design to Increase Sales" and *JQH5*, pages 6.16–6.19.

Processes evolve into various phases of maturity. At General Electric, five maturity classes are ad hoc, consistent, organized, systematic, and statistically stable (McNamara, 1997). McNamara reports that most important business processes are not statistically stable.

To summarize, business process management includes some similarities to and some differences from the concepts discussed earlier in this book. The similarities include an emphasis on the customer concept and the use of techniques from quality planning, quality control, and quality improvement. But important additions are provided by the business process management concept:

1. Emphasis is placed on the overall effectiveness, efficiency, and adaptability of a cross-functional *process*, rather than on the output of individual functional departments.
2. The process is analyzed in an integrated manner to correct defects, errors, and other problems and also to identify and meet customer needs.
3. Responsibility for the process is defined in terms of a process owner and a process team. The owner and the team are permanent. The process team may set up quality teams that operate temporarily to address specific problems within the total process.

Finally, the special features of business process management require us to consider its impact on an organization.

7.8 IMPACT OF BUSINESS PROCESS MANAGEMENT ON AN ORGANIZATION

This chapter presents a concept of business process management that identifies the key processes for organizational success and selects process owners and process teams that are responsible for planning, controlling, and improving all aspects of processes. These teams are permanent, not ad hoc quality planning or quality improvement teams.

TABLE 7.1
AT&T process quality management and improvement methodology

Steps	Activities	Tools
1. Establish process management responsibilities	Review owner selection criteria Identify owner and process members Establish/review responsibilities of owner and process members	Nominal group technique
2. Define process and identify customer requirements	Define process boundaries and major groups, outputs and customers, inputs and suppliers, and subprocesses and flows Conduct customer needs analysis Define customer requirements and communicate your own requirements to suppliers	Block diagram Survey Customer/supplier relations checklist Interview Benchmarking Affinity diagram Tree diagram
3. Define and establish measures	Decide on effective measures Review existing measures Install new measures and reporting system Establish customer satisfaction feedback system	Brainstorming Nominal group technique Survey Interview
4. Assess conformance to customer requirements	Collect and review data on process operations Identify and remove causes of abnormal variation Compare performance of stable process to requirements and determine chronic problem areas	Control chart Interview Survey Pareto diagram Cause-and-effect diagram Brainstorming Nominal group technique Trend chart
5. Investigate process to identify improvement opportunities	Gather data on process problems Identify potential process problem areas to pursue Document potential problem areas Gather data on subprocess problems Identify potential subprocess problems to pursue	Interview Flowcharting Brainstorming Pareto diagram Nominal group technique
6. Rank improvement opportunities and set objectives	Review improvement opportunities Establish priorities Negotiate objectives Decide on improvement projects	Pareto diagram Nominal group technique Trend chart
7. Improve process quality	Develop action plan Identify root causes Test and implement solution Follow through Perform periodic process review	Pareto diagram Nominal group technique Brainstorming Cause-and-effect diagram Cause-and-effect/force-field analysis Control chart Survey

Source: Shaw et al. (1988).

Although only a minority of organizations apply this full concept of business process management the application is growing. The changes are indeed profound: Work activities are focused in cross-functional process teams, not functional departments; the organizational structure changes from hierarchical to flat; managers change from supervisors to coaches and leaders; employees change from people who perform narrow tasks and rigid procedures to empowered individuals who are trained in a broader work scope; and more.

The impact on the functional department organization can be dramatic. If an organization is run by processes, the processes will drive the organization, including the structure. When most work is done by permanent process teams, the role of functional departments becomes one of serving processes; i.e., a process is the internal customer of the functional departments. Functional departments will still be needed to do internal consulting, training, measurement, and research.

Perhaps business process management will bury the silos, break down the walls between functional departments, and replace command and control management with empowered employees to whom management provides capable processes. Perhaps.

SUMMARY

- A process is a collection of activities that convert inputs into outputs or results.
- A primary process is a collection of cross-functional activities that are essential for external customer satisfaction and achieving the mission of the organization.
- Business process management is an approach for planning, controlling, and improving the primary processes in an organization by using permanent process teams.
- Business process management is most effective when a process owner and a permanent process team manage and operate a process.

PROBLEMS

7.1. For an organization with which you are familiar, discuss the critical success factors, i.e., the few events that must occur for the organization to be successful. Then identify three cross-functional processes and rank each of them on the basis of (a) their importance with regard to the critical factors and (b) relative need for improvement.

7.2. For one of the processes in problem 7.1, prepare a process mission statement and a statement of process goals.

7.3. For one of your key business processes, propose a process owner and members of a process team.

7.4. For the process in problem 7.3, develop a flowchart of the current process.

7.5. For the process in problem 7.3, propose at least three process measurements that could be collected for process analysis and improvement.

7.6. For the process in problem 7.3, analyze the process and propose a redesign.

REFERENCES

Batson, R. G. and T. K. Williams (1998). "Process Simulation in Quality and BPR Teams," *Annual Quality Congress Proceedings,* ASQ, Milwaukee, WI, pp. 368–381.

Camp, R. C. (1995). *Business Process Benchmarking,* Quality Press, ASQ, Milwaukee, WI.

Carr, N. G. (1999). "A New Way to Manage Process Knowledge," *Harvard Business Review,* September–October, pp. 24–25.

Hammer, M. and J. Champy (1993). *Reengineering the Corporation,* HarperCollins, New York.

Hardaker, M. and B. K. Ward (1987). "Getting Things Done: How to Make a Team Work," *Harvard Business Review,* November–December, pp. 112–119.

Hildreth, S. W. (1993). "Rolling out BPQM in the Core R & D of Lederle–Praxis Biologicals, American Cyanamid," *Proceedings of R & D Quality Symposium,* Juran Institute, Inc., Wilton, CT, pp. 2A-1 to 2A-9.

Juran Institute, Inc. (1989). *Quality Improvement Tools—Flow Diagrams,* Wilton, CT.

Juran Institute, Inc. (1990). *Business Process Quality Management,* Wilton, CT.

McNamara, D. M. (1997). "Process Maturity Classes," *Annual Quality Congress Proceedings,* ASQ, Milwaukee, pp. 891–897.

Salvendy, G. (1992). *Handbook of Industrial Engineering,* 2nd ed., John Wiley & Sons, New York.

Shaw, G. T., E. Leger, and J. C. MacDorman (1988). "Process Quality at AT&T," *Impro Conference Proceedings*, Juran Institute, Inc., Wilton, CT, pp. 4D-5 to 4D-9.

Snee, R. D. (1993). "Creating Robust Work Processes," *Quality Progress,* February, pp. 37–41.

Van Aken, E. M. and S. K. Hacker (1997). "Enhancing Conventional Process Improvement with Systems Mapping," *Proceedings Annual Quality Congress,* ASQ, Milwaukee, pp. 84–94.

CHAPTER 8

Organization Roles to Support a Quality Culture

8.1 EVOLUTION OF ORGANIZATION FOR QUALITY

Many organizations, particularly in manufacturing industries, have centralized quality-related activities in a "quality department." Over the decades, the names and the scope of activities have changed—inspection, statistical quality control, quality control, reliability engineering, total quality control, quality assurance, total quality management (TQM).

From the 1980s until the turn of the century, five major trends emerged in organizing for quality:

1. Quality management tasks were assigned (or transferred) to functional line departments rather than to quality departments. For example, process capability studies were transferred from a quality department to a process engineering department.
2. The scope of quality management was broadened from operations only (little q) to all activities (big Q) and from external customers to external and internal customers. Most organizations now train personnel in the functional departments in the tools of quality management and make employees responsible for implementing modern concepts of quality.
3. A major expansion occurred in the use of design and improvement teams.
4. Authority to make decisions was delegated to lower levels closest to where work is done.
5. Many companies are including key suppliers and customers in quality activities—improvement, planning, and control. The term employed is *partnering*. Sometimes partnering results in a "virtual corporation," i.e., a temporary network of functions from different companies that come together for a specific business purpose; the network then disbands when the purpose has been met. For elaboration, see Christie and Levary (1998).

These trends led to new approaches to coordinating and organizing quality activities. Becker et al. (1994) conducted research at 30 companies to learn the impact of TQM on the organization. The findings: TQM led to (1) flatter organizations, (2) an increase in liaison devices such as cross-functional teams, (3) a decrease in size and an increase in output, (4) a shift to group reward systems, (5) participation in planning by lower level employees, and (6) changing organizational boundaries.

The Quality Office: Leading the Way Forward

The "quality office" is a term the authors used to describe the full-time function with responsibilities of managing the quality system (little q) and the performance excellence (big Q) system across an organization. The little q quality management system is the traditional role of the quality office. The big Q system defines the application of methods, tools, and structure to link all functions and processes in a "total organization" approach.

The quality office of today versus the quality assurance function of the past is being transformed to recognize and reinforce the current view that quality is not separate and distinct from business but its principles are central to achieving performance excellence. Your organization may not use the term *quality office*. The function may be named quality assurance, quality department, performance excellence program office, strategic quality, or customer satisfaction office, to name a few. The quality office of the future (and some are doing it today) will play a dual role. It will administer the quality management system to control and assure that the policies and procedures are in place to avoid product and service failures and reduce risk. The second role is to lead the performance excellence drive within the organization. Today, there may be two separate offices administering and carrying out these two roles. In the future, we see the quality office doing both.

✧ 8.2 ROLE OF THE QUALITY OFFICE

The approach used to coordinate quality activities throughout an organization takes two major forms:

1. Coordination for *control* is achieved by the regular line and staff departments, primarily through employment of formal procedures and use of the feedback loop. Feedback loops take such forms as audit of execution versus plans, sampling to evaluate process and product quality, control charts, and reports on quality.
2. Coordination for *creating change* is achieved primarily through the use of quality project teams and other organizational forms for creating change.

Coordination for control is often the focus of a quality department; sometimes such a focus is so preoccupying that the quality department is unable to make major strides in coordination for change. As a result, some "parallel organizations" for creating change have evolved.

Parallel Organizations for Creating Change

All organizations are engaged in creating beneficial change as well as in preventing adverse change ("control"). Much of the work of creating change consists of processing small, similar changes. An example is the continuing introduction of new products consisting of new colors, sizes, shapes, and so forth. Coordination for this level of change can often be handled by carefully prepared procedures.

Nonroutine, unusual programs of change usually require new organizational forms. These new forms are called "parallel organizations." *Parallel* means that these organizational forms exist in addition to and simultaneously with the regular "line" organization.

Examples of parallel organizations for achieving change in quality management are process teams, (see Chapter 7), quality councils, and various types of quality project teams (discussed later). Parallel organizations may be permanent or ad hoc (a process team versus a quality improvement team) and may be mandatory or voluntary (a quality council versus a workforce team). An early form of a parallel organization for product reliability is described by Gryna (1960).

The *quality officer,* the leader of this function, should be part of a community of practitioners that must act as the surrogate conscious of the customer for the organization. This chapter will deal with the development of the quality and performance excellence office needed to support both little q and big Q. It will include how to establish an effective function, how to develop and maintain the skills and behaviors of its staff, and how to establish credibility with upper managers.

Computer Sciences Corporation, one of the largest technology solutions and outsourcing organizations in the world, understands the quality office of the future. Darryl W. Bonadio, Master Black Belt and Director of MSS Quality Management and Improvement, describes quality and the company's quality office this way: "Quality is important to everyone at CSC. Quality must become a cultural attitude that results in personalization of our five standards: Positive Talk, Confident Perspective, Outcomes Oriented, Be Accountable, and Respect our Client. CSC's Managed Services Sector manages quality through its Global Process Innovation and Quality Excellence organization. This is our Quality Office. The Enterprise Excellence Program we have implemented meets this challenge by addressing three important tenets. The first is Delivery Excellence, which is the integration of quality principles, tools and approaches that enable achievement of business objectives and promote client advocacy. Second, Passionate Delivery where engaged employees are the most effective quality assurance factor in our service-oriented environment. And finally, Enterprise Performance Management that provides an end-to-end analysis and a feedback loop for continuous improvement."

Discussion over the changing role of quality professionals has also occurred by the likes of Spichiger (2002) and Wescott (2004). The role of the quality office is best discussed by considering it at two levels—tactical and strategic.

"Successful strategy execution depends on both satisfying the customer today and achieving excellence in the future. We continue to use the Juran Trilogy to establish a solid foundation for achieving and sustaining breakthroughs throughout the entire organization of the Builder Cabinet Group at Masco Corporation. This has enabled us to continually meet or exceed the expectations of customers today, develop innovative breakthrough products and processes for tomorrow, and most importantly provided the means for developing team members throughout my company to execute the strategy to action. As the executive of the Quality Office, we will continue to drive our performance towards sustainability" (Wittig, 2009).

At the tactical level, the traditional role was and still is to provide an independent evaluation of product quality or service quality. Inspections, testing, and product or service audits are examples. In this role, the quality office is often viewed as being limited to supporting operations—providing an independent evaluation to ensure that productivity targets are not pursued at the expense of not meeting quality specifications. Examples include:

- Inspection and checking
- Product testing
- Supplier quality
- Root cause corrective action
- Quality system audits

As technology and the needs of the organization have changed, the role of the quality office must expand beyond this traditional view. At the tactical level, the quality office should play a role in enabling others in the organization to carry out the management of quality and performance excellence either independently of or in collaboration with other functions. Examples of expanded tasks include:

- Calculate the costs due to poor quality and use the costs to marshal the right resources to reduce these costs.
- Conduct process capability studies in nonmanufacturing processes.
- Participate in design reviews of new products as early as possible in the development cycle.
- Work with the supplier chain, and provide evaluations of supplier selection and ongoing performance.
- Monitor customer satisfaction, and create and manage the corrective action process.
- Be involved with the engineering change process.
- Reduce the inspection by creating in-process checking and self-inspection by the workforce.
- Work with functions to implement error-proofing efforts.
- Participate in kaizen or rapid improvement events.
- Use data-driven methods such as statistical process control to monitor performance.
- Identify improvement projects such as lean and six sigma projects.
- Integrate with environmental, health, and safety programs.
- Develop a skilled and competent function of experts that can consult with all functions.

Strategic Level

In many enterprises, the quality office is often viewed as having only a tactical role. That the quality office can perform a useful strategic role is not well recognized by many organizations. However, enlightened organizations have recognized the strategic value of the quality office. They view the quality office as a strategic asset with a key role in shaping, planning, and enabling the deployment of strategies, strategic goals, and business plans of the organization. A strategic approach for the quality office of the future is to think of their role as providing "enterprise assurance" (Juran Institute White Paper, 2008).

The following is a list of strategic activities in which the quality office needs to play an important role:

- Vision, mission, and policy development
- Assisting upper management with strategic planning and goal-setting
- Recommending to upper managers how to reduce the costs of poor quality
- Providing the organization with the most effective methods and/or tools to reduce these costs
- Being involved in annual business planning to incorporate improvement goals
- Demonstrating how quality can affect social responsibility and the environment
- Organizationwide assessment and planning to close the gaps
- Organizationwide transformation and improvement initiatives such as lean six sigma
- Innovation of major business processes (such as demand creation, product development, order fulfillment, supply chain, and HR management processes)
- Development and deployment of balanced scorecards and data systems to support all key business processes, policy deployment, and improvement efforts

These roles are consistent with the views presented by Gryna, Chua, and DeFeo (2007), where the quality office should have both tactical and strategic roles.

✦ 8.3 ORGANIZING THE QUALITY OFFICE OF THE FUTURE

To better understand responsibilities, we need to make an important distinction. There are direct responsibilities and indirect responsibilities. Direct responsibilities are those activities and results over which the Quality Office has control because they are executed by full-time personnel who report to the Quality Executive. Indirect responsibilities are those over which the Quality Office has an influence, but with little or no direct control. Indirect responsibilities are those which the Quality Office enables others (non-quality personnel) to do. The Quality Office provides the infrastructure and the means for these indirect responsibilities to be carried out by personnel in other functions. For example, the execution of process control plans to assure quality compliance is one where the Quality Office enables others (in Operations) to carry out this important task. Another important example is the development and deployment of a change management program to attain performance excellence. This is one where the Quality Office has both direct and indirect responsibility. Direct because Quality personnel are performance excellence specialists. Indirect because selected non-Quality personnel from other functions are also performance excellence specialists. In both cases, the Quality Office plays an integral part in developing and enabling all performance excellence specialists (such as six sigma and lean six sigma green belts, black belts, and lean experts) to contribute to the improvement of the enterprise.

Consider Its Mission

Who should lead the Quality Office? At what level of management? To whom is the Quality Office accountable and to whom should the Quality Office report? These are important questions to ask when setting up the Quality Office and are best answered by first considering the mission of the Quality Office. Ideally, we want to ensure that the mission and associated responsibilities are supported by the corresponding management level and the appropriate level of authority is given to the Quality Office. Authority must be consistent with responsibility otherwise, we have a paralyzed office not able to carry out its duties.

For example, at an organization that provides customer care and call center services outsourced by major corporations, the mission of the Quality Office is to contribute to the organization's financial growth by providing services that:

  A. Enable the efficient delivery of exceptional customer experiences
  B. Foster a culture of fact based leadership and continuous improvement

Not surprisingly, in this organization the Quality Office is led by the Quality Director, who reports to the Chief Financial Officer. To support that mission, responsibilities of this office reflect both the tactical and strategic role of the Quality Office discussed earlier. Functions that are the responsibility of this Quality Office are:

1. Process improvement (six sigma and lean)
2. Workforce management (contact center volume forecasting, staffing, and intra-day staff management)
3. Customer contact quality (call and email quality)

Note: The Quality Office should always be part of the executive management team or a direct report to it to be effective at leading performance excellence. If not there will be little credibility of the Office.

The Quality Office should plan to enable others as much as possible to drive toward achieving what Dr. Juran calls "Self-Control" where (a) what is expected is clearly known, (b) actual performance is known through short or immediate feedback loops, and (c) the means and ability to regulate is available so that actual performance meet expectations.

Example: Organizing the Quality Office in a Multinational Global Enterprise

The scope, size, and structure become more challenging when the enterprise is global, with multiple divisions and regions operating around the world. To what extent should the Quality Office be centralized (versus decentralized)? How much of the structure should be matrix? To what extent should there be direct reporting relationships versus indirect dotted-line reporting relationships to the head of the Quality Office? Should a plant quality manager report directly or indirectly to the local plant's general manager?

To illustrate how the Quality Office can be organized, we will look at an example of a multinational global enterprise. Given the nature of the responsibilities, the Chief Quality Officer (who is the Vice President of Quality) reports to the President and COO (who in turn reports to the CEO), at a high enough level of authority that is consistent with the global responsibilities of the Quality Office.

The Quality Office is structured to drive global standards in areas with significant opportunity (such as reducing the number of suppliers and to increase global coordination for key customer quality improvements), to leverage best practices (such as by maintaining a library of lean six sigma projects, and FMEA libraries to capture engineering experience) and revitalize use of quality audits to highlight continued improvements.

Responsibilities of Global Functional Quality are:

- Customer Quality
  - To coordinate consistent proactive and reactive customer support.
- Supplier Quality
  - To enforce one global standard for suppliers and contract manufacturers.
- Lean Six Sigma
  - To eliminate waste.
  - To improve product and process quality.
- Audit
  - Continuous evolvement of quality systems.
  - Global utilization of IQRS audit and related improvement plans.
- Project Management
  - To coordinate global development and implementation of systems, tools, and quality process improvements.
- Rules of Engagement
  - The manufacturing plant has responsibility for manufacturing related ERA (Emergency Response Action), containment, Customer Complaints, and 8D actions and responses.
  - The selling entity has responsibility for communicating the ERA from seller to customer and to enter the complaint in SAP and inform the manufacturing plant.
  - Escalation above the plant quality manager goes first to the division quality head for the manufacturing division.
  - When a customer requires an on site visit to discuss a quality problem, it is the responsibility of the closest (geographic) division quality head.

The Quality Director (The Leadership)

The Quality Director (or Quality Vice President or Quality Manager, depending on organization size and titles) is someone who leads the Quality Office and works with senior management to ensure that enterprise quality is planned, assured, controlled, and improved. The competencies of a Quality Director are:

1. All of the competencies of a Quality Engineer PLUS the following
2. Leadership and Management: Motivating and influencing senior management and others, managing departments and leading the Quality Office
3. Strategic Planning and Deployment: Developing and aligning strategies and goals, and incorporating Quality and Performance Excellence into strategic planning and deployment
4. Customer Relationship Management: Identifying customers and needs, utilizing customer feedback, and improving customer satisfaction and loyalty
5. Supply Chain Management: Evaluation and selection of suppliers, management and improvement of suppliers, and supplier certification and partnering

6. Quality Information System: Establish metrics, monitoring, and evaluation protocols
7. Training and Development: Skills assessment, training needs analysis, and development of personnel

As mentioned above, the Quality Director should also posses all six of the taxonomy levels.

The Quality Engineer: The Technical Experts

The Quality Engineer is a person who studies the management of quality and is an expert at the deployment of it. The Quality Engineers should have the requisite knowledge, skills, and experience to be a critical enabler, capable of carrying out the three universal processes for managing quality (quality planning, improvement, and control) within the context of a quality system. The competencies of the Quality Engineer are listed under seven categories in the Competency Matrix in the Appendix. The seven categories for the Quality Engineer's competencies are:

1. Basic concepts of quality
2. Quality systems
3. Organization-wide assessment of quality
4. Quality planning
5. Quality control
6. Quality improvement
7. Change management

The Competencies for the Quality Officer Engineer require at a bare minimum the first three taxonomy levels: Remembering, Understanding, and Applying knowledge. Most competencies however require the addition of the next two, Analyzing and Evaluating.

For specific examples of duties, desired attributes, skills, experience, and educational backgrounds required for different quality officer positions, the reader is referred to the example job profiles in the Appendix. These examples showcase the descriptions and requirements for a quality vice president, director, manager, and engineer in a multinational global enterprise.

The Performance Excellence Practitioners: The Belts and Change Agents

In addition to the Quality Director and Engineers the Quality Office of the future will include full-time practitioners: Master and Black Belts, Lean Experts, and Change Agents. These are not new to many organizations. What is new is that their role should be tied to the Quality Office to assure these positions continue to keep up with the needs of the organization.

The following is a list of perquisite attributes and capabilities that can be used to identify candidates for development into future quality office and performance excellence experts (such as Black Belts, Master Black Belts, Lean Experts, Lean Masters, and various other titles).

The candidate:

1. Has a managerial or technical specialist position
2. Has deep knowledge of business practice and understands your organization's business plan
3. Has high-level or university level math and reading skills; has basic education in data analysis and statistics
4. Has been trained in the Management of Quality Methods and Tools
5. Has a track record of superior performance, including leadership of new initiatives or change initiatives or has demonstrated they are capable of it
6. Welcomes accountability and challenges and is willing to take prudent risks
7. Has solid technical skills and knowledge in target environment
8. Makes decisions based on facts and data and searches for "best practices"
9. Has perseverance, stability, and is creative yet pragmatic
10. Good coach, mentor, and teacher; ideally has had experience leading teams
11. Has credibility at all levels of the organization
12. Has experience using common technology and software; not averse to learning new tools

In addition to the skills and experience prerequisites, there are a number of characteristics and traits that will naturally enhance the candidate's ability to drive change. It is beneficial if the candidate:

1. Is a clear communicator
2. Manages stress effectively
3. Learns new concepts quickly
4. Has a "can-do" attitude and the ability to manage multiple assignments
5. Is goal driven and plans ahead
6. Is solution oriented
7. Able to work effectively with all levels of the organization
8. Stays on top of key aspects and is timely and efficient

Professional Certification for Key Quality Officers

The introduction of six sigma in the past decade led to an insurgence in certification of Belts. This was largely due to a lesson learned from the total quality management era. During TQM many so called experts were trained in the 'methods of TQM.' Unfortunately few were trained in the tools to collect and analyze data. As a result, numerous organizations did not benefit from the TQM programs. Motorola introduced a core curriculum that all six sigma practitioners needed to learn. That evolved into a certification program that went beyond the borders of Motorola. As a result there are many 'certifiers' that will provide a certification as a Master Black, Black Belt, Green Belt and so on. Most certifications state that the person certified is an 'expert' in the skills of six sigma or lean or both. Certification did lead to improved performance but also to some weak experts due to no oversight of the certifiers, many of which were consulting companies or universities, not well versed in the methods or tools of six sigma and lean.

The American Society for Quality (ASQ) has, for many years, offered certification for quality technicians, quality auditors, quality engineers, and quality managers. As the six sigma movement grew, the ASQ and its affiliates around the world began to certify Black Belts. Although not perfect, the ASQ is in a better position to monitor the certifications than self-serving firms. Certification must be based on legitimacy to be effective. Having too many firms certifying Belts will only lead to a weaker certification process.

ASQ's Certified Quality Engineer (CQE) program is for people that want to understand the principles of product and service quality evaluation and control (ASQ, 2009). For a detailed list of the CQE body of knowledge, the reader is referred to the certification requirements for Certified Quality Engineer at www.asq.org.

ASQ also offers a certification for quality officers at the quality management level, called Certified Manager of Quality/Organizational Excellence. ASQ views the Certified Manager of Quality/Organizational Excellence as "a professional who leads and champions process-improvement initiatives—everywhere from small businesses to multinational corporations—that can have regional or global focus in a variety of service and industrial settings. A Certified Manager of Quality/Organizational Excellence facilitates and leads team efforts to establish and monitor customer/supplier relations, supports strategic planning and deployment initiatives, and helps develop measurement systems to determine organizational improvement. The Certified Manager of Quality/Organizational Excellence should be able to motivate and evaluate staff, manage projects and human resources, analyze financial situations, determine and evaluate risk, and employ knowledge management tools and techniques in resolving organizational challenges" (ASQ, 2009).

Note: No matter what organization you use to certify your experts, here are some lessons learned about certification.

1. One project is not enough to make someone an expert.
2. Passing a written test that is not proctored is no guarantee the person that is supposed to be taking the test is actually taking it.
3. Getting your organization to sign-off on the success of the Belt project is no guarantee unless someone in the organization is knowledgeable about the methods of six sigma.
4. Select a reputable certifying body.

Responsibilities for Development of Capable Experts

The training and development of quality and performance excellence experts can succeed only if there is accountability and responsibility for its implementation and effectiveness. This accountability and responsibility lies with the same group that it does in any other key competitive or developmental strategy—with the leadership team. It is their responsibility to agree on the strategy and assure that it will support the other operational, cultural, and financial corporate strategies. They are not responsible for the planning, design, and execution of the development strategy; this responsibility generally lies with a component of the human resource function, with technical support provided by the Quality Office. The responsible parties are executive leadership, human resources, and the quality office.

Executive leadership: The executive team bears the responsibility for creating a quality and performance excellence culture in the enterprise. A quality and performance excellence culture is a product of behaviors, skills, tools, and methods as they are applied to the work. These changes don't come about without showing people "how" to implement and sustain this culture. Therefore the executive team must become educated in quality and performance excellence, and stimulate their professional development team to offer options for training and development for quality and performance excellence. On the basis of these options, the executive team will then develop and approve a strategy and strategic goals for the training and development effort. This effort may be enterprise-wide and long term (3 to 5 years), or very narrowly focused on a particular segment of the organization or product/service line, and planned for a relatively short duration.

Human resources: The human resources (HR) function (or sub-function) bears the responsibility for implementing the quality and performance excellence training and development strategy. The implementation activities include the selection of subject matter, training design and delivery, and establishing an evaluation process. This is integrated with other corporate training and development activities, and follows the same implementation process. The subject matter may be internally sourced, or may be outsourced to external quality and performance excellence training providers. The major difference between how this is approached now compared to the past is that there is a strong trend to seamlessly integrate the quality and performance excellence training into the professional development curriculum and to include a high degree of customization to reflect the organization's culture. This is especially true for organizations that have a mature quality and performance excellence system in place.

The Quality Office: The Quality Office bears the responsibility to collaborate with the HR professionals to share their technical expertise on quality and performance excellence, much the same as key sales professionals would share their expertise in identifying and developing the curriculum for sales training. This is a departure from the past, when organizations had elaborate (and sometimes very large) Quality departments that identified, developed, and delivered quality and performance excellence training, separate from the training department. This created barriers in the implementation of performance excellence as an integral part of all activities (big Q) and contributed to the "quality versus real work" dilemma of the late 1980s and early 1990s.

An underlying principle of quality and performance excellence is to have an unwavering focus on the customer. Training and development for quality and performance excellence demands the same. A clear understanding of who the customers are, what their needs are, and what the features should be of a training and development strategy, the subsequent subject matter that responds to those needs are critical components.

A clear understanding of the customer means that all of those who will participate or benefit from the training must be considered in the design and delivery. Responsive organizations carefully approach this identification of customers and their objectives, and communicate how the training can help achieve those objectives. Many times, for lack of such clear definition, organizations waste huge amounts of time and money providing training on tools and techniques that they will never use. For example, providing training on advanced statistical tools to champions or team members of lean six sigma projects is wasteful. It was not uncommon in the past for organizations to measure success in quality and performance excellence in terms of the number of individuals they trained and the number of subjects in which they were trained!

The Quality Office

Traditionally, the Quality Office is given the responsibility to ensure that the functions required for establishing and producing quality products and services meet the customers' specified requirements. This Office is called by various names however, it is most commonly known as the Quality or Quality Assurance department. In regulated industries such as the pharmaceutical or the medical device industry, it is often called QA/RA, short for Quality Assurance/Regulatory Assurance. Within defense industries it may be known as Mission Assurance or Quality and Mission Assurance. In other industries, it might be called Quality and Safety Compliance. In this chapter, the traditionally limited role of the Quality Office as mentioned above is challenged. The modern definition of the Quality Office is expanded and to a business level and is called the Performance Excellence Office, discussed below.

Quality functions are the actions or activities which are carried out on a daily basis according to the three universal processes of the Juran Trilogy—quality planning, improvement, and control.

Quality planning activities include joint supplier planning, designing or re-designing processes, new development of products or services, design reviews, toll-gate reviews, and quality plans.

Quality improvement activities include problem-solving, root-cause analysis, and projects to remove waste or improve process capability.

Quality control activities include implementing quality standards, carrying our source inspection, testing, in-process inspection, final inspection, and audits.

These functions may or may not be performed by Quality personnel alone and usually require the participation and input of employees (the community of practitioners) throughout an organization.

ISO 9000 identifies eight quality management principles that sum up what can be used by the Quality Office in order to lead the organization toward performance excellence.

1. Customer focus: Organizations depend on their customers and therefore should understand current and future customer needs, should meet their customer requirements and strive to exceed customer expectations.
2. Leadership: Leaders establish unity of purpose and direction of the organization. They should create and maintain the internal environment in which people can become fully involved in achieving the organization's objectives.
3. Involvement of people: People at all levels are the essence of an organization and their full involvement enables their abilities to be used for the organization's benefit.
4. Process approach: A desired result is achieved more efficiently when activities and related resources are managed as a process.

5. System approach to management: Identifying, understanding and managing interrelated processes as a system contributes to the organization's effectiveness and efficiency in achieving its objectives.
6. Continual improvement: Continual improvement of the organization's overall performance should be a permanent objective of the organization.
7. Factual approach to decision making: Effective decisions are based on the analysis of data and information.
8. Mutually beneficial supplier relationships: An organization and its suppliers are interdependent and a mutually beneficial relationship enhances the ability of both to create value.

Implement the Quality System (little q)

A formal definition of a Quality System can be found in the ISO 9000 series of standards. ISO 9000 defines a Quality Management System as a "management system to direct and control an organization with regard to quality." Since quality is an organization-wide function, the Quality System is therefore organization-wide. While the Quality and Performance Excellence Office plays a major role, the Quality System is much larger in scope and therefore may be 'directed and controlled' from multiple offices.

Next we examine the roles of various constituencies in an overall quality effort. The key role is that of upper management.

→ 8.4 ROLE OF UPPER MANAGEMENT

A company president said this: "My people have disappointed me. I clearly told them that quality is our first priority; I provided training; now two years later there is little evidence of improvement." He missed the point of active leadership. Of all the ingredients for successfully achieving quality superiority, one stands out: active leadership by upper management. Commitment to quality is assumed but is not enough.

Certain roles can be identified:

- Establish and serve on a quality council (discussed in a later section).
- Establish quality strategies (see Chapter 2).
- Establish, align, and deploy quality goals (see Chapter 2).
- Provide the resources.
- Provide training in quality methodology (see Chapters 4, 5, and 6).
- Serve on upper management quality improvement teams that address chronic problems of an upper management nature (see Chapter 9).
- Review progress and stimulate improvement (see Chapter 9).
- Provide for reward and recognition (see Chapter 9).

In brief, upper management develops the strategies for quality and ensures their implementation through personal leadership.

One effective example is the action taken by the head of a manufacturing division. He personally chairs an annual meeting at which improvement projects are proposed and discussed. By the end of the meeting, a list of approved projects for the coming year is finalized, and responsibility and resources are assigned for each project.

Unfortunately, there is a price to be paid for this active leadership. The price is time. Upper managers will need to spend at least 10% of their time on quality activities—with other managers, with front-line employees, with suppliers, and with customers.

Providing Resources for Quality Activities

The modern approach to quality requires an investment of time and resources throughout the entire organization—for many people, the price is about 10% of their time. In the long run, this investment yields time savings that then become available for quality activities or other activities; in the short run, the investment of resources can be a problem.

Upper management has the key role in providing resources for quality activities. One alternative is to add resources, but in highly competitive times, this approach may not be feasible. Often, time and resources can be found only by changing the priorities of both line and staff work units. Thus some work must be eliminated or delayed to make personnel available for quality activities.

People who are assigned to quality teams should be aware of the amount of time that will be required while they are on a team. If time is a problem, they should be encouraged to propose changes in their other priorities (before the team activity starts). Resources for project teams will be made available only if the pilot teams demonstrate benefits by achieving tangible results—thus the importance of nurturing those pilot teams to yield results. As a track record builds for successful teams, the resource issue becomes less of a problem. One outcome can be for the annual budgeting process of an organization to include a list of proposed projects and necessary resources on the agenda for discussion.

In a dramatic action to provide resources for a comprehensive quality effort, upper management at a modem manufacturer assigned 3% of the total workforce (of 3000 people) to work full time on quality. The team included six upper level managers who were responsible for designing and overseeing the effort. These managers returned to their original jobs after one year.

Quality Council

A *quality council* (sometimes called a "leadership team") is a group of upper managers who develop the quality strategy and guide and support the implementation. Councils may be established at several levels—corporate, division, business unit. When multiple councils are established, they are usually linked together, i.e., members of high-level councils serve as chairpersons of lower level councils. For any level, membership consists of upper managers—both line and staff. The chairperson is the manager who has overall responsibility and authority for that level, e.g., the president for the corporate council and the division and site managers for their levels. One member of the council is the quality director whose role in the company is discussed later.

Each council should prepare a charter that includes responsibilities such as

- Formulating quality strategies and policies.
- Estimating major dimensions of the quality issue.
- Establishing an infrastructure for selecting quality projects and assigning project team leaders and members.
- Providing resources including support for teams.
- Planning for training for all levels.
- Establishing strategic measures of progress.
- Reviewing progress and remove any obstacles to improvement.
- Providing for public recognition of teams.
- Revising the reward system to reflect progress in quality improvement.

At Kelly Services, Inc. (McCain, 1995), the quality council identified its key responsibilities as

- Setting and deploying vision, mission, shared values, quality policy, and quality goals.
- Reviewing progress against goals.
- Integrating quality goals into business plans and performance management plans.

Upper managers often ask, Isn't the quality council identical in membership to the regular top-management team? Usually, yes. If so, instead of having a separate council, why not add quality issues to the agenda of the periodic upper management meetings? Eventually (when quality has become "a way of life"), the two can be combined, *but not at the start.* The seriousness and complexity of quality issues require a focus that is best achieved by meetings that address quality alone.

As a council works on its various activities, it often designates one or more full-time staff people to assist the council by preparing draft recommendations for council review. In another approach, several council members serve on ad hoc task forces to investigate various issues for the council.

In addition, individual council members often serve as advocates ("champions") for key quality-related projects. In this role, they continually monitor the project, use their executive position to remove obstacles to completing the project, and recognize project teams for their efforts.

Providing evidence of upper management leadership is clearly important in establishing a positive quality culture. For elaboration, see Section 9.6.

Finally, note that leadership is the first of the Baldrige criteria. The leadership criterion covers organizational leadership, public responsibility, and citizenship. Examination of the complete criteria reveals that leadership should be linked to other Baldrige criteria such as strategic planning, customer and market focus, information and analysis, and human resources.

❖ 8.5 ROLE OF THE QUALITY DIRECTOR

The quality director of the future is likely to have two primary roles—administering the quality department and assisting upper management with strategic quality management (Gryna, 1993).

The Quality Department of the Future

What will the future role of a quality department be? The major activities are indicated in Table 8.1. The table indicates some traditional activities of the quality department and some important departures from the current norm.

Note, for example, "transferring activities to line departments." Recent decades have shown that the best way by far to *implement* quality methodologies is through line organizations rather than through a staff quality department. (Isn't it a shame that it took us so long to understand this point?) A few organizations have stressed this approach for several years, and some have been eminently successful in transferring many quality activities to the line organization.

To achieve success in such a transfer, the line departments must clearly and fully understand the activities for which they are responsible. In addition, the line departments must be trained to execute these newly acquired responsibilities.

Examples of such transfer include shifting the sentinel (sorting) type inspection activity from a quality department to the workforce itself, transferring reliability engineering work from a quality department to the design engineering department, and transferring supplier quality activities from a quality department to the purchasing department.

A key role for the quality department is to help its internal customers achieve their quality objectives. In most organizations, the key internal customer is the operations department (called "operations" in the service sector and "manufacturing" or "production" in the manufacturing sector). Clearly, quality and operations must be partners, not adversaries. As a quality director, the author each year had to obtain 50% of his departmental budget from operations departments. This experience vividly causes a manager to become customer oriented. If a quality director truly wants to assist internal customers, a constructive step is to conduct internal "market research" on these customers. Bourquin (1995) describes how a unit of AT&T conducted a survey of internal customers (mainly manufacturing but also marketing, design, and other departments). The survey contained 10 questions reflecting both the relative

**TABLE 8.1
Functions of the quality department of the future**

Companywide quality planning

Setting up quality measurement at all levels

Auditing outgoing quality

Auditing quality practices

Coordinating and assisting with quality projects

Participating in supplier partnerships

Training for quality

Consulting for quality

Developing new quality methodologies

Transferring activities to line departments

importance of certain services and the degree of satisfaction with the services. The survey required little time or money but provided valuable quantitative feedback from the main internal customers.

The author believes that there will always be a need for a quality department to provide an independent evaluation of product quality and to furnish services to internal customers (see Table 8.1). Changing business conditions, however, make it essential to review the role of the department periodically and make appropriate changes. These changing business conditions include company mergers, changes in customer (external and internal) expectations, outsourcing, the information explosion, new communication technologies, the impact of global business and cultures, and the maturity of the quality effort within the organization. In addition, the emphasis on continuous improvement within organizations has resulted in internal competitors to a quality department for coordinating improvement activities. Such competitors include industrial engineering, human resources, financial audit, and information technology.

Assisting Upper Management with Strategic Quality Management (SQM)

A wonderful opportunity exists for a quality director to assist upper management to plan and execute the many activities of SQM. Some of these activities are shown in Table 8.2. (For elaboration, see Chapter 2.) Can a quality director achieve such an exalted role? It is useful to cite an analogy to the area of finance (see Table 8.3). Many organizations today have a CFO. This officer is concerned with broad financial planning, addressing questions such as, Where should our company be going with respect to finance? Other financial managers direct and manage detailed financial processes, such as accounts payable, accounts receivable, cash management, acquisitions, and budgeting. These roles are also vital in the organization but different from the broader role of the CFO. Finally, of course, line managers throughout the entire organization have specific activities that help meet the financial objectives of the company. Note that the quality function comprises both technical quality activities and strategic quality management activities. Provision must be made to staff a quality department with skills in these two areas. Larger organizations may benefit from having both a

TABLE 8.2
Assisting upper management with strategic quality management

Assessing quality

Formulating goals and policies

Developing quality strategies to increase sales revenue and reduce internal costs

Delegating organizational responsibilities for quality

Carrying out reward and recognition

Reviewing progress

Determining personal roles for upper management

Acting as facilitator to the quality council

Integrating quality during strategic business planning cycle

**TABLE 8.3
A contrast of roles**

Finance (today)	Quality (the future)
Chief financial officer	Quality director
Other financial managers	Other quality managers
Line managers	Line managers

quality engineering function and a quality management function (just as accounting and financial planning are separate functions). These functions would then report to the quality director.

The quality director of the future could act as the right hand of upper management on quality in the same way that the chief financial officer acts as the right hand of upper management on finance. (Note that other essential quality activities such as inspection, audit, and quality measurement must be directed and managed by other quality managers.)

Some of the scars of experience that have been accumulated within financial circles apply also to the quality function. Years ago, there was no such role as chief financial officer. The detailed financial processes were handled by one or several managers. As time went on, the need for a person with a broader financial viewpoint became apparent. In some companies, the person who was then the "controller" was promoted to become the chief financial officer. Other companies felt that the controller did not have the breadth of vision and scope to assist upper management in broad financial planning, even though the controller was excellent in administering some of the detailed financial processes. Thus not every controller became a chief financial officer.

A similar issue arises in regard to the broad scope necessary for the quality director of the future. Is the present quality director prepared, or is that person willing to become prepared, for the broader business role with respect to quality? Quality directors who wish to assume this broader role in the future need to learn from the lessons of the financial controllers.

Table 8.4 shows some ingredients for success as a quality director of the type indicated here.

Clearly, this list goes far beyond the current scope of quality directors in many companies. In larger companies, quality managers, including those who will manage some of the technical quality activities, will be needed at various levels. A quality director is likely to be involved in administering technical activities but will increasingly be called on to assist upper management as upper management—not the quality director—*leads* the company on quality. For elaboration of the future role of the quality department, see Gryna (1993).

Cinelli and Schein (1994) present the results of a survey on the profile of the U.S. senior quality executive. The survey examines the professional background, day-to-day activities, and responsibilities of quality executives at 223 *Fortune* 500 firms. For the results of a survey of 38 U.K. companies on organizing for quality, see Groocock (1994).

TABLE 8.4
Ingredients for success as a quality director

Focus on customer orientation and customer advocacy
Ability to build collaborative relationships across functions
Strong oral and written communication skills to encourage sharing of information
Goal orientation
Ability to analyze complex issues and generate innovative solutions
Initiative, persistence, and self-confidence in gaining acceptance of new ideas
Ability to organize activities
Ability to provide for self-development of subordinates

"Ingredients for success as a quality director" adapted from G. Watson, "The Emancipation of Quality: Building Bridges and Closing Gaps" as appeared in *Quality Progress,* August, 1998, Vol. 31, No. 8. Reprinted by permission of Gregory H. Watson.

8.6 ROLE OF MANAGEMENT

Middle managers, supervisors, professional specialists, and the workforce are the people who execute the quality strategy developed by upper management.

The roles of middle managers, supervisors, and specialists include

- Nominating quality problems for solutions.
- Serving as leaders of various types of quality teams.
- Serving as members of quality teams.
- Serving on task forces to assist the quality council in developing elements of the quality strategy.
- Leading the quality activities within their own area by demonstrating a personal commitment and encouraging their employees.
- Identifying customers and suppliers and meeting with them to discover and address their needs.

Increasingly, middle managers are asked to serve as team leaders as a continuing part of their job. For many of them, their roles as team leaders require special managerial skills. For a manager of a department directing the people in that department, a traditional hierarchical approach is common. The leader of a cross-functional quality improvement team faces several challenges, e.g., the leader usually has no hierarchical authority over anyone on the team because members come from various departments, serve part time, and have priorities in their home departments. The success of a team leader depends on technical competence, ability to get people to work together as a team, and a personal sense of responsibility for arriving at a solution to the assigned problem. Leading a team calls for quite an array of talents and willingness to assume responsibility. For some middle managers, the required change in managerial style is too much of a burden; for others, the role presents an opportunity. Dietch et al. (1989) identified and studied 15 characteristics of team leaders at the

Southern California Edison Company. The research concluded that team leaders (as compared to team members) have a higher tolerance for handling setbacks, believe that they have great influence over what happens to them, exhibit a greater tolerance for ambiguity, are more flexible, and are more curious (see Section 9.6, "Provide Evidence of Management Leadership").

8.7 ROLE OF THE WORKFORCE

By workforce, we mean all employees except those in management and professional specialists.

Recall that most quality problems are management or system controllable. Therefore management must (1) direct the steps necessary to identify and remove the causes of quality problems (see Chapter 3) and (2) provide the system that places workers in a state of self-control (see Section 6.3, "The Ultimate Inspector: Self-Managed Processes"). Inputs from and cooperation by the workforce are essential. The roles of the workforce include

- Nominating quality problems for solution
- Serving as members of various types of quality teams
- Identifying elements of their own jobs that do not meet the three criteria of self-control
- Becoming knowledgeable as to the needs of their customers (internal and external)

At last we are starting to tap the potential of the workforce by using its experience, training, and knowledge. One plant manager says this: "No one knows a workplace and a radius of 20 feet around it better than the worker." Quality goals cannot be achieved unless we use the hands and the heads of the workforce. Period. Some of the workforce roles on teams are discussed next.

8.8 ROLE OF TEAMS

The "organization of the future" will be influenced by the interaction of two systems that are present in all organizations: the technical system (equipment, procedures, etc.) and the social system (people, roles, etc.); thus the name "sociotechnical systems" (STS).

Much of the research on sociotechnical systems has concentrated on designing new ways of organizing work, particularly at the workforce level. Team concepts play an important role in these new approaches. Some organizations now report that, within a given year, 40% of their people participate on a team; some organizations have a goal of 80%. A summary of the most common types of quality teams is given in Table 8.5. Quality project teams are discussed in Section 4.7, "Define Phase"; business management quality teams are described in Chapter 7; quality project teams, workforce teams, and self-managing teams are discussed in this chapter.

Aubrey and D. Gryna (1991) describe the experiences of more than 1000 quality teams during four years at 75 banking affiliates of Banc One. This effort yielded significant results: $18 million in cost savings and revenue enhancement; a 10 to 15% improvement

TABLE 8.5
Summary of types of quality teams

	Improvement Quality project team	Workforce team	Business process quality team	Self-directed team
Purpose	Solve cross-functional quality problems	Solve problems within a department	Plan, control, and improve the quality of a key cross-functional process	Plan, execute, and control work to achieve a defined output
Membership	Combination of managers, professionals, and workforce from multiple departments	Primarily workforce from one department	Primarily managers and professionals from multiple departments	Primarily workforce from one work area
Basis and size of membership	Mandatory; 4–8 members	Voluntary; 6–12 members	Mandatory; 4–6 members	Mandatory; all members in the work area (6–18)
Continuity	Team disbands after project is completed	Team remains intact, project after project	Permanent	Permanent
Other names	Quality improvement team	Employee involvement group	Business process management team; process team	Self-supervising team; semiautonomous team

in customer satisfaction; and a 5 to 10% reduction in costs, defects, and lost customers. On some teams, membership was assigned; on other teams, membership was voluntary. A summary of some of the organizational results is given in Table 8.6. The focus is to improve customer satisfaction, reduce cost and increase revenue, and improve communication between front-line employees and management.

TABLE 8.6
Observations on organization of quality teams at a bank

Feature	Results of research
Team size	Average of 7 members, with a range of 2–11
Project selection	75% by management, 15% by quality council, 10% by individual team
Average savings related to project selection	Projects selected by management or the quality council achieved savings about twice as high as projects selected by the team
Project duration	Average of 3 months; 24 worker-hours per team member (excluding time spent outside team meetings)
Factors to maximize team success	Ideal team size of 4–5 employees; 75% officer/staff level, 25% nonexempt employees; members elected by management; project selected by management or quality council; project duration of 3–4 months with weekly 90-minute team meetings

Next we examine three types of teams: quality project teams, workforce teams, and self-directed teams (see Figure 8.1 for reference).

8.9 TRAINING AND CERTIFICATION OF PERSONNEL

The principles of selection, training, and retention of personnel are known but are not always practiced with sufficient intensity in many functions.

Selection of Personnel

Norrell, a human resource company, provides client companies with traditional temporary help, managed staffing, and outsourcing services. A survey of more than 1000 of their clients clarified the client definition of *quality* as "excellence of personnel." *Excellence* was defined in terms of eight criteria for clerical positions and nine criteria for technical-industrial positions. These criteria are used in selecting and training personnel assigned to the client companies. In another example based on a survey of five service organizations and nine manufacturers, Jeffrey (1995) identified 15 competencies that these organizations and their customers viewed as important in customer service activities by front-line employees. Many of these competencies apply to both front-line personnel and back-office personnel.

A human resources firm that provides temporary personnel to clients performed research to select a series of questions for prospective employees. A battery of questions was developed and given to all employees. Employees who are rated superior by clients (based on market research studies) answer certain questions differently from employees who are not superior. (Other questions result in the same response from most employees.) Responses to these "differentiating questions" help to select new employees.

Personality is one important attribute for many (but not all) positions in the operations function. This attribute is increasingly important as organizing by teams becomes more prevalent. A chemical manufacturer even places job applicants in a team problem-solving situation as part of the selection process. Larry Silver of Raymond James Financial Services Inc. states the case well: "We need to recruit people who play well in the sandbox with others, i.e., don't throw sand in people's face and do work together to build castles."

One tool for evaluating personality types is the Myers-Briggs Type Indicator. This personality test describes 16 personality types that are based on four preference scales: extrovert or introvert, sensing or intuition, thinking or feeling, and judgment or perception. Thus one personality type is an extrovert, sensing, thinking, judgment person. Analyzing responses to test questions from prospective or current employees helps to determine the personality types of individuals. Organizations need many personality types, and the Myers-Briggs approach describes the contributions of each of the 16 types to the organization. By understanding the types and making job assignments accordingly, an organization can take advantage of all personality types to achieve high performance in the workplace. McDermott (1994) explains the 16 types and how the tool can help in recruiting new personnel and assigning current personnel. Anderson and Anderson (1997) describe a candidate assessment tool that uses four areas of abilities, three areas of motivation and interests, and 13 areas of personality.

Training

One essential ingredient of a broad quality program is extensive training. Table 8.7 identifies constituencies and subject matter.

Experience in training has identified the reasons that some training programs fail:

- *Failure to provide training when it will be used.* In too many cases, training is given to employees who have little or no opportunity to use it until many months later (if ever). A much better approach schedules training for each group when it is needed—"just in time" training.
- *Lack of participation by line managers in designing training.* Without this participation, training is often technique oriented rather than problem and results oriented.
- *Reliance on the lecture method of training.* Particularly in the industrial world, training must be highly interactive, i.e., it must enable a trainee to *apply* the concepts during the training process.

TABLE 8.7
Who needs to be trained in what

Subject matter	Top management	Quality managers	Other middle managers	Specialists	Facilitators	Workforce
Quality awareness	X		X	X	X	X
Basic concepts	X	X	X	X	X	X
Strategic quality management	X	X				
Personal roles	X	X	X	X	X	X
Three quality processes	X	X	X	X	X	
Problem-solving methods		X	X	X	X	X
Basic statistics	X	X	X	X	X	X
Advanced statistics		X		X		
Quality in functional areas		X	X	X		
Motivation for quality	X	X	X		X	

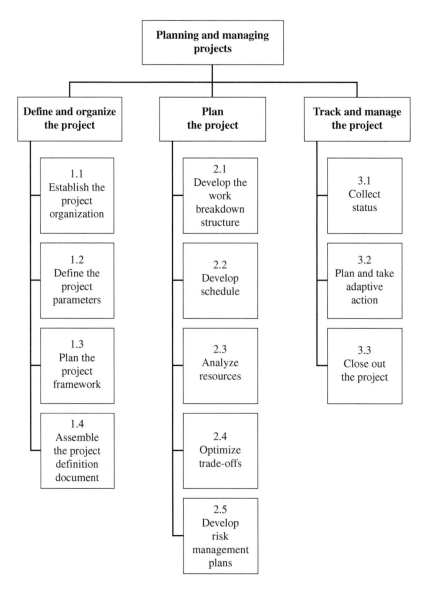

FIGURE 8.1
Project management process model. From *Project Management Manual.* Boston, MA, 1996. Copyright © 1996 by the Harvard Business School Publishing Corporation; all rights reserved. Reprinted by permission of Harvard Business School Press.

- *Poor communication during training.* The technology of quality, particularly statistical methodology, can be mystifying to some people. Many benefits are possible if we emphasize simple language and graphical techniques.

Training programs are a failure if they do not result in a change in behavior. Applying these lessons can help to prevent such failures.

Retention

Investing increased resources in selection and training helps to retain skilled employees. Compensation, of course, is also an essential contributor to employee retention. Other factors, however, are also essential, including

1. Providing for career planning and development.
2. Designing jobs for self-control (see Chapters 13 and 14).
3. Providing sufficient empowerment and other means for personnel to excel.
4. Removing sources of job stress and burnout.
5. Providing continuous coaching for personnel.
6. Providing for participation in departmental planning.
7. Providing the opportunity to interact with customers (both external and internal).
8. Providing various forms of reward and recognition.

Performance appraisal has become a subject of intense debate (and confusion). The author believes that, when properly conducted, performance appraisal is a useful tool. For further discussion, see Section 9.7.

Retention of key employees depends on many issues. An information technology services firm conducted a survey of more than 200 employees to determine their reasons for leaving—with some surprising results. In rank order (based on frequency and intensity of their feeling), employees gave the following reasons for leaving: communicate with me more, train my boss, train me more, help me set goals and give me feedback, provide a higher salary, and provide better benefits. Many of these issues are addressed in Chapter 9, "Creating a Quality Culture."

Morgan and Smith (1996) is a useful reference for selection, training, and retention of personnel.

SUMMARY

- Coordination of quality activities throughout an organization requires two efforts: coordination for control and coordination for creating change.
- Coordination for control is often the focus of a quality department; coordination for creating change often involves "parallel organizations" such as a quality council and quality project teams.
- New forms of organization aim to remove the barriers, or walls, between functional departments.
- To achieve quality excellence, upper management must lead the quality effort. The roles in this leadership can be identified.
- A quality council is a group of upper managers who develop quality strategy and guide and support its implementation.
- Middle management executes the quality strategy through a variety of roles.
- Inputs from the workforce are essential to identify the causes of quality problems and to design work systems for self-control.

- Quality teams create change. Four important types of teams are quality project teams, workforce teams, process teams, and self-managing teams.
- The implementation of the quality strategy must occur through the line organization rather than through a staff quality department.
- The quality director of the future will have two roles: administering the quality department and assisting upper management with strategic quality management.

PROBLEMS

8.1. Conduct a research study and report on how any of the following institutions was organized for quality:
  (a) The Florentine Arte Della Lana (wool guild) of the 12th, 13th, and 14th centuries.
  (b) Venetian shipbuilding in the 14th century.
  (c) Construction of cathedrals in medieval Europe.
  (d) The Gobelin tapestry industry in the 16th and 17th centuries.

8.2. Study the plan of organization for quality of any of the following institutions and report your conclusions:
  (a) A hospital.
  (b) A university.
  (c) A chain supermarket.
  (d) A chain of motels.
  (e) A restaurant.
  (f) A manufacturing company.

8.3. For any organization, make a list of the activities assigned to the quality department. Compare your list with Table 8.1 and explain why the two lists might be different.

8.4. For any organization, study the methods in use to attain
  (a) Coordination for control.
  (b) Coordination for change.
  Report your findings.

8.5. Review the three elements of self-control in Chapter 6. Explain how each of these elements can be helpful in setting up a self-managing team.

8.6. An often-asked question is, Who is responsible for quality? As worded, the question is too vague. The question must be restated in terms of actions and decisions concerning quality. Choose one functional area of an organization. For this area, list some key actions or decisions in the rows of a table. Then set up a few columns showing organizational units that are candidates for responsibility. Fill in the table to show which unit should be responsible for each action or decision.

REFERENCES

Anderson, D. N. and T. Anderson (1997). "Finding the Right People with the Right Tools," *IIE Solutions Conference Proceedings,* Institute of Industrial Engineers, Norcross, GA.

Aubrey, C. A. II and D. S. Gryna (1991). "Revolution through Effective Improvement Projects," *ASQ Quality Congress Transactions,* Milwaukee, pp. 8–13.

Becker, S., W. A. J. Golomski, and D. C. Lory (1994). "TQM and Organization of the Firm," *Quality Management Journal*, January, pp. 18–24.

Bourquin, C. R. (1995). "A Quality Department Surveys Its Customers (or, Shoes for the Cobbler's Children)," *Annual Quality Congress Proceedings*, ASQ, Milwaukee, pp. 970–976.

Christie, P. M. J. and R. R. Levary (1998). "Virtual Corporations: Recipe for Success," *Industrial Management*, July–August, pp. 7–11.

Cinelli, D. L. and L. Schein (1994). *A Profile of the U.S. Senior Quality Executive*, The Conference Board, New York.

Dietch, R., S. Tashian, and H. Green (1989). "Leadership Characteristics and Culture Change: An Exploratory Research Study," *Impro Conference Proceedings*, Juran Institute, Inc., Wilton, CT, pp. 3C-21 to 3C-29.

Groocock, J. M. (1994). "Organizing for Quality—Including a Study of Corporate-Level Quality Management in Large U.K.-Owned Companies," *Quality Management Journal*, January, pp. 25–35.

Gryna, F. M. (1960). "Total Quality Control through Reliability," *ASQ Quality Congress Transactions*, Milwaukee, pp. 746–754.

Gryna, F. M. (1993). "The Role of the Quality Director—Revisited," *Impro Conference Proceedings*, Juran Institute, Inc., Wilton, CT, pp. 11.12–11.14.

Jeffrey, J. R. (1995). "Preparing the Front Line," *Quality Progress*, February, pp. 79–82.

McCain, C. (1995). "Successfully Solving the Quality Puzzle in a Service Company," *Impro Conference Proceedings*, Juran Institute, Inc., Wilton, CT, pp. 6A.1-1 to 6A.1-13.

McDermott, R. (1994). "The Human Dynamics of Total Quality," *Quality Congress Transactions*, ASQ, Milwaukee, pp. 225–233.

Morgan, R. B. and J. E. Smith (1996). *Staffing the New Workplace*, ASQ Quality Press, Milwaukee, and CCH Inc., Chicago.

Watson, G. H. (1998). "The Emancipation of Quality: Building Bridges and Closing Gaps," *Quality Progress*, August, pp. 110–114.

CHAPTER 9

Creating a Quality Culture

→ 9.1 CORPORATE CULTURE

A *quality culture* is the pattern of human habits, beliefs, values, and behavior concerning quality.

Corporate culture consists of habits, beliefs, values, and behavior. Management needs to define and create the culture necessary for business success. In a classic book, Miller (1984) identifies eight "primary values" that promote employee loyalty, productivity, and innovation. His ideas reflect the thinking of many behavioral scientists and managers regarding the roles facing management. The eight values are

1. *Purpose.* Purpose is the vision stated in terms of product or service and benefit to the customer.
2. *Consensus.* Three decision-making styles—command, consultative, and consensus—should be matched to particular situations.
3. *Excellence.* Management creates an environment in which the pursuit of knowledge for improvement is pervasive.
4. *Unity.* The emphasis here is on employee participation and ownership of work.
5. *Performance.* Individual and team rewards are the focus along with performance measurements to tell individuals how they are doing.
6. *Empiricism.* Management by fact and the use of the scientific method form the basis of this value.
7. *Intimacy.* Intimacy relates to sharing ideas, feelings, and needs in an open and trusting manner without fear of punishment.
8. *Integrity.* The norm here is for managers to act as role models for ethical practices.

→ 9.2 QUALITY CULTURE

Quality culture is an integral part of corporate culture. Our discussion here concentrates on the aspects of corporate culture that relate to the quality of products and services.

Differences in quality culture can have extreme illustrations, both negative and positive. Two examples:

- *Negative quality culture ("hide the scrap" scenario).* The culture in a paint manufacturing plant put pressure on supervisors to avoid reporting any batch of paint that did not meet specifications. One supervisor resorted to hiding some paint; he directed his workers to dig a hole in the yard and bury the paint. In the service industries, reporting on the processing of transactions (e.g., insurance claims, mishandled baggage) is sometimes subject to data manipulation to yield good performance reports.
- *Positive quality culture ("climb the ladders to delight the customer" scenario).* The culture in a hotel resulted in taking an extraordinary step to please a customer—on short notice. About an hour before a seminar was to be presented, the seminar leader heard a steady, clear tinkling sound. The cause—small glass prisms in two chandeliers colliding due to air movement from the air conditioning system. He mentioned this distraction to the hotel management. Action was immediate—two work crews were assembled, ladders were set up, and every second prism was removed (about 100 in all).

Cameron and Sine (1999) discuss research on four different quality cultures and the quality tools associated with each culture. The four cultures are absence of quality emphasis, error detection, error prevention, and creative quality. Analysis of responses from managers in 68 organizations revealed that the more advanced levels of quality culture are associated with higher levels of organizational effectiveness.

An important starting point is to determine the current quality culture. This topic is discussed in Section 3.7. Watson and Gryna (1998) report on research assessing quality culture in small manufacturing and service organizations. Knowing the quality culture enables us to implement a strategy in a way that encourages people to embrace the quality strategy and make it successful.

Culture can be changed. We need to provide goals and measurements, evidence of upper management leadership, self-development and empowerment, participation, and recognition and rewards—critical success factors for achieving a positive quality culture. These paths are the means to drive changes in actions that lead to changes in attitude and finally to changes in quality culture. Above all, there must be a sense of urgency and initiative on quality throughout all levels of the organization. Instilling such a sense is easier said than done because "talk doesn't cook rice."

The paths for quality culture must be integrated with the methodologies and structure for quality (see Figure 9.1). The three elements of self-control are really prerequisites for achieving a quality culture. Thus we must provide people with the knowledge of what they are supposed to do, provide feedback on how they are doing, and provide a means of regulating a capable process. Actions to "motivate" people will not be successful unless these basics of self-control are in place. Readers are urged to review

FIGURE 9.1
Technology and culture.

Section 6.3, "The Ultimate Inspector: Self-Managed Processes." Further elaboration is given in Chapters 13 and 14.

We proceed then to examine five key drivers for developing a quality culture.

9.3 CREATING A CULTURE OF QUALITY

Employees in an organization have opinions, beliefs, traditions, and practices concerning quality. We call this set of characteristics the "organization quality culture." Gaining an understanding of this culture should be part of a company assessment of quality for two reasons: (1) The culture clearly has a major impact on quality results, and (2) knowing the present culture can identify barriers for developing a strategy and implementing an action plan based on the companywide assessment of quality.

Changing a culture is difficult and usually unsuccessful unless a comprehensive approach exists to achieve and sustain it. The Juran performance excellence change model describes five separate and unique types of breakthroughs that must occur in an organization before a quality sustainable culture is attained. Without them, an organization attains superior results, but the results may not be sustained for long

268 Quality Management and Analysis

FIGURE 9.2
Juran performance excellence change model.
(Juran Institute, 2009)

periods of time. If performance excellence is the state in which an organization attains superior results through the application of the universal quality management methods, then an organization must ensure these methods are successfully used. The journey from where your organization is to where it wants to go may require a transformational change. This change will result in the ability of the organization to sustain its performance and attain world-class status and market leadership.

The five breakthroughs are:

1. A breakthrough in leadership and management, which leads to
2. A breakthrough in organization and structure, which leads to
3. A breakthrough in current performance, which leads to
4. A breakthrough in culture, which leads to
5. A breakthrough in adaptability and sustainability (see Figure 9.2).

→ 9.4 THE PERFORMANCE EXCELLENCE CHANGE MODEL

The performance excellence change model (Figure 9.2) is based on more than 60 years of experience and research from Juran and the Juran Institute. The five breakthroughs, when complete, help produce the state of performance excellence. Each breakthrough addresses a specific organizational subsystem that must change. Each is essential for supporting organizational life; none by itself is sufficient. In effect, they all empower the operational subsystem, whose mission is to achieve technological proficiency in producing the goods, services, and information for which customers will pay or use.

Among the different breakthrough types, there is some overlap and duplication of activities and tasks. This overlap is expected because each subsystem is interrelated with all the others, and each is affected by activities in the others. The authors acknowledge that some issues in each type of breakthrough may have already been addressed by the reader's organization. All the better! If so, if you did not start your organization's performance excellence journey from the beginning, pick up the journey from where your organization finds itself at present. Closing the gaps will likely be part of your organization's next strategic business planning cycle. To close the gaps, design strategic and operational goals and projects to reach those goals and deploy to all functions and levels.

Breakthrough and Transformational Change

Breakthroughs can occur in an organization at any time, usually as the result of a specific initiative like a specific improvement project, e.g., a six sigma DMAIC improvement project, a design of a new service, or the invention of a new technology. These changes can produce sudden explosive bursts of beneficial change for an organization and society. But they may not be enough to cause the culture to change or sustain itself to the changes that occurred. This is because it may not have happened for the right reason. It was not purposeful. It came about through chance. Change by "chance" is not predictable. What an organization needs is predictable change.

Today's organizations operate in a state of perpetual, unpredictable change, which requires continuous adaptive improvements as pressure mounts for new improvements from the outside. These improvements may take months or even years to accomplish because it is the cumulative effect of many coordinated and interrelated organizational plans, policies, and breakthrough projects. Taken together, these diligent efforts gradually transform the organization.

Organizations that do not intend to change usually do when a crisis—or a fear of impending crisis—triggers a need for change within an organization. Consider the following scenario:

> Two of the largest competitors have introduced new products that are better than ours. Consequently, sales of product X and Y are heading steadily down, and taking our market share along. Our new-product introduction time is much slower than the competition, making the situation even worse. The new plant can't seem to do anything right. Some equipment is down often, and even when operating, produces too many costly defective items.
>
> Too many of our invoices are returned because of errors, with resulting postponement of revenue and a growing number of dissatisfied customers, not to mention the hassle and costs of rework. Accounts receivable have been much too high, and are gradually increasing. We are becoming afraid that the future may offer additional threats we need to ward off or, more importantly, plan for so they can be prevented altogether. Leadership must take action, or the organization is going to experience a loss of market share, customer base, and revenue.

Breakthroughs are Essential to Organizational Vitality

There are three important reasons an organization cannot survive very long without the medicinal renewing effects of continual breakthrough:

1. The costs of poorly performing processes (or COP^3) continue to increase if they are not tackled. They are too high. One obvious reason is that organizations are plagued by a continuous onslaught of crises precipitated by mysterious sources of chronic high costs of poorly performing processes. As stated previously the total chronic levels of COP^3 have been reported to be as high as 20 percent or more of the costs of goods sold. The number varies by type of industry and organization. It is not unusual for these costs at times to exceed profit or be the major contributor to losses. In any case, the average overall level is appalling (because it is substantial and avoidable), and the toll it takes on the organization can be devastating. COP^3 is a major driver of many cost-cutting initiatives, not only because it can be so destructive if left unaddressed, but also because savings realized by reducing COP^3 go directly to the bottom line. Furthermore, the savings continue, year after year, so long as the remedial improvements are irreversible, or controls are placed on reversible improvements.

    It makes good business sense that the mysterious, chronic causes of waste must be discovered, removed, and prevented from returning. Breakthrough improvement becomes a preferred initial method of attack because of its ability to uncover and remove specific root causes—and hold the gains. It is designed to do just that. We could describe breakthrough improvement methodology as the application of the scientific method to the solution of performance problems. It closely resembles the medical model of diagnosis and treatment.

2. Another reason breakthrough is required for organizational survival is the state of chronic, accelerating change found in today's business environment. Unrelenting change has become so powerful and so pervasive that no constituent part of an organization finds itself immune from its effects for long. Because any or all components of an organization can be threatened by changes in the environment, if an organization wishes to survive, it likely will be forced into creating basic changes that are powerful enough to bring about accommodation with new conditions. Performance breakthrough, consisting as it does of several specific types of breakthrough in various organization functions, is a powerful approach that is capable of determining countermeasures sufficiently effective to prevail against the inexorable forces of change. An organization may have to re-invent itself. It may even be driven to re-examine and perhaps modify its core products, business, service, or even customers.

3. Without continuous improvement, organizations die. An additional reason breakthrough is essential for organizational survival is found in knowledge derived from scientific research into the behavior of organizations. Leaders can learn valuable lessons about how organizations function and how to manage them by examining open systems theory. Among the more important lessons taught by open systems theory is the notion of negative entropy. Negative entropy refers to characteristics that human organizations share with biological systems such as the living cell, or

the living organism (which is a collection of cells). Entropy is the tendency of all living things—and all organizations—to head toward their own extinction. Negative entropy consists of countermeasures that living systems and social systems take in order to stave off their own extinction. Organisms replace aging cells, heal wounds, and fight disease. Organizations build up reserves of energy (backlogs and supplies) and constantly replace expended energy by acquiring more energy (sales and raw materials) from their environment. Eventually, living organisms lose the race. So do organizations if they do not continually adapt, heal "wounds" (make performance breakthrough improvements), and build up reserves of cash and goodwill. The Juran performance excellence model is a means by which organizations can stave off their own extinction.

System Thinking and Transformation Change

Organizations are like living organisms. They consist of a number of subsystems, each of which performs a vital specialized function that makes specific, unique, and essential contributions to the life of the whole. A given individual subsystem is devoted to its own specific function such as design, production, management, maintenance, sales, procurement, and adaptability. One cannot carry the biological analogy very far because living organisms separate subsystems with physical boundaries and structures (e.g., cell walls, the nervous system, the digestive system, the circulatory system, etc.). Boundaries and structure of subsystems in human organizations, on the other hand, are not physical; they are repetitive events, activities, and transactions. The repetitive patterns of activities are, in effect, the work tasks, procedures, and processes carried out by organizational functions. Open systems theorists call these patterns of activities *roles*. A role consists of one or more recurrent activities out of a total pattern of activities that, in combination, produce the organizational output.

Roles are maintained and carried out in a repetitive, relatively stable manner by means of mutually understood sets of expectations and feedback loops, shown in Figure 9.3. Open systems theory and Juran's model focus particularly on the technical

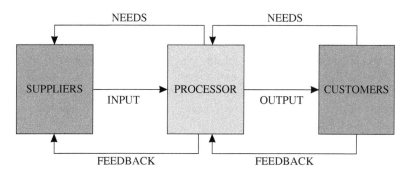

FIGURE 9.3
The triple role.
(Juran Institute, 2009)

methods, human relationships, organization structures, and interdependence of functional roles associated with these activities and transactions. A detailed knowledge of the repetitive transactions between the organization and its environment, and also within the organization itself, is essential to accomplishing breakthroughs because these transactions determine the effectiveness and efficiency of performance.

Figure 9.3 represents a model that applies equally to an organization as a whole, to individual subsystems and organizational functions (departments, work stations, etc.) within the organization, and to individual organizational members performing tasks in any function or level. All these entities perform three more or less simultaneous roles: They act as customer, processor, and supplier. As a processor, charged with the duty of transforming imported energy, they receive raw materials—goods, information, and/or services—from their suppliers, who may be located inside or outside the organization. The processor's job consists of transforming the received things into a new product of some kind—goods, information, or service. In turn, the processor supplies the product to his or her customers, who may be located within or outside the organization.

Each of these roles requires more than merely the exchange of things. Each role is linked by mutually understood expectations (i.e., specifications, work orders, and procedures) and feedback as to how well the expectations are being met (i.e., complaints, quality reports, kudos, and rewards). Note that in the diagram, the processor must communicate (shown by arrows) to the supplier a detailed description of his or her needs and requirements. In addition, the processor provides the supplier with feedback on the extent to which the expectations are being met. This feedback is part of the control loop and helps to ensure consistent adequate performance by the supplier. The customer bears the same responsibilities to his or her processors who, in effect, are also suppliers (not of the raw materials, but of the product).

When defects, delays, errors, or excessive costs occur, the causes can be found somewhere in the activities performed by suppliers, processors, and customers; in the set of transactions between them; or perhaps in gaps in the communication of needs and feedback. Breakthrough efforts must uncover the precise root causes by deep probing and exploration. If the causes are really elusive, discovering them may require placing the offending repetitive process under a microscope of unprecedented power and precision, as is done in six sigma. Performance excellence initiatives require that all functions and levels be involved, at least to some extent, because each function's performance is interrelated and dependent to some degree on all the other functions. Moreover, a change in the behavior of any one function will have some effect on all the others, even though it may not be apparent at the time. This interrelatedness of all functions has practical day-to-day implications for the leader at any level, namely, the imperative of employing "systems thinking" when making decisions, particularly decisions to make changes.

Because an organization is an open system, its life depends on (1) successful transactions with the organization's external environment and (2) proper coordination of the organization's various specialized internal functions and their outputs. The proper coordination and performance of the various internal functions is dependent on the management processes of planning, controlling, and improving and human factors such as leadership, organizational structure, and culture. To manage in an open system (such as an organization), management at all levels must think and act in systems terms. They must consider the impact of any proposed change not only upon the whole organization, but also the impact on the interrelationships of all the parts. Failure to do so, even when changing seemingly little things, can make some pretty big messes. Leaders need to reason as follows: "If there is to be a change in X, what is required (inputs) from all functions to create this change, and how will X affect each of the other functions, and the total organization as well (ultimate output/results)?"

There are three important lessons learned from the experience of the authors:

1. All organizations need a systematic approach to ensure the change happens. The problems that appear in one function or step in a process often have their origin upstream from that function or step in the process. People in a given work area cannot necessarily solve the problem in their own work area by themselves—they need to involve others in the problem solving process. Without systematic involvement of the other functions, suboptimization will occur. Suboptimization results in excess costs and internal customer dissatisfaction—the exact opposite of what is intended.
2. Change can only be created with active participation of all employees top down and over time. This includes not only those who are the source of a problem, but also those affected by the problem and those who will be the source of the changes to provide a remedy to the problem (usually those who are the source of the problem, and perhaps others).
3. Functional change alone is not sufficient to transform an organization. Breakthroughs attempted in isolation or within a silo from the whole organization and without systems thinking can easily create more problems than existed at the start of the breakthrough attempt.

Attempts to bring about substantial organizational change such as performance excellence require not only changing the behavior of individuals (as might be attempted by training), but also redefining their roles in the social system. This requires, among other things, changing the expectations that customers have for their processors and those that processors have for their suppliers. In other words, performance breakthroughs require organizational design capability in order to produce consistent coordinated behavior in support of specific organizational goals. Modifications will likely also be made to other elements that define roles such as job descriptions, job fit, work procedures, control plans, other elements of the quality system, training, etc. To achieve breakthrough, it is not sufficient simply to train a few black belts or experts and complete a few projects. Although this will probably result in some improvement, it is unlikely to produce long-term culture change and sustainability. The authors believe too many organizations are settling for simple improvements when they should be striving for breakthroughs.

As we have seen, attaining a performance excellence state consists of achieving and sustaining the beneficial changes. It is noteworthy that creating a bright idea for a change does not, by itself, make change actually happen. People must understand why the change is needed and see the impact it will have on them before they can change what they do, and perhaps how they do it. Beneficial change is often resisted, sometimes by the very persons who could benefit most from it, especially if they have been successful in doing things the current way. Leading change can be a perplexing and challenging undertaking. Accordingly, those trying to implement change should acquire know-how in how to do it.

Breakthrough in Leadership and Management

Breakthrough in leadership occurs when managers respond to two basic questions:

1. How does management set performance goals for the organization and motivate the people in the organization to reach them?
2. How do they best utilize the power of the workforce and other resources in the organization and how should they best manage them?

Issues with leadership are found at all levels, not just at the top of an organization. A breakthrough in leadership and management results in an organization characterized by unity of purpose and shared values as well as a system that enables participation of the workforce.

Each work group knows what its goals are, and specifically what performance is expected from the team and the individual. Each individual knows specifically what he or she is to contribute to the overall organizational mission and how his or her performance will be measured. Little erratic or counterproductive behavior occurs. Should it occur, or should conflict arise, guidelines to behavior and decision making are in place to enable relatively quick and smooth resolution of the problem. There are two major elements to leadership: (1) Leaders must decide and clearly communicate where they want everyone to go, and (2) leaders must entice them to follow the path. In this text, the words *leader* and *manager* do not necessarily refer to different persons. Indeed, most leaders are managers, and managers should be leaders. The distinctions are matters of intent and activities, not players. Leadership can and should be exercised by managers; leaders also need to manage. If leadership consists of influencing others in a positive manner that attracts others, it follows that those at the top of the managerial pyramid (CEOs and chairpersons) can be the most effective leaders because they possess more formal authority than anyone else in an organization. In fact, those at the top usually are the most influential leaders. If dramatic change, such as introducing lean six sigma into an organization is to be undertaken, by far the most effective approach is for the CEO to lead the charge. A lean six sigma launch is helped immensely if other leaders, such as union presidents, also lead the charge. The same can be said if senior and middle managers, first-line supervisors, and nonmanagement work crew leaders "follow the leader" and support a performance excellence program by word and deed. Leadership is not dictatorship because dictators make people afraid of behaving in "incorrect" ways and perhaps occasionally provide public treats (e.g., free gasoline, for example, as has happened in Turkmenistan); freeing prisoners; or staging public spectacles that, together with propaganda, are designed to make people follow the great leader. Dictators don't really get people to want to behave "correctly" (what the dictator says is correct); the people merely become afraid not to.

❖ 9.5 THE ROLES OF LEADERS TO ATTAIN A BREAKTHROUGH IN LEADERSHIP

Demonstrate Strategic Planning and Deployment

The first step in strategic planning is to determine the organization's mission. (What business are we in? What services do we provide?) Next, a vision for the desired future state of the organization is formulated and published (e.g., "We will become the supplier of choice, worldwide, of product X or service Y."). After proclaiming the basic reason for the organization's existence, and the overall general goal the organization seeks to achieve in the future, the upper management team generates a few key strategies the organization must implement in order to fulfill the mission and realize the vision (e.g., assure ourselves of a reliable source of high-quality raw materials; assure ourselves of a stable well-qualified work force at all levels for the foreseeable future; reduce by the end of this year our overall costs of poor quality by 50% of last year's annual cost). Now the process becomes more precise. For each key strategy, a small number of quantified strategic goals (targets) are listed, few enough to be accomplished by the resources and people available. These quantified strategic goals are further divided into goals for this year, goals for the next two years, etc. Finally, for each quantified strategic goal, a practical number of operational goals are established that describe exactly who is to do exactly what to reach each specific strategic goal. Normally, operational goals are specific projects to be accomplished (such as six sigma projects)—specific performance targets to be reached by each function or work group, etc.

Strategic deployment is a process of converting goals into specific precise actions, each action designed to realize a specific goal. Deployment occurs in two phases: one phase is during the strategic planning process; the other phase is after the strategic plan is completed. During the strategic planning process, after the management team determines key strategies, it circulates these to others in the organization: department heads, functional heads, process owners, etc. They, in turn, may circulate the strategies further out: to supervisors, working forepersons, etc. They, in turn, may circulate the strategies to everyone they supervise. Each party is asked to contribute ideas and suggestions concerning what activities could be undertaken to carry out the strategies, what the specific quantified strategic goals should be, what resources would be required, etc. These responses go back up to the upper management team, who utilize the responses to promulgate more specific strategic and operational goals. This exchange of proposed activities to reach goals may take place several times. Some call these iterations and reiterations "catch ball." With each cycle, the various goals are refined, becoming more specific, more practical, and quantified. Finally, a set of precise strategic and operational goals emerges, each with owners. In addition, metrics are devised by which to measure performance toward goals and to provide managers at all levels with a scorecard of progress. Most significantly, these goals have been established with the participation of those who will be responsible for carrying them out.

Emerging from this event is an organization united in its commitment to reaching the same goals. All functions and levels have been included. This is highly significant because leadership is not considered to be something exercised by one person at the top of an organization. It is a function performed at any level and in any function, by anyone who influences others. With a well-deployed strategic plan, specific acts of leadership (attempts to influence others) should be relatively consistent from leader to leader, function to function, and time to time. Decisions made at different levels or in different functions should not conflict with one another very often. So, at least, is the ideal.

When managers do everything they can to provide the means for everyone to be empowered or attain a state of self-control, it greatly enhances their credibility and the level of trust followers feel toward them. This happens because when we are in self-control, we have at our disposal all of the elements necessary to be successful on our job. When a leader does this, followers feel gratitude and respect toward that leader and are inclined to follow that leader because, "My leader comes through for me. My leader doesn't just talk; she delivers!"

A brief review of them follows here because they can be so instrumental in demonstrating leadership. A person is in a state of self-control, if he or she meets the following criteria:

1. Knows exactly what is expected: the standard of performance for the process; who does what and who decides what, knows how he or she is doing compared to the standards
2. Receives timely feedback to have the ability to regulate the process
3. Has a capable process that includes the necessary tools, equipment, materials, maintenance, time, and authority to adjust it when it is nonconforming

A person in a state of self-control has at his or her disposal all the means necessary to perform his or her work tasks successfully. Management must provide the means because only management controls the required resources needed to put a person in self-control. Persons who have long been suffering from lack of self-control and its associated inability, through no fault of their own, are especially grateful to a leader/manager who relieves them from the suffering by making self-control possible. They come to respect and trust such a manager and tend to become an enthusiastic follower of him or her, mindful of the good things—including enhanced self-confidence and self-esteem—that have flowed from that manager.

Perform Periodic Audits

Conducting periodic audits performed by leaders and managers is a superb method of demonstrating commitment to and support for an effort to change. Leaders and managers, especially the senior executive managers, enhance their credibility and power to lead by personally walking around the organization and talking to the people about what they do and how they do it. The management audit has both formal and informal aspects. The formal aspect consists of asking each person being audited to answer certain specific written questions and to produce data and other evidence of performance that conforms to the formal controls. The informal aspect is simply talking with the people being audited about what's on their mind and sharing with them what's on the manager's mind. The management audit is roughly the equivalent of senior generals visiting the troops in the field. It is a chance for managers to demonstrate their interest in how things are going: what's going well and what needs corrective action. It is a splendid opportunity to listen to what people have to say and to show respect for them. If the managers follow up on the suggestions and complaints they hear, that is yet another way to demonstrate that they care enough about "the troops" to provide them with needed support and assistance. It grants to anyone in the organization a direct line of communication with the top, something that makes many people feel important and motivates them to keep performing at their best. Importantly, the managers' ability to lead is reinforced.

Provide Resources to Continually Innovate and Improve

Breakthroughs are produced by teams, project by project. These teams are assigned goals to attain by means of project charters. Each project is formally chartered, in writing, by the executive council. The executive council also provides the project teams with the people and other resources the teams need to carry out their missions.

The manager's role is one of managing the organization so that high standards are met, proper behavior is rewarded (or individuals are held accountable), facilities and processes are maintained, and employees are motivated and supported. Performance toward goals is measured and tracked for all functions and levels (i.e., overall organization, function, division, department, work group, and individual). Performance metrics are regularly summarized, reported, and reviewed to compare actual performance with goals. Management routinely initiates corrective action to address poor performance or excessively slow progress toward goals. Actions may include establishing performance breakthrough improvement projects, providing additional training or support, clearing away resistance, providing needed resources, implementing disciplinary action, etc. Leaders and management must do the following:

- Create and maintain systems and procedures that ensure the best, most efficient and effective performance of an organization, in all functions and levels.
- Reward (and holds people accountable, if necessary) appropriate behavior.
- Consistently uphold and demonstrate high standards.
- Focus on stability.

Breakthroughs in Organization Structure

Creating a breakthrough in organizational structure accomplishes the following:

- Designs and puts into place the organization's operational systems (i.e., quality system, orientation of new employees, training, communication processes, supply chains, etc.).
- Designs and puts into practice a formal structure that integrates each function with all the others and sets forth relative authority levels and reporting lines (e.g., the organization chart and the means to manage across it).
- Aligns and coordinates the respective interdependent individual functions into a smooth functioning, integrated organization.

Creating a breakthrough in organization structure is a response to the basic question: "How do I set up organizational structures and processes to reap the most effective and efficient performance toward our goals?"

Trends in this area are clear. More and more work is performed by project teams. Job tasks may be described by team project descriptions rather than, or in addition to, individual job descriptions. Performance evaluation is often related to the accomplishments of one's team instead of or in addition to, one's individual accomplishments.

Management structure consists of cross-functional processes that are managed by process owners, as well as vertical functions that are managed by functional managers. Where there is both vertical and horizontal responsibility, potential conflicts are resolved by matrix mechanisms that require negotiated agreements by the function manager and the cross-functional (horizontal) process owner.

Unity and consistency in the operation of both cross-functional processes and vertical functions are essential to creating performance breakthroughs and are essential to continued organizational survival. All members of leadership teams at all levels simply must be in basic agreement as to goals, methods, priorities, style, etc. This is especially vital when attempting performance breakthrough improvement projects because the causes of so many performance problems are cross-functional, and the remedies to these problems must be designed and carried out cross-functionally. Consequently, in a lean or six sigma implementation, for example, quality or executive councils, steering committees, champions (who periodically meet as a group), cross-functional project teams, project team leaders, black belts, and master black belts are seen. All of these roles involve dealing with change and teamwork issues. There is also a steady trend toward fewer authority or administrative levels and shorter reporting lines.

> The rate of change in the business world is not going to slow down anytime soon. If anything, competition in most industries will probably speed up over the next few decades. Enterprises everywhere will be presented with even more terrible hazards and wonderful opportunities, driven by the globalization of the economy along with related technological and social trends. (John P. Kotter, *Leading Change,* 1996)

There are three accepted basic types of organization for managing any function work, and one newer, emerging approach. The most traditional and accepted organization types are functional, process, and matrix. They are important design baselines because these organizational structures have been tested and studied extensively, and their advantages and disadvantages are well known. The newer, emerging organizational designs are network organizations.

Function-Based Organization

In a *function-based organization*, departments are established based on specialized expertise. Responsibility and accountability for process and results are usually distributed piecemeal among departments. Many firms organize around functional departments having a well-defined management hierarchy. This applies both to the major functions (e.g., human resources, finance, operations, marketing, product development) and also to sections within a functional department. Organizing by function has certain advantages—clear responsibilities and efficiency of activities within a function.

A function-based organization typically develops and nurtures talent, and fosters expertise and excellence within the functions themselves.

Therefore, it offers several long-terms benefits, and so on. But this organizational form also creates "walls" between the departments. These walls—sometimes visible, sometimes invisible—often cause serious communications barriers. However, function-based organizations can result in a slow, bureaucratic decision-making apparatus, as well as the creation of functional business plans and objectives that may be inconsistent with overall strategic business unit plans and objectives. The outcome can be efficient operations within each department, but with less-than-optimal results delivered to external (and internal) customers.

Business Process Managed Organizations

Many organizations are beginning to experiment with an alternative to the function-based organization in response to today's "make it happen fast" world. Businesses are constantly redrawing their lines, work groups, departments, divisions, even entire companies, trying to enable productivity increases, cycle-time reduction, revenue enhancement, or an increase in customer satisfaction. Increasingly, organizations are being rotated 90 degrees into *process-based organizations.*

In a process organization, reporting responsibilities are associated with a process, and accountability is assigned to a process owner. In a process-based organization, each process is provided with the functionally specialized resources necessary. This has the effect of eliminating the barriers associated with the traditional function-based organization, making it easier to create cross-functional teams to manage the process on an ongoing basis.

Process-based organizations are usually accountable to the business unit or units that receive the benefits of the process under consideration. Therefore, process-based organizations are usually associated with responsiveness, efficiency, and customer focus.

However, over time, pure process-based organizations run the risk of diluting and diminishing the skill level within the various functions. Furthermore, a lack of process standardization can evolve, which can result in inefficiencies and organizational redundancies. Additionally, such organizations frequently require a matrix reporting structure, which can result in some confusion if the various business units have conflicting objectives. The matrix structure is a hybrid combination of the functional and divisional archetypes.

Merging Functional Excellence with Process Management

What is required, however, is an organization that identifies and captures the benefits of supply chain optimization in a responsive, customer-focused manner, while promoting and nurturing the expertise required to manage and improve continuously the key processes on an ongoing basis: the *matrix structure.* This organization will likely be a hybrid of the functional and process-based organizations, with the business unit accountable for objectives, priorities, and results, and the functional department accountable for process management and improvement and resource development.

The organization of the future, according to the late Dr. Frank Gryna of the Center for Quality at the University of Tampa, will be influenced by the interaction of two systems that are present in all organizations: the technical system (equipment, procedures, etc.) and the social system (people, roles, etc.)—thus the name sociotechnical systems (STSs).

Much of the research on sociotechnical systems has concentrated on designing new ways of organizing work, particularly at the workforce level. For example, supervisors are emerging as "coaches"; they teach and empower rather than assign and direct. Operators are becoming "technicians"; they perform a multiskilled job with broad decision making, rather than a narrow job with limited decision making. Team concepts play an important role in these new approaches. Some organizations now report that within a given year, 40% of their people participate on a team; some organizations have a goal of 80%. Permanent teams (e.g., process team, self-managing team, etc.) are responsible for all output parameters, including quality; ad hoc teams (e.g., a quality project team) are typically responsible for improvement in quality. The literature on organizational forms in operations and other functions is extensive and increases continuously. For a discussion of research conducted on teams, see Katzenbach and Smith (1993). Mann (1994) explains how managers in process-oriented operations will need to develop skills as coaches, developers, and "boundary managers." The attributes associated with division managers, functional managers, process managers, and network managers of customer services are summarized in Table 9.1. There is emerging evidence that divisional and functional organizations may not have the flexibility to adapt to rapidly changing marketplace or technological changes.

Design a System That Promotes Employee Engagement and Self-Management

Traditional management was based on Frederick Taylor's teachings of specialization. At the turn of the 20th century, Taylor recommended that the best way to manage

TABLE 9.1
Attributes of various roles

Attributes of Roles	Division Manager	Function Manager	Process Manager	Network Leader
Strategic orientation	Entrepreneurial	Professional	Cross-functional	Dynamic
Focus objectives	Customer adaptability	Internal efficiency	Customer effectiveness	Variable adaptability, speed
Operational responsibility	Cross-functional	Narrow, parochial	Broad, pan-organizational	Flexible
Authority	Less than responsibility	Equal to responsibility	Equal to responsibility	Ad hoc, based on leadership
Interdependence	May be high	Usually high	High	Very high
Personal style	Initiator	Reactor	Active	Proactive
Ambiguity of task	Moderate	Low	Variable	Can be high

Sources: The first two columns are adapted from the work of Financial Executive Research Foundation, Morristown, NJ. The last two columns represent the work of Edward Fuchs.

manufacturing organizations was to standardize the activity of general workers into simple, repetitive tasks and then closely supervise them (Taylor, 1947). Workers were "doers"; managers were "planners." In the first half of the 20th century, this specialized system resulted in large productivity increases and a very productive economy. As the century wore on, workers became more educated, and machinery and instruments more numerous and complicated. Many organizations realized the need for more interaction among employees. The training and experience of the work force was not being used. Experience in team systems, where employees worked together, began in the latter half of the 20th century, though team systems did not seriously catch on until the mid-1970s as pressure mounted on many organizations to improve performance. Self-directed teams began to catch on in the mid-1980s (Wellins et al., 1991). For maximum effectiveness, the work design should require a high level of employee involvement.

Empowerment and Commitment

Workers who have been working under a directive command management system, in which the boss gives orders and the worker carries them out, cannot be expected to adapt instantly to a highly participative, high-performance work system. There are too many new skills to learn; too many old habits to overcome. According to reports from numerous organizations that have employed high-performance work systems, such systems must evolve. This evolution is carefully managed, step-by-step, to prepare team members for the many new skills and behaviors required of them.

The first stage of involvement is the consultative environment, in which the manager consults the people involved, asks their opinions, discusses their opinions, then takes unilateral action. A more advanced state of involvement is the appointment of a special team or project team to work on a specific problem, such as improving the cleaning cycle on a reactor. This involvement often produces in team member's pride, commitment, and a sense of ownership.

An example of special quality teams is the blitz team from St. Joseph's Hospital in Paterson, New Jersey. Teams had been working for about a year as a part of the TQM effort there. Teams were all making substantial progress, but senior management was impatient because TQM was moving too slowly. Recognizing the need for the organization to produce quick results in the fast-paced marketplace, the team developed the "blitz team" method (from the German word for lightning). The blitz team approach accelerated the standard team problem-solving approach by adding the services of a dedicated facilitator. The facilitator reduced elapsed time in three areas: problem-solving focus, data processing, and group dynamics.

Because the facilitator was very experienced in the problem-solving process, the team asked the facilitator to use that experience to provide more guidance and direction than is normally the style on such teams. The result was that the team was more focused on results and took fewer detours than usual. In the interest of speed, the facilitator took responsibility for the processing of data between meetings, thus enabling reduction of the time elapsed between team meetings. Further, the facilitator managed the team dynamics more skillfully than might be expected of an amateur in training within the organization. The team went from first meeting to documented root causes in one week. Some remedies were designed and implemented within the next few weeks.

The team achieved the hospital's project objectives by reducing throughput delays for emergency room (ER) patients. ER patients are treated more quickly, and worker frustrations have been reduced (Niedz, 1995). Special teams can focus sharply on specific problems. The success of such a team depends on assigning to the team people capable of implementing solutions quickly.

Project Teams

Employees need time to organize work that should be accomplished by the team. Time is necessary to organize and make sure the team knows what its objective is, why they are doing it, how to organize the work, and who will be involved. However, the schedules are so short; there is never time to organize the work team.

Many teams start working, believing they know what to do, and so go into action directly. They don't have a minute to find the support from others, determine the right goals for the team, or create and implement a plan that allows them to achieve the proposed goals or plan how to work together. Nevertheless, what they all have in common is that no one understands in the same way what they are doing, why, how, and with whom. Linking the effort of the team to five critical success factors can solve the problem.

Leadership Style

Members of empowered teams share leadership responsibilities, sometimes willingly and sometimes reluctantly. Decision making is more collaborative, with consensus as the objective. Teams work toward win–win agreements. Teamwork is encouraged. Emphasis is more on problem solution and prevention, rather than on blame. During a visit to Procter & Gamble's plant in Foley, Florida, the host employee commented that a few years before he would not have believed he would ever be capable of conducting this tour. His new leadership roles had given him confidence to relate to customers and other outsiders.

Citizenship

Honesty, fairness, trust, and respect for others are more readily evident. In mature teams, members are concerned about each other's growth in the job (i.e., members reaching their full potential). Members become more willingly to share their experiences and coach each other because their goal is focused on the team's success, rather than on their personal success. Members more readily recognize and encourage each other's (and the team's) successes.

Reasons for High Commitment

As previously stated, empowered team members have authority, capability, and desire and an understanding of the organization's goals. In many organizations, they believe that this makes members feel and behave as owners and makes them more willing to accept greater responsibility. They also have greater knowledge, which further enhances their motivation and willingness to accept responsibility.

Means of Achieving High Performance

It has been observed that as employees accept more responsibility, have more motivation, and have greater knowledge, they freely participate more toward the interests of the business. They begin to truly act like owners, displaying greater discretionary effort and initiative. As previously stated, empowered team members have authority, capability, and desire and understand the organization's direction. Consequently, members feel and behave as owners, and are willing to accept greater responsibility. They also have greater knowledge, which further enhances their motivation and willingness to accept responsibility.

Enough progress has been made with various empowered organizations that we can now observe some key features of successful efforts. These have come from experiences of various consultants, visits by the authors to other companies, and published books and articles. These key features can help us learn how to design new organizations or redesign old ones to be more effective. The emphasis is on key features, rather than a prescription of how each is to operate in detail. This list is not exhaustive, but it is a helpful checklist, useful for a variety of organizations. The focus is on the external customers, their needs, and the products or services that satisfy those needs.

- The organization has the structure and job designs in place to reduce variation in process and product.
- The organizational layers are few.
- There is a focus on the business and customers.
- Boundaries are set to reduce variances at the source.
- Networks are strong.
- Communications are free-flowing and unobstructed.
- Employees understand who the critical customers are, what their needs are, and how to meet the customer needs with their own actions. Thus, all actions are based on satisfying the customer. The employees (operator, technicians, plant manager, etc.) understand that they work for the customer rather than for the plant manager.
- Supplier and customer input are used for managing the business.

In empowered organizations, managers create an environment to make people great, rather than control them. Successful managers are said to "champion" employees and make them feel good about their jobs, their organization, and themselves. Marvin Runyon, when head of the Nissan plant in Smyrna, Tennessee, stressed that "management's job is to provide an environment in which people can do their work" (Bernstein, 1988).

Breakthroughs in Current Performance

Breakthroughs in current performance (or improvement) do the following:

1. Significantly improve on the levels of results an organization is currently attaining. It happens when a systematic project-by-project improvement system discovers the root causes of current chronic problems and implements solutions to eliminate them.
2. Devise changes to the "guilty" processes and reduce the costs of poorly performing processes.
3. Install new systems and controls to prevent the return of these causes.

A system to attain breakthroughs in current performance addresses the question: "How do we reduce or eliminate things that are wrong with our products or processes, and the associated customer dissatisfaction and high costs (waste) that consume the bottom line?" A breakthrough improvement program addresses quality problems—failures to meet specific important needs of specific customers, internal and external. (Other types of problems are addressed by other types of breakthrough). Lean, six sigma, lean six sigma, root cause corrective action, and so on need to be part of a systematic approach to improve current performance. These methods address a few specific types of things that always go wrong:

- Excessive number of defects
- Undue number of delays
- Unnecessary long cycle times
- Unwarranted costs of the resulting rework, scrap, late deliveries, dissatisfied customers, replacement of returned goods, loss of customers, loss of goodwill, etc.

Lean and six sigma teams are all methods to improve performance. They are all project based and require multi-functional teams to improve current levels of performance. Each of them requires a systematic approach to complete the projects. A systematic approach to improving performance of processes is to:

1. Define the problem (the champions and executive council do this)
2. Measure (the project team does this)
3. Analyze (the project team does this)
4. Improve (the project team, often with help of others, does this)
5. Control (the project team and the operating forces do this)

Breakthroughs in current levels of performance problems are attained using these methods. The Lean and six sigma methods place your ailing processes under a microscope of unprecedented precision and clarity and make it possible to understand and control the relationships between input variables and desired output variables.

Your organization does have a choice as to what "system" to bring to bear on your problems: a "conventional" weapon system (quality improvement) or a "nuclear" system (six sigma). The conventional system is perfectly effective on many problems and much cheaper than the more elaborate and demanding nuclear system. The return on investment is considerable from both approaches, but especially so from six sigma if your customers are demanding maximum quality levels.

Breakthroughs in current performance solve problems such as excessive number of defects, excessive delays, excessively long time cycles, and excessive costs.

Breakthroughs in Culture

The result of completing many improvements creates a habit of improvement in the organization. Each of these improvements starts to create a quality culture because collectively it:

- Creates a set of new behavior standards and social norms that best supports organizational goals and climate.
- Instills in all functions and levels the values and beliefs that guide organizational behavior and decision making.

- Determines organizational cultural patterns such as style (informal versus formal, flexible versus rigid, congenial versus hostile, entrepreneurial/risk-taking versus passive/risk adverse, rewarding positive feedback versus punishing negative feedback, etc.), extent of internal versus external collaboration, high energy/morale versus low energy/morale, etc. Performance breakthrough in culture is a response to the basic question: "How do I create a social climate that encourages organization members to align together eagerly toward the organization's performance goals?"

As employees continue to see their leadership "sticking to it," culture change happens. An organization is not at a sustainable level yet or transformation change yet. There are still a number of issues that must be addressed. Among them are:

- Reviewing the organization's vision, mission, and values
- Orientation of new employees and training practices
- Reward and recognition policies and practices
- Human resource policies and administration
- Quality and customer satisfaction policies
- Fanatic commitment to customers and their satisfaction
- Commitment to continuous improvement
- Standards and conduct codes, including ethics
- No "sacred cows" regarding people, practices, and core business content
- Community benefit and public relations

An organization's culture exerts an extraordinarily powerful impact on organizational performance. The culture determines what is right or wrong, legitimate or illegitimate, and acceptable or unacceptable. Consequently, breakthrough in culture is profoundly influential in achieving performance breakthrough. It is also probably the most difficult and time-consuming type of breakthrough to make happen. In addition, it is so widely misunderstood that attempts to pull it off often fail.

Breakthroughs in culture (1) create a set of behavior standards and a social climate that supports organizational goals; (2) instill in all functions and levels the values and beliefs that guide organizational behavior and decision-making; and (3) determine organizational cultural patterns such as style (informal versus formal, flexible versus rigid, authoritarian top-down versus participative collaboration, management-driven versus leadership-driven, and the like), the organization's caste system (the relative status of each function), and the reward structure (who is rewarded for doing what).

➔ 9.6 RESISTANCE TO CHANGE

Curiously, even with such reinforcement, change, even beneficial change, will often be resisted. The would-be agent of change needs to understand the nature of this resistance and how to prevent or overcome it.

The control chart case example drew the conclusion that the main resistance to change is due to the disturbance of the cultural pattern of the shop when a change is proposed or attempted. People who are successful—and therefore comfortable—functioning in the current social or technical system don't want to have their comfortable existence disrupted, especially by an "illegitimate" change.

When a technical or social change is introduced into a group, group members immediately worry that their secure status and comfort level under the new system may be very different (worse) than under the current system. Threatened with the frightening possibility of losing the ability to perform well or losing status, the natural impulse is to resist the change. They have too much at stake in the current system. The new system will require them not only to let go of the current system willingly, but also to embrace the uncertain, unpredictable new way of performing. This is a tall order. It is remarkable how profoundly even a tiny departure from cultural norms will upset society members.

Norms are Helpful in Achieving a Cultural Transformation

Transforming a culture requires a highly supportive workforce. Certain cultural norms appear to be instrumental in providing the needed support. If these norms are not now part of your culture, some breakthroughs in culture may be required to implant them. Here are some of the more enabling norms:

- A belief that the quality of a product or process is at least of equal importance, and probably of greater importance than the mere quantity produced. This belief results in decisions favoring quality: Defective items do not get passed on down the line or out the door; chronic errors and delays are corrected, etc.
- A fanatic commitment to meeting customer needs. Everyone knows who his or her customers are (those who receive the results of their work), and how well he or she is doing at meeting those needs (they ask). Organization members, if necessary, drop everything and go out of their way to assist customers in need.
- A fanatic commitment to stretch goals and continuous improvement. There is always an economic opportunity for improving products or processes. Those who practice continuous improvement keep up with, or become better than competitors.

  Those organizations that do not practice continuous improvement fall behind and become irrelevant or worse. Six sigma product design and process improvement is capable, if executed properly, of producing superb economical designs, and nearly defect-free processes to produce them. The result is very satisfied customers and sharply reduced costs. The resulting sales and savings show up directly on the bottom line.
- A customer-oriented code of conduct and code of ethics. This code is published, taught in new employee orientations, and taken into consideration in performance ratings and in distributing rewards. Everyone is expected at all times to behave and make decisions in accordance with the code. It is enforced, if needed, by managers at all levels. The code applies to everyone, even board members—perhaps especially to them considering their power to influence everyone else.
- A belief that continuous adaptive change is not only good, but necessary. To keep alive, you must develop a system for discovering social, governmental, international, or technological trends that could have an impact on your organization. In addition, you will need to create and maintain structures and processes that enable quick effective response to these newly discovered trends.

Given the difficulty of predicting trends in the fast-moving contemporary world, it becomes vital for organizations to have such processes and structures in place and operating. If you fail to learn and appropriately adapt to what you learn, your organization can be left behind very suddenly and unexpectedly and could end up in the scrap heap. Many former e-commerce companies come to mind. Many rusting abandoned factories—the world over—testify to the consequences of not keeping up and consequently being left behind.

9.7 POLICIES AND CULTURAL NORMS

Policies are guides for managerial action and decision making. Organization manuals typically begin with a statement of the organization's quality policy. This statement rates the relative worth that organization members should place on producing high-quality products, as distinguished from the mere quantity of products produced. ("High-quality products" are goods, services, or information that meets important customer needs at the lowest optimum cost with few if any defects, delays, errors, etc.) High-quality products produce customer satisfaction, sales revenue, repeat demand or sales, and low costs of poor quality (unnecessary waste). Here, in that one sentence, are the reasons for attempting quality improvement. The inclusion of the value statement in your organization's quality manual reinforces some of the instrumental cultural norms and patterns essential for achieving a "quality culture" and, ultimately, performance breakthrough.

Keep in mind that if the value statement, designed to be a guide for decision making, is ignored and not enforced, it becomes worthless, except perhaps as a means of deceiving customers and employees in the short term. But you can be sure that customers and employees will soon catch on to the truth. They will dismiss the quality policy, waving it away as a sham that diminishes the whole organization and degrades the credibility of the management.

9.8 HUMAN RESOURCES AND CULTURAL PATTERNS

The human resources function plays a significant role in reinforcing cultural norms. It does so by several means that include:

- *Recruiting*. Advertisements contain descriptions of desirable traits (e.g., dependable, energetic, self-starter, creative, analytic, etc.), as well as characterizations of the organization (e.g., service-oriented, customer-oriented, committed to being a world leader in quality, progressive, world-class, equal opportunity, etc.). Organizational values are often featured in these messages.
- *Orientation and training*. It is customary when providing new employees with an introduction to an organization to review with them expected modes of dress, behavior, attitudes, traditional styles of working together, etc.
- *Publishing employee handbooks*. The handbooks distributed to new employees, and to everyone annually, are replete with descriptions of organizational history, traditional policies and practices, and expectations for organization members. All of these topics express directly or indirectly detailed elements of the official culture.

- *Reward and recognition practices.* In our rapidly changing world, management teams find themselves agonizing over what kind of employee behavior should be rewarded. Whatever the behavior is, when it is rewarded, the reward reinforces the cultural norms embodied in that behavior, and it should induce more of the same behavior from the rewarded ones, as well as attract others to do the same.
- *Career path and promotion practices.* If you track the record of those promoted in an organization, you are likely to find either behavior that conforms to the traditional cultural norms in their background or behavior that resembles desired new cultural norms required for a given organizational change, such as launching a six sigma effort. In the former case, management wants to preserve the current culture; in the latter case, management wants to create breakthroughs in culture and bring about a new culture that is at least somewhat altered. In both cases, the issue of the relationship of the person being promoted to the organizational culture is a significant factor in granting the promotion.

Breakthroughs in Adaptability and Sustainability

Creating a breakthrough in adaptability and sustainability require:

- Creating structures and processes that uncover and predict changes or trends in the environment that are potentially promising or threatening to the organization
- Creating processes that evaluate information from the environment and refer it to the appropriate organizational person or function
- Participation in creating an organizational structure that facilitates rapid adaptive action to exploit the promising trends or avoid the threatening disasters
- A response to the question: "How do I prepare my organization to respond quickly and effectively to unexpected change?"

The survival of an organization, like all open systems, depends on its ability to detect and react to threats and opportunities that present themselves from within and from outside. To detect potential threats and opportunities, an organization must not only gather data and information about what's happening, but it must also discover the (often) elusive meaning and significance the data hold for the organization. Finally, it must take appropriate action to minimize the threats and exploit the opportunities gleaned from the data and information.

To do all this will require appropriate organizational structures, some of which may already exist (an intelligence function, using an adaptive cycle, an information quality council) and a data quality system. The information quality council acts, among other things, as a "voice of the market." Dates are defined as "facts" (such as name, address, age) or "measurements of some physical reality, expressed in numbers and units of measure that enable our organization to make effective decisions by." These measurements are the raw material of information, which is defined as "answers to questions," or the "meaning revealed by the data, when analyzed." The typical contemporary organization appears to the authors to be awash in data but bereft of useful information. Even when in possession of multiple databases, much doubt exists regarding the quality of the data and therefore its ability to tell the truth about the question it is supposed to answer.

Managers dispute the reliability of reports, especially if the messages contained in the data are unfavorable. Department heads question the accuracy of financial statements, and sales figures, especially when they bring bad tidings.

Often, multiple databases will convey incongruent or contradictory answers to the same question, because each individual database has been designed to answer questions couched in the unique dialect or based on the unique definitions of terms used by one particular department or function, but not all functions. Data often are stored (hoarded?) in isolated unpublicized pockets, out of sight of the very people in other functions who could benefit from them if they knew they existed. Anyone who relies on data for making strategic or operational decisions is rendered almost helpless if the data are not available or are untrustworthy. How can a physician decide on a treatment if X-rays and test results are not available? How can the sales team plan promotions when it does not know how its products are selling compared with the competition? What if these same sales people knew that the very database that could answer their particular questions already exists but is used for the exclusive benefit of another part of the organization? It is clear that making breakthroughs in adaptability is difficult if we cannot get necessary data and information or if we cannot trust the truthfulness of the information we do get. Some organizations for which up-to-date and trustworthy data are absolutely critical go to great lengths to get useful information. And yet, in spite of their considerable efforts, many nevertheless remain plagued by chronic data quality problems.

The Route to Adaptability—The Adaptive Cycle and Its Prerequisites

Creating a breakthrough in adaptability creates structures and processes that:

1. Detect changes or trends in the internal or external environment that are potentially threatening or promising to the organization.
2. Interpret and evaluate the information.
3. Refer the distilled information to empowered functions or persons within the organization.
4. Take action to ward off the threats and exploit the opportunities. This is a continuous, perpetual cycle.

The cycle might more precisely be conceptualized as a spiral because it goes round and round, never stopping (see Figure 9.4). Several prerequisite actions are needed to set the cycle in motion and create breakthroughs in adaptability. Although each prerequisite is essential, and all are sufficient, perhaps the most crucial is the information quality council and data quality system. All else flows from timely trustworthy data—data that purport to truthfully describe aspects of reality vital to your organization.

Prerequisites for the Adaptive Cycle: Breakthroughs

1. Leadership and management
2. Organization structure
3. Current performance
4. Culture

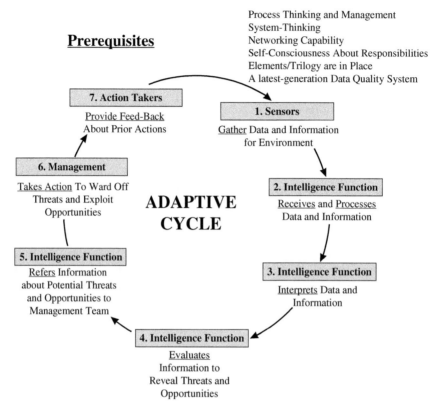

FIGURE 9.4
Adaptive cycle—to detect and react to organizational threats and opportunities. (*Juran Institute*, Six Sigma Breakthrough, and Beyond, *McGraw-Hill, 2003*)

A Journey around the Adaptive Cycle

An intelligence function gathers data and information from the internal and external environment. What do we need to know? Several of the following basic things, at a minimum:

From the internal environment:

- Process capability of our measurement and data systems
- Process capability of our key repetitive processes
- Performance of our key repetitive processes (human resources, sales, design, engineering, procurement, logistics, production, storage, transportation, finance, training, etc.; yields, defect types and levels, and time cycles)
- Causes of our most important performance problems
- Management instrument panel information: score cards (performance toward goals)
- Internal costs and costs of poor quality (COP^3)
- Characteristics of our organizational culture (how much does it support or subvert our goals)

- Employee needs
- Employee loyalty

From the external environment:

- Customer needs, now and in the future (what our customers or clients and potential customers or clients want from us or our products)
- Ideal designs of our products (goods, services, and information)
- Customer satisfaction levels
- Customer loyalty levels
- Scientific, technological, social, and governmental trends that can affect us
- Market research and benchmarking findings (us compared to our competition; us compared to best practices)
- Field intelligence findings (how well our products or services perform in use)

You may add to this list other information of vital interest to your particular organization. This list may seem long. It may seem expensive to get all this information. (It can be.) You may be tempted to wave it away as excessive or unnecessary. Nevertheless, if your organization is to survive, there appears to be no alternative but to gather this kind of information—on a regular, periodic basis. Fortunately, as part of routine control and tracking procedures already in place, your organization probably gathers much of this data and information. Gathering the rest of the information is relatively easy to justify, given the consequences of being unaware of, or deaf or blind to, vital information.

Information about internal affairs is gathered from routine production and quality reports, sales figures, accounts receivable and payable reports, monthly financial reports, shipment figures, inventories, and other standard control and tracking practices. In addition, specially designed surveys—written and interviews—can be used to gain insights into such matters as the state of employee attitudes and needs. A number of these survey instruments are available off the shelf in the marketplace. Formal studies to determine the capability of your measurement systems and your repetitive processes are routinely conducted if you are utilizing six sigma in your organization. Even if you don't use six sigma, such studies are an integral part of any contemporary quality system. Scorecards are very widely utilized in organizations that carry out annual strategic planning and deployment. The scores provide management with a dashboard, or instrument panel, which indicates warnings of trouble in specific organizational areas. Final reports of operational projects from quality improvement teams, six sigma project teams, and other projects undertaken as part of executing the annual strategic business plan are excellent sources of "lessons learned" and ideas for future projects. The tools and techniques for conducting COP^3 studies on a continuing basis are widely available. The results of COP^3 studies become powerful drivers of new breakthrough projects because they identify specific areas in need of improvement. In sum, the materials and tools for gathering information about your organization's internal functioning are widely available and easy to use.

Gathering information about conditions in the external environment is somewhat more complex. Some approaches require considerable know-how and great care. Determining customer needs is an example of an activity that sounds simple but actually requires some know-how to accomplish properly. Firstly, it is proactive.

Potential and actual customers are personally approached and asked to describe their needs, in terms of benefits they want from a product, services, or information. Many interviewees will describe their needs in terms of a problem to be solved or a product feature. Responses like these must be translated to describe the benefits the interviewee wants, not the problem to be solved or the product feature they'd like. Tools and techniques for determining ideal designs of current and future products or services are also available. They require considerable training to acquire the skills, but the payoffs are enormous. The list of such approaches includes quality planning, design for six sigma (DFSS), and TRIZ, a technique developed in Russia for projecting future customer needs and product features. Surveys are typically used to get a feel for customer satisfaction. A "feel" may be as close as you can get to knowledge of customer feelings and perceptions. These glimpses can be useful if they reveal distinct patterns of perceptions whereby large proportions of a sample population respond very favorably or very unfavorably to a given issue. Even so, survey results can hardly be considered "data," although they have their uses if suitable cautions are kept in mind. The limitations of survey research methodology cloud the clarity of results from surveys. (What really is the precise difference between a rating of "2" and a rating of "3"? A respondent could answer the same question different ways at 8:00 a.m. and at 3:00 p.m., for example. A satisfaction score increase from one month to another could be meaningless if the group of individuals polled in the second month is not the exact same group that was polled the first month. Even if they were the same individuals, the first objection raised would still apply to confound the results.)

A more useful approach for gauging customer "satisfaction"—or, more precisely, their detailed responses to the products or services they get from you—is the customer loyalty study, which is conducted in person with trained interviewers every six months or so on the same people. The results of this study go way beyond the results from a survey. Results are quantified and visualized. Customers and former customers are asked carefully crafted standard questions about the organization's products and performance. Interviewers probe the responses with follow-up questions, clarifying questions, etc. From the responses, a number of revealing pieces of information are obtained and published graphically. Not only do you learn the features of your products or services that cause the respondents happiness and unhappiness, but also such things as how much improvement of defect X (e.g., late deliveries) it would take for former customers to resume doing business with you.

Here's another example. You can graphically depict the amount of sales (volume and revenue) that would result from given amounts of specific types of improvements. You can also learn what specific "bad" things you'd better improve, and the financial consequences of doing so or not doing so. Results from customer loyalty studies are powerful drivers of strategic and tactical planning and breakthrough improvement activity.

Discovering scientific, technological, social, and governmental trends that could affect your organization simply requires plowing through numerous trade publications, journals, news media, websites, and the like—and networking as much as possible. Regular searches can be subcontracted so you receive, say, published weekly summaries of information concerning very specific types of issues of vital concern to you. Although there are numerous choices of sources of information concerning trends, there appears to be little choice of whether to acquire such information. The trick is to sort out the useful from the useless information.

A basic product of any intelligence function is to discover how the sales and performance of our organizations' products, services, and sales compare with our competitors and potential competitors. Market research and field intelligence techniques are standard features in most commercial businesses, and books on those topics proliferate.

Many organizations undertake benchmarking studies to gather information on world-class best practices. They study the inner workings of repetitive processes such as design, warehousing, operating oil wells, mail order sales, and almost anything else. The processes studied are not necessarily those of your competitors; they need only be the very best (efficient, effective, most economical, etc.). Benchmarking studies are classic intelligence detective work and are often conducted on a subcontract basis with organizations that specialize in benchmarking. The results are typically published and shared with all participants. When you have discovered best practices, you can compare your performance with them and describe gaps between theirs and yours, thus identifying breakthrough opportunities.

Completing the adaptive cycle will enable the organization to attain a breakthrough in adaptability and lead to sustainability. Skipping a breakthrough may not indicate a problem in the short term, only in the long term. Consider the economic crisis that hit the global economy in 2008. There were many global organizations that we considered leaders in their markets—when business was good. During the crisis, many top performers of the past went out of business, merged with others, or went into bankruptcy only to emerge a different organization. Why did so many organizations have trouble? Our theory is that although the organizations were good at responding to their customer needs, they were not watching society's needs. This led to a lack of information that, if it had been available, would have provided enough time to "batten down the hatches" to ride the crisis out. To keep this from happening, creating a high performing, adaptable organization may lead to better performance when things are not so good.

Sustainability

The second part of this breakthrough is sustainability. Sustainability has two important meanings. The first is to sustain the benefits of the transformational changes that took place. The second is to ensure the organization is sustainable from an environmental point of view. We felt at the time of this publication we would only focus on the long-term results. As more organizations take on the environmental issues that will plague us in the future, sustainability will focus on both. The next chapter elaborates on eco-quality.

Sustainability is the return to evaluating performance annually based on the findings of the information quality council. Through this information, the leaders can adjust the organization to ensure it stays ahead of its customers and can sustain itself for the long term.

→ 9.9 A TRANSFORMATION ROADMAP

The Juran roadmap has five phases each corresponding to the breakthroughs that are described in this chapter. Each phase is independent, but the beginning and end of each phase are not clearly delineated. Each organization reacts differently to changes. This means that one business unit in an organization may remain in one phase longer than another unit. These phases once again are a managerial guide to change, not a prescription.

The five phases of the transformation roadmap are shown in the Figure 9.5. The road starts at the Decide Phase. This phase begins when someone on the executive team decides that something must be done or else the organization will not meet shareholder expectations or will not meet its plan, etc. It ends with a clear plan for change.

In the Decide Phase, the organization will need to create new information or better information than it may have had about itself. This information can come from a number of reviews or assessments. Our experience shows that the more new information an organization has, the better its planning for change. There are a number of areas that should be reviewed. Here are some of the important ones:

- Conduct a customer loyalty assessment to determine what customers like or dislike about your products and services.
- Identify the areas of strength and uncover possible problems in the organization's performance.
- Understand employee attitudes toward the proposed changes.
- Understand the key business processes and how the changes will affect them.
- Conduct a cost analysis of poorly performing processes to determine the financial impact of these costs on the bottom line.
- Conduct a world-class quality review of all business units to understand the level of improvement needed in each unit.

A comprehensive review of the organization prior to launch is essential for success. Here we show a typical review that we recommend to all organizations embarking

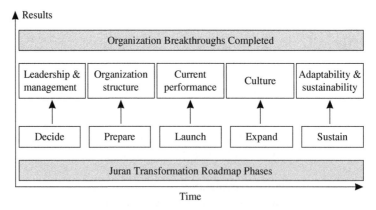

FIGURE 9.5
The Juran roadmap and breakthroughs.

upon a six sigma transformational initiative. From these assessments and reviews, the executive team now has qualitative and quantitative information to define the implementation plan for its organization.

The deployment plan must include:

1. The infrastructure needed to steer the changes
2. The methodology and tools that will be used throughout the implementation
3. The goals and objectives of the effort
4. The detailed milestones for achieving results

The conclusion of this phase results in the breakthrough in leadership and management.

The second phase is the Prepare Phase. In this phase, the executive team begins to prepare for the changes that will take place. It focuses on developing a pilot effort to try the change in a few business units before carrying it out in the total organization.

This phase begins by deploying the plan created in the Decide Phase, and it ends after a successful launch of pilot projects in the Launch Phase. From here, the organization begins to identify the improvement projects that must be carried out to meet the desired goals established in the Decide Phase. In this phase, the organization launches the pilot projects, reviews the pilots' progress, and enables success. Upon completion of the pilot projects, the executives evaluate what has worked and what has not. Then they either abandon their efforts or change the plan and expand it throughout the organization.

For your organization:

1. Identify the areas of strength and uncover possible problems in the organization's performance from the Decide Phase.
2. Identify value streams and key business processes that need improvement.
3. Select multifunctional pilot or demonstration projects and create project charters.
4. Create a training plan and set of learning events to train the teams.
5. Communicate the steps taken in this phase to the workforce.

The conclusion of this phase results in the breakthrough in organization structure.

The third phase is the Launch Phase. In this phase, the executive team launches the demonstration projects to a few business units before carrying them out in the total organization. Each project will require a project charter, a team and an effective launch, a review of progress, and holding the gains before results are attained. The length of this phase depends on the number of projects and results expected. For most organizations, this phase completion takes less than one year. As each project is completed and results are attained leaders can then evaluate the lessons learned and expand by launching more projects.

The conclusion of this phase results in the breakthrough in current performance.

Expansion can take months or years, depending on the size of the organization. An organization of 500 employees will require less time to deploy a plan across the organization than an organization of 50,000. The Expand Phase may take three to five years. Note that positive financial results will occur long before cultural changes take place. Staying in the Expand Phase is not a bad thing. An organization must continue to implement its plan, business unit by business unit, organization by organization, until all of the organization has had enough time to implement the desired changes. The final phase is the Sustain Phase when the organization has a fully integrated operation. All improvement and six sigma goals are aligned with the strategy of the organization. Key business processes are defined and well managed, and process owners are assigned to manage them. Employee performance reviews and compensation are in line with the changes required. Those who comply with the change are rewarded. The executives and business unit heads conduct regular reviews and audits of the change process. This may result in a discussion or even a change in the strategy of the organization.

The organization may have learned more about its capabilities and more about its customers that may lead to a change in strategy.

The conclusion of this phase results in the breakthrough in culture.

The Sustain Phase also lasts as long as the organization is meeting its strategic and financial goals. Deviations from expected results, possibly due to macroeconomic events outside the organization, require a review of the scorecard to determine what has changed. When this is determined, the organization makes the changes, continues, and sustains itself at the current level.

The conclusion of this phase results in the breakthrough in adaptability and sustainability.

SUMMARY

- Breakthrough was defined as the purposeful creation of significant, sustainable beneficial change. It is often associated with process improvement targets. Transformation requires that the organization attain breakthroughs. The required breakthroughs are leadership and management, organization structure, current performance, culture, and adaptability and sustainability.
- Organizational change is important for three reasons (each of which can destroy an organization): (1) costs of poorly performing processes are too high; (2) continuous societal changes must be dealt with; and (3) without change, organizations die.
- All organizations must be thought of as open systems. They depend on successful transactions with an organization's external environment, and proper coordination of the organization's various specialized internal functions.
- In attempting to create these breakthroughs, problems that appear in one work area often have their origin upstream in the process. Therefore, people in a given work location suffering from a performance problem can't necessarily solve it by themselves.
- Performance excellence can only be attained with the active participation not only of those who are the source of the problems, but also of those affected by the problem and those who will be the sources of remedial changes to the problem (usually those who are the source of the problem and perhaps others).
- Organizational change attempted in isolation from the whole organization and without systems thinking can easily create more problems than existed. The Juran performance excellence change model can provide the model and roadmap for transformational change.

- To become superior in quality, we need to (1) develop technologies to create products and processes that meet customer needs and (2) stimulate a culture that continually views quality as a primary goal.
- The culture for quality *can* be changed. We need to provide
    Goals and measurements.
    Evidence of upper management leadership.
    Self-development and empowerment.
    Participation.
    Recognition and rewards.
- These elements must be integrated with the methodologies and structure for quality.
- To change culture requires years, not months; to change quality requires trust, not techniques.

PROBLEMS

9.1. Consider an individual sport such as golf or racquetball. Apply the three principles of self-control to such a sport and decide whether individuals are in a state of self-control. How does your conclusion relate to the design of jobs in an organization?

9.2. Prepare two lists to analyze the forces that have an impact on implementing quality improvement teams. One list should contain the restraining forces that are obstacles to implementing teams (e.g., lack of time); the other list should show the driving forces that can help (e.g., support by upper management). (This approach is part of a "force field analysis" or a "barriers and aids analysis.")

9.3. Look about your community to identify some of the ongoing drives or campaigns, such as for traffic safety, fund-raising, political election, or to keep your city clean. For any campaign, analyze and report on
(a) The methods used to secure attention.
(b) The methods used to secure interest and identification with the program.
(c) The methods used to secure action.

9.4. For any organization known to you, study the prevailing continuing program of motivation for quality. Report your findings on (1) the ingredients of the program and (2) the organization's effectiveness in carrying out the various aspects of the program.

9.5. Develop three quality indicators for the performance of a buyer in a purchasing department: a lagging indicator, a concurrent indicator, and a leading indicator.

9.6. Tektronix, Inc. (1985) describes "people involvement" as moving through nine types of management (a "continuum"): Autocratic authoritarian management, directive command, selective information sharing, employee input, problem-solving groups, ad hoc task forces, participative decision making, work redesign and goal setting, and semiautonomous teams. For each type, describe in a few sentences the respective roles of managers and nonmanagers.

9.7. Speak with at least three people, preferably from different organizations. Identify three specific situations that contribute to a negative quality culture. Identify three situations that contribute to a positive quality culture.

REFERENCES

Cameron, K. and W. Sine (1999). "A Framework for Organizational Quality Culture," *Quality Management Journal,* vol. 6, no. 4, pp. 7–25.

Miller, L. M. (1984). *American Spirit: Visions of a New Corporate Culture,* William Morrow, New York.

Tektronix, Inc. (1985). "People Involvement: A Continuum," Tektronix, Inc., Beaverton, OR. All rights reserved. Reproduced with permission.

Watson, M. A. and F. M. Gryna (1998). "Assessment of Quality Culture in Small Business: Exploratory Research," Report No. 905, College of Business, University of Tampa, Tampa, FL.

CHAPTER 10

Inspection, Test, and Measurement

→ 10.1   THE TERMINOLOGY OF INSPECTION

Inspection and test typically include measurement of an output and comparison to specified requirements to determine conformity. Inspection and test activities ensure that manufactured products, individual components, and multicomponent systems are adequate for their intended purpose. Inspection and testing are the operational parts of quality control, which is the most important factor to the survival of any manufacturing company. Quality control directly supports the other factors of cost, productivity, on-time delivery, and market share. Therefore, all quality standards needed to produce the components of a product and perform its assembly must be specified in a manner such that customers' expectations are met. Inspection is performed for a wide variety of purposes, e.g., distinguishing between good and bad product, determining whether a process is changing, measuring process capability, rating product quality, securing product design information, rating the inspectors' accuracy, and determining the precision of measuring instruments. Each of these purposes has its special influence on the nature of the inspection and on the manner of doing it.

The distinction between "inspection" and "test" has become blurred. Whereas inspection is the activity of examining the product or its components to determine if they meet the design standards, testing is a procedure in which the item is observed during operation in order to determine whether it functions properly for a reasonable period of time. Inspection, typically performed under static conditions on items such as components, can vary from simple visual or destructive examination to a series of complex measurements. The emphasis in inspection is to determine conformance to a standard. Test, on the other hand, is performed under either static or dynamic conditions and is typically performed on more complex items such as subassemblies or systems. Test results determine conformance and can also be used as input for other analyses such as evaluating a new design, diagnosing problems, or making physical adjustments on products. Some industries have their own terms for inspection or test, e.g., assay is used in the mining and pharmaceutical industries.

Although the terms *inspection* and *test* usually refer to manufacturing industries, the concepts also apply to other industries. In service industries, different terms are used, e.g., *review, checking, reconciliation, examination.* The evaluation of the correctness of an income tax return, the cleanliness of a hotel room, or the accuracy of a bank teller's closing balance are forms of inspection—a measurement, a comparison to a standard, and a decision.

✢ 10.2 CONFORMANCE TO SPECIFICATION AND FITNESS FOR USE

Of all the purposes of inspection, the most ancient and the most extensively used is product acceptance, i.e., determining whether a product conforms to standard and therefore should be accepted. *Product* can mean a discrete unit, a collection of discrete units (a "lot"), a bulk product (a tank car of chemicals), or a complex system.

Product can also mean a service, such as a transaction at a bank; an inquiry to an agency about tax regulations; or the performance of personnel before, during, and after an airline flight. In these examples, inspection characteristics can be identified, standards set, and conformance judged.

Product acceptance involves the disposition of product based on its quality. This disposition involves several important decisions:

1. *Conformance.* Judging whether the product conforms to specification.
2. *Fitness for use.* Deciding whether nonconforming product is fit for use.
3. *Communication.* Deciding what to communicate to insiders and outsiders.

The Conformance Decision

Except in small companies, the number of conformance decisions made each year is huge. There is no possibility that the supervisory body can become involved in the details of so many decisions. Hence the work is organized so that inspectors or production workers can make these decisions themselves. To this end, they are trained to understand the products, the standards, and the instruments. Once trained, they are given the jobs of inspecting and judging conformance. (In many cases, the delegation is to automated instruments.)

Associated with the conformance decision is the disposition of conforming product. The inspector is authorized to identify the product ("stamp it up") as acceptable product. This identification then serves to inform packers, shippers, etc., that the product should proceed to its next destination (further processing, storeroom, customer). Strictly speaking, this decision to "ship" is made not by inspectors but by management. With some exceptions, product that conforms to specification is also fit for use. Hence company procedures (which are established by the managers) provide that conforming products should be shipped as a regular practice.

The Fitness-for-Use Decision

In the case of nonconforming products, a new question arises: Is this nonconforming product fit or unfit for use? Product features are said to possess "fitness-for-use" if they are able to meet customer demands, protect human safety, and protect the environment. Unfit product is disposed of in various ways: scrap, sort, rework, return to supplier, sell at discount, etc. In some cases, the answer is obvious—the nonconformance is so severe that the product is clearly unfit. Hence it is scrapped or, if economically repairable, brought to conformance. However, in many cases, the answer as to fitness for use is not obvious. In such cases, if enough is at stake, a study is made to determine fitness for use. This study involves securing inputs such as the following:

- *Who will the user be?* A technologically sophisticated user may be able to deal successfully with the nonconformance; a consumer may not. A nearby user may have easy access to field service; a distant or foreign user may lack such easy access.
- *How will the product be used?* For many materials and standard products, the specifications are broad enough to cover a variety of possible uses, and the actual use to which the product will be put is not known at the time of manufacture. For example, sheet steel may be cut up to serve as decorative plates or as structural members; a television receiver may be stationed at a comfortable range or at an extreme range; chemical intermediates may be employed in numerous formulas.
- *Are there risks to human safety or to structural integrity?* Where such risks are significant, all else is academic.
- *What is the urgency?* For some applications, the client cannot wait because the product in question is critical to putting some broader system into operation. Hence the product may demand delivery now and cause repairs in the field.
- *What are the company's and the users' economics?* For some nonconformances, the economics of repair are so forbidding that the product must be used as is, although at a price discount. In some industries, e.g., textiles, the price structure formalizes this concept by use of a separate grade—"seconds."
- *What are the users' measures of fitness for use?* These may differ significantly from those available to the manufacturer. For example, a manufacturer of abrasive cloth used a laboratory test to judge the *ability* of the cloth to polish metal; a major client evaluated the *cost* per 1000 pieces polished.

These and other inputs may be needed at several levels of fitness for use, i.e., the effects on the economics of subsequent processors, the marketability requirements of the merchants, the qualities that determine fitness for the ultimate user, and the qualities that influence field maintenance. The internal costs can be estimated to arrive at an economic optimum. However, the effects can go well beyond money: Schedules are disrupted, people are blamed, etc.

The job of securing such inputs is often assigned to a staff specialist, e.g., a quality engineer who "makes the rounds," contacting the various departments that are able to provide pertinent information. There may be a need to contact the customer and even to conduct an actual tryout. A typical list of sources is shown in Table 10.1.

Once all the information has been collected and analyzed, the fitness-for-use decision can be made. If the amount of money at stake is small, this decision will be

TABLE 10.1
Sources of information

Input	Usual sources
Who will the user be?	Marketing
How will this product be used?	Marketing; client
Are there risks to human safety or to structural integrity?	Product research and design
What is the urgency?	Marketing; client
What are the company's and users' economics?	All departments; client
What are the users' measures of fitness for use?	Market research; marketing; client

delegated to a staff specialist, to the quality manager, or to some continuing decision-making committee such as a material review board. If the amount at stake is large, the decision will usually be made by a team of upper managers.

Deliberations on the fitness-for-use decision are often a dramatic blend of voices—some balanced and judicious, others bowing to the pressures of delivery deadlines, even if it means tossing up gems of earnest nonsense.

The Communication Decision

The conformance and fitness-for-use decisions are a source of essential information, although some of the data is not well communicated.

Data on nonconforming products are usually communicated to the producing departments to aid them in preventing a recurrence. More elaborate data collection systems may require periodic summaries to identify "repeaters," which then become the subject of special studies.

When nonconforming products are sent out as fit for use, the need for two additional categories of communication arises:

1. *Communication to "outsiders"* (usually customers) who have a right and a need to know. All too often, manufacturing companies neglect or avoid informing their customers when shipping nonconforming products. Such avoidance can be the result of bad experience, i.e., some customers will seize on such nonconformances to secure a price discount despite the fact that use of the product will not add to their own costs. Neglect is more usually a failure even to face the question of what to communicate. A major factor here is the design of the forms used to record the decisions (nonconforming material control as part of the quality management system). With rare exceptions, these forms lack provisions that force those involved to make recommendations and decisions on (a) whether to inform the outsiders and (b) what to communicate to them.
2. *Communication to insiders.* When nonconforming goods are shipped as fit for use, the reasons for doing so are not always communicated to inspectors and especially not to production workers. The resulting vacuum of knowledge has been known to breed some bad practices. When the same type of nonconformance has been shipped

several times, an inspector may conclude (in the absence of knowing why) that it is a waste of time to report such nonconformances in the first place. Yet in some future case, the special reasons that were the basis for the decision to ship the nonconforming goods may not be present. In like manner, a production worker may conclude that it is a waste of time to exert all that effort to avoid some nonconformance that will be shipped anyway. Such reactions by well-meaning employees can be minimized if the company squarely faces the question, What shall we communicate to insiders?

➔ 10.3 DISPOSITION OF NONCONFORMING PRODUCT

Once an inspector finds that a lot of product is nonconforming, he or she prepares a report to that effect. Copies of this report are sent to the implicated departments. This action sets a planned sequence of events into motion. The lot is marked "hold" and is often sent to a special holding area to reduce the risk of mixups. The product is put into quarantine. Schedulers look into the possibility of shortages and the need for replacement. An investigator is assigned to collect the type of information needed as inputs for the fitness-for-use decision, as discussed earlier.

Decision Not to Ship

The investigation may conclude that the lot should not be shipped as is. In that event, the economics are studied to find the best disposition: sorting, repairing, downgrading, scrapping, etc. Supplemental accounting efforts may charge the costs to the responsible source, especially if supplier responsibility is involved. Some degree of action to prevent a recurrence also occurs.

Decision to Ship

This decision may come about in several ways:

- *Waiver by the designer.* Such a waiver is a change in specification as to the lot in question that thereby puts the lot into a state of conformance. These decisions should be managed through systems such as engineering change notices or engineering change requests.
- *Waiver by the customer* or by the marketing department on behalf of the customer. Such a waiver in effect supersedes the specification. (The waiver may have been "bought" by a change in warranty or by a discount in price.)
- *Waiver by the quality department* under its delegation to make fitness-for-use decisions on noncritical matters. The criteria for "noncritical" may be based on prior seriousness classification of characteristics, on the low cost of the product involved, or on still other bases. For minor categories of seriousness, the delegation may even be made by the quality engineers or by inspection supervisors. However, for major and critical defects, the delegation is typically by the technical manager, the quality manager, or some team of managers.

- *Waiver by a formal material review board.* This board concept was originally evolved by military buyers of defense products to expedite decisions on nonconforming lots. Membership on the board includes the military representative plus the designer and the quality specialist. A unanimous decision is required to ship nonconforming product. The board procedures provide for formal documentation of the facts and conclusions, thereby creating a data source of great potential value.
- *Waiver by upper managers.* This part of the procedure is restricted to cases of a critical nature involving risks to human safety, marketability of the product, or risk of loss of large sums of money. For such cases, the stakes are too high to warrant decision making by a single department. Hence, the managerial team takes over. Waivers, however, have an insidious way of becoming part of a culture. It is valuable to continuously track the amount of product shipped under the waiver of specifications, e.g., the percentage of lots shipped each month under waiver.

The service sector also has a variety of actions for disposition of a nonconforming service. Peach (1997, page 435) provides some examples:

- *For a cable television distribution failure,* the disposition is to repair by reinitializing the connection.
- *For an incorrect crediting in an account by a bank,* the disposition is to rework by crediting the correct account and debiting the incorrect amount.
- *For tainted food in a restaurant,* the disposition is to scrap by returning the food to the supplier, disposing in the trash, and notifying local health authorities.

Corrective Action

Aside from a need to dispose of the nonconforming lot, there is a need to prevent a recurrence. This prevention process is of two types, depending on the origin of the nonconformance.

1. Some nonconformances originate in some isolated, sporadic change that took place in an otherwise well-behaved process. Examples are a mix-up in the materials used, an instrument that is out of calibration, or a human mistake in turning a valve too soon. For such cases, local supervision is often able to identify what went wrong and to restore the process to its normal good behavior. Sometimes, this troubleshooting may require the assistance of a staff specialist. In any case, no changes of a fundamental nature are involved because manufacturing planning has already established an adequate process. Documentation of these decisions should at best follow simple investigation and resolutions such as through an 8D process.
2. Other nonconformances are "repeaters." They arise over and over again, as evidenced from their recurring need for disposition by the material review board or other such agency. Such recurrences point to a chronic condition that must be diagnosed and remedied if the problem is to be solved. Local supervision is seldom able to find the cause of these chronic nonconformances, mainly because the responsibility for diagnosis is vague. Lacking agreement on the cause, the

problem goes on and on amid earnest debates about who or what is to blame—unrealistic design, incapable process, poor motivation, etc. The need is not for troubleshooting to restore the normal good behavior because the normal behavior is bad. Instead, the need is to organize for an improvement project, as discussed in Chapter 4. In these instances, one should use more sophisticated techniques such as root cause analysis or six sigma methodologies.

✣ 10.4 INSPECTION PLANNING

Inspection planning is the activity of (1) designating the "stations" at which inspection should take place and (2) providing those stations with the means for knowing what to do plus the facilities for doing it. For simple, routine quality characteristics, the planning is often done by the inspector. For complex products made in large multidepartmental companies, the planning is usually done by specialists such as quality engineers.

Locating the Inspection Stations

The basic tool for choosing the location of inspection stations is the flowchart (see, e.g., Figure 13.5). The most usual locations are

- At receipt of goods from suppliers, usually called "incoming inspection" or "supplier inspection."
- Following the setup of a production process to provide added assurance against producing a defective batch. In some cases, this "setup approval" also becomes approval of the batch.
- During the running of critical or costly operations, usually called "process inspection."
- Prior to delivery of goods from one processing department to another, usually called "lot approval" or "tollgate inspection."
- Prior to shipping completed products to storage or to customers, usually called "finished-goods inspection."
- Before performing a costly, irreversible operation, e.g., pouring a melt of steel.
- At natural "peepholes" in the process.

The concept of inspection stations also applies to the service sector. For example, in arranging travel for customers, "receiving inspection" often includes verifying information on a customer credit card; "process inspection" includes verifying that the customer has a passport, visa, and driver's license; "final inspection" includes confirming that tickets match agreements (Peach, 1997, page 432).

The inspection station is not necessarily a fixed zone where the work comes to the inspector. In some cases, the inspector goes to the work by patrolling a large area and performing inspections at numerous locations, including the shipping area, at the supplier's plant, or on the customer's premises (typically known as roving inspection).

Process inspection is often within the responsibility of production operators. An adjunct to this approach is the use of "poka-yoke" devices as part of the inspection.

These devices are installed in the machine to inspect process conditions and product results and provide immediate feedback to the operator (see Section 13.8). See also Section 6.3 for a discussion of the self-adjusting concept where operators employ simple, direct methods (including inspection) continuously. Increasingly, inspection is built into the process rather than being placed at the end of the process (see Durkee and Gookins, 1999).

Choosing and Interpreting Quality Characteristics

The planner prepares a list of which quality characteristics are to be checked at which inspection stations. The planner should use classification of characteristics from engineering (critical, major, minor, or incidental) to help determine those vital few product or process characteristics for inspection. For some of these characteristics, the planner may find it necessary to provide information that supplements the specifications. Product specifications are prepared by comparatively few people, each generally aware of the needs of fitness for use. In contrast, these specifications must be used by numerous inspectors and operators, most of whom lack such awareness. The planner can help bridge this gap in a number of ways:

- By providing inspection and test environments that simulate the conditions of use. This principle is widely used, for example, in testing electrical appliances. It is also extended to applications such as the type of lighting used for inspecting textiles.
- By providing supplementary information that goes beyond the specifications prepared by product designers and process engineers. Some of this information is available in published standards—company, industry, and national. Other information is specifically prepared to meet the specific needs of the product under consideration. For example, in an optical goods factory, the generic term *beauty defects* was used to describe several conditions that differed widely as to their effect on fitness for use. A scratch on a lens surface in the focal plane of a microscope made the lens unfit for use. A scratch on the large lens of a pair of binoculars, although not functionally serious, was visible to the user and hence was not acceptable. Two other species of scratches were neither adverse to fitness for use nor visible to the user and hence were unimportant. These distinctions were clarified through planning analysis and woven into the procedures.
- By helping to train inspectors and supervisors to understand the conditions of use and the "why" of the specification requirements.
- By providing seriousness classification (see Section 10.5).

Detailed Inspection Planning

For each quality characteristic, the planner determines the detailed work to be done. This determination covers matters such as

- The type of inspection or test to be done. This area may require a detailed description of the testing environment, testing equipment, testing procedure, and tolerances for accuracy.

- The number of units to be inspected or tested (sample size).
- The method of selecting the samples to be inspected or tested.
- The type of measurement (attributes, variables, other).
- Conformance criteria for the units, usually the specified product tolerance limits.

Beyond this detailed planning for the characteristics and units is further detailed planning applicable to the product, the process, and the data system:

- Conformance criteria for the lot, usually consisting of the allowable number of nonconforming units in the sample.
- The physical disposition to be made of the product—the conforming lots, the nonconforming lots, and the units tested.
- Criteria for decisions on the process—should it run or stop?
- Data to be recorded, forms to be used, reports to be prepared.

This planning is usually included in a formal document that must be approved by the planner and the inspection supervisor. For an example from Baxter Travenol, see Figure 10.1.

Sensory Characteristics

Sensory characteristics are those for which we lack measuring instruments and for which the senses of human beings must be used as measuring instruments. Sensory qualities may involve technological performance of a product (e.g., adhesion of a protective coating), aesthetic characteristics (e.g., odor of a perfume), taste (e.g., food), or human services characteristics (e.g., the spectrum of hotel services).

An important category of sensory characteristics is the visual quality characteristic. Typically, written specifications are not clear because of their inability to quantify the characteristics. Among the approaches employed to describe the limits for characteristics are the following:

1. Providing photographs to define the limits of acceptability of the product.

    EXAMPLE 10.1. A fast-food enterprise has the problem of defining quality standards for suppliers of hamburger buns. The solution is photographs showing the ideal and the maximum and minimum acceptable limits for "golden brown" color, symmetry of bun, and distribution of sesame seeds.

2. Providing physical standards to define the limits of acceptability.

    EXAMPLE 10.2. A government agency needed to define the lightest and darkest acceptable shades of khaki for suppliers of uniforms. Color swatches of cloth were prepared for the limiting shades and issued to inspectors. Imagine the follow-up required to periodically replace the swatches when fading was imminent!

3. Specifying the *conditions* of inspection instead of trying to explicitly define the limits of acceptability.

    EXAMPLE 10.3. Riley (1979) describes a special inspection procedure for cosmetic (appearance) defects of electronic calculators. Part drawings indicate the relative importance

Part Number: XXXX							Part Name: YYYY	
Process	Characteristics	C_p[1] Index	C_{pk}[1] Index	Frequency[2]	Sample size[2]	Analysis methods	Out-of-control conditions are encountered[4]
Incoming inspection	Stock thickness	1.6	1.0	Every shipment	—	Review control charts provided with each lot	Impound lot—contact supplier for resolution
In-process inspection	Thickness	1.9	1.1	Every 1000 parts	2 pieces	Micrometers/$\bar{X}$ and s chart	Correct process
	Width	1.5	1.4	Every 10,000 parts	5 pieces	Micrometer/median chart	Correct process
	Length	1.6	1.2	Every 4 hours	75 pieces	Tapered ring gage/p chart	Correct process
Assembly area	Thickness	2.0	1.8	Hourly	30 pieces	Special gage/p chart	Correct process
	Width	2.2	1.9	Chart hourly	100%	Automatic tester/u chart	Repair by responsible operator
Outgoing[3]	Complete assembly	2.8	1.9	Hourly	20 pieces	Automatic tester/$\bar{X}$ and s chart	Correct process
	Complete assembly	NA	1500 DPM	Each lot	50 pieces	Complete visual inspection plus gage and test stand/c chart	Reject lot and sort for identified nonconformance

[1]Explanations and formulas are contained in the SPC Guideline.
[2]The frequencies and sample size are determined from the performance study of the stability of each process. They are periodically reviewed and updated as required.
[3]After 6 months production experience, the process control and inspection records will be reviewed to determine if outgoing inspection can be reduced.
[4]If any nonconforming products are found in the process samples, then there will be performed 100% inspection of all products produced since the last in control point.

FIGURE 10.1
Control plan. (*From Baxter Travenol Laboratories, 1986.*)

of different surfaces, using a system of category numbers and class letters. Three categories identify the surface being inspected:

I. Plastic window (critical areas only)
II. External
III. Internal

Three classes indicate the frequency with which the surface will be viewed by the user:

A. Usually seen by the user.
B. Seldom seen by the user.
C. Never seen by user (except during maintenance).

For example, a sheet-metal part that will seldom be seen carries a grade of Coating IIB.

The conditions of inspection are stated in terms of viewing distance, viewing time, and lighting conditions. The distance and time are specified for each combination of surface being inspected and the frequency of viewing by the user. Lighting conditions must be between 75 and 150 foot-candles from a nondirectional source.

The guidelines help to establish cosmetic gradings on parts drawings. However, a judgment must still be made by the inspector whether or not the end user would consider the flaw(s) objectionable, using the specified time and distance.

Elaboration on sensory characteristics is provided in *JQH5,* pages 23.25–23.29.

✦ 10.5 SERIOUSNESS CLASSIFICATION

Quality characteristics are decidedly unequal in their effect on fitness for use. A relative few are "serious," i.e., of critical importance; many are of minor importance. Clearly, the more important the characteristic, the greater the attention it should receive in matters such as extent of quality planning; precision of processes, tooling, and instruments; sizes of samples; strictness of criteria for conformance; etc. However, making such discrimination requires that the relative importance of the characteristics be made known to the various decision makers involved: process engineers, quality planners, inspection supervisors, etc. To this end, many companies use formal systems of seriousness classification. The resulting classification is used in inspection and quality planning and also in specification writing, supplier relations, product audits, executive reports on quality, etc. This multiple use of seriousness classification dictates that the system be prepared by an interdepartmental committee which then:

1. Decides how many classes or strata of seriousness to create (usually three or four).
2. Defines each class.
3. Classifies each characteristic into its proper class of seriousness.

Characteristics and Defects

There are actually two lists that need to be classified. One is the list of quality characteristics derived from the specifications. The other is the list of "defects," i.e., symptoms of nonconformance during manufacture and of field failure during use. There is a good deal of commonality between these two lists, but there are differences as well. (For example, the list of defects found on glass bottles has little resemblance to the list of characteristics.) In addition, the two lists do not behave alike. The design characteristic "diameter," for example, gives rise to two defects—oversize and undersize. The amount by which the diameter is oversize may be decisive as to seriousness classification.

Normally, it is feasible to make one system of classification applicable to both lists. However, the uses for the resulting classifications are sufficiently varied to make it convenient to publish the lists separately.

Definitions for the Classes

Most sets of definitions show the influence of the pioneering work of the Bell System in the 1920s. Study of numerous such systems reveals an inner pattern that is a useful guide to any committee faced with applying the concept to its own company. Table 10.2 shows the nature of this inner pattern applied to a company in the food industry.

Classification

Classification is a long and tedious but essential task. However, it yields some welcome by-products by pointing out misconceptions and confusion among departments and thereby opening the way to clear up vagueness and misunderstandings. Then, when the final seriousness classification is applied to several different purposes, it is subjected to several new challenges that provide still further clarification of vagueness.

A problem often encountered is the reluctance of the designers to become involved in seriousness classification of characteristics. They may offer plausible reasons: All characteristics are critical, the tightness of the tolerance is an index of seriousness, etc. Yet the real reasons may be unawareness of the benefits, a feeling that other matters have higher departmental priority, etc. In such cases, it may be worthwhile to demonstrate the benefits of classification by working out a small-scale example. In one company, the classification-of-characteristics program reduced the number of dimensions that had to be checked from 682 to 279 and reduced inspection time from 215 minutes to 120 minutes.

✣ 10.6 AUTOMATED INSPECTION

Automated inspection and testing are widely used to reduce inspection costs, reduce error rates, alleviate personnel shortages, shorten inspection time, avoid inspector monotony, and provide still other advantages. Applications of automation have successfully been made to mechanical gauging, electronic testing (for high volumes of components as well as system circuitry), nondestructive tests of many kinds, chemical analyses, color discrimination, visual inspection (e.g., of large-scale integrated circuits), etc. In addition, automated testing is extensively used as a part of scheduled maintenance programs for equipment in the field.

TABLE 10.2
Composite definitions for seriousness classification in the food industry

Defect	Effect on consumer safety	Effect on usage	Consumer relations	Loss to company	Effect on conformance to government regulations
Critical	Will surely cause personal injury or illness	Will render the product totally unfit for use	Will offend consumers' sensibilities due to odor, appearance, etc.	Will lose customers and will result in losses greater than value of product	Fails to conform to regulations for purity, toxicity, identification
Major A	Very unlikely to cause personal injury or illness	May render the product unfit for use and may cause rejection by the user	Will likely be noticed by consumers and will likely reduce product salability	May lose customers and may result in losses greater than the value of the product; will substantially reduce production yields	Fails to conform to regulations on weight, volume, or batch control
Major B	Will not cause injury or illness	Will make the product more difficult to use, e.g., removal from package, or will require improvisation by the user; affects appearance, neatness	May be noticed by some consumers and may be an annoyance if noticed	Unlikely to lose customers; may require product replacement; may result in loss equal to product value	Minor nonconformance to regulations on weight, volume, or batch control, e.g., completeness of documentation
Minor	Will not cause injury or illness	Will not affect usability of the product; may affect appearance, neatness	Unlikely to be noticed by consumers and of little concern if noticed	Unlikely to result in loss	Conforms fully to regulations

Examples in nonmanufacturing activities range from the spelling check provided within word processors to the checking of bank transactions for errors.

A company contemplating the use of automated inspection first identifies the few tests that dominate the inspection budgets and use of personnel. The economics of automation are computed, and trials are made on some likely candidates for a good return on investment. As experience is gained, the concept is extended further and further.

With the emphasis on defect levels in the parts-per-million range, many industries are increasingly accepting on-machine automated 100% inspection and testing. Orkin and Olivier (1999) provide an extensive table that identifies seven categories of potential applications of automated inspection ranging from dimensional gauging to nondestructive testing.

A dramatic example of automated inspection is the concept of "machine vision" where electronic eyes inspect and guide an array of industrial processes. The applications include steering robots to place doors on cars, finding blemishes on vegetables in a frozen food processing line, examining wood for knots in veneer panels, and checking that the right color drug capsule goes into correctly labeled packages before being shipped to pharmacies (*Fortune,* February 16, 1998, page 104B). High-speed visual inspection devices are either integrated or are slightly off-line from manufacturing operations. Products can automatically undergo multicharacteristic checks. At the end of the inspection cycle, the computer monitor tells the operator whether the product is acceptable and then updates quality process statistics. Human inspection methods usually detect only 80 to 90% of the defects (see later discussion); with machine visual inspection, essentially all defects are detected. For elaboration, see *JQH5,* page 23.15.

A critical requirement for all automated test equipment is precision measurement, i.e., repeated measurements on the same unit of product should yield the "same" test results within some acceptable range of variation. This repeatability is inherent in the design of the equipment and can be quantified by the methods discussed in Section 10.8. In addition, means must be provided to keep the equipment "accurate," i.e., in calibration with respect to standards for the units of measure involved.

Still another aspect of automated test equipment is the problem of processing the data that are generated by the tests. Modern systems of electronic data processing allow entering these test data directly from the test equipment into the computer without the need for intermediate documents. Such direct entry supports the prompt preparation of data summaries, conformance calculations, comparisons with prior lots, etc. In turn, it is feasible to program the computer to issue instructions to the test equipment with respect to frequency of test, disposition of units tested, alarm signals relative to improbable results, etc.

Cooper (1997) discusses how to design, test, implement, and maintain a paperless inspection system.

➔ 10.7 INSPECTION ACCURACY

Inspection accuracy depends on (1) the completeness of the inspection planning (see earlier discussion), (2) the bias and precision of the instruments (see later in this section), and (3) the level of human error.

High error rates are particularly prevalent in inspection tasks having a high degree of monotony, e.g., viewing jars of a food product for foreign particles, screening luggage at an airport security gate. Surprisingly, monotony that causes an inspector to miss defects can build up in a short time. With monotonous inspection, inspectors detect about 80 to 90% of the defects and miss the remainder. Thus 100% inspection that is monotonous is *not* 100% effective in detecting defects. One of the advantages of automated inspection is the elimination of human error.

Human errors in inspection arise from multiple causes, of which four are most important: technique errors, inadvertent errors, conscious errors, and communication errors. The nature of these errors is similar to the same categories for personnel in other activities (see Section 4.9 under "Test of Theories Involving Human Error"). For specific elaboration on inspection errors, see *JQH6,* pages 595–602.

Measure of Inspector Accuracy

Some companies carry out regular evaluations of inspector accuracy as part of the overall evaluation of inspection performance. The plans employ a check inspector who periodically reviews random samples of work previously inspected by the various inspectors. The check inspection findings are then summarized, weighted, and converted into some index of inspector performance. *JQH5,* pages 23.51–23.53, explains this procedure.

Harris and Chaney (1969) provided some early research on inspector accuracy. Among their findings were that inspection accuracy decreases with reductions in defect rates, inspection accuracy increases with repeated inspections (up to a total of six), inspection accuracy decreases with additional product complexity, and the effect cannot be overcome by increasing the allowable inspection time. These conclusions are sobering.

✦ 10.8 ERRORS OF MEASUREMENT

Variation in a process has two sources: variation of the process making the product and variation of the measurement process (see Figure 10.3). Particularly with the low defect levels demanded under the six sigma approach, we must understand the capability of the manufacturing process and also the capability of the measurement process.

Even when correctly used, a measuring instrument may not give a true reading of a characteristic. The difference between the true value and the measured value can be due to one or more of five sources of variation (Figure 10.2).

There is much confusion as to terminology. *Bias* is sometimes referred to as "accuracy." Because *accuracy* has several meanings in the literature (especially in measuring instrument catalogs), its use as an alternative for "bias" is not recommended.

The distinction between repeatability and bias is illustrated in Figure 10.3. *Repeatability* is often referred to as "precision."

Any statement of bias and repeatability (precision) must be preceded by three conditions:

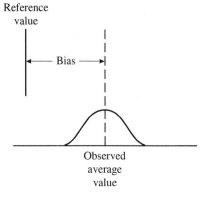

Bias

Bias is the difference between the observed average of measurements and the reference value. The reference value, also known as the accepted reference value or master value, is a value that serves as an agreed-upon reference for the measured values.[1] A reference value can be determined by averaging several measurements with a higher level (e.g., metrology lab or layout equipment) of measuring equipment.

[1]ASTM D 3980-88.

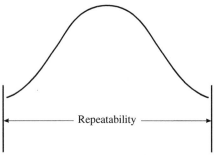

Repeatability

Repeatability is the variation in measurements obtained with one measurement instrument when used several times by an appraiser while measuring the identical characteristic on the same part.

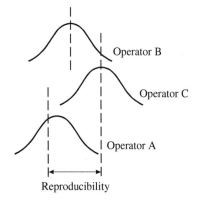

Reproducibility

Reproducibility is the variation in the average of the measurements made by different appraisers using the same measuring instrument when measuring the identical characteristic on the same part.

FIGURE 10.2

Five sources of measurement variation. (*Reprinted with permission from the MSA Manual. DaimlerChrysler, Ford, General Motors Supplier Quality Requirements Task Force.*)

1. *Definition of the test method.* This definition includes the step-by-step procedure, equipment to be used, preparation of test specimens, test conditions, etc.
2. *Definition of the system of causes of variability,* such as material, analysts, apparatus, laboratories, days, etc. ASTM recommends that modifiers of the word *precision* be used to clarify the scope of the precision measure. Examples of such modifiers are single operator, single analyst, single-laboratory-operator-material-day, and multilaboratory.

Inspection, Test, and Measurement 315

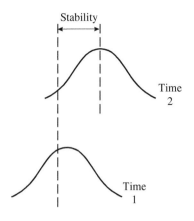

Stability
Stability (or drift) is the total variation in the measurements obtained with a measurement system on the same master or parts when measuring a single characteristic over an extended time period.

Linearity
Linearity is the difference in the bias values through the expected operating range of the gauge.

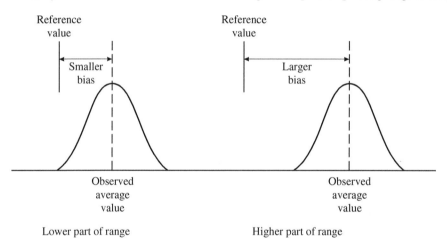

FIGURE 10.2 (*Continued*)
Five sources of measurement variation.

3. *Existence of a statistically controlled measurement process.* The measurement process must have stability for the statements on bias and precision to be valid. This stability can be verified by a control chart (see Section 20.6).

Effect of Measurement Error on Acceptance Decisions

Error of measurement can cause incorrect decisions on (1) individual units of product and on (2) lots submitted to sampling plans.

In one example of measuring the softening point of a material, the standard deviation of the test precision is 2°, yielding two standard deviations of ±4°. The specification

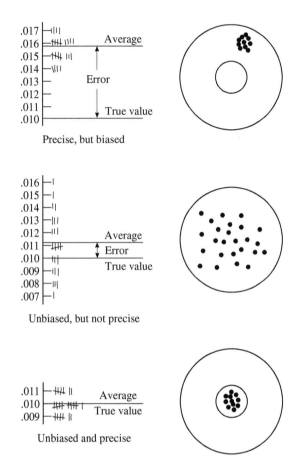

FIGURE 10.3
Distinction between bias and repeatability (precision).

limits on the material are ±3°. Imagine the incorrect decisions that are made under these conditions.

Two types of errors can occur in the classification of a product: (1) a nonconforming unit can be accepted (the consumer's risk) and (2) a conforming unit can be rejected (the producer's risk). In a classic paper, Eagle (1954) showed the effect of precision on each of these errors.

The probability of accepting a nonconforming unit as a function of measurement error (called test error, σ_{TE}, by Eagle) is shown in Figure 10.4. The abscissa expresses the test error as the standard deviation divided by the plus-or-minus value of the specification range (assumed equal to two standard deviations of the product).

For example, if the measurement error is one-half of the tolerance range, the probability is about 1.65% that a nonconforming unit will be read as conforming (due to the measurement error) and therefore will be accepted.

Figure 10.5 shows the percentage of *conforming* units that will be *rejected* as a function of the measurement error. For example, if the measurement error is one-half

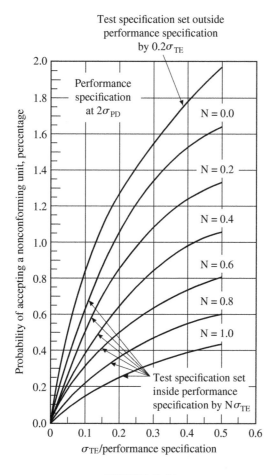

FIGURE 10.4
Probability of accepting a nonconforming unit.
(*From Eagle, 1954.*)

of the plus-or-minus tolerance range, about 14% of the units that are within specifications will be rejected because the measurement error will show that these conforming units are outside specification.

The test specification can be adjusted with respect to the performance specification (see Figures 10.4 and 10.5). Moving the test specification inside the performance specification reduces the probability of accepting a nonconforming product but increases the probability of rejecting a conforming product. The reverse occurs if the test specification is moved outside the performance specification. Both risks can be reduced by increasing the precision of the test, i.e., by reducing the value of σ_E (see "Reducing and Controlling Errors of Measurement").

Hoag et al. (1975) studied the effect of inspector errors on type I (α) and type II (β) risks of sampling plans. For a single sampling plan and an 80% probability of the inspector detecting a defect, the real value of β is two to three times that specified, and the real value of α is about one-fourth to one-half of that specified.

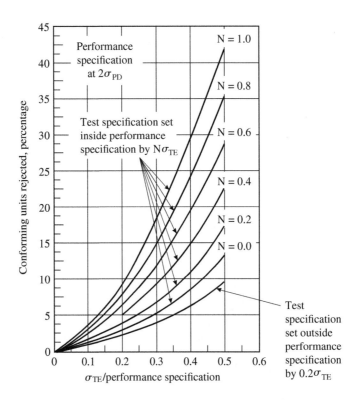

FIGURE 10.5
Conforming units rejected, percentage. (*From Eagle, 1954.*)

Case et al. (1975) investigated the effect of inspection error on the *average outgoing quality* (AOQ) of an attribute sampling procedure. They concluded that the AOQ values change and also significant changes can occur in the shape of the AOQ curve.

The Automotive Industry Action Group (1995, page 77) presents the concept of a gauge performance curve to determine the probability of accepting or rejecting a part when the gauge repeatability and reproducibility (R&R) are unknown.

All these investigations concluded that measurement error can be a serious problem.

Components of Variation

In drawing conclusions about measurement error, it is worthwhile to study the causes of variation in observed values. The relationship is:

$$\sigma_{observed} = \sqrt{\sigma^2_{cause\ A} + \sigma^2_{cause\ B} + \cdots + \sigma^2_{cause\ N}}$$

The formula assumes that the causes act independently.

It is valuable to find the numerical values of the components of observed variation because the knowledge may suggest where effort should be concentrated to reduce variation in the product. A separation of the observed variation into product variation plus other causes of variation may indicate important factors other than the manufacturing process. Thus, if it is found that the *measurement* error is a large percentage of the total variation, this finding must be analyzed before proceeding with a quality improvement program. Finding the components (e.g., instrument, operator) of this error may help to reduce the measurement error, which in turn may completely eliminate a problem.

Observations from an instrument used to measure a series of different units of product can be viewed as a composite of (1) the variation due to the measuring method and (2) the variation in the product itself. This value can be expressed as:

$$\sigma_O = \sqrt{\sigma_P^2 + \sigma_E^2}$$

where $\sigma_O = \sigma$ of the observed data
$\sigma_P = \sigma$ of the product
$\sigma_E = \sigma$ of the measuring method

Solving for σ_P yields

$$\sigma_P = \sqrt{\sigma_O^2 - \sigma_E^2}$$

The components of measurement error often focus on repeatability and reproducibility (R&R). *Repeatability* concerns variation due to measurement gauges and equipment; *reproducibility* concerns variation due to human "appraisers" who use the gauges and equipment. Studies to estimate these components are often called "gauge R&R" studies.

A gauge R&R study can provide separate numerical estimates of repeatability and reproducibility. Two methods are usually used to analyze the measurement data. Each method requires a number of appraisers, a number of parts, and repeat trials of appraisers measuring different parts. For example, an R&R study might use three appraisers, 10 parts, and two trials.

One method analyzes averages and ranges of the measurement study data. This method requires minimum statistical background and does not require a computer. The second method is the analysis of variance, ANOVA (see Section 18.14). Compared to the first method, ANOVA requires a higher level of statistical knowledge for interpretation of the results but can evaluate the data for possible interaction between appraisers and parts. ANOVA is best done on a computer using MINITAB or other software. Overall, the ANOVA method is preferred to analyzing the averages and ranges. Detailed illustrations of each method are provided in the Automotive Industry Action Group booklet *Measurement Systems Analysis* (1995). Also, see Tsai (1988) for an example using ANOVA and considering both no interaction and interaction of operators and parts. Burdick and Larsen (1997) provide methods for constructing confidence intervals on measures of variability in R&R studies.

When the total standard deviation of repeatability and reproducibility is determined from ANOVA, a judgment must then be made on the adequacy of the measurement process. A common practice is to calculate 5.15σ ($\pm 2.575\sigma$) as the total spread of the measurements that will include 99% of the measurements. If 5.15σ is equal to or less than 10% of the specification range for the quality characteristic, the measurement process is viewed as acceptable for that characteristic; if the result is greater than 10%, the measurement process is viewed as unacceptable. Engel and DeVries (1997) examine how the practice of comparing measurement error with the specification interval relates to making correct decisions in product testing.

Reducing and Controlling Errors of Measurement

Steps can be taken to reduce and control errors for all sources of measurement variation. The systematic errors that contribute to bias can sometimes be handled by applying a numerical correction to the measured data. If an instrument has a bias of -0.001, then, on the average, it reads 0.001 too low. The data can be adjusted by adding 0.001 to each value of the data. Of course, it is preferable to adjust the instrument as part of a calibration program.

In a calibration program, the measurements made by an instrument are compared to a reference standard of known accuracy. If the instrument is found to be out of calibration, an adjustment is made.

A calibration program can become complex for these reasons:

1. The large number of measuring instruments.
2. The need for periodic calibration of many instruments.
3. The need for many reference standards.
4. The increased technological complexity of new instruments.
5. The variety of types of instruments, i.e., mechanical, electronic, chemical, etc.

A calibration program should include provisions for periodic audits. These follow the general approach for quality audits (see Chapter 16).

Precision in measurement can be improved through either or both of the following procedures:

- *Discovery of the causes of variation and remedy of these causes.* A useful step is to resolve the observed values into components of variation (see earlier discussion). This process can lead to the discovery of inadequate training, perishable reagents, lack of sufficient detail in procedures, and other such problems. This fundamental approach also points to other causes for which the remedy is unknown or uneconomic, i.e., basic redesign of the test procedure.
- *Multiple measurements and statistical methodology to control the error in measurement.* The use of multiple measurements is based on the following relationship (see Section 18.3):

$$\sigma_{\bar{x}} = \frac{\sigma}{\sqrt{n}}$$

The formula states that halving the error in measurement requires quadrupling (not doubling) the number of measurements.

As the number of tests grows larger and larger, a significant reduction in the error in measurement can be achieved only by taking a still *larger* number of additional tests. Thus the cost of the additional tests versus the value of the slight improvement in measurement error becomes an issue. The alternatives of reducing the causes of variation (by control charts or other techniques) must also be considered.

For reducing other forms of measurement error, see Automotive Industry Action Group (1995).

Example of Measurement System Analysis (MSA) in a Six Sigma Improvement Project (DMAIC) (Courtesy of Steve Wittig and Chris Arquette at a Juran Institute client)

Background

The paint line has a first run yield of 74%. This means that 26% of all frames need to be reworked at least once. Defects due to finish issues account for 15% and material (wood) issues account for 11%. This project looks only at finish defects because this is readily within our control. Any rework is non-value added and contributes to wasted paint/primer, labor, utilities, w.i.p., capacity, and more hazardous waste. Our goal is to improve first run yield to 90% only for finish defects.

Summary of MSA Effort

The paint line is very old and somewhat neglected. Our first MSA results were expectedly poor and the appraisers were contributing to the defect rate by rejecting good frames. We improved this by continued training of the appraisers by Q.C. We did two more MSAs with acceptable results. This will need to be an ongoing test/train routine. Figures 10.6, 10.7, and 10.8 are attribute MSA, and Figures 10.9 and 10.10 are variable MSA.

Validate Measurement System
Attribute Data Analysis — MSA 1

Sample #	Expert	Operator 1		Operator 2		Operator 3	
		Try 1	Try 2	Try 1	Try 2	Try 1	Try 2
1	Blister	Blister	Blister	Blister	Blister	Blister	Blister
2	Good	Light ed.	Good	Good	Good	Good	Good
3	Drip	Drip	Drip	Dirt	Dirt	Drip	Drip
4	Dirt	Contam	Dirt	Dirt	Dirt	Dirt	Dirt
5	Contam	Contam	Contam	Good	Good	Overrun	Overrun
6	Blister	Blister	Blister	Blister	Blister	Dirt	Blister
7	Good	Good	Good	Good	Good	Good	Good
8	Dirt	Light ed.	Contam	Good	Good	Dirt	Dirt
9	Good	Good	Drip	Dirt	Good	Drip	Drip
10	Good	Orange p.	Good	Good	Good	Good	Good
11	Dirt	Dirt	Good	Dirt	Dirt	Dirt	Dirt
12	Good	Contam	Light ed.	Good	Good	Overrun	Overrun
13	Good	Light ed.	Light ed.	Good	Good	Light ed.	Overrun
14	Contam	Contam	Contam	Good	Good	Good	Good
15	Drip	Drip	Light ed.	Dirt	Good	Drip	Drip
16	Light ed.	Light ed.	Light ed.	Good	Good	Good	Good
17	Dirt	Contam	Contam	Good	Dirt	Dirt	Dirt
18	Dirt	Contam	Contam	Dirt	Dirt	Dirt	Dirt
19	Blister	Blister	Good	Good	Blister	Blister	Blister
20	Good	Good	Good	Good	Good	Orange p.	Orange p.

FIGURE 10.6
Baseline attribute MSA on appraisers' accept/reject decisions.

Data Collection Plan for MSA

Ref.	List of questions to be answered	Tools to be used	Description data type	Sample size, number of samples	Data To Be Collected			Remarks
					Where to collect data	Who will collect data	How will data be recorded	
1.	How accurate are the appraisers at classifying defects?	MSA discrete	Appraiser accuracy/	20 frames	Paint line	QC inspectors/ Green Belts	Data sheet	
2.	How accurate are the appraisers at accept/reject decisions?	MSA discrete	Appraiser accuracy/	20 frames	Paint line	QC inspectors/ Green Belts	Data sheet	

Attribute Data Analysis — MSA 1 Results

Within appraiser

Assessment agreement

Appraiser	# inspected	# matched	Percent (%)	95.0% CI
1	20	11	55.0	(31.5, 76.9)
2	20	16	80.0	(56.3, 94.3)
3	20	18	90.0	(68.3, 98.8)

matched: Appraiser agrees with him/herself across trials.

Each appraiser vs standard

Assessment agreement

Appraiser	# inspected	# matched	Percent (%)	95.0% CI
1	20	8	40.0	(19.1, 63.9)
2	20	11	55.0	(31.5, 76.9)
3	20	12	60.0	(36.1, 80.9)

matched: Appraiser's assessment across trials agrees with standard.

Between appraisers

Assessment agreement

# inspected	# matched	Percent (%)	95.0% CI
20	2	10.0	(1.2, 31.7)

matched: All appraisers' assessments agree with each other.

All appraisers vs standard

Assessment agreement

# inspected	# matched	Percent (%)	95.0% CI
20	2	10.0	(1.2, 31.7)

matched: All appraisers' assessments agree with standard

Note: 38% were called bad that were good. 22% were called good that were bad. This could yield an improvement in the defect rate by 16%

FIGURE 10.7
Results of baseline attribute MSA—Results are not acceptable.

Attribute Data Analysis — MSA 2 Results

Within appraiser

Assessment agreement

Appraiser	# inspected	# matched	Percent (%)	95.0% CI
1	20	16	80.0	(56.3, 94.3)
2	20	19	95.0	(75.1, 99.9)
3	20	20	100.0	(86.1, 100.0)

matched: Appraiser agrees with him/herself across trials.

Each appraiser vs standard

Assessment agreement

Appraiser	# inspected	# matched	Percent (%)	95.0% CI
1	20	15	75.0	(50.9, 91.3)
2	20	19	95.0	(75.1, 99.9)
3	20	18	90.0	(68.3, 98.8)

matched: Appraiser's assessment across trials agrees with standard.

Between appraisers

Assessment agreement

# inspected	# matched	Percent (%)	95.0% CI
20	13	65.0	(40.8, 84.6)

matched: All appraisers' assessments agree with each other.

All appraisers vs standard

Assessment agreement

# inspected	# matched	Percent (%)	95.0% CI
20	13	65.0	(40.8, 84.6)

matched: All appraisers' assessments agree with standard.

Conclusion: MSA is greatly improved over first. Continue Q.C. training of inspectors to bring up agreement between all appraisers and standard.

FIGURE 10.8
Attribute MSA after improvement.

Measurement System Analysis
Sheen Gauge Study

Gauge R&R

Source	VarComp	%Contribution (of VarComp)
Total Gauge R&R	0.0519	0.69
Repeatability	0. 0298	0. 39
Reproducibility	0.02 21	0.29
Operator	0.0028	0. 04
Operator*measurement	0. 0193	0. 26
Part-to-part	7.4851	99.31
Total variation	7.5370	100.00

Source	StdDev (SD)	Study Var (5.15*SD)	%Study Var (%SV)
Total Gauge R&R	0.22776	1.1730	8.30
Repeatability	0.17248	0.8883	6 .28
Reproducibility	0.14874	0.7660	5 .42
Operator	0.05294	0.2726	1 .93
Operator*measurement	0.13900	0. 7159	5 .06
Part-to-part	2.73590	14.0899	99. 66
Total variation	2.74536	14.1386	100.00

Number of distinct categories = 17

FIGURE 10.9
Results of baseline variable data MSA on sheen gauge—results are acceptable.

Gauge R & R — Sheen Gauge

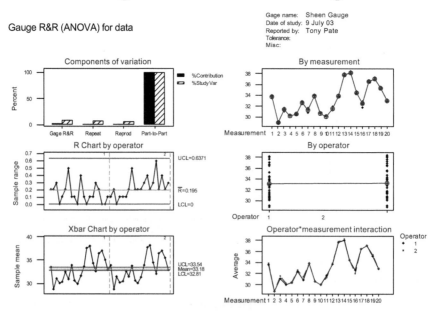

FIGURE 10.10

→ 10.9 HOW MUCH INSPECTION IS NECESSARY?

The amount of inspection to decide the acceptability of a lot can vary from no inspection to a sample to 100% inspection. The decision is governed mainly by the amount of prior knowledge available as to quality, the *homogeneity* of the lot, and the allowable degree of risk.

Prior knowledge that is helpful in deciding on the amount of inspection includes:

- Previous quality history on the product item and the supplier (internal or external).
- Criticality of the item on overall system performance.
- Criticality on later manufacturing or service operations.
- Warranty or use history.
- Process capability information. A process that is in statistical control with good uniformity around a target value (e.g., a 6σ process, see Section 20.13) will require minimum inspection.
- Measurement capability information, e.g., the availability of accurate and precise instruments.
- The nature of the manufacturing process. For example, some operations depend primarily on the adequacy of the setup.
- Inspection of the first few and the last few items in a production run. This is usually sufficient.
- Product homogeneity. For example, a fluid product is homogeneous and reduces the need for large sample sizes.
- Data on process variables and process conditions, e.g., as provided by automatic recording charts.
- Degree of adherence to the three elements of self-control for the personnel operating the process (see Section 6.3).

Competition to reduce costs has resulted in pressures to reduce the amount of inspection. The concept of inspection by the producers (self-inspection) has added to the focus of reducing inspection. Opportunities do exist for cost reduction in inspection activities. First, however, the causes of the high failure costs must be diagnosed and removed, and the prerequisites for self-inspection must be met.

> **EXAMPLE 10.4.** The Datapoint Corporation manufactures office and computer products (Adams, 1987). Part of the operation was 100% in-line inspection of visual characteristics by the quality staff. A dramatic shift was planned—production personnel would do their own visual inspection; the quality staff would perform an audit inspection and do diagnostic work on the causes of nonconformities. But a number of steps were required: a quality education process for first-line management, supervisors, and line personnel; special training in workmanship standards to help people recognize nonconformances; an 18-month implementation plan to phase in the new approach; use of data from functional acceptance tests for process yield reports; analysis of data; and a process audit system to review documentation, tools, materials, and people.
>
> The results were dramatic: The staff of 35 in-line inspectors was reduced to five process auditors, and scrap and rework plunged from 15 to 2%.

Economics of Inspection

We have several alternatives for evaluating lots:

1. *No inspection.* This approach is appropriate if the same lot has already been inspected by qualified laboratories, e.g., in other divisions of the same company or in supplier companies. Prior inspections by qualified production workers have the same effect.
2. *Small samples.* Small samples can be adequate if the process is inherently uniform and the order of production can be preserved. For example, in some punch press operations, the stamping dies are made with a high degree of stability. As a result, the successive pieces stamped out by such dies exhibit a high degree of uniformity for certain dimensional characteristics. For such characteristics, if the first and last pieces are correct, the remaining pieces are also correct, even for lot sizes running to many thousands of pieces. In its generalized form, the press example is one of a high degree of process capability combined with "stratified" sampling—sampling based on knowledge of the order of production.

    Small samples can also be used when the product is homogeneous due to its fluidity (gases, liquids) or to prior mixing operations. This homogeneity need not be assumed—it can be verified by sampling. Even solid materials may be homogeneous due to *prior* fluidity. Once homogeneity has been established, the sampling needed is minimal.

    For a product with a continuing history of good quality, sampling can be periodic, e.g., "skip-lot" or "chain sampling" (see *JQH5*, Section 46).
3. *Large samples.* In the absence of prior knowledge, the information about lot quality must be derived solely from sampling, which means random sampling and hence relatively large samples. The actual sample sizes depend on two main variables: (a) the tolerable percentage of defects and (b) the risks that can be accepted. Once values have been assigned to these variables, the sample sizes can be determined scientifically in accordance with the laws of probability (see Section 17.15). However, the choices of defect levels and risks are based largely on empirical judgments.

    Random sampling is clearly needed in cases where there is no ready access to prior knowledge, e.g., purchases from certain suppliers. However, there remain many cases in which random sampling is used despite the availability of inputs such as process capability, order of manufacture, fluidity, etc. A major obstacle is a lack of publications which show how to design sampling plans in ways which use these inputs. In the absence of such publications, quality planners are faced with creating their own designs. This means added work amid the absence of protection derived from the use of recognized, authoritative published materials.

    See *JQH5*, pages 46.15–46.17, for a discussion of the formation of inspection lots and the selection of samples.
4. *One hundred percent inspection.* This technique is used when the results of sampling show that the level of defects present is too high for the product to go on to the users. In critical cases, added provisions may be needed to guard against inspector fallibility, e.g., automated inspection or redundant 200% inspection.

An economic evaluation of these alternatives requires a comparison of *total* costs under each one.

Let N = number of items in lot
   n = number of items in sample
   p = proportion defective in lot
   A = damage cost incurred if a defective slips through inspection
   I = inspection cost per item
   P_a = probability that lot will be accepted by sampling plan

A and I are sometimes denoted as k_1, and k_2, respectively.

Consider the comparison of sampling inspection versus 100% inspection. Suppose it is assumed that no inspection errors occur and the cost to replace a defective found in inspection is borne by the producer or is small compared to the damage or inconvenience caused by a defective. The total costs are summarized in Table 10.3. These costs reflect both inspection costs and damage costs and recognize the probability of accepting or rejecting a lot under sampling inspection. The expressions can be equated to determine a break-even point. If it is assumed that the sample size is small compared to the lot size, the break-even point, p_b, is:

$$p_b = \frac{I}{A}$$

If the lot quality (p) is less than p_b, the total cost will be lowest with sampling inspection or no inspection. If p is greater than p_b, 100% inspection is best. This principle is often called the Deming *kp* rule.

For example, a microcomputer device costs $.50 per unit to inspect. A damage cost of $10.00 is incurred if a defective device is installed in the larger system. Therefore:

$$p_b = \frac{0.50}{10.00} = 0.05 = 5.0\%$$

If the percentage defective is expected to be greater than 5%, then 100% inspection should be used. Otherwise, use sampling or no inspection.

The variability in quality from lot to lot is important. If past history shows that the quality level is much better than the break-even point and is stable from lot to lot, little if any inspection may be needed. If the level is much worse than the break-even point and consistently so, it will usually be cheaper to use 100% inspection rather than sampling. If the quality is at neither of these extremes, a detailed economic comparison of no inspection, sampling, and 100% inspection should be made. Sampling is usually

TABLE 10.3
Economic comparison of inspection alternatives

Alternative	Total cost
No inspection	NpA
Sampling	$nI + (N - n)pAP_a + (N - n)(1 - P_a)I$
100% inspection	NI

best when the product is a mixture of high-quality lots and low-quality lots or when the producer's process is not in a state of statistical control.

The high costs of component failures in complex electronic equipment coupled with the development of automatic testing equipment for components has resulted in the economic justification of 100% inspection for some electronic components. The cost of finding and correcting a defective can increase by a ratio of 10 for each major stage that the product moves to from production to the customer, i.e., if it costs $1 at incoming inspection, the cost increases to $10 at the printed circuit board stage, $100 at the system level, and $1000 in the field.

→ 10.10 THE CONCEPT OF ACCEPTANCE SAMPLING

Acceptance sampling is the process of evaluating a portion of the product in a lot for the purpose of accepting or rejecting the entire lot. It involves the application of specific sampling plans to a designated lot or sequence of lots. Any acceptance sampling application must distinguish whether the purpose is to accumulate information on the immediate product being sampled or on the process which produced an immediate product at hand.

The main advantage of sampling is economy. Despite some added costs for designing and administering sampling plans, the lower costs of inspecting only part of the lot result in an overall cost reduction.

In addition to this main advantage, there are others:

- The smaller inspection staff is less complex and less costly to administer.
- There is less damage to the product, i.e., handling incidental to inspection is itself a source of defects.
- The lot is disposed of in shorter (calendar) time so that scheduling and delivery are improved.
- The problem of monotony and inspector error induced by 100% inspection is minimized.
- Rejection (rather than sorting) of nonconforming lots tends to dramatize the quality deficiencies and to urge the organization to look for preventive measures.
- Proper design of the sampling plan commonly requires study of the actual level of quality required by the user. The resulting knowledge is a useful input to the overall quality planning.

The disadvantages are sampling risks, greater administrative costs, and less information about the product than provided by 100% inspection.

Acceptance sampling is used when (1) the cost of inspection is high in relation to the damage cost resulting from passing a defective product, (2) 100% inspection is monotonous and causes inspection errors, or (3) the inspection is destructive. Acceptance sampling is most effective when it is preceded by a prevention program that achieves an acceptable level of quality of conformance.

We must also emphasize what acceptance sampling does not do. It does not provide refined estimates of lot quality. (It does determine, with specified risks, an acceptance or rejection decision on each lot.) Also, acceptance sampling does not provide judgments whether or not rejected product is fit for use. (It does give a decision on a lot with respect to the defined quality specification.)

In recent years, the emphasis on statistical process control has led some practitioners to conclude that acceptance sampling is no longer a valid concept. Their belief, stated here in oversimplified terms, is that only two levels of inspection are valid—no inspection or 100% inspection. This text takes the viewpoint that the concept of prevention (using statistical process control and other statistical and managerial techniques) is the foundation for meeting product requirements. Acceptance sampling procedures are, however, important in a program of *acceptance control.* Under this latter approach, described at the end of this chapter, sampling procedures are continually matched to process history and quality results. This step ultimately leads to phasing out acceptance sampling in favor of supplier certification and process control.

This chapter presents examples of specific acceptance sampling plans.

For some perceptive discussions of the modern role of acceptance sampling, see Schilling (1994) and Taylor (1994). *JQH5* provides specifics on the various types of sampling plans.

→ 10.11 SAMPLING RISKS: THE OPERATING CHARACTERISTIC CURVE

Neither sampling nor 100% inspection can guarantee that every defective item in a lot will be found. Sampling involves a risk that the sample will not adequately reflect the conditions in the lot; 100% inspection has the risk that monotony and other factors will result in inspectors missing some of the defectives (see Section 10.7). Both of these risks can be quantified.

Sampling risks are of two kinds:

1. Good lots can be rejected (the producer's risk). This risk corresponds to the α risk.
2. Bad lots can be accepted (the consumer's risk). This risk corresponds to the β risk.

The α and β risks are discussed in Section 18.7.

The operating characteristic (OC) *curve* for a sampling plan quantifies these risks. The OC curve for an attribute plan is a graph of the percentage defective in a lot versus the probability that the sampling plan will accept a lot. Because p is unknown, the probability must be stated for all possible values of p. It is assumed that an infinite number of lots are produced. Figure 10.11 shows an "ideal" OC curve where it is desired to accept all lots 1.5% defective or less and reject all lots having a quality level greater than 1.5% defective. All lots less than 1.5% defective have a probability of acceptance of 1.0 (certainty); all lots greater than 1.5% defective have a probability of acceptance of zero. Actually, however, no sampling plan exists that can discriminate perfectly; there always remains some risk that a "good" lot will be rejected or that a "bad" lot will be accepted. The best that can be achieved is to make the acceptance of good lots more likely than the acceptance of bad lots.

An acceptance sampling plan basically consists of a sample size (n) and an acceptance criterion (c). For example, a sample of 125 units is to be randomly selected from

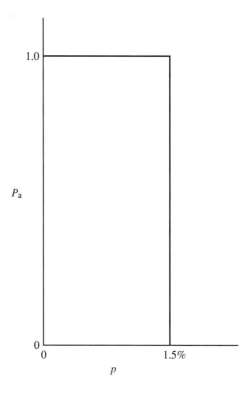

FIGURE 10.11
Ideal OC curve.

the lot. If five or fewer defectives are found, the lot is accepted. If six or more defectives are found, the lot is rejected.

The sample of 125 could, by the laws of chance, contain 0, 1, 2, 3, even up to 125 defectives. It is this *sampling variation* that causes some good lots to be rejected and some bad lots to be accepted. The OC curve for $n = 125$ and $c = 5$ is curve A, Figure 10.12. (The other curves will be discussed later.) A 1.5% defective lot has about a 98% chance of being accepted. A much worse lot, say 6% defective, has a 23% chance of being accepted. With the risks stated quantitatively, the adequacy of the sampling plan can be judged.

The OC curve for a specific plan states *only* the chance that a lot having p percent defective will be accepted by the sampling plan. The OC curve does *not:*

- Predict the quality of lots submitted for inspection. For example (Figure 10.12), it is incorrect to say that there is a 36% chance that the lot quality is 5% defective.
- State a "confidence level" in connection with a specific percentage defective.
- Predict the final quality achieved after all inspections are completed.

These and other myths about the OC curve require careful explanation of the concept to those using it. (Acceptable quality level is explained in Section 10.12.)

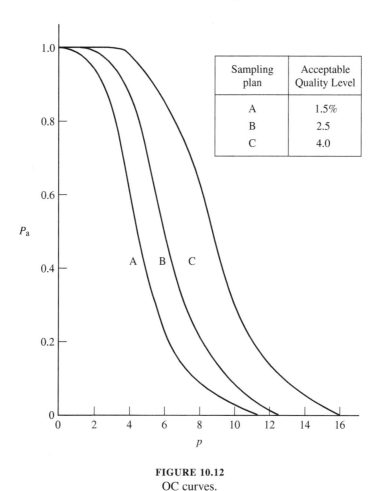

FIGURE 10.12
OC curves.

Constructing the Operating Characteristic Curve

An OC curve can be developed by determining the probability of acceptance for several values of incoming quality, p. The probability of acceptance is the probability that the number of defectives in the sample is equal to or less than the acceptance number for the sampling plan. Three distributions can be used to find the probability of acceptance: the hypergeometric, binomial, and Poisson distributions. When its assumptions can be met, the Poisson distribution is preferable because of the ease of calculation.

Grant and Leavenworth (1996, pages 183–193) describe the use of hypergeometric and binomial distributions.

The Poisson distribution yields a good approximation for acceptance sampling when the sample size is at least 16, the lot size is at least 10 times the sample size, and p is less than 0.1. The Poisson distribution function as applied to acceptance sampling is

$$P\begin{pmatrix} \text{exactly} \\ r \text{ defectives} \\ \text{in sample of } n \end{pmatrix} = \frac{e^{-np}(np)^r}{r!}$$

The equation can be solved by using a calculator or by using Table C in Appendix II. This table gives the probability of *r or fewer* defectives in a sample of n from a lot having a fraction defective of p. To illustrate Table C, consider the sampling plan previously cited: $n = 125$ and $c = 5$. To find the probability of accepting a 4% defective lot, calculate np as $125(0.04) = 5.00$. Table C then gives the probability of five or fewer defectives as 0.616. Figure 10.12 (curve A) shows this probability as the value of P_a for 4% defective lot quality.

The preceding discussion of sampling risks assumes that the proportion defective of incoming lots is reasonably constant. This assumption is often made in practice. Chun and Rinks (1998) derive modified producer's and consumer's risks when incoming quality is not constant.

→ 10.12 QUALITY INDEXES FOR ACCEPTANCE SAMPLING PLANS

Many of the published plans can be categorized in terms of one of several quality indexes:

1. *Acceptable quality level (AQL)*. The units of quality level can be selected to meet the particular needs of a product. Thus ANSI/ASQC Z1.4 (1993) defines AQL as "the maximum percent nonconforming (or the maximum number of nonconformities per hundred units) that, for purposes of sampling inspection, can be considered satisfactory as a process average." If a unit of product can have a number of different defects of varying seriousness, then demerits can be assigned to each type and product quality measured in terms of demerits. Because an AQL is an *acceptable* level, the probability of acceptance for an AQL lot should be high (see Figure 10.13).
2. *Limiting quality level (LQL)*. LQL defines *unsatisfactory* quality. Different titles are sometimes used to denote an LQL; for example, Dodge-Romig plans use the term *lot tolerance percentage defective (LTPD)*. Because an LQL is an *unacceptable* level, the probability of acceptance for an LQL lot should be low (see Figure 10.13). In some tables, this probability, known as the consumer's risk, is designated P_c, and has been standardized at 0.1. The consumer's risk is not the probability that the consumer will actually receive product at the LQL. The consumer will, in fact, not receive 1 lot in 10 at LQL fraction defective. What the consumer actually gets depends on the actual quality in the lots *before* inspection and on the probability of acceptance.
3. *Indifference quality level (IQL)*. IQL is a quality level somewhere between the AQL and LQL. It is frequently defined as the quality level that has a probability of acceptance of 0.5 for a given sampling plan (see Figure 10.13).

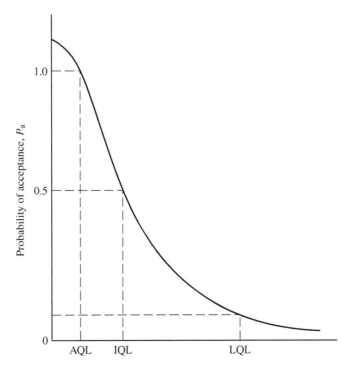

FIGURE 10.13
Quality indexes for sampling plans.

It should be emphasized to both internal and external suppliers that all product submitted for inspection is expected to meet specifications. An acceptable quality level does not mean that submitting a certain amount of nonconforming product is approved. The AQL simply recognizes that, under sampling, some nonconforming product will pass through the sampling scheme.

4. *Average outgoing quality limit (AOQL).* An approximate relationship exists between the fraction defective in the material before inspection (incoming quality p) and the fraction defective remaining after inspection (outgoing quality AOQ): AOQ = pP_a. When incoming quality is perfect, outgoing quality must also be perfect. However, when incoming quality is bad, outgoing quality will also be perfect (assuming no inspection errors) because the sampling plan will cause all lots to be rejected and inspected in detail. Thus at either extreme—incoming quality excellent or terrible—the outgoing quality will tend to be good. Between these extremes is the point at which the percentage of defectives in the outgoing material will reach its maximum. This point is the average outgoing quality limit (AOQL). For a sample calculation, see *JQH5,* page 46.11.

These indexes apply primarily when the production occurs in a continuing series of lots. The LQL concept is recommended for isolated lots. The indexes were originally developed by statisticians to help describe the characteristics of sampling plans.

Misinterpretations (particularly of the AQL) are common and are similar to those mentioned in Section 10.3. For example, a sampling plan based on AQL *will* accept some lots that have a quality level worse than the AQL.

10.13 TYPES OF SAMPLING PLANS

Sampling plans are of two types:

1. *Attributes plans.* A random sample is taken from the lot, and each unit is classified as acceptable or defective. The number defective is then compared with the allowable number stated in the plan, and a decision is made to accept or reject the lot. This chapter illustrates attributes plans based on AQL.
2. *Variables plans.* A sample is taken, and a *measurement* of a specified quality characteristic is made on each unit. These measurements are then summarized into a sample statistic (e.g., sample average), and the observed value is compared with an allowable value defined in the plan. A decision is then made to accept or reject the lot.

The key advantage of a variables sampling plan is the additional information provided in each sample that, in turn, results in smaller sample sizes compared with an attributes plan that has the same risks. However, if a product has several important quality characteristics, each must be evaluated against a separate variables acceptance criterion (e.g., numerical values must be obtained and the average and standard deviation for each characteristic calculated). In a corresponding attributes plan, the sample size required may be higher, but the several characteristics can be treated as a group and evaluated against one set of acceptance criteria. *JQH5,* Section 46, provides examples of variables plans.

Single, Double, and Multiple Sampling

Many published sampling tables give a choice among single, double, and multiple sampling. In single-sampling plans, a random sample of n items is drawn from the lot. If the number of defectives is less than or equal to the acceptance number (c), the lot is accepted. Otherwise, the lot is rejected. In double-sampling plans (Figure 10.14), a smaller initial sample is usually drawn, and a decision to accept or reject is reached on the basis of this smaller first sample if the number of defectives is either quite large or quite small. A second sample is taken if the results of the first are not decisive. Because it is necessary to draw and inspect the second sample only in borderline cases, the average number of pieces inspected per lot is generally smaller in double sampling. In multiple-sampling plans, one, two, or several still smaller samples are taken, usually continuing as needed until a decision to accept or reject is obtained. Thus, double- and multiple-sampling plans may mean less inspection but are more complicated to administer.

In general, it is possible to derive single-, double-, or multiple-sampling schemes with essentially identical OC curves (see *JQH5,* page 46.17 and Table 46.6).

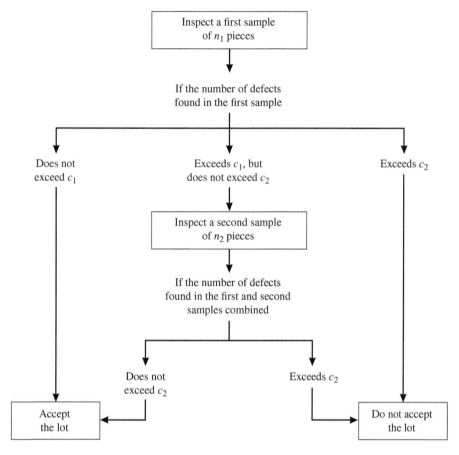

FIGURE 10.14
Schematic operation of double sampling.

✧ 10.14 CHARACTERISTICS OF A GOOD ACCEPTANCE PLAN

An acceptance sampling plan should have these characteristics:

- The index (AQL, AOQL, etc.) used to define "quality" should reflect the needs of the consumer and the producer and not be chosen primarily for statistical convenience.
- The sampling risks should be known in quantitative terms (the OC curve). The producer should have adequate protection against the rejection of good lots; the consumer should be protected against the acceptance of bad lots.
- The plan should minimize the *total* cost of inspection of all products. This requires careful evaluation of the pros and cons of attributes and variables plans, as well as single, double, and multiple sampling. It should also reflect product priorities, particularly from the fitness-for-use viewpoint.

- The plan should use other knowledge, such as process capability, supplier data, and other information.
- The plan should have built-in flexibility to reflect changes in lot sizes, quality of product submitted, and any other pertinent factors.
- The measurements required by the plan should provide information useful in estimating individual lot quality and long-run quality.
- The plan should be simple to explain and administer.

See *JQH5*, Section 46, for elaboration of these characteristics. Fortunately, published tables are available that meet many of these characteristics. We now proceed to a discussion of AQL plans.

→ 10.15 ANSI/ASQC Z1.4

ANSI/ASQC Z1.4 (1993) is an attributes sampling system. Its quality index is the acceptable quality level (AQL). The AQL is the maximum percentage nonconforming (or the maximum number of nonconformities per 100 units) that, for purposes of sampling inspection, can be considered satisfactory as a process average. (The standard uses the term *nonconformity* rather than *defective unit*.) The probability of accepting material of AQL quality is always high but not exactly the same for all plans. For lot quality just equal to the AQL, the "percentage of lots expected to be accepted" ranges from about 89 to 99. The choice may be made from 26 available AQL values ranging from 0.010 to 1000.0. (AQL values of 10.0 or less may be interpreted as percentage nonconforming or nonconformities per 100 units; values above 10.0 are interpreted as nonconformities per 100 units.)

The tables specify the relative amount of inspection to be used as "inspection level" I, II, or III; level II is regarded as normal. The inspection-level concept permits the user to balance the cost of inspection against the amount of protection required. The three levels involve inspection in amounts roughly in the ratio 0.4:1.0:1.6. (Four additional inspection levels are provided for situations requiring "small-sample inspection.")

A plan is chosen from the tables as follows:

1. The following information must be known:
    - AQL.
    - Lot size.
    - Type of sampling (single, double, or multiple).
    - Inspection level (usually level II).
2. Knowing the lot size and inspection level, a code letter is obtained from Table 10.4.
3. Knowing the code letter, AQL, and type of sampling, the sampling plan is read from Table 10.5. (Table 10.5 is for single sampling; the standard also provides tables for double and multiple sampling.)

For example, suppose that a purchasing agency has contracted for a 1.5% AQL. Suppose also that the parts are bought in lots of 1500 pieces. The table of sample-size code letters (Table 10.4) shows that K plans are required for inspection level II. Table 10.5

Inspection, Test, and Measurement

TABLE 10.4
Sample-size code letters

Lot or batch size	Special inspection levels				General inspection levels		
	S-1	S-2	S-3	S-4	I	II	III
2–8	A	A	A	A	A	A	B
9–15	A	A	A	A	A	B	C
16–25	A	A	B	B	B	C	D
26–50	A	B	B	C	C	D	E
51–90	B	B	C	C	C	E	F
91–150	B	B	C	D	D	F	G
151–280	B	C	D	E	E	G	H
281–500	B	C	D	E	F	H	J
501–1200	C	C	E	F	G	J	K
1201–3200	C	D	E	G	H	K	L
3201–10,000	C	D	F	G	J	L	M
10,001–35,000	C	D	F	H	K	M	N
35,001–150,000	D	E	G	J	L	N	P
150,001–500,000	D	E	G	J	M	P	Q
500,001 and above	D	E	H	K	N	Q	R

states that the sample size is 125. For AQL = 1.5, the acceptance number is given as five and the rejection number as six. Therefore, the entire lot of 1500 articles may be accepted if five or fewer nonconforming articles are found but must not be accepted (rejected) if six or more are found.

Sampling risks are defined by the OC curve published in the standard. The curve for this plan is shown as curve A in Figure 10.12.

The standard provides single, double, and multiple plans for each code letter (i.e., lot-size category). The plans for code letter K are shown in Table 10.6. Thus, the three plans can be found under the AQL column of 1.5. For example, the double-sampling plan calls for a first sample of 80 units. If two or fewer nonconforming are found, the lot is accepted. If five or more nonconforming are found, the lot is not accepted. If three or four nonconforming are found in the sample of 80, then a second sample of 80 is taken, giving a cumulative sample size of 160. If the total number of nonconforming in both samples is six or less, the lot is accepted; seven or more nonconforming means lot rejection.

Switching Procedures in ANSI/ASQC Z1.4

ANSI/ASQC Z1.4 includes provision for tightened inspection if quality deteriorates. If two out of five consecutive lots are not acceptable (rejected) on original inspection, a tightened inspection plan is imposed. The sample size is the same as usual, but the acceptance number is reduced. (The tightened plans do require larger sample sizes if the probability of acceptance for an AQL lot is less than 0.75.) For the example

TABLE 10.5

Master table for normal inspection (single sampling)

Sample-size code letter	Sample size	\multicolumn{24}{c}{Acceptable quality levels (normal inspection)}

Sample-size code letter	Sample size	0.010 Ac Re	0.015 Ac Re	0.025 Ac Re	0.040 Ac Re	0.065 Ac Re	0.10 Ac Re	0.15 Ac Re	0.25 Ac Re	0.40 Ac Re	0.65 Ac Re	1.0 Ac Re	1.5 Ac Re
A	2	↓	↓	↓	↓	↓	↓	↓	↓	↓	↓	↓	↓
B	3	↓	↓	↓	↓	↓	↓	↓	↓	↓	↓	↓	↓
C	5	↓	↓	↓	↓	↓	↓	↓	↓	↓	↓	↓	↓
D	8	↓	↓	↓	↓	↓	↓	↓	↓	↓	↓	↓	0 1
E	13	↓	↓	↓	↓	↓	↓	↓	↓	↓	↓	0 1	↓
F	20	↓	↓	↓	↓	↓	↓	↓	↓	↓	0 1	↓	↑
G	32	↓	↓	↓	↓	↓	↓	↓	↓	0 1	↓	↑	1 2
H	50	↓	↓	↓	↓	↓	↓	↓	0 1	↓	↑	1 2	2 3
J	80	↓	↓	↓	↓	↓	↓	0 1	↓	↑	1 2	2 3	3 4
K	125	↓	↓	↓	↓	↓	0 1	↓	↑	1 2	2 3	3 4	5 6
L	200	↓	↓	↓	↓	0 1	↓	↑	1 2	2 3	3 4	5 6	7 8
M	315	↓	↓	↓	0 1	↓	↑	1 2	2 3	3 4	5 6	7 8	10 11
N	500	↓	↓	0 1	↓	↑	1 2	2 3	3 4	5 6	7 8	10 11	14 15
P	800	↓	0 1	↓	↑	1 2	2 3	3 4	5 6	7 8	10 11	14 15	21 22
Q	1250	0 1	↓	↑	1 2	2 3	3 4	5 6	7 8	10 11	14 15	21 22	↑
R	2000	↓	↑	1 2	2 3	3 4	5 6	7 8	10 11	14 15	21 22	↑	↑

Notes: ↓, use first sampling plan below arrow. If sample size equals, or exceeds, lot of batch size, do 100 percent inspection.
↑, use first sampling plan above arrow.
Ac, acceptance number.
Re, rejection number.

previously cited, the tightened plan can be read from Table 10.6 as a sample size of 125 and an acceptance number of three.

ANSI/ASQC Z1.4 also provides for reduced inspection when the supplier's record has been good. The preceding 10 lots must have had a normal inspection with all lots accepted. A table of lower limits for the process average is provided to help decide whether the supplier's record has been good enough to switch to reduced inspection. The plan does, however, provide an option of switching to reduced inspection without using the table of lower limits. Under reduced sampling, the sample size is usually 40% of the normal sample size.

These switching rules apply when production is submitted at a steady rate. The plan provides other rules for using normal, tightened, and reduced inspection.

Other Provisions of ANSI/ASQC Z1.4

The standard provides OC curves for most of the individual plans along with "limiting quality" values for a probability of acceptance of 10 and 5%. Average sample-size curves for double and multiple sampling are also included. The latter curves show the average sample sizes expected as a function of the product quality submitted. Although the OC curves are roughly the same for single, double, and multiple

Inspection, Test, and Measurement

TABLE 10.5

Master table for normal inspection (single sampling) (*cont.*)

					Acceptable quality levels (normal inspection)							
2.5	4.0	6.5	10	15	25	40	65	100	150	250	400	650	1000
Ac Re	Ac Re	Ac Re	Ac Re	Ac Re	Ac Re	Ac Re	Ac Re	Ac Re	Ac Re	Ac Re	Ac Re	Ac Re	Ac Re

```
                    0  1              1  2    2  3    3  4    5  6    7  8   10 11  14 15  21 22  30 31
             0  1                     1  2    2  3    3  4    5  6    7  8   10 11  14 15  21 22  30 31  44 45
  0  1                    1  2   2  3    3  4    5  6    7  8   10 11  14 15  21 22  30 31  44 45
             1  2   2  3    3  4    5  6    7  8   10 11  14 15  21 22  30 31  44 45
       1  2   2  3    3  4    5  6    7  8   10 11  14 15  21 22  30 31  44 45
  1  2   2  3    3  4    5  6    7  8   10 11  14 15  21 22
  2  3   3  4    5  6    7  8   10 11  14 15  21 22
  3  4   5  6    7  8   10 11  14 15  21 22
  5  6   7  8   10 11  14 15  21 22
  7  8  10 11  14 15  21 22
 10 11  14 15  21 22
 14 15  21 22
 21 22
```

sampling, the average sample-size curves vary considerably because of the inherent differences among the three types of sampling. The standard also states the AOQL that would result if all rejected lots were screened for nonconforming units.

In ANSI/ASQC Z1.4, *a sampling scheme* is defined as "a combination of sampling plans with switching rules and possibly a provision for discontinuance of inspection." For the sampling schemes associated with the individual plans, the standard provides OC curves and information on AOQL, limiting quality, and average sample sizes—all for single sampling.

Dodge-Romig Sampling Tables

Dodge and Romig (1959) provide four sets of attributes plans emphasizing either lot-by-lot quality (LTPD) or long-run quality (AOQL);

Lot tolerance percentage defective (LTPD):	single sampling
	double sampling
Average outgoing quality limit (AOQL):	single sampling
	double sampling

These plans differ from those in ANSI/ASQC Z1.4 in that the Dodge-Romig plans assume that all rejected lots are 100% inspected and the defectives are replaced with acceptable items. Plans with this feature are called *rectifying inspection plans*. The tables provide protection against poor quality on either a lot-by-lot basis or

TABLE 10.6
Sampling plan for sample-size code letter K

Type of sampling plan	Cumulative sample size	Acceptable quality levels (normal inspection)											
		Less than 0.10		0.10		0.15		×	...	1.0		1.5	...
		Ac	Re	Ac	Re	Ac	Re	Ac	Re	Ac	Re	Ac	Re
Single	125	∇		0	1	0				3	4	5	6
Double	80	∇		*		Use letter J		Use letter M		1	4	2	5
	160									4	5	6	7
Multiple	32	∇		*						#	3	#	4
	64									0	3	1	5
	96									1	4	2	6
	128									2	5	3	7
	160									3	6	5	8
	192									4	6	7	9
	224									6	7	9	10
		Less than 0.15		0.15		×		0.25		1.5		2.5	

Acceptable quality levels (tightened inspection)

Notes: Δ, use next preceding sample-size code letter for which acceptance and rejection numbers are available.
∇, use next subsequent sample-size code letter for which acceptance and rejection numbers are available.
Ac, acceptance number.
Re, rejection number.
*, use single-sampling plan above (or alternatively use letter N).
#, acceptance not permitted at this sample size.

average long-run quality. The LTPD plans assure that a lot of poor quality will have a low probability of acceptance, i.e., the probability of acceptance (or consumer's risk) is 0.1 for a lot with LTPD quality. LTPD values range from 0.5 to 10.0% defective. The AOQL plans assure that, after all sampling and 100% inspection of rejected lots, the *average* quality over many lots will not exceed the AOQL. The AOQL values range from 0.1 to 10.0%. Each LTPD plan lists the corresponding AOQL, and each AOQL plan lists the LTPD.

→ 10.16 SELECTION OF A NUMERICAL VALUE OF THE QUALITY INDEX

The problem of selecting a value of the quality index (e.g., AQL, AOQL, or lot tolerance percentage defective) is one of balancing the cost of finding and correcting a defective against the loss incurred if a defective slips through an inspection procedure.

TABLE 10.6

Sampling plan for sample-size code letter K (*cont.*)

Acceptable quality levels (normal inspection)												
×		4.0		×		...		×		10		Higher than 10		Cumulative sample size
Ac	Re	Ac	Re	Ac	Re	Ac	Re	Ac	Re	Ac	Re	
8	9	10	11	12	13	18	19	21	22	Δ	10	125
3	7	5	9	6	10	9	14	11	16	Δ		80
11	12	12	13	15	16	23	24	26	27			160
0	4	0	5	0	6	1	8	2	9	Δ		32
2	7	3	8	3	9	6	12	7	14			64
4	9	6	10	7	12	11	17	13	19			96
6	11	8	13	10	15	16	22	19	25			128
9	12	11	15	14	17	22	25	25	29			160
12	14	14	17	18	20	27	29	31	33			192
14	15	18	19	21	22	32	33	37	38			224
4.0		×		6.5		10		×		Higher than 10		

Acceptable quality levels (tightened inspection)

Enell (1954), in a classic paper, has suggested using the break-even point (see Section 10.9) in the selection of an AQL. The *break-even point* for inspection is defined as the cost to inspect one piece divided by the damage done by one defective. For the example cited, the break-even point was 5% defective.

Because a 5% defective quality level is the break-even point between sorting and sampling, the appropriate sampling plan should provide for a lot to have a 50% probability of being sorted or sampled, i.e., the probability of acceptance for the plan should be 0.50 at a 5% defective quality level. The OC curves in a set of sampling tables such as ANSI/ASQC Z1.4 can now be examined to determine an AQL. For example, suppose that the device is inspected in lots of 3000 pieces. The OC curves for this case (code letter K) are shown in ANSI/ASQC Z1.4 and Figure 10.12. The plan closest to having a P_a of 0.50 for a 5% level is the plan for an AQL of 1.5%. Therefore, this is the plan to adopt.

Some plans include a classification of defects to help determine the numerical value of the AQL. Defects are first classified as critical, major, or minor according to definitions provided in the standard. Different AQLs may be designated for groups of defects considered collectively or for individual defects. Critical defects may have a 0% AQL, whereas major defects may be assigned a low AQL, say 1%, and minor defects a higher AQL, say 4%. Some manufacturers of complex products specify quality in terms of the number of defects per million parts.

In practice, quantification of the quality index is a matter of judgment based on the following factors: past performance in quality, effect of nonconforming product on later production steps, effect of nonconforming product on fitness for use, urgency of delivery requirements, and cost of achieving the specified quality level.

A thorough discussion of this difficult issue is provided by Schilling (1982, pages 571–586) in the classic, comprehensive book on acceptance sampling.

➔ 10.17 HOW TO SELECT THE PROPER SAMPLING PROCEDURES

The methods of acceptance sampling are many and varied. It is essential to select a sampling procedure appropriate to the acceptance sampling situation to which it is applied. Sampling procedures can serve different purposes. As itemized by Schilling (1982), they include:

- Guaranteeing quality levels at stated risks.
- Maintaining quality at AQL level or better.
- Guaranteeing an AOQL.
- Reducing inspection after good history.
- Checking inspection.
- Ensuring compliance with mandatory standards.
- Reliability sampling.
- Checking inspection accuracy.

For each purpose, Schilling recommends specific attributes or variables sampling plans. Selection of a plan depends on the purpose, the quality history, and the extent of knowledge of the process.

The steps involved in the selection and application of a sampling procedure are shown in Figure 10.15. Emphasis is on the feedback of information necessary for the proper application, modification, and evolution of sampling to encourage continuous improvement and reduced inspection costs. This can be achieved by moving from a system of acceptance sampling to acceptance control.

Moving from Acceptance Sampling to Acceptance Control

Acceptance sampling is the process of evaluating a portion of the product in a lot for the purpose of accepting or rejecting the entire lot as either conforming or not conforming to a quality specification. *Acceptance control* is a "continuing strategy of selection, application, and modification of acceptance sampling procedures to a changing inspection environment" (Schilling, 1982, page 564). This evaluation of a sampling plan application is shown in the life cycle of acceptance control (Table 10.7). The cycle is applied over the lifetime of a product to achieve (1) quality improvement (using process control and process capability concepts) and (2) reduction and elimination of inspection (using acceptance sampling).

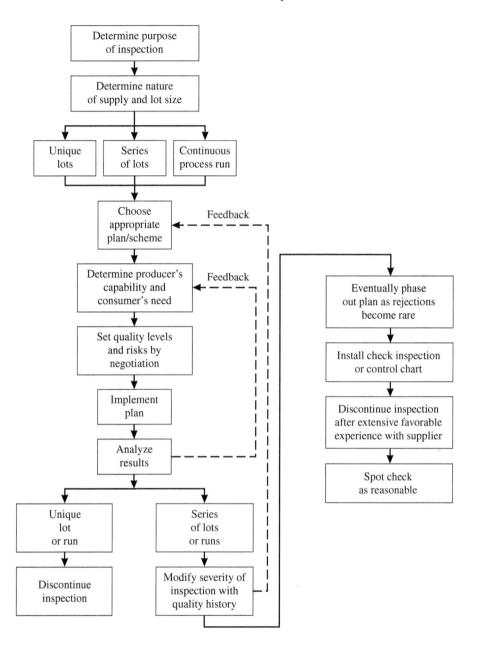

FIGURE 10.15
Check sequence for implementation of sampling procedure. From *Acceptance Sampling in Quality Control* by Edward G. Schilling. Copyright 1982 by Taylor & Francis Group LLC. Reproduced with permission of Taylor & Francis Group LLC in the format Textbook via Copyright Clearance Center.

TABLE 10.7
Life cycle of acceptance control application

Stage	Step	Method
Preparatory	Choose plan appropriate to purpose	Analysis of quality system to define the exact need for the procedure
	Determine producer capability	Process performance evaluation using control charts
	Determine consumer needs	Process capability study using control charts
	Set quality levels and risks	Economic analysis and negotiation
	Determine plans	Standard procedures if possible
Initiation	Train inspector	Include plan, procedure, records, and action
	Apply plan properly	Ensure random sampling
	Analyze results	Keep records and control charts
Operational	Assess protection	Periodically check quality history and OC curves
	Adjust plan	When possible, change severity to reflect quality history and cost
	Decrease sample size if warranted	Modify to use appropriate sampling plans taking advantage of credibility of supplier with cumulative results
Phase out	Eliminate inspection effort where possible	Use demerit rating or check inspection procedures when quality is consistently good
		Keep control charts
Elimination	Spot check only	Remove all inspection when warranted by extensive favorable history

From *Acceptance Sampling in Quality Control* by Edward G. Schilling. Copyright 1982 by Taylor & Francis Group LLC. Reproduced with permission of Taylor & Francis Group LLC in the format Textbook via Copyright Clearance Center.

SUMMARY

- Product acceptance involves three decisions: conformance, fitness for use, and communication.
- In deciding whether or not nonconforming product is fit for use, input must be secured from several sources.
- The communication decision involves both outsiders (customers) and insiders.
- Inspection planning includes the designation of inspection stations and spelling out the instructions and facilities required.
- A classification of characteristics is a list of quality characteristics derived from the specifications; a classification of defects is a list of symptoms of nonconformance during manufacture and field use.
- The amount of inspection necessary depends mainly on the amount of prior knowledge about product quality, homogeneity of the lot, and the allowable risk.

- Human errors in inspection arise from technique errors, inadvertent errors, conscious errors, and communication errors.
- Errors of measurement have two parts: precision and bias. Both parts can be quantified.
- Acceptance sampling is the process of evaluating a portion of the product in a lot for the purpose of accepting or rejecting the entire lot.
- We have four alternatives for evaluating lots: no inspection, small samples, large samples, or 100% inspection.
- Sampling risks are of two kinds: good lots can be rejected and bad lots can be accepted. An operating characteristic (OC) curve quantifies these risks.
- Published sampling plans categorize quality in terms of acceptable quality level (AQL), limiting quality level (LQL), indifference quality level (IQL), or average outgoing quality limit (AOQL).
- Attributes sampling plans evaluate each unit of product as simply acceptable or defective: ANSI/ASQC Z1.4 (1993) and the Dodge-Romig sampling tables are examples of attributes plans.
- Variables sampling plans involve measurements on product units.

PROBLEMS

10.1. Discuss the inspection of new homes with the appropriate municipal department. Determine the purpose of its inspection, the specification used, and how the inspection is conducted. Comment on this inspection from the viewpoint of the purchaser of a home.

10.2. In one large company making consumer durable products, the chief inspector operated on the principle that disposition of nonconforming lots of components must be done in only three ways: (1) by scrapping them, (2) by repairing them to bring them into conformance, or (3) by securing a waiver from the design department. What she did not permit was a tryout to see whether the components were usable despite nonconformance. Her stated reason was that if she authorized such tryouts, the production people would devote their energies to tryouts rather than to making the components right in the first place. What do you think of this approach?

10.3. Certain plates stamped out in a stamping press include holes for which there are close tolerances for diameter and for the distance between holes. In the discussion of how many pieces to gauge for these dimensions, one proposal is to measure the first and the last piece for each lot and to accept the lot if both pieces conform to specification. A statistician objects to this proposal on the grounds that the sample is only two pieces and that, if the lot were 50% defective, it could easily be accepted due to statistical variation. What is your opinion?

10.4. A large manufacturer of minicomputers is incurring high costs due to the need for assembling and testing the computers to discover and eliminate defects before the computers are installed in clients' computer rooms. After this assembly, testing, and repair, the computers are disassembled, shipped to the clients' facilities, reassembled, and checked out. How would you go about reducing the cost of preassembly and pretest?

10.5. You are a quality manager engaged in a seminar to discuss common problems with other quality managers. A lively discussion has developed over some problems associated with pressures applied to quality managers. The pressures concern the shipment of nonconforming, unfit, or even unsafe products. In addition, the pressures concern the matter of the quality manager signing a test certificate or other document that puts him on record as having approved something when he was really against it.

Here are several of the problem categories identified by the group:

- A lot of product has been made with a nonconformance to specification. All company managers (including the quality manager) are convinced that the nonconforming product is fit for use. They are not agreed on whether to tell the client about the nonconformance. The marketing manager is against informing the client on the ground that some clients may use such information to wring a price concession out of the manufacturer.

- A product lot contains a small percentage of units that are clearly unfit for use. There is a debate on whether to sort the lot to remove the defective units or to ship the lot as is and pay any claims as they arise. The production manager (who wants to ship the product without sorting) contends that the problem is purely economic and that the quality considerations are secondary.

- A large electromechanical system has been made under a contract that includes a penalty clause for late delivery. The system has already met its test requirements and is being crated for shipment. At this point, it is discovered that the test equipment used to test one of the subsystems was out of calibration at the time of the test. Under the accepted practice in the industry, such a discovery throws suspicion on the quality of the subsystem and hence on the quality of the system. Unfortunately, the subsystem is not easily accessible. It is buried deep within the system so that it would involve a serious delay as well as a large expense to take the system apart, test the subsystem, and then put it all back together again. The manufacturing people take the position that the subsystem is okay despite the condition of the test equipment. They note that another subsystem built by the same people using the sample process has just tested okay. They urge that the system be shipped based on this evidence of a reliable process and work force.

- A product with a good safety record has resulted in serious injury to a user. The injury involved a most unusual combination of unlikely events plus an obvious misuse by the user. The design manager defends the design on the record—the only known serious injury was associated with misuse.

What are your conclusions as to the position to be taken by the quality manager in the foregoing cases with respect to (a) shipping the product and (b) signing the documents?

10.6. An instrument has been used to measure the length of a part. The result was 6.70052 cm. An error-of-measurement study on the instrument had the following results:

Bias: +0.00254 cm (on the average, the instrument reads 0.00254 cm high).

Repeatability (precision): 0.001016 cm (1σ).

Make a statement concerning the true value of the part just measured. State all assumptions needed.

10.7. The precision of a specific mechanical gauge is indicated by a standard deviation (of individual repeat measurements) of 0.00254 cm. Investigate the effect on precision of making multiple measurements. Consider 2, 3, 4, 5, 10, 20, and 30 as multiples. Graph the results.

10.8. One ball bearing was measured by one inspector 13 times with each of two vernier micrometers. The results are shown here:

Measurement number	Model A	Model B
1	0.6557	0.6559
2	0.6556	0.6559
3	0.6556	0.6559
4	0.6555	0.6559
5	0.6556	0.6559
6	0.6557	0.6559
7	0.6556	0.6559
8	0.6558	0.6559
9	0.6557	0.6559
10	0.6557	0.6559
11	0.6556	0.6559
12	0.6557	0.6560
13	0.6557	0.6560

Suppose that the true diameter is 0.65600. Calculate measures of bias and precision for each micrometer. What restrictions must be placed on the applicability of the numbers you determined?
Answer: A: bias = -0.00035, precision = 1σ of 0.000075; B: bias = -0.000085, precision = 1σ of 0.000036.

10.9. A large sample of product has been measured. The mean was 2.506 inches and the standard deviation was 0.002 inches. A separate error-of-measurement study yielded the following results:

Bias = $+0.001$ in.

Precision: 0.0005 in. (1σ). The product has only a minimum specification limit. What should this limit be to take into account bias and precision and reject only 5% of the product?
Answer: 2.502.

10.10. A sample of measurements shows a mean and standard deviation of 2.000 in. and 0.004 in., respectively. These results are for the *observed* values. A separate error-of-measurement study indicates a precision of 0.002 in. (1σ). There is no bias error. The dimension has a specification of 2.000 ± 0.006 in. What percentage of the population has *true* dimensions outside the specification?

10.11. A large gray-iron foundry casts the base for precision grinders. These bases are produced at the rate of 18 per day and are 100% inspected at the foundry for flaws in the metal. The castings are then stored and subsequently shipped in lots of 300 to the grinder manufacturer. The grinder manufacturer has found that these lots are 10% defective. Upon receiving a lot, the manufacturer inspects 12 of them and rejects the lot if two or more defectives are found. What is the chance that the manufacturer will reject a given lot?

10.12. The following double-sampling plan has been proposed for evaluating a lot of 50 pieces:

Sample	Sample size	Acceptance number	Rejection number
1	3	0	3
2	3	2	3

Using only the addition and multiplication rules for probability, calculate the probability of accepting a lot that is 10% defective.
Answer: 0.984.

10.13. Prepare the operating characteristic curves for a single-sampling plan with an acceptance number of zero. Use sample sizes of 2, 5, 10, and 20.

10.14. You are quality manager for a company receiving large quantities of materials from a supplier in lots of 1000. The cost of inspecting the lots is $0.76/unit. The cost that is incurred if bad material is introduced into your product is $15.20/unit. A sampling plan of 75 with acceptance number equal to two has been submitted to you by one of your engineers. In the past, lots submitted by the supplier have averaged 3.4% defective.
(a) Is a sampling plan economically justified?
(b) Prepare an operating characteristic curve.
(c) If you want to accept only lots of 4% defective or better, what do you think of the sampling plan submitted by the engineer?
(d) Suppose that rejected lots are 100% inspected. If a supplier submits many 4% defective lots, what will be the average outgoing quality of these lots?

10.15. Refer to ANSI/ASQC Z1.4 with the following conditions:
- Lot size = 10,000.
- Inspection level II.
- Acceptable quality level = 4%.
(a) Find a single-sampling plan for normal inspection.
(b) Suppose that a lot is sampled and accepted. Someone makes the statement, "This means the lot has 4% or less defective." Comment on this statement. (Assume that the sample was randomly selected and no inspection errors were made.)
(c) Calculate the probability of accepting a 4% defective lot under normal inspection.
Answer: (c) 0.98.

10.16. A manufacturer sells his product in large lots to a customer who uses a sampling plan for incoming inspection. The plan calls for a sample of 200 units and an acceptance number of two. Rejected lots are returned to the manufacturer. If a lot is rejected and returned, the manufacturer has decided to gamble and send it right back to the customer without screening it (and without telling the customer it was a rejected lot). The manufacturer hopes that another random sample will lead to an acceptance of the lot. What is the probability that a 2% defective lot will be accepted on either one or two submissions to the customer?
Answer: 0.42.

10.17. A customer is furnished resistors under ANSI/ASQC Z1.4. Inspection level II has been specified with an AQL of 1.0%. Lot sizes vary from 900 to 1200.
(a) Which single-sampling plan will be used?
(b) Calculate the quality (in terms of percentage defective) that has an equal chance of being accepted or rejected.
(c) What is the chance that a 1% defective lot will be accepted?
Answer: (b) 3.4 percent; (c) 0.95.

10.18. A rule-of-thumb sampling plan states that, for any lot size, the sample size should be 10% of the lot and the acceptance number should be zero. It is believed that this procedure will hold the sampling risks constant. Prepare the operating characteristic curves using this plan for lot sizes of 100, 200, and 1000. Calculate points for quality levels of 0, 2, 4, 6, 10, and 14% defective. Compare the three curves and draw conclusions about the sampling risks.

REFERENCES

Adams, R. (1987). "Moving from Inspection to Audit," *Quality Progress,* January, pp. 30–31.
ANSI/ASQC Z1.4 (1993). *Sampling Procedures and Tables for Inspection by Attributes,* American Society for Quality Control, Milwaukee.
Automotive Industry Action Group (1995). *Measurement Systems Analysis,* Southfield, MI.
Baxter Travenol Laboratories (1986). *Statistical Process Control Guideline,* Baxter Travenol Laboratories, Deerfield, IL, p. 23.
Burdick, R. K. and G. A. Larsen (1997). "Confidence Intervals on Measures of Variability in R&R Studies," *Journal of Quality Technology,* vol. 29, no. 3, pp. 261–273.
Case, K. E., G. K. Bennett, and J. W. Schmidt (1975). "The Effect of Inspection Error on Average Outgoing Quality," *Journal of Quality Technology,* vol. 7, no. 1, pp. 1–12.
Chun, Y. H. and D. B. Rinks (1998). "Three Types of Producer's and Consumer's Risks in the Single Sampling Plan," *Journal of Quality Technology,* vol. 30, no. 3, pp. 254–268.
Cooper, J. (1997). "Implementing a Paperless Inspection System," *Annual Quality Congress Proceedings,* ASQ, Milwaukee, pp. 231–235.
Dodge, H. F. and H. G. Romig (1959). *Sampling Inspection Tables,* 2nd ed., John Wiley & Sons, New York.
Durkee, D. and B. Gookins (1999). "The Inspector's Role in Organizational Quality," *Annual Quality Congress Proceedings,* ASQ, Milwaukee, pp. 117–121.
Eagle, A. R. (1954). "A Method for Handling Errors in Testing and Measuring," *Industrial Quality Control,* March, pp. 10–14.
Enell, J. W. (1954). "What Sampling Plan Shall I Choose?" *Industrial Quality Control,* vol. 10, no. 6, pp. 96–100.
Engel, J. and B. DeVries (1997). "Evaluating a Well-Known Criterion for Measurement Precision," *Journal of Quality Technology,* vol. 29, no. 4, pp. 469–476.
Grant, E. L. and R. S. Leavenworth (1996). *Statistical Quality Control,* 7th ed., McGraw-Hill, New York.
Harris, D. H. and F. B. Chaney (1969). *Human Factors in Quality Assurance,* John Wiley and Sons, New York, pp. 77–85.
Hoag, L. L., B. L. Foote, and C. Mount-Campbell (1975). "The Effect of Inspector Accuracy on Type I and II Errors of Common Sampling Techniques," *Journal of Quality Technology,* vol. 7, no. 4, pp. 157–164.

Orkin, F. I. and D. P. Olivier (1999). In *JQH5,* Table 10.2.
Peach, R. W., ed. (1997). *The ISO 9000 Handbook,* 3d ed., McGraw-Hill, New York.
Riley, F. D. (1979). "Visual Inspection—Time and Distance Method," *ASQC Annual Technical Conference Transactions,* Milwaukee, p. 483.
Schilling, E. G. (1982). *Acceptance Sampling in Quality Control,* Marcel Dekker, New York.
Schilling, E. G. (1994). "The Importance of Sampling in Inspection," *Annual Quality Congress Proceedings,* ASQ, Milwaukee, pp. 809–812.
Taylor, W. A. (1994). "Acceptance Sampling in the 90s," *Annual Quality Congress Proceedings,* ASQ, Milwaukee, pp. 591–598.
Tsai, P. (1988). "Variable Gauge Repeatability and Reproducibility Study Using the Analysis of Variance Method," *Quality Engineering,* vol. 1, no. 1, pp. 107–115.

CHAPTER 11

Understanding Customer Needs

➔ 11.1 QUALITY AND COMPETITIVE ADVANTAGE

The preceding chapters in this book present the key concepts in understanding the dimensions of quality, determining current status, and starting to achieve quality excellence. All departments in an organization (see Section 1.4, "The Quality Function") have roles in the quality effort. Starting with this chapter on understanding customer needs, the remainder of this book discusses the roles of these various departments.

In developing new products or services (or modifying existing ones), the quality by design method or design for six sigma method is used. This chapter elaborates on two of the early steps, identify the customers and discover customer needs. These two steps are part of the process of designing for quality.

In the competitive world, all organizations aspire to have a unique competitive advantage. Such an advantage can be achieved by price, by ability to meet customer needs on short notice, and by quality. This chapter starts the journey to show how quality—both product features and freedom from deficiencies—can lead to a unique competitive advantage. By identifying customers, analyzing their needs, and understanding our quality status relative to competition, we can establish new product quality goals that will lead to a competitive advantage. We start with customers.

➔ 11.2 IDENTIFY THE CUSTOMERS

We define a *customer* as anyone who is affected by the product or process. Three categories of customers then emerge:

1. *External customers, both current and potential.* A multiplicity of these customers gives rise to a variety of influences, depending on whether the customer

is economically powerful and on its technological sophistication. For service organizations, the list of external customers may extend far. For example, customers of the Internal Revenue Service include taxpayers and also the Treasury Department, Office of the President, Congress, accountants, lawyers, etc. Each customer has needs that must first be determined and then addressed in planning a product.

2. *Internal customers or processors.* These customers include all functions affected by the product at both the managerial and workforce levels. Internal suppliers often view their internal customers as "captive" customers. Not so. Internal customers may have an alternative source; they may be able to purchase the product from an external supplier. For example, an engineering department procures research services from the company research department. The engineering department is an internal customer that may decide to use an external consultant to provide the services. An assembly plant purchases components from a sister plant within the company. That assembly department must be viewed as an internal customer that, to meet its own goals, could decide to go outside the company family to obtain the required quality in components.

3. *Suppliers as customers.* Suppliers should be viewed as extensions of internal customer departments such as manufacturing. Thus their needs must be understood and addressed during planning for quality.

In identifying customers, some are obvious and some are not. An important tool for identifying all those who are affected (i.e., customers) is the flow diagram. Often no one individual or department is able to describe the total process; a cross-functional team can create the flow diagram. A flow diagram also helps to understand linkages that can reveal hidden customer needs. These linkages are information flows, work flows, and other interdependencies among operations.

For external customers, a "cast of characters" is often the "customer." Thus in selling products and services to a hospital, a supplier must understand the needs of the hospital purchasing manager, a quality assurance manager, heads of hospital departments, physicians, nurses, and (of no little consequence) the patient. One manufacturer learns the needs of four levels of customers: those who approve the purchase, those who influence the decision, those who sign the purchase order, and those who are end users.

In practice, we must recognize that some customers are more important than others. It is typical that about 80% of the total sales volume comes from about 20% of the customers; these are the "vital few" customers who command priority. Within these key customer organizations are individual customers who also may have a hierarchy of importance; e.g., a surgeon at a hospital is a key customer for surgical needles.

Note that customers include both current and potential customers and that it is often useful to identify segments of customers (for an example of customer segmentation, see Section 5.10, "Spectrum of Customers"). Finally, a difficult but important decision involves which potential customers *not* to pursue. Restricting customer segments enables an organization to concentrate resources on the vital few customers and perform those activities that will lead to customer loyalty.

11.3 CUSTOMER BEHAVIOR

We will first define certain terms concerning customer behavior. The terms are *needs, expectations, satisfaction,* and *perception.*

Customer needs are the basic physiological and psychological requirements and desires for survival and well-being. A. H. Maslow is a primary source of information on both physiological and psychological needs. He identifies a hierarchy of such needs as physiological, safety, social, ego, and self-fulfillment (for elaboration, see Section 9.2).

Customer expectations are the anticipated characteristics and performance of the goods or service. Kano and Gitlow (1995) suggest that three levels of customer expectation are related to product attributes. The "expected" level of quality represents the minimum or "must be" attributes. We cannot drive up satisfaction with these attributes because they are taken for granted, but if performance of the basic attributes is poor, strong dissatisfaction will result. At the "unitary" (or desired) level, better performance leads to greater satisfaction but (in a limited time period) usually in small increments. For the "attractive" (or surprising) level, better performance results in delighted customers because the attributes or the level of performance are a pleasant surprise to the customers. (For an application to hospitals, see Section 11.4.) Discovering and understanding customer needs and expectations is necessary in defining specific product attributes for market research and product development.

Customer satisfaction is the degree to which the customer believes that the expectations are met or exceeded by the benefits received. Customer expectation has a strong influence on satisfaction. Thus a customer staying at a luxury hotel expects perfection, and even a minor inconvenience will result in low satisfaction. A budget hotel can have poor features, but if customers get a reasonable night's sleep, they will have high satisfaction because their expectation is low.

Customer perception is the impression made by the product. Perception occurs after a customer selects, organizes, and interprets information on the product. Customer perceptions are heavily based on previous experience. But other factors influence perception, and these factors can occur before the purchase, at the point of purchase, and after the purchase. For elaboration, see Onkvisit and Shaw (1994).

11.4 SCOPE OF HUMAN NEEDS AND EXPECTATIONS

In planning to collect information on customer needs, we must go beyond the search for obvious needs to the more subtle ones that present opportunities for innovative new-product designs.

First, let us focus on the distinction between stated needs and real needs. A consumer states a need for a "clothes dryer," but the real need is "to remove moisture"; a consumer wants a "lawn mower," but the real need is "to maintain the height of the lawn." In both cases, expressing the need in terms of a basic verb and noun can spawn new product ideas. One historical example is the replacement of hair nets by hair spray to satisfy the basic need of "secure hair." When a customer states: "I need an X," perhaps we should ask: "What would you use the X for?"

Some needs are disguised or even unknown to the customer at the time of purchase. Such needs often lead to the customer using the product in a manner different from that intended by the supplier—a telephone number intended for emergencies is used for routine questions, a hair dryer is used in winter weather to thaw a lock, a tractor is used in unusual soil conditions. Designers view such applications as misuse of the product but might better view them as new applications for their products.

Some of these applications are misuse, but such needs must be understood and, in some cases, alternative design concepts considered. Then there are other needs that go far beyond the utilitarian. Some needs may be perceptional (e.g., the now classic example of Stew Leonard's supermarket customers, who believe that only unwrapped fish could be fresh); some needs may be cultural (e.g., a process such as computer-aided design threatens to reduce the need for human expertise, and thus design engineers resist it).

The Hospital Corporation of America identifies several levels of customer expectations. At level 1, a customer *assumes* that a basic need will be met; at level 2, the customer will be *satisfied;* at level 3, the customer will be *delighted* with the service. For example, suppose that patient must receive 33 radiation treatments. Waiting time in the therapy area is one attribute of this outpatient service. At level 1, the patient *assumes* that the radiation equipment will be functioning each day. At level 2, the patient will be *satisfied* if the waiting time in the area is moderate, say, 15 minutes. At level 3, the patient will be *delighted* if the waiting time is short, say, one minute. To achieve a unique competitive advantage, we must focus on level 3.

Customer needs may be clear or disguised; they may be rational or less than rational. To create customers, we must discover and serve their needs.

Anderson and Narus (1998) view needs in terms of the worth, in monetary units, of the technical, economic, service, and social benefits a customer receives in exchange for the price paid.

→ 11.5 SOURCES OF MARKET QUALITY INFORMATION

Market quality information includes quality alarm signals arising from a decline in sales and also from field failure reports, customer complaints, claims, lawsuits, etc.

Most alarm signals are poor measures of quality—rather, they are measures of expressed product dissatisfaction. A low level of alarm signals does not necessarily mean a high level of quality. Particularly for inexpensive products, complaint rates are a poor indicator of customer satisfaction. If customers are not satisfied, they simply switch brands—without submitting a complaint. Neither does the absence of product dissatisfaction mean that there is product satisfaction, since these two terms are not opposites. A product may be failure free and yet not be salable because a competitor's design is superior or has a lower price.

A second source of market quality information is the vast array of published data available relative to quality. Some of these data are internal to the company. Many are published external to the company. Such "field intelligence" includes databases on sales volume, price changes, success rate on bids, complaints, spare parts usage, salespersons' reports, ratings from customers and consumer journals, government reports, etc.

These two major sources of market quality information, while necessary, are not sufficient. A good deal of information is missing, and data can be provided only through special market research studies.

➔ 11.6 MARKET RESEARCH IN QUALITY ("VOICE OF THE CUSTOMER")

The American Marketing Association defines *marketing research* (we use the term *market research*) as "the function which links the consumer, customer, and public to the marketer, through information—information used to identify and define marketing opportunities and problems; generate, refine, and evaluate marketing actions; monitor marketing performance; and improve understanding of marketing as a process."

Although market research includes the compilation and analysis of existing information (field failure reports, customer complaints, government reports), the exciting aspect concerns information that does not yet exist, i.e., the voice of the customer (VOC). VOC is a continuous process of collecting customer views on quality and can include customer needs, expectations, satisfaction, and perception. The emphasis is on in-depth observing, listening, and learning. This process addresses the three main purposes of market research for quality.

Purposes of Market Research in Quality

The broad purposes are mainly to

- Determine customer needs.
- Develop new features.
- Measure current customer satisfaction.
- Analyze customer retention and loyalty issues.

Determination of customer needs has several facets. These include both short-term needs to perform surgery on current products and longer term needs to integrate quality into new-product development. We ask customers directly what their needs are and also methodically study how customers currently use the product; then we analyze their total system of use to identify hidden needs. This process provides specific ideas for modifying current products and helps to discover opportunities for new products. Needs must be periodically evaluated because today's new needs become routine expectations tomorrow.

Measurement of current customer satisfaction also involves several elements. First, the term *quality* must be translated into specific attributes that customers say are important. These attributes should not be restricted to attributes of the core and auxiliary products provided to customers, but should also include quality attributes for the full life cycle from initial contact with a salesperson to customer service after the sale. We also need to learn the relative importance of these attributes to the customer and how our product compares to the competition (see Section 3.6 for an example).

Finally, we must ask customers whether they would purchase from us again or recommend us to friends.

Using market research for customer retention and loyalty starts with the distinction between customer satisfaction and customer loyalty—two different concepts (see Section 5.4). Some levels of customer satisfaction do not achieve a high level of customer loyalty. The research must include finding the reasons for losing customers, i.e., customer defections. A framework for achieving high customer retention and loyalty is presented in Section 5.5.

Thus market research in quality looks for answers to some cardinal questions:

- What is the relative importance of various product qualities, as seen by the user? The answers provided by market research are typically different from the prior beliefs of the manufacturer. Sometimes the difference is dramatic.
- For the more important qualities, how does our product compare with competitors' products, as seen by users?
- What is the effect of these competing qualities (including our own) on users' costs, well-being, and other aspects of fitness for use?
- What are users' problems about which they do not complain but which we might nevertheless be able to remedy?
- Do users have ideas that we might be able to use for their benefit?

In conducting market research, certain principles are basic:

1. Customer views on quality are most useful when they include views on competitors' products.
2. Market research should also capture the views of "noncustomers," i.e., prospective customers and lost customers.
3. In designing market research, operating managers within the organization should be asked to which questions they need answers to improve their operations.
4. The occasional survey is better than none, but periodic surveys provide data for analysis of trends and other matters.
5. The concept of sampling is important. Surveying a carefully designed sample of customers in depth is more useful than superficially surveying a large sample of customers.
6. The customer desires product performance during the entire life of a product, not just during the warranty period. Market research should go beyond the warranty period.

A useful reference for market research is Churchill (1991). Goodman et al. (1996) discuss eight common pitfalls that undermine the integrity and value of customer feedback.

Critical Incident Technique

The technique was developed by J. C. Flanagan during World War II to improve the performance of fighter pilots.

An "incident" is best thought as "any observable human activity that is sufficiently complete in itself to permit inferences and predictions to be made about the person performing the act." For the incident to be considered "critical," it "must occur in a situation where the purpose or intent of the act seems fairly clear to the observer and where its consequences are sufficiently definite to leave little doubt concerning its effects" (Flanagan, 1954).

Since then, critical incident surveys have been used to establish competence-based education in various professions (such as medicine, dentistry, pharmacy, teaching, and law enforcement).

End users (customers or potential customers) are asked to identify specific incidents which they experienced personally and which had an important effect on the final outcome. The emphasis is on incidents rather than on vague opinions. The context of the incident may also be elicited. Data from many users are collected and analyzed.

A critical incident allows us to determine the competencies that were present or lacking in that particular situation. By collecting a sufficient number of such critical incidents, it is possible to build a profile of the competencies that are required for satisfactory performance. One of the co-authors of this book (Chua) has adapted and applied the critical incident technique to the determination of customer needs in various design for six sigma (DFSS) projects.

The CIT is an open-ended retrospective method of finding out what users feel are the critical features of the product or service being studied. It focuses on user behavior. It is more flexible than a questionnaire or survey and is recommended when the only alternative is to develop a questionnaire or survey from the start.

In the following discussion, the market research addresses customer needs for both components of quality—product features and freedom from deficiencies.

✣ 11.7 NEEDS RELATED TO PRODUCT FEATURES

To start, we need to identify the attributes that customers say are important in their purchasing decision. Table 11.1 shows the results of research in which customers (and potential customers) of floppy disks were asked to select important attributes. The relative importance was measured as the percentage of customers who selected each attribute. Note that the list of attributes focuses on both the disks and on specific and general aspects of service.

An important next step is to learn how our product compares with the competition's. This task can be accomplished by using a multiattribute study. Customers are asked to consider several product attributes and, for each attribute, to state the relative importance and a rating for our product and competing products. Section 3.6 presents three examples to illustrate customers' role in assessment. Additional examples of market research are given later.

In another approach for determining relative importance of service attributes, a bank asks customers to select the five most important services from a list of 12 (e.g., processing transactions without error, greeting you with a smile, knowledge of bank products and services).

Sometimes the list of attributes is lengthy, and the identification of the vital few can be extremely helpful (Mitsch, 1998). Thus at Unigroup (United Van Lines and

TABLE 11.1
Relative importance of attributes of floppy disks

Attribute	Relative importance, %
Quality of delivered products	98
Meeting delivery schedules	95
In-field reliability	95
Price	72
Provision of technical assistance	66
Handling large orders	64
Product sophistication	62
Meeting changing customer needs	51
Short delivery time	48
Capability of field service	45
Accessibility of suppliers' management	44
Specialization in floppy disk drive products	34

Mayflower Transit), from a list of 71 attributes in moving household goods, customers selected 14 as most important (e.g., ability of salesperson, handling your goods with care—loading crew, keeping promises).

The relative importance of product attributes can be determined by several methods. The simplest is to present a list to customers and ask them to select the most important. In another approach, customers are asked to allocate 100 points over various attributes. In a more complicated method ("conjoint measurement"), customers are presented with combinations of product attributes and asked to indicate their preferences. The importance rating can then be calculated. Churchill (1991), Appendix 9B, describes this method. Peña (1999) presents a similar method for estimating the relative importance of service quality attributes with applications to the measurement of quality at a university and to the Spanish railroad system.

The Shaving System Case

A leading manufacturer of razors and razor blades was confronted by a new competing model that featured an ingenious, improved blade-changing mechanism. When the competitor promoted the new model aggressively, the company marketing managers became apprehensive. They then succeeded in initiating a costly new-product development to bring out a model that could match the competition's. However, the top managers also authorized a field market research study to discover the reaction of consumers to the new development.

The planners of this classic research study identified the seven major quality characteristics of a shaving system. Next (through an intermediary market research company), the company provided each of several hundred consumers with all three principal shaving systems then on the market. These consumers were asked to use each shaving system for a month and to report:

- The ranking of the seven characteristics in their order of importance to the user.
- The ranking of the performance of each of the three systems for each of the seven quality characteristics.

TABLE 11.2
Market research study—the shaving system case

Quality characteristics	Users' rankings			Importance
	Gillette	Gem	Schick	
1. Remove beard				
2. Safety				
3. Ease of cleaning				
4. Ease of blade changing				
5. . . .				
6. . . .				
7. . . .				

Table 11.2 shows the plan of the market research study. The resulting data showed that, as seen by the users

1. Ease of blade changing was the least important characteristic.
2. The competitor's new blade-changing mechanism had failed to create a user preference over competing forms of blade changing.
    These findings denied the beliefs of the company marketing managers and came as a welcome surprise. They enabled the company to terminate the costly new-product development then already in progress.
3. On one of the most critical qualities (product safety), the company's shaving system was inferior to both competing systems. This news came as an unwelcome surprise and stimulated the company to take steps to eliminate this weakness. Note that *this inferiority could not have been discovered from field complaints* because no normal user would conduct such a comparative study on his or her own initiative. It was "an alarming situation for which our present alarm signals are silent."

Discovering Customer Needs and Marketing Opportunities

Market research in the field provides access to realities that cannot be discovered in the laboratory. The conditions of use can involve environments, loads, user training levels, misapplication, etc., all of which may be different from the conditions prevailing in the laboratory. The laboratory does provide a relatively prompt, inexpensive simulation that is most helpful for making many decisions. However, this simulation cannot fully disclose the needs of fitness under the actual conditions of use.

Field studies can provide access to the realities of the conditions of use and also to the users themselves. Through access to the user, it becomes possible to understand problems that are not a matter of poor quality but for which the manufacturer may nevertheless be able to provide a solution. For example, most users prefer to avoid disagreeable, time-consuming chores. The food-processing industry has successfully transferred many such chores from the household kitchen to the factory (e.g., soluble coffee, precooked foods) and incidentally greatly increased their sales.

The study of the user's operation can be aided by dissecting the process of use. This job is done by documenting all of the steps, analyzing them, and identifying opportunities for new-product development. Such information is valuable input to the product development function as that group embarks on a new design project.

> **EXAMPLE 11.1.** For a certain health care product, a clamp controlled the amount of flow of fluid from an external source to a patient during dialysis. An analysis using a flow diagram showed that the proposed clamp was excellent in controlling the flow of fluid but was inconvenient in a later step that required the patient to wear the product under clothing. The original focus was on performance of the product, but the flow diagram pinpointed a later problem of inconvenience to the patient. See Section 11.4 for further discussion.

This approach gathers information by observation of customers rather than inquiry of customers. The approach is not new, but Leonard and Rayport (1997) suggest a useful framework of five steps, using a team to do the research:

1. *Observation.* Customers are observed carrying out normal routines.
2. *Capturing data.* Relatively few data are gathered through questions. Instead, photography and videography are often used.
3. *Reflection and analysis.* After gathering the data, the team reviews the visual and other data to identify customers' possible problems and needs.
4. *Brainstorming for solutions.* The observations are transformed into visual representations of possible solutions.
5. *Developing prototypes of possible solutions.* Prototypes clarify the concept of the new product or service and stimulate discussion with colleagues and customers.

Some researchers call this approach "stapling ourselves to the product."

Opportunities for improvement span the full range of customer use from initial receipt through operation. The primary stages and examples of ideas for improving fitness for use are given in Table 11.3. (Also see Table 5.9 for examples of hidden customer needs that presented opportunities for improvement.) These opportunities

TABLE 11.3
Opportunities for improving fitness for use of an industrial product

Stage	Opportunities
Receiving inspection	Provide data so incoming inspection can be eliminated
Material storage	Design product and packaging for ease of identification and handling
Processing	Do preprocessing of material (e.g., ready-mixed concrete); design product to maximize productivity when it is used in customer's manufacturing operation
Finished goods storage, warehouse and field	Design product and packaging for ease of identification and handling
Installation, alignment, and checkout	Use modular concepts and other means to facilitate setups by customer rather than manufacturer
Maintenance, preventive	Incorporate preventive maintenance in product (e.g., self-lubricated bearings)
Maintenance, corrective	Design product to permit self-diagnosis by user

TABLE 11.4
Adding value through customer service

Possible ways to add value (basic product)	Example from financial services (checking account)	Example from retailing (CDs)
Be flexible	Let customers design their own checks	Allow customers to return CDs
Tolerate customer errors	Cover overdrafts without charging a fee	Extend credit when customers forget to bring money
Give personal attention	Help customers with individual tax questions	Learn customers' musical tastes and suggest new CDs they might enjoy
Provide helpful information	Publish a brochure on financial planning	Distribute a newsletter that reviews new stereo equipment
Increase convenience	Install ATMs	Let customers order by phone

Source: Bovée and Thill (1992).

must then be translated into specific product quality goals, which, in turn, help to create a unique competitive advantage.

In another approach, Bovée and Thill (1992) suggest five ways of adding value through enhanced customer service with an application to financial services and retailing of compact discs (Table 11.4).

Sometimes information on customer needs is too broad to translate into specific product attributes and features. To obtain more precision, we can break down broadly stated needs into secondary and tertiary needs. This process provides a detailed quality function deployment matrix, i.e., a customer needs analysis matrix. An example is given in Table 11.5 for scheduling medical resources.

The methods for collecting information on customer needs are many and varied. Some of the common methods include focus groups (discussed later), observations at customer sites, executive interactions with customers, special customer surveys, analysis of complaints, participation at trade shows, and comparing products with those of competitors. A recent method is the monitoring of Internet messages to find out what customers are saying about a product (Finch, 1997).

JQH5, Section 3, provides a list of 13 methods for collecting information on customer needs. Plsek (1997) is an excellent reference on the application of creativity concepts to customer needs analysis.

Focus Groups

A focus group consists of about 8 to 14 current or potential customers who meet for about two hours to discuss a product. Here are some key features:

1. The discussion has a focus, hence the name.
2. The discussion can focus on current products, proposed products, or future products.

TABLE 11.5
Needs analysis spreadsheet

Primary needs	Secondary needs	Tertiary needs
Efficient use of clinical resources	No wasted time between patients	Procedure length adjustable to MD practice pattern
		No incompatible procedures scheduled
	Maximum number of appointments booked	Can easily find appointment records across facilities
		Accurate inventory of available resources—people, equipment, supplies
Good clinical care	Right patient gets right procedure	Correct association of patient with MD orders
		Medical record information available
	Special patient needs known ahead and attended to	
	All locations have patient and MD information	

Source: Juran Institute, Inc.

3. A moderator who is skilled in group dynamics guides the discussion.
4. The moderator has a clear goal for the information needed and a plan for guiding the discussion.
5. Often company personnel observe and listen in an adjacent room shielded by a one-way mirror.

Focus groups can cover many facets of a product or can discuss only quality. A discussion can be broad (e.g., obtaining views on customer needs) or can have a narrower scope (e.g., determining customer sensitivity to various degrees of surface imperfections on silverware). The groups might consist of customers from various market segments, or groups may represent noncustomer segments the firm wants to penetrate. Sessions can provide depth of information on customer needs, expectations, perceptions, satisfaction, intentions, and reactions to new concepts or ideas. One useful approach is to ask customers to dream about the ideal product, e.g., the product that would overcome customer frustrations. The responses provide input for new-product development. Typically, focus group respondents are quite candid as they receive reinforcement by like-minded people in the group.

A key to a successful focus group is the qualification of the focus group moderator. Being a moderator is not a task for a well-intentioned amateur. The moderator must be a quick learner, a friendly leader, knowledgeable but not all-knowing, a good listener, a facilitator not a performer, flexible, empathic, a big-picture thinker, and a good writer; he or she must also have an excellent memory. For elaboration, see Greenbaum (1988).

Electronic meeting support (EMS) tools can help to record and summarize views during a focus group meeting. Typically, the participants meet in a room, with each participant sitting at a computer. Using the computers, the participants—simultaneously and anonymously—respond to the moderator's questions. Responses are collected and summarized by the computer network for all to see and discuss.

Mass Customization

There are many commercial examples where products are customized to a customer order (for example, in the retail computer industry). Davis (1987) first coined the term and refers to mass customization when "the same large number of customers can be reached as in mass markets of the industrial economy, and simultaneously they can be treated individually as in the customized markets of pre-industrial economies." Pine (1993) provides a good treatment of mass customization in his book. In his definition, the goal is to detect customers needs first and then to fulfill these needs with an efficiency that almost equals that of mass production. Tseng and Jiao (2001) provide a working definition, stating that the objective of mass customization is "to deliver goods and services that meet individual customers' needs with near mass production efficiency."

Mass customization is applicable only to those products for which the value of customization, to the extent that customers are willing to pay for it, exceeds the cost of customizing (Piller, 2003; Tseng and Piller, 2003). This implies that even while the price of a to-be-customized product may increase, the same group of customers that was buying a standard (mass) product before is now heading toward customized products. An important indicator of the extent of value being created is the level of customization. An interesting project in mass customization was recently started in Europe. The European Commission launched the EuroShoe project, a large scale European project dedicated to mass customization in the footwear industry (for more details, see www.euro-shoe.net). Feet of each individual customer can be scanned biometrically, his/her specific habits analyzed and used to make an individual last, insole, and sole for each customer. Shoes and lasts are then produced when an order is placed by an end-consumer.

✥ 11.8 NEEDS RELATED TO PRODUCT DEFICIENCIES

Understanding customer needs must also include the deficiencies side of the quality definition. Clearly, the emphasis during planning must be on *prevention* of deficiencies (defects, failures, errors, etc.).

With respect to prevention of deficiencies, the following chapters discuss the concepts and tools necessary in various functional areas. An important input to such prevention efforts are the data on field complaints. Most organizations have systems of collecting and analyzing information on customer complaints. A Pareto analysis of field failures, complaints, product returns, etc., identifies the vital few quality

TABLE 11.6
Pareto analysis of complaints on fuel nozzles

Failure mode	Ranking based on various measures		
	Frequency	Warranty charges	Effect on user costs
G	1	4	3
A	2	1	6
D	3	2	1
F	4	5	4
N	5	7	8
L	6	3	2
P	7	6	5
R	8	8	7

problems to be addressed in both current and future product development. This type of analysis is in wide use. Table 11.6 shows a Pareto analysis of field complaints on fuel nozzles using several different measures. A conventional Pareto analysis would rank the various types of complaints according to the frequency of occurrence, but when those complaints are evaluated on an economic basis, the ranking changes. Note also that there are several possible economic indexes such as warranty costs and effect on user costs. The priorities for future product development can be quite different depending on the measure used in the Pareto analysis. Similarly, data on defects found internally must be analyzed on the proper basis to prioritize prevention efforts, many of which must be addressed during product development.

To prevent recurrence of specific complaints in future products requires careful analysis of the words and phrases used to describe complaints.

> **EXAMPLE 11.2.** An annual mail survey of automobile customers revealed a significant number of complaints on "fit of doors"; within the company, this term referred to the "margin" and "flushness." Margin is the space between the front door and fender or between the front and rear doors. Sometimes the margin was not uniform, e.g., a wider space at the top compared to a narrower space at the bottom of the door. Flushness refers to the smoothness of fit of the door with the body of the car after the door is shut. Steps were taken (at a significant cost) during processing and assembly to correct the margin and flushness problems. But subsequent surveys again reported complaints on fit of doors.
>
> Fortunately, other market research input became available. The company held periodic focus group meetings at which customers answered a questionnaire and discussed selected issues in more detail. Fit of doors again was proclaimed to be a problem by the customers. On the spot, they were asked, "What do you mean by fit of doors?" Their answer was *not* margin or flushness. First, fit meant the amount of effort required to close the door. Customers complained that they had to "slam the door hard in order to get it to close completely the first time." Second, *fit* meant "sound." When they left the car and closed the door, they wanted to hear a solid, "businesslike" sound instead of "a metallic, loose sound" (that told them the door was not closed). The market research provided a correct understanding of the complaint, which subsequently required changes in the product design and the manufacturing process. Earlier, the company had acted on the complaint, but it had solved the wrong problem. It was a waste of resources—like spraying scrap with perfume.

When customer needs are determined, they must be related to internal business processes by defining metrics for each need. For an example, see Figure 14.3.

In analyzing customer needs related to both product features and to product deficiencies, an essential input is information on the current level of customer satisfaction.

➔ 11.9 MEASURING CUSTOMER SATISFACTION

In measuring customer satisfaction, terms like *customer satisfaction* and *quality* are too foggy. Instead, we must identify the attributes of the product that collectively define satisfaction. Examples are the accuracy of transactions at a bank or specific attributes of workmanship in a residential home. A sample of customers should be asked (in preliminary research) which attributes they think constitute high quality. Inputs from company managers, trade journals, and other sources should also be included. With these inputs, the list of attributes is then finalized for research with the larger sample of customers. The list of attributes should span the entire cycle of customer contact from initial contact with a salesperson through use and servicing of the product and handling of complaints. Examples of customer satisfaction studies from both manufacturing and service industries are presented in Section 3.6.

It is possible to use a generic template of attributes. For example, in the area of customer service, the SERVQUAL model (Zeithaml et al., 1990) identifies five dimensions: tangibles, reliability, responsiveness, assurance, and empathy. Such broad terms need to be defined. Market research for the five areas is conducted by presenting customers with 22 statements that examine the difference between customer expectations and perceptions. These statements can be customized for each company.

Customer satisfaction measurement should also include asking customers about the relative importance of the various attributes (see earlier discussion). When this information is combined with customer satisfaction data, the resulting quadrant diagram helps to direct future action on modifying a product to meet customer needs. For an example, see Figure 3.4.

Information about customer satisfaction relative to the competition can be sobering; the data may spur necessary action on a product. Data on competitor quality can be obtained from laboratory testing, direct questions to customers, mystery shoppers, focus groups, and benchmarking. When a noncustomer is purchasing from a competitor, it is useful to ask an open-ended question, e.g., what are the chief reasons that you buy from company X?

Another essential question to ask customers is, Would you purchase from us again, or would you recommend us to your friends? An example of this approach, along with numerical results at United Van Lines (UVL), is shown in Table 11.7.

Benchmarking sources report that world-class companies earn "definitely would recommend" scores 60 to 65% of the time. For quality superiority, organizations need to score in the top (not top two) response category.

Data on customer satisfaction can be collected by various methods, including market surveys, posttransaction interviews, focus groups, mystery shoppers, employee reports, surveys and interviews, call centers for complaints and other comments, customer advisory groups, and customer segment surveys. For further discussion, see *JQH5*, Section 18. Vavra (1997) and Churchill (1991) are also useful references.

TABLE 11.7
Results of loyalty survey

	Current 4 months	
Willingness to recommend UVL	**Responses**	**Percentage**
Definitely would recommend	37	55.2
Would recommend	14	20.9
Might or might not recommend	9	13.4
Would not recommend	4	6.0
Definitely would not recommend	3	4.5

Source: United Van Lines, Inc. customer service survey.

The Web can also be used to collect customer satisfaction data through online customer surveys. Suppliers specializing in electronic surveys have assembled panels of ready respondents. Online research is faster than phone or mail surveys and can have higher response rates because it is more convenient and less intrusive than other methods. On the downside, online polling may not be representative because many consumers do not use the Internet regularly.

Customer satisfaction and needs data from various sources can be integrated into a database, and customer relationship management software can then be used to analyze and make the results available immediately to all functions in an organization. The potential for electronic commerce is exciting, e.g., data can be sliced by ZIP code to discover local needs.

Customer satisfaction data must be linked to customer loyalty analysis.

Collecting information on customers and their needs provides the input for the quality function deployment matrix on customers and customer needs (see Section 5.12). The data then help us to translate customer needs into product features (discussed in Chapter 12).

Rhey and Gryna (2001) discuss the application of market research for quality to small business firms.

➔ 11.10 MARKET RESEARCH FOR INTERNAL CUSTOMERS

Market research concepts apply to both internal and external customers. In the service sector, a support department of AT&T Bell Laboratories conducted a survey of the needs of internal customers (Weimer, 1995). The result was a multilevel hierarchy of needs. Service quality was defined as one of the primary needs. Customers defined service quality with 10 secondary (tactical) needs such as "no red tape" and "access to service information." These needs were further defined in terms of 42 tertiary (operational) needs such as "deal with person providing service directly" and "know who supplies a service or owns a process."

For an application using EMS tools with a focus group of financial managers, see Section 14.2.

In another example, the central engineering staff of a large automotive manufacturer applied marketing research concepts to internal customers (Stevens, 1987).

A survey questionnaire sent to internal customers first requested information about the frequency of contact with each of the seven staff areas. Respondents were then asked to state their degree of satisfaction with eight attributes of service: services meeting requirements, follow-up required, assistance available, competent people, cooperation, input requested, problem notification, and help in seeking contacts. An overall satisfaction rating was also requested. Finally, respondents were asked for details on any low ratings for the eight attributes and suggestions how engineering could improve its service to help achieve customer quality goals.

Spending the time and effort to understand customer needs clearly provides the foundation for planning and executing the total work of the organization—product development, supplier relations, operations, marketing, field service, and support functions. That's the scenario for the rest of this book.

SUMMARY

- A customer is anyone who is affected by a product or process.
- Customers fall into three categories—external, internal, and suppliers.
- Customer needs may be clear or disguised; they may be rational or less than rational. But these needs must be discovered and served.
- A low level of customer dissatisfaction (e.g., complaints) does *not* necessarily mean that customers are satisfied.
- Market research for quality analyzes aspects of quality that are affected by forces in the marketplace.
- Customer satisfaction should be measured relative to the competition and should address both product features and freedom from deficiencies.
- Market research can help to discover opportunities for achieving a unique competitive advantage.

PROBLEMS

11.1. For a specific output of your department in an organization, sketch a simple flow diagram of the journey of the output throughout the organization. Then prepare a list of the internal and external customers.

11.2. Identify your own customers and suppliers for a representative 24-hour day.

11.3. For one of your important customers, define the steps you should take to discover the customer's primary needs. Emphasize the discovery of real needs rather than stated needs.

11.4. For any *consumer* product, prepare a plan of market research aimed at
   (a) Identifying the principal product qualities.
   (b) Ranking these qualities in their order of importance to users.

(c) Discovering the relative performance of competing companies with respect to the principal product qualities.

11.5. For any *industrial* product, prepare a plan of market research aimed at securing data similar to those set out in problem 11.4.

11.6. Walk through a supermarket and identify some products for which work has been transferred from the household kitchen to food-processing factories. In addition, identify some potential products for such transfer. Report your findings, with special emphasis on the quality problems involved.

11.7. You are participating in the preparation of a bid on a military optical system used for range finding. The military specification requires that the subsystems be interchangeable (so that in case of damage, the failed subsystem can be unplugged and replaced with a spare). You know from experience that this provision is wise. However, you are disturbed by the further requirement that the optical elements (lenses, prisms, etc.) within each subsystem should also be interchangeable. To your knowledge, this requirement is foolish because the military has never had the special repair and test facilities needed to reassemble optical elements in the field. To your knowledge, the military also lacks the means of testing for this interchangeability.

A procedure is available for proposing a change in such unrealistic specifications, but you once went through this procedure and found it to be shockingly long and difficult. In addition, your company has had poor success in securing military business because it has been underbid in price by competitors. You have suspected for some time that one reason for this lack of success is that your company does its bidding on the basis of meeting all specification requirements (sensible or not), whereas the competitors may be doing their bidding on the basis of meeting only the requirements that make sense.

You are now faced with making a recommendation on how to structure the bid with respect to the requirement for the unneeded interchangeabilty. What do you propose? Why? If you propose to bid on the basis of not meeting the "foolish" requirement, what do you propose to do if the military later finds out what has happened?

REFERENCES

Anderson, J. C. and J. A. Narus (1998). "Business Marketing: Understand What Customers Value," *Harvard Business Review,* November–December, pp. 53–65.
Bovée, C. L. and J. V. Thill (1992). *Marketing,* McGraw-Hill, New York.
Churchill, G. A. Jr. (1991). *Marketing Research Methodological Foundations,* Dryden Press, Chicago.
Davis, S. (1987). *Future Perfect,* Addison-Wesley, Reading, Massachusetts.
Finch, B. J. (1997). "A New Way to Listen to the Customer," *Quality Progress,* May, pp. 73–76.
Flanagan J. C. (1954). "The Critical Incident Technique," *Psychological Bulletin,* 51.4, 327–359.
Goodman, J., D. DePalma, and S. Broetzmann (1996). "Maximizing the Value of Customer Feedback," *Quality Progress,* December, pp. 35–39.
Greenbaum, T. L. (1988). *The Practical Handbook and Guide to Focus Group Research,* Lexington Books, D.C. Heath, Lexington, MA, pp. 50–54.

Kano, N. and H. Gitlow (1995). "The Kano Program," notes from seminar on May 4 and 5, University of Miami.

Leonard, D. and J. F. Rayport (1997). "Spark Innovation through Empathic Design," *Harvard Business Review,* November–December, pp. 102–113. Reprinted by permission of Harvard Business Review.

Mitsch, G. J. (1998). "Moving Experiences: How System-Wide and Singular Customer Satisfaction Measurement Drives Quality at United and Mayflower," Customer Satisfaction and Quality Measurement Conference, American Marketing Association and American Society for Quality, Atlanta.

Onkvisit, S. and J. J. Shaw (1994). *Consumer Behavior,* Macmillan, New York.

Peña, D. (1999). "A Methodology for Building Service Quality Indices," *Annual Quality Congress Proceedings,* ASQ, Milwaukee, pp. 550–558.

Piller, F. (2003). *Mass Customization,* 3rd edition, Gabler, Wiesbaden.

Pine, B. J. (1993). *Mass Customization,* Harvard Business School Press, Boston.

Rhey, W. L. and F. M. Gryna (2001). *Market Research for Quality in Small Business, Quality Progress,* January.

Stevens, E. R. (1987). *Implementing an Internal-Customer Satisfaction Improvement Process,* Juran Report No. 8, pp. 140–145.

Tseng, M. M. and J. Jiao (2001). Mass Customization, in G. Salvendy (Ed.) *Handbook of Industrial Engineering,* 3rd edition, Wiley, New York, pp. 684–709.

Tseng, M. M. and F. Piller (2003). "Towards the Customer Centric Enterprise," in M. Tseng and F. Piller (Ed.) *The Customer Centric Enterprise: Advances in Mass Customization and Personalization,* Springer, New York, pp. 1–27.

Vavra, T. G. (1997). *Improving Your Measurement of Customer Satisfaction,* Quality Press, ASQ, Milwaukee.

Weimer, C. K. (1995). "Translating R & D Needs into Support Measures," *Annual Quality Congress Proceedings,* ASQ, Milwaukee, pp. 113–119.

Zeithaml, V. A., A. Parasuraman, and L. L. Berry (1990). *Delivering Quality Service,* Free Press, New York.

CHAPTER 12

New-Product Development's Role in Designing for Quality

➔ 12.1 OPPORTUNITIES FOR IMPROVEMENT IN DESIGN

This step on the quality spiral translates the needs of the user into a set of product design requirements for manufacturing (or the operations function in a service organization). This process is called product development, research and development, engineering, or product design.

Many problems encountered by both external and internal customers can be traced to the design of the product.

> **EXAMPLE 12.1.** In a classic study of 850 field failures of relatively simple electronic equipment, 43% of the failures were due to engineering design deficiencies.
>
> **EXAMPLE 12.2.** In one chemical company, a startling 50% of the product shipped was out of specification. Fortunately, the product was fit for use. A review concluded that many of the specifications were obsolete and had to be changed.
>
> **EXAMPLE 12.3.** A study of health care products revealed that 34% of the product recalls were caused by faulty product or software design.
>
> **EXAMPLE 12.4.** A manufacturer of terminals, modems, and other computer products analyzed the reasons for design changes. Local "wisdom" said that (1) about 10% of the changes were due to errors in the design and (2) the remaining changes were related to cost-reduction projects, requests from manufacturing, and changing customer requirements. But an in-depth analysis of design changes on four product lines reached a surprising conclusion: 78% of the changes were due to design errors.

For mechanical and electronic products of at least moderate complexity, errors during product development cause about 40% of fitness-for-use problems. When product development is responsible for both creating the formulation (design) of the product and developing the manufacturing process, as in chemicals, about 50% of the problems are due to development.

Product design people are highly educated and highly competent, particularly in quantitative skills. With respect to quality, a product design must provide the necessary product features to meet customer needs, and these features must be free of deficiencies. Poor designs are not the fault of individual designers—the cause is the *process* of design and development. We need, then, to identify some key issues in that process.

→ 12.2 DESIGN AND DEVELOPMENT AS A PROCESS

A *process* is a collection of activities that produce an output or result. This concept is in contrast to viewing design and development as the job of a functional department, often called "Design and Development." When design and development is viewed as a process, several key issues emerge:

1. Design and development has "customers" (i.e., anyone affected by the design). Design and development departments frequently function as the brain trust for new product ideas with an emphasis on "come up with the new product and let other departments manufacture it and sell it." That thinking is becoming obsolete, but some residue remains and needs to be recognized. Clearly, design and development has customers—external customers who purchase the product and also internal customers who make, sell, and service the product. Thus a product design must use modern design concepts to satisfy the needs of customers—as defined by the customers.
2. For external customers, the needs must be studied in detail and translated in a structured way into product features and design parameters—the goal being to knock a competitor off its pedestal. These needs are defined by customers, not by product development personnel. Competitive forces now dictate that customer needs must be analyzed in more detail than in the past. An example of a new need is products that are uniquely designed for one customer—in high or low volumes. Thus the emergence of "mass customization." This customization is often associated with faster order fulfillment. Feitzinger and Lee (1997) discuss the impact of mass customization on product design and process design.
3. For internal customers, the development process must recognize that the design becomes a major determinant of costs in areas such as manufacturing, purchasing, and servicing. Thus the development function must share some of the responsibility for these costs—at least to the extent of including these other functions as part of the development process. A helpful approach in this regard is the use of quality by design and the associated spreadsheets (see Chapter 5, "Quality by Design to Increase Sales").
4. Designers must recognize that product development is a cross-functional process (with subprocesses) involving the design department and other internal departments. A flow diagram of the development process (and key subprocesses) can be a useful starting point in identifying and understanding the roles of these departments. The leadership role of the design department in product development requires that designers become knowledgeable about the impact of the design on these other activities and that designers actively view these activities as part of the development process.

It also means that the design function must gather and document information on the lessons learned during development for use in future designs.
5. Suppliers must become part of the development team. In this age of specialization, suppliers can provide continuously updated expertise in designing certain subsystems of a system.
6. The product development process must address the issue of minimizing variation of product around the nominal values of design parameters. Traditionally, the design analysis sets the nominal (or desired average) value of a design parameter and then assigns (with less than a scientific approach) upper and lower specification limits around the nominal value. The implication is that all values within these limits are acceptable for performance. In practice, there are significant advantages to minimizing the variation around a nominal value (see Section 20.2). Design of experiments and other techniques can determine the optimum values of product and process parameters to minimize variation. Such approaches can help to achieve a six sigma defect goal of 3.4 parts per million. This is elaborated under "Application of Design of Experiments (DOE) to Product and Process Design."
7. The product development process (for physical as well as service products) can benefit from the application of quality management concepts. Quality by design (Chapter 5) steps can provide a road map for developing a new or modified product; quality control (Chapter 6) can provide the measurements and feedback for the various steps in product development; quality improvement (Chapter 4) can provide the methodology for continuous improvement of the development process. See Section 12.14 for details.

This integrated view of the product development process is sometimes called "quality by design" (QbD). For elaboration on product development, see *JQH6,* Section 14 ("Continuous Innovation Using Design for Six Sigma") and Section 28 ("Research & Development: More Innovation, Scarce Resources"). Endres (1997) explains the use of the Juran approach in improving research and development. Sobek et al. (1998) explain how Toyota integrates product development using (1) "social processes" among engineers to foster in-depth technical knowledge and efficient communication and (2) design standards to speed development, allow flexibility, and build the company's base of knowledge. Brocato and Potocki (1998) explain a system model for quality management at the Johns Hopkins Applied Physics Laboratory.

We proceed, then, to examine the phases of product development and then specific techniques for design effectiveness.

✦ 12.3 PHASES OF PRODUCT DEVELOPMENT AND THE EARLY WARNING CONCEPT

The process of developing modern products involves an evolution through distinct phases of development (see Table 12.1 for an example of phases).

The frequency and severity of problems caused by design have stimulated companies to develop more and better forms of early warning for impending troubles. These early warnings are available in various forms (Table 12.1). Special quality-oriented

TABLE 12.1
Forms of early warning of new-product problems

Phases of new-product progression	Forms of early warning of new-product troubles
Concept and feasibility study	Concept review
Prototype design	Design review, reliability and maintainability prediction, failure mode, effect and criticality analysis, safety analyses, value analysis and engineering
Prototype construction	Prototype test, environmental test, overstressing
Preproduction	Pilot production lots, design verification test, evaluation of specifications
Early full-scale production	In-house testing (e.g., kitchen, road), consumer use panels, limited marketing area
Full-scale production, marketing, and use	Employees as test panels, special provisions for prompt feedback
All phases	Failure analysis, data collection and analysis

tools described in this chapter help to evaluate designs and to improve the design process itself. Collectively, these early warnings and quality-oriented tools provide added assurance that the new designs will not create undue trouble as they progress around the spiral. Many forms of early warning are administered by specialists in reliability, maintainability, and other fields. The timing of their inputs is critical. Early timing can provide constructive help; late timing causes resistance to the warnings and often creates an atmosphere of blame. The cost of changes in a design can be huge; e.g., a design *change* during pilot production of a major electronics product can cost more than $1 million.

Next, we examine some techniques that help to ensure overall design effectiveness. These design assurance techniques address functional performance, reliability, maintainability, safety, manufacturability, and other attributes.

➜ 12.4 DESIGNING FOR BASIC FUNCTIONAL REQUIREMENTS; QUALITY BY DESIGN SPREADSHEETS

Product development translates customer needs for functional requirements into specific engineering and quality characteristics. For traditional products, this process was not complicated and could be achieved by experienced design engineers without using any special techniques. For modern products, it is necessary to document and analyze the design logic. This means starting with the desired product functionality—that is, what it should do—and then identifying the necessary characteristics for raw materials, parts, assemblies, and process steps. Such an approach goes under a variety of names, such as systems engineering, functional analysis systems technique, structured product/process analysis, quality function deployment, and quality by design.

Quality by Design

One technique for documenting overall design logic is quality by besign (QbD), sometimes called quality function deployment (QFD). QbD is a cross-functional structured process that systematically ties together the customer, the customer needs, the functional characteristics of the product, the product features, the production process features, and the control features. It uses a series of interlocking spreadsheets that translates customer needs into product and process characteristics. (See Figure 12.1; it's identical to Figure 5.6a but is repeated here for convenience.)

In addition to tying the needs with all the features and controls, the spreadsheets document the relative importance of various needs and features and include specific measureable goals, including quality goals, for all features.

An example of a QbD spreadsheet for high-level product design at a bank is shown in Table 12.2. This "high-level product design matrix" translates customer needs into product features and product feature goals for service provided by an automated teller machine.

Parameter Design and Robust Design

The most basic product feature is performance, i.e., the output—the color density of a television set, the turning radius of an automobile. To create such output, engineers

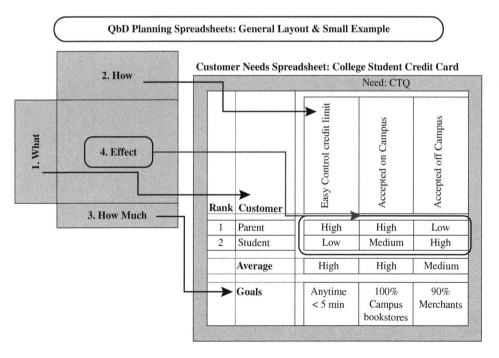

FIGURE 12.1
QbD spreadsheets: General layout and small example.

376 Quality Management and Analysis

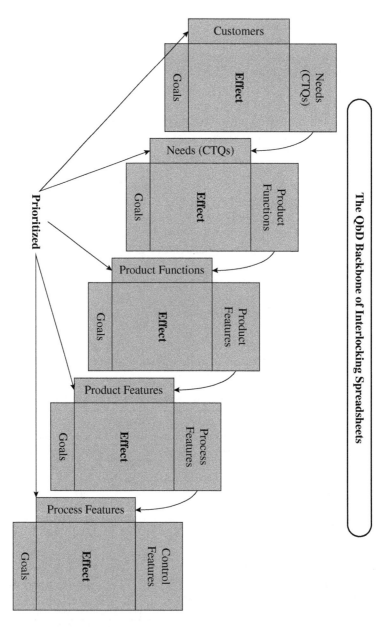

FIGURE 12.2
The QbD backbone of interlocking structures.

use engineering principles to combine inputs of materials, parts, components, assemblies, liquids, etc. For each input, the engineer identifies parameters and specifies numerical values to achieve the required output of the final product. For each parameter, the specifications state a target (or nominal) value and a tolerance range around the target. The process is called "parameter and tolerance design."

TABLE 12.2
Product design matrix for automated teller machine

Customers' needs	Product features	Product feature goals
Convenience	Hours of use	Available 24 hours a day, 99% of the time
Use by non-English speaking	Available in different languages	Customer can choose from four languages—English, Spanish, Japanese, and French
Easy access	Access to all accounts	Capable of accessing accounts and performing the following functions: get cash, make deposit, get information, transfer money between accounts
Speed/no lines	Multiple locations/machines available	Located at all 300 branches, with at least two machines available at all times
Sense of control	Customers can direct all transactions themselves	Directions can be followed by anyone with more than a sixth-grade education 90% of the time
Privacy and safety	Location of machines; well lighted and secure	Machines securely located indoors with card access required

Source: Juran Institute, Inc.

In selecting these target values, it is useful to set values so that the performance of the product in the field is not affected by variability in manufacturing or field conditions. Then the design is said to be "robust." Robust designs provide optimum performance simultaneously with variation in manufacturing and field conditions.

Designers have always tried to create robust designs. But as products become more complex, with many factors affecting performance, it becomes difficult to know (1) what factors do affect performance and (2) what nominal values to set for each factor. Furthermore, some factors affect the mean value of an output parameter, whereas others affect the variation about the mean. One purpose of development testing is to investigate these matters. A powerful aid to planning development testing is the statistical design of experiments (see Section 18.11, "The Design of Experiments").

Application of Design of Experiments (DOE) to Product and Process Design

G. Taguchi has developed a method for determining the optimum target values of product and process parameters that will minimize variation while keeping the mean on target. This method is known as "parameter design"; setting limits around a target value is known as "tolerance design." The emphasis on setting target values on design parameters to minimize variation is an important contribution of this approach.

In the Taguchi approach, the design of experiments is employed to investigate control factors in the presence of noise factors. *Control factors* are factors that can be controlled in the design of a product or process. During design, we want to set their values for optimum product performance and minimum variation. *Noise factors* are factors that cannot be easily controlled (e.g., ambient temperature, variation and deterioration of materials), but we want to minimize them. Taguchi recommends using fractional factorial experiments (see Section 18.15) to identify the critical control factors and set their levels to minimize variation.

Phadke et al. (1983) describe an application of this approach to integrated circuits in the Bell System. The example concerns the dimensions of the contact window of a large-scale integrated chip. Windows that do not open or are too small result in loss of contact to the devices, whereas excessively large windows lead to shortened device features. The steps in window forming and the factors critical to each step are shown in Table 12.3.

Levels of each of the nine critical factors were selected. Six of the factors have three levels each; three of the factors have only two levels each. A full factorial experiment to explore all possible factor-level combinations would require $3^6 \times 2^3$, or 5832, observations. (Twelve experiments were originally planned.) Instead, a fractional factorial design was chosen to investigate 18 combinations with a total of 34 measurements. Three dimensions of each window were measured. Transformed variables instead of absolute values were used in the data analysis. The variables selected were the mean and the signal-to-noise ratio (S/N). The signal-to-noise ratio is defined as

$$\text{S/N} = \log_{10}\left(\frac{\text{mean}}{\text{standard deviation}}\right).$$

The problem was to determine the factor levels that yield a maximum S/N while keeping the mean on target. Two steps were needed:

1. Determining the factors that have a significant effect on the S/N. These are the factors that control process variability and are referred to as control factors. For each control factor, the level chosen is that with the highest S/N, thus maximizing the overall ratio.
2. Selecting the control factor that has the smallest effect on the S/N. This factor is called a "signal factor." The levels of the factors that are neither control factors nor signal factors are set at nominal levels prior to the experiment. Finally, the level of the signal factor is set so that the mean response is on target.

TABLE 12.3
Fabrication steps and critical factors in window forming

Fabrication step	Critical factors
Applying photoresist	Photoresist viscosity *(B)* and spin speed *(C)*
Baking	Baking temperature *(D)* and baking time *(E)*
Exposing	Mask dimension *(A)*, aperture *(F)*, and exposure time *(G)*
Developing	Developing time *(H)*
Plasma etching	Etching time *(I)*
Removing photoresist	Does not affect window size

TABLE 12.4
Optimum factor levels

Label	Factor	Standard level	Optimum level
A	Mask dimension, μm	2.0	2.5
B	Viscosity	204	204
C	Spin speed, r/min	3000	4000
D	Bake temperature, °C	105	105
E	Bake time, min	30	30
F	Aperture	2	2
G	Exposure, PEP setting	Normal	Normal
H	Developing time, s	45	60
I	Plasma etch time, min	13.2	13.2

Source: Phadke et al. (1983, p. 1303).

The analysis of variance revealed that the control factors were factors A, B, C, F, G, and H. Based on an analysis of the data and engineering judgment, the exposure time was chosen as the signal factor.

The levels selected for each factor are shown in Table 12.4. Using these levels, a sample of chips was manufactured. The resulting benefits were as follows:

	Old conditions	Optimum conditions
Standard deviation	0.29	0.14
Visual defects (window per chip)	0.12	0.04

After observing these improvements, the process engineers eliminated a number of in-process checks, thereby reducing the overall time spent by the wafers in photolithography by a factor of 2.

Ross (1996) provides a complete discussion of the Taguchi approach. The Taguchi methodology has spawned some controversy. For a rigorous and practical account of the "triumphs and tragedies" of the methodology, see Pignatiello and Ramberg (1992). Montgomery (1997) and others agree that the concept of robust parameter design is good but are critical of the experimental designs and the data analysis methods (e.g., the signal-to-noise ratio) used in the Taguchi approach.

↦ 12.5 DESIGNING FOR TIME-ORIENTED PERFORMANCE (RELIABILITY)

Design engineers realize that a product should have a long service life with few failures. As products become more complex, failures increase with operating time. Traditional efforts of design, although necessary, are often not sufficient to achieve both the functional performance requirements and a low rate of failures with time. To prevent these failures, specialists have created a collection of tools called reliability engineering. Reliability is quality over time.

Reliability is the ability of a product to perform a required function under stated conditions for a stated period of time. (More simply, reliability is the chance that a product will work for the required time.) If this definition is dissected, four implications become apparent:

1. The quantification of reliability in terms of a probability.
2. A statement defining successful product performance.
3. A statement defining the environment in which the equipment must operate.
4. A statement of the required operating time between failures. (Otherwise, the probability is a meaningless number for time-oriented products.)

To achieve high reliability, it is necessary to define the specific tasks required. This task definition is called the *reliability program.* The early development of reliability programs emphasized the *design* phase of the product life cycle. However, it soon became apparent that the manufacturing and field-use phases could not be handled separately. The result was reliability programs that spanned the full product life cycle, i.e., "cradle to grave."

A reliability program typically includes the following activities: setting overall reliability goals, apportionment of the reliability goals, stress analysis, identification of critical parts, failure mode and effect analysis, reliability prediction, design review, selection of suppliers (see Chapter 15), control of reliability during manufacturing (see Chapters 4 and 20), reliability testing, and failure reporting and corrective action system. Except as indicated, these activities are discussed in this chapter and Chapter 19.

Some elements of a reliability program are old (e.g., stress analysis, part selection). The significant new aspect is the *quantification* of reliability. The act of quantification makes reliability a design parameter just like weight and tensile strength. Thus reliability can be submitted to specification and verification. Quantification also helps to refine certain traditional design tasks such as stress analysis and part selection.

Before embarking on a discussion of reliability tasks, a remark on the application of the tasks is in order.

- These tasks are clearly not warranted for simple products. However, many products that were originally simple are now becoming more complex. In this case, the various reliability tasks should be examined to see which, if any, might be justified.
- Reliability techniques were originally developed for electronic and space products, but their adaptation to mechanical products has now achieved success.
- These techniques apply to products a company markets and also to capital equipment a company purchases, e.g., numerically controlled machines. A less obvious application is to chemical-processing equipment.

For an overview of key reliability activities, see Figure 12.3.

Setting Overall Reliability Goals

The original development of reliability quantification consisted of a probability and a mission time along with a definition of performance and use conditions. This definition proved confusing to many people, so the index was abbreviated (using a mathematical relationship) to mean time between failures (MTBF). Many people

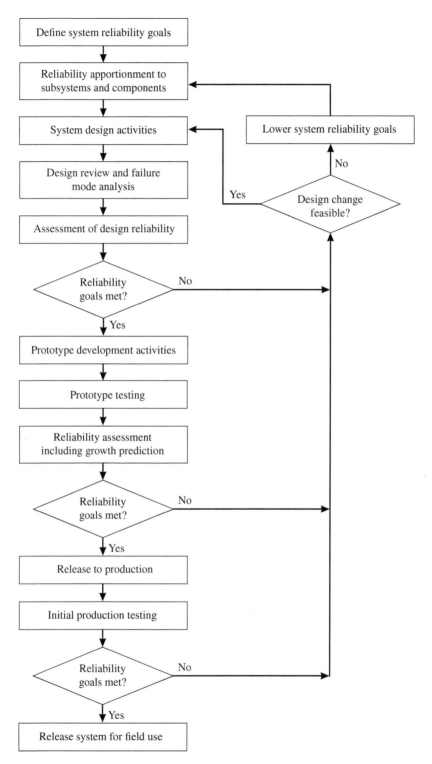

FIGURE 12.3
Some key reliability activities. (*From Ireson et al., 1996, p. 24.18.*)

TABLE 12.5
Reliability metrics of merit

Figure of merit	Meaning
Mean time between failures (MTBF)	Mean time between successive failures of a repairable product
Failure rate	Number of failures per unit time
Mean time to failure (MTTF)	Mean time to failure of a nonrepairable product or mean time to first failure of a repairable product
Mean life	Mean value of life ("life" may be related to major overhead, wear-out time, etc.)
Mean time to first failure (MTFF)	Mean time to first failure of a repairable product
Mean time between maintenance (MTBM)	Mean time between a specified type of maintenance action
Longevity	Wear-out time for a product
Availability	Operating time expressed as a percentage of operating and repair time
System effectiveness	Extent to which a product achieves the requirements of the user
Probability of success	Same as reliability (but often used for "one shot" or non-time-oriented products)
b_{10} life	Life during which 10% of the population would have failed
b_{50} life	Median life, or life during which 50% of the population would have failed
Repairs/100	Number of repairs per 100 operating hours

believe that MTBF is the only reliability index, but no single index applies to most products. A summary of common indexes (often called *figures of merit*) is presented in Table 12.5.

As experience is gained in quantifying reliability, many companies are learning that it is best to create an index that uniquely meets the needs of those who will use the index. Users of the index include internal technical personnel and also marketing personnel and users of the product. Two examples of reliability indexes and goals follow:

- *For a telephone system.* The downtime of each switching center should be a maximum of 24 hours per 40 years.
- *For an engine manufacturer.* Seventy percent of the engines produced should pass through the warranty period without generating a claim. The number of failures per failed engine should not exceed one.

Note that both examples quantify reliability.

Setting overall reliability goals requires a meeting of the minds on (1) reliability as a number, (2) the environmental conditions to which the numbers apply, and (3) a definition of successful product performance. This task is not easily accomplished.

However, the act of requiring designers to define both environmental conditions and successful product performance with precision forces the designers to understand the design in greater depth.

Reliability Apportionment, Prediction, and Analysis

The process of reliability quantification involves three phases:

1. *Apportionment (or budgeting).* The process of allocating (aligning) reliability objectives among various elements that collectively make up a higher level product.
2. *Prediction.* The use of prior performance data plus probability theory to calculate the expected failure rates for various circuits, configurations, etc.
3. *Analysis.* The identification of the strong and weak portions of the design to serve as a basis for improvements, trade-offs, and similar actions.

These phases are illustrated in Table 12.6, one of the first published examples.

In the top section of the table, an overall reliability requirement of 95% for 1.45 hours is apportioned to the six subsystems of a missile. The second section of the table apportions the budget of the explosive subsystem to the three units within the subsystem. The allocation for the fusing circuitry is 0.998 or, in terms of MTBF, 725 hours. In the final section of the table, the proposed design for the circuitry is analyzed and a reliability prediction made, using the method of adding failure rates. As the prediction indicates an MTBF of 1398 hours as compared to a budget of 725 hours, the proposed design is acceptable. The prediction technique provides a quantitative evaluation of a design or a design change and also can identify design areas having the largest potential for reliability improvement. Thus the vital few will be obvious by noting the components with the highest failure rates. In this example the transistors, diodes, and tantalum capacitors account for about 70% of the unreliability.

The approach of adding failure rates to predict system reliability is analogous to the control of weight in aircraft structures, where a running record is kept of weight as various parts are added to the design.

Reliability prediction is a continuous process starting with paper *predictions* based on a design analysis, plus historical failure-rate information (see Section 12.5). The evaluation ends with reliability *measurement* based on data from customer use of the product. Table 12.7 lists some characteristics of the various phases.

Although the visible result of the prediction procedure is to quantify the reliability numbers, the *process* of prediction is usually as important as the resulting numbers because the prediction cannot be made without obtaining rather detailed information on product missions, environments, critical component histories, etc. Acquiring this information often gives the designer knowledge previously unavailable. Even if the designer is unable to secure the needed information, this inability nevertheless identifies the areas of ignorance in which the designer is forced to work.

The use of computer software improves the effectiveness of reliability predictions and reliability analyses. Such software handles the detailed calculations and also makes it feasible to identify and assess many alternatives before finalizing a design. The "Directory of Software" published periodically in *Quality Progress* includes a description of software for reliability.

TABLE 12.6
Establishment of reliability objectives*

System breakdown

Subsystem	Type of operation	Reliability	Unreliability	Failure rate per hour	Reliability objective*
Air frame	Continuous	0.997	0.003	0.0021	483
Rocket motor	One shot	0.995	0.005		1/200 operations
Transmitter	Continuous	0.982	0.018	0.0126	80.5 h
Receiver	Continuous	0.988	0.012	0.0084	121 h
Control system	Continuous	0.993	0.007	0.0049	207 h
Explosive system	One shot	0.995	0.005		1/200 operations
System		0.95	0.05		

Explosive subsystem breakdown

Unit	Operating mode	Reliability	Unreliability	Reliability objective
Fusing circuitry	Continuous	0.998	0.002	725 h
Safety and arming mechanism	One shot	0.999	0.001	1/1000 operations
Warhead	One shot	0.998	0.002	2/1000 operations
Explosive subsystem		0.995	0.005	

Unit breakdown

Fusing circuitry component part classification	Number used, n	Failure rate per part, λ (%/1000 h)	Total part failure rate, $n\lambda$ (%/1000 h)
Transistors	93	0.30	27.90
Diodes	87	0.15	13.05
Film resistors	112	0.04	4.48
Wirewound resistors	29	0.20	5.80
Paper capacitors	63	0.04	2.52
Tantalum capacitors	17	0.05	8.50
Transformers	13	0.20	2.60
Inductors	11	0.14	1.54
Solder joints and wires	512	0.01	5.12
			71.51

$$\text{MTBF} = \frac{1}{\text{failure rate}} = \frac{1}{\Sigma n\lambda} = \frac{1}{0.0007151} = 1398 \text{ h}$$

Source: Adapted by F. M. Gryna, Jr., from Beaton (1959, p. 65).
*For a mission time of 1.45 hours.

Parts Selection and Control

Table 12.6 shows how system reliability rests on a base of the reliability of component parts.

The vital role played by parts reliability has resulted in programs for thorough selection, evaluation, and control of parts. These programs include parts application studies, approved parts lists, identification of critical components, and use of derating.

TABLE 12.7
Stages of reliability prediction and measurement

	1. Start of design	2. During detailed design	3. At final design	4. From system tests	5. From customer use
Basis	Prediction based on approximate part counts and part failure rates from previous product use; little knowledge of stress levels, redundancy, etc.	Prediction based on quantities and types of parts, redundancies, stress levels, etc.	Prediction based on types and quantities of part failure rates for expected stress levels, redundancies, external environments, special maintenance practices, special effects of system complexity, cycling effects, etc.	Measurement based on the results of tests of the complete systems: appropriate reliability indexes are calculated from the number of failures and operating time	Same as step 4 except calculations are based on customer use data
Primary uses	Evaluate feasibility of meeting a proposed numerical requirement Help in establishing a reliability goal for design	Evaluate overall reliability Define problem areas	Evaluate overall reliability Define problem areas	Evaluate overall reliability Define problem areas	Measure achieved reliability Define problem areas Obtain data for future designs

Note: System tests in steps 4 and/or 5 may reveal problems that result in a revision of the "final" design. Such changes can be evaluated by repeating steps 3, 4, and 5.

Critical components list

A component part is considered "critical" if any of the following conditions apply:

- It has a high population in the equipment.
- It has a single source of supply.
- It must function to special, tight limits.
- It has not been proved to the reliability standard, i.e., no test data are available, or use data are insufficient.

The critical components list should be prepared early in the design effort. It is common practice to formalize these lists, showing, for each critical component, the nature of the critical features, the plan for quantifying reliability, the plan for improving reliability, etc. The list becomes the basic planning document for (1) test programs for qualifying parts; (2) design guidance in application studies and techniques; and (3) design guidance for application of redundant parts, circuits, or subsystems.

Derating practice

Derating is the assignment of a product to operate at stress levels below its normal rating, e.g., a capacity rated at 300 V is used in a 200-V application. For many components, data are available showing failure rate as a function of stress levels. The conservative designer will use such data to achieve reliability by using the parts at low power ratios and low ambient temperatures.

Some companies have established internal policies with respect to derating. Derating is also a form of quantifying the factor of safety and hence lends itself to setting guidelines as to the design margins to be used. Derating may be considered a method of determining more scientifically the factor of safety that engineers have long provided empirically. For example, if the calculated load of a structure is 20 tons, engineers might design the structure to withstand 100 tons as a protection against unanticipated loads, misuse, hidden flaws, deterioration, etc.

Failure Mode, Effect, and Criticality Analysis

Two techniques provide a methodical way of examining a design for possible ways in which failures can occur. In the *failure mode, effect, and criticality analysis* (FMECA), a product is examined at the system and/or lower levels for all the ways in which a failure may occur. For each potential failure, an estimate is made of its effect on the total system and of its seriousness. In addition, a review is made of the action being taken (or planned) to minimize the probability of failure or to minimize the effect of failure. Figure 12.4 shows a portion of an FMECA for a traveling lawn sprinkler. Each hardware item is listed on a separate line. Note that the failure "mode" is the symptom of the failure, as distinct from the cause of failure, which consists of the proved reasons for the existence of the symptoms. The analysis can be elaborated to include matters such as

- *Safety.* Injury is the most serious of all failure effects. In consequence, safety is handled through special programs.
- *Effect on downtime.* Must the system stop until repairs are made, or can repairs be made during off-duty time?
- *Access.* What hardware items must be removed to get at the failed component?

1 = Very low (<1 in 1000)
2 = Low (3 in 1000)
3 = Medium (5 in 1000)
4 = High (7 in 1000)
5 = Very high (>9 in 1000)

T = Type of failure
P = Probability of occurrence
S = Seriousness of failure to system
H = Hydraulic failure
M = Mechanical failure
W = Wear failure
C = Customer abuse

Product HRC-1
Date Jan. 14, 2000
By S.M.

Component part number	Possible failure	Cause of failure	T	P	S	Effect of failure on product	Alternatives
Worm bearing 4224	Bearing worn	Not aligned with bottom housing	M	1	4	Spray head wobbles or slows down	Improve inspection
Zytel 101		Excessive spray head wobble	M	1	3	Spray head wobbles or slows down	Improve worm bearing
Bearing stem 4225	Excessive wear	Poor bearing/ material combination	M	5	4	Spray head wobbles and loses power	Change stem material
Brass		Dirty water in bearing area	M	5	4	Spray head wobbles and loses power	Improve worm seal area
		Excessive spray head wobble	M	2	3	Spray head wobbles and loses power	Improve operating instructions
Thrust washer 4226	Excessive wear	High water pressure	M	2	5	Spray head will stall out	Inform customer in instructions
Fulton 404		Dirty water in washers	M	5	5	Spray head will stall out	Improve worm seal design
Worm 4527	Excessive wear in bearing area	Poor bearing/ material combination	M	5	4	Spray head wobbles and loses power	Change bearing stem material
Brass		Dirty water in bearing area	M	5	4	Spray head wobbles and loses power	Improve worm seal design
		Excessive spray head wobble	M	2	3	Spray head wobbles and loses power	Improve operating instructions

FIGURE 12.4
Failure mode, effect, and criticality analysis.

- *Repair planning.* What is the anticipated repair time? What special repair tools are needed?
- *Recommendations.* What changes in designs or specifications should be made? What tests should be added? What instructions should be included in manuals of inspection, operation, or maintenance?

In Figure 12.4, a ranking procedure has been applied to assign priorities to the failure modes for further study. The ranking is twofold: (1) the probability of occurrence of the failure mode and (2) the severity of the effect. For each of these, a scale of 1 to 5 is used. If desired, a risk priority number can be calculated as the product of the ratings. Priority is then assigned to investigating failure modes with high risk priority numbers.

In this example, the analysis revealed that about 30% of the expected failures were in the worm-and-bearing-stem area and that a redesign could easily be justified.

For most products, conducting the analysis of failure mode and failure effect for each component is not economical. Instead, engineering judgment is used to single out items that are critical to the operation of the product. As the FMECA proceeds for these selected items, the designer will discover that ready answers for some of the failure modes are lacking and that further analysis is necessary.

Generally, FMECA on one item is helpful to designers of other items in the system. In addition, the analyses are useful in planning for inspection, assembly, maintainability, and safety.

Although the FMECA was originally developed for analyzing the design of physical products, the concept also applies to the service sector. Figure 12.5 shows an application at a bank in which the "product" is a new checking account and the "component" is printing new checks. This example also shows the additional factor of probability of detection of the defect before it affects the customer.

Fault tree analysis (FTA) is another method of studying potential failures in a product. Whereas FMECA usually studies all potential modes of failure, FTA studies are usually applied only to failures that are considered serious enough to warrant detailed analysis. For an example, see "Fault Tree Analysis" section.

The FMECA should not be developed by an individual designer, but by a team of designers, operations, quality, and other personnel who examine the proposed design from the several perspectives necessary for external and internal customer satisfaction. Perkins (1996) discusses this team approach in a resource-limited design community. Hatty and Owens (1994) explain how training and a multidisciplined team are important prerequisites for an effective analysis. The FMECA can also be applied to analyze proposed processes (rather than products). See Section 13.6.

Evaluating Designs by Testing

Although reliability prediction, design review, FMECA, and other techniques are valuable as early-warning devices, they cannot be a substitute for the ultimate proof, i.e., use of the product by the customer. However, field experience comes too late and must be preceded by a substitute—various forms of testing the product to simulate field use.

Long before reliability technology was developed, several types of tests (performance, environmental, stress, life) were done to evaluate a design. The advent of reliability, maintainability, and other parameters resulted in additional types of tests.

A summary of types of tests for evaluating a design is given in Table 12.8. Test programs often use one type of test for more than one purpose, e.g., evaluating both performance and environmental capabilities.

Product: _____ New checking account
Component: _____ Printing new checks

1	2	3	4	5	6	7	8	9
Mode of failure	Cause of failure	Effect of failure	Frequency of occurrence (1–10)	Degree of severity (1–10)	Chance of detection (1–10)	Risk priority (1–1000) [4] × [5] × [6]	Design action	Design validation
Checks being printed incorrectly	Incorrect information on application form	Checks have to be reissued	4	6	8	192	Clerk reviews information with customer	
	Data entry error	Checks have to be reissued	8	6	5	240	Review step in software	
	Information entered in wrong field on application form	Checks have to be reissued	5	6	2	60	—	
Incorrect account number								

FIGURE 12.5
Failure mode and effect analysis: checking account. (*From Juran Institute, Inc.*)

TABLE 12.8
Summary of tests used to evaluate a design

Type of test	Purpose
Performance	Determine ability of product to meet basic performance requirements
Environmental	Evaluate ability of product to withstand defined environmental levels; determine interval environments generated by product operation; verify environmental levels specified
Stress	Determine levels of stress that a product can withstand to determine the safety margin inherent in the design; determine modes of failure that are not associated with time
Reliability	Determine product reliability and compare to requirements; monitor for trends
Maintainability	Determine time required to make repairs and compare to requirements
Life	Determine wear-out time for a product and failure modes associated with time
Pilot run	Determine whether fabrication and assembly processes are capable of meeting design requirements; determine whether reliability will be degraded
Customer ("beta")	Determine whether product functions properly under customer use conditions

All tests provide some degree of design assurance. They also involve a risk of leading one astray. The principal sources of risk follow.

Intended use versus actual use. The designer typically aims to attain fitness for the intended use. However, actual use can differ from the designer's concept because of variations in the environment and other conditions of use. In addition, some users will misapply or misuse the product.

Model construction versus subsequent production. Models are usually built by skilled specialists under the supervision of designers. Subsequent production is carried out by less skilled factory workers under supervisors who must meet standards of productivity as well as quality. In addition, factory processes seldom possess the adaptive flexibility available in the model shop.

Variability due to small numbers. The number of models built is usually small. (Often there is only one.) Yet tests on these small numbers are used to judge the adequacy of the design for making many production units, sometimes running into the thousands and even millions.

Evaluation of test results. Pressures to release a design for production can result in test plans and evaluations that do not objectively evaluate conformance to performance requirements, let alone complete fitness for use. One organization studied the process of qualifying a new design by having an independent review team analyze a sample of "qualification test" results on designs that had been approved for release. Two conclusions were that

- Pressures to release designs caused design approvals that later resulted in field problems. The situation was traced to (1) inadequate initial testing or (2) lack of verification of design changes made to correct failures that surfaced on the initial test.

- More than 50% of the approved test results were rejected because the test procedure was unable to evaluate the requirements set by the product development team.

As a homework problem, readers are asked to recommend action to avoid these risks.

Methods for Improving Reliability during Design

The general approach to quality improvement (see Chapter 4) is widely applicable to reliability improvement as far as economic analysis and managerial tools are concerned. The differences are in the technological tools used for diagnosis and remedy. Projects can be identified through reliability prediction, design review, FMECA, or other reliability evaluation techniques.

Action to improve reliability during the design phase is best taken by the designer. The designer understands best the engineering principles involved in the design. The reliability engineer can help by defining areas needing improvement and by assisting in the development of alternatives. The following actions indicate some approaches to improving a design:

1. *Review the users' needs* to see whether the *function* of the unreliable parts is really necessary to the user. If not, eliminate those parts from the design. Alternatively, determine whether the reliability index (figure of merit) correctly reflects the real needs of the user. For example, availability (discussed in detail in the next section) is sometimes more meaningful than reliability. If so, a good maintenance program might improve availability and hence ease the reliability problem.
2. *Consider trade-offs* of reliability for other parameters, e.g., functional performance, weight. Here again, you may find that customers' real needs may be better served by such a trade-off.
3. *Use redundancy* to provide more than one means of accomplishing a given task so that all the means must fail before the system fails. This technique is discussed in Section 19.4, "The Relationship between Part and System Reliability."
4. *Review the selection of any parts* that are relatively new and unproven. Use standard parts whose reliability has been proven by actual field use. (However, be sure that the conditions of previous use are applicable to the new product.)
5. *Use derating* to ensure that the stresses applied to the parts are lower than the stresses the parts can normally withstand.
6. *Use "robust" design methods* that enable a product to handle unexpected environments.
7. *Control the operating environment* to provide conditions that yield lower failure rates. Common examples are potting electronic components to protect them against climate and shock and the use of cooling systems to keep down ambient temperatures.
8. *Specify replacement schedules* to remove and replace low-reliability parts before they reach the wear-out stage. In many cases, the replacement is made contingent

on the results of checkouts or tests that determine whether degradation has reached a prescribed limit.
9. *Prescribe screening tests* to detect "infant mortality" failures and to eliminate substandard components. The tests take various forms, e.g., bench tests, "burn-in," and accelerated life tests.
10. *Conduct research and development* to improve the basic reliability of components that contribute most of the unreliability. Although such improvements avoid the need for subsequent trade-offs, they may require advancing the state of the art and hence making an investment of an unpredictable size.

Although none of the foregoing actions provides a perfect solution, the range of choice is broad. In some instances, the designer can arrive at a solution independently. More usually, collaboration with other company specialists will be necessary. In still other cases, the customer and/or company management will have to be adaptable because of the broader considerations involved.

12.6 AVAILABILITY

One major parameter of fitness for use is availability. *Availability* is the ability of a product, when used under given conditions, to perform satisfactorily when called upon. The total time in the operative state (also called uptime) is the sum of the time spent in active use and in the standby state. The total time in the nonoperative state (also called downtime) is the sum of the time spent under active repair and waiting for spare parts, paperwork, etc. The quantification of both availability and unavailability dramatizes the extent of the problems and areas for potential improvements. Formulas for quantifying availability are presented in Section 19.9, "Availability."

The proportion of time that a product is available for use depends on (1) freedom from failures, i.e., reliability, and (2) the ease with which service can be restored after a failure. The latter factor brings us to the subject of maintainability.

Designing for Maintainability

The tools for ensuring maintainability follow the same basic pattern as those for ensuring reliability, i.e., there are tools for specifying, predicting, analyzing, and measuring maintainability. The tools apply to both preventive maintenance (to reduce the number of failures) and corrective maintenance (to restore a product to operable condition). Some of the basic tools are discussed later.

Maintainability is often specified quantitatively, such as the mean time to repair (MTTR). MTTR is the mean time needed to perform repair work assuming that a spare part and technician are available, e.g., a test set may have a specified MTTR of 2.5. As with reliability, no one maintainability index applies to most products. Other examples of indexes are percentage of downtime due to hardware failures, percentage of downtime due to software errors, and mean time between preventive maintenance.

Following the approach used in reliability, a maintainability goal for a product can be apportioned to the various components of the product. Also, maintainability can be predicted based on an analysis of the design. For elaboration, see Ireson et al. (1996), Chapter 19.

Approaches to improving maintainability of a design are both general and specific. General approaches include

- *Reliability versus maintainability.* For example, given an availability requirement, should the response be an improvement in reliability or in maintainability?
- *Modular versus nonmodular construction.* Modular design requires added design effort but reduces the time required for diagnosis and remedy in the field. The fault need only be localized to the module level, after which the defective module is simply unplugged and replaced. This concept also applies to consumer products such as television sets.
- *Repairs versus throwaway.* For some products or modules, the cost of field repair exceeds the cost of making new units in the factory. In such cases, design for throwaway is an economic improvement in maintainability.
- *Built-in versus external test equipment.* Built-in test features reduce diagnostic time but usually at an added investment.
- *Person versus machine.* For example, should the operation/maintenance function be highly engineered with special instrumentation and repair facilities, or should it be left to skilled technicians with general-use equipment?

Specific approaches are based on detailed checklists that are used as guides to good design for maintainability.

Maintainability can also be demonstrated by testing. The demonstration consists of measuring the time needed to locate and repair malfunctions or to perform selected maintenance tasks.

✣ 12.7 DESIGNING FOR SAFETY

Safety analysis tools include hazard quantification, designation of safety-oriented characteristics and components, fault tree analysis, fail-safe concepts, in-house and field testing, and publication of product ratings.

Quantification of safety

Generally, quantification of safety has been time related. Industrial injury rates are quantified on the basis of lost-time accidents per million labor-hours of exposure. (Note that this value expresses the frequency of occurrence, but does not indicate the severity of the accidents.) Motor vehicle injury rates are on the basis of injuries per 100 million miles. School injury rates are on the basis of injuries per 100,000 student days.

Product designers have tended to quantify safety in two ways:

1. *Hazard frequency.* A hazard is any combination of parts, components, conditions, or changing set of circumstances that present an injury potential. Hazard frequency takes the form of frequency of occurrence of an unsafe event and/or injuries per unit of time, e.g., per million hours of exposure. MIL-STD-882D has established categories of probability levels for hazards ranging from frequent to improbable. Such probabilities are sometimes referred to as "risk."
2. *Hazard severity.* MIL-STD-882D recognizes four levels of severity:

    - Severity level 1—catastrophic: Could result in one or more of the following: death, permanent total disability, irreversible significant environmental impact, or loss exceeding $10 million.
    - Severity level 2—critical: Could result in one or more of the following: permanent partial disability, injuries or occupational illness that may result in hospitalization of at least three personnel, reversible significant environmental impact, or loss exceeding $1 million but less than $10 million.
    - Severity level 3—marginal: Could result in one or more of the following: injury or occupational illness resulting in 10 or more lost work days, reversible moderate environmental impact, or loss exceeding $100,000 but less than $1 million.
    - Severity level 4—negligible: Could result in one or more of the following: injury or illness resulting in less than 10 lost work days, minimal environmental impact, or loss less than $100,000.

    MIL-STD-882C identifies 22 tasks to eliminate hazards.

Hazard analysis

Hazard analysis is similar to FMECA—but the failure event is one that causes an injury. Three forms of hazard analysis can be prepared: design concept, operating procedures, and hardware failures.

Fault tree analysis

This top-down approach starts by supposing that an accident takes place. It then considers the possible direct causes which could lead to this accident. Next, it looks for the origins of these causes. Finally, it looks for ways to avoid these origins and causes. The branching out of origins and causes is what gives the technique the name of "fault tree" analysis. The approach is the reverse of FMECA, which starts with origins and causes and looks for any resulting bad effects.

Hammer (1980) presents a fault tree analysis for an interlock safety circuit (see Figures 12.6a and 12.6b).

Based on field experience with specific products, detailed checklists are often developed to provide the designer with information on potential hazards, the injuries that can result, and specific types of design actions that can be taken to minimize the risk.

As products have become more complex, the interaction of products with the human beings operating the product has assumed increasing importance. The evaluation of the design of a product to ensure compatibility with the capabilities of human beings is referred to as "ergonomics" or "human-centered design." For elaboration, see Ireson et al. (1996).

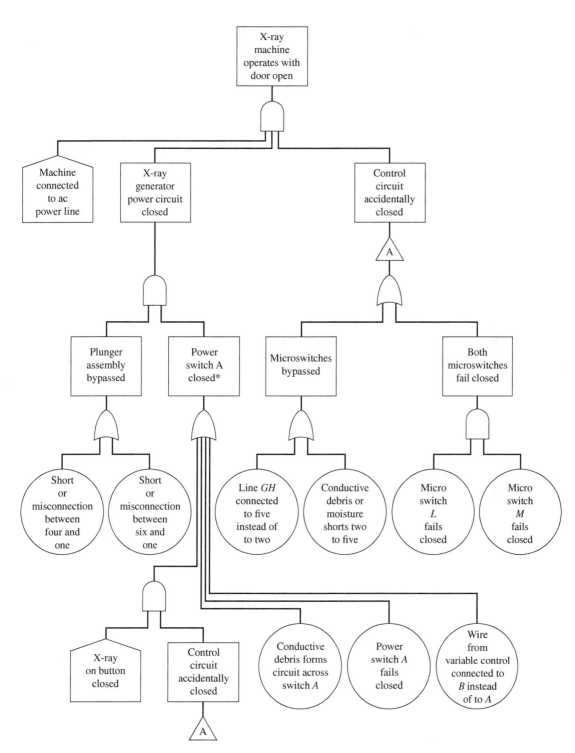

*Fault indicator lamp will light anytime this condition exists.

FIGURE 12.6a
Fault tree analysis of an interlock safety circuit. (*Modified from Hammer, 1980.*)

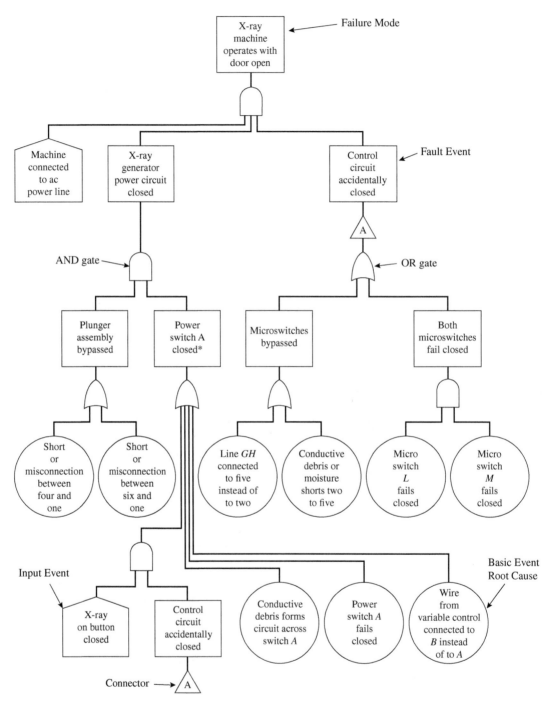

FIGURE 12.6b
Fault tree analysis of safety interlock on X-ray machine.

→ 12.8 DESIGNING FOR MANUFACTURABILITY

Decisions made during design are the dominant influence on product costs, the ability to meet specifications, and the time required to bring a new product to the marketplace. Furthermore, once these decisions are made (in hardware or software), the cost of changes in the design can be huge, e.g., a design *change* during pilot production of a major electronics product can cost more than $1 million.

An important set of decisions is the selection of specifications for product characteristics to be controlled during production. *Specification limits* specify the allowable limits of variability above and below the nominal value set by the designer.

The selection of limits has a dual effect on the economics of quality. The limits affect

- Fitness for use and hence the salability of the product.
- Costs of manufacture (facilities, tooling, productivity) and quality (equipment, inspection, scrap, rework, material review, etc.).

Chapter 19 explains some quantitative techniques for setting specification limits.

A technique called design for manufacturability focuses on simplifying a design to make it more producible. The emphasis is on reducing the total number of parts, the number of different parts, and the total number of manufacturing operations. This type of analysis is not new—"value engineering" tools have been useful in achieving design simplification (see Section 12.9). What is new, however, is the computer software available for analyzing a design and identifying opportunities for simplifying assembly products. Such software dissects the assembly step by step; poses questions concerning parts and subassemblies; and provides a summary of the number of parts, the assembly time, and the theoretical minimum number of parts or subassemblies. Use of such software enables designers to learn the principles for ease of manufacturing analogous to reliability, maintainability, and safety analyses. In one example, the proposed design of a new electronic cash register was analyzed with design for manufacturability (DFM) software. As a result, the number of parts was reduced by 65%. A person using no screws or bolts can assemble the register in less than two minutes—blindfolded. This simplified terminal was put onto the marketplace in 24 months—a record. Such design simplification reduces assembly errors and other sources of quality problems during manufacture.

In organizations that use the six sigma approach, the extremely low defect levels desired (3.4 defects per million) have an impact on product development. Product designers must understand and study manufacturing process capability before the design is released. Kodak designers use an approach that starts with the identification of critical parts and critical characteristics. The measurement process is then defined, including establishing measurement capability. The manufacturing process is then studied first for stability and then for process capability. For elaboration, see Turmel and Gartz (1997). Process stability and process capability are covered in this book in Chapter 20, "Statistical Process Control."

→ 12.9 COST AND PRODUCT PERFORMANCE

Designing for reliability, maintainability, safety, and other parameters must be done with the simultaneous objective of minimizing cost. Formal techniques of achieving an optimum balance between performance and cost include both quantitative and qualitative approaches.

The quantitative approach uses a ratio relating performance and cost. Such a ratio tells "what we get for each dollar we spend." The ratio is particularly useful in comparing alternative design approaches to accomplish a desired function.

A cost-effectiveness comparison of four alternative designs is shown in Table 12.9. Note that design 3 is the optimum design, even though design 4 has higher availability.

Another approach to comparing several different designs over a number of attributes is shown in Table 12.10. A design for a food-waste disposer is compared to the designs of two competitor models for 10 attributes describing fitness for use. For each combination of attribute and design, an effectiveness cost ratio is calculated. For example, for design G and the grinding-time characteristic, the ratio is 6/2.14, or 2.8. The value of 6 is the product of a weighting factor of 3 for grinding time and a score of 2 for design G in grinding time. The value of $2.14 is the estimated cost of achieving the grinding time of design G using the design G concept in production. The total of the ratios for each company provides a cost-effectiveness type of index.

Several approaches to achieving a balance between performance and cost have been developed. Value engineering is a technique for evaluating the design of a product to ensure that the essential functions are provided at minimal overall cost to the manufacturer or user. A useful reference is Park (1999). A complementary technique is the "design to cost" approach. It starts with a definition of (1) a cost target for the product and (2) the function desired. Alternative design concepts are then developed and evaluated.

During the development cycle, the product undergoes several reviews. One form is a business review in which the current results of the development effort are summarized and a decision is made whether to proceed. Another type of review is technical and is usually called "design review."

TABLE 12.9
Cost-effectiveness comparison of alternative designs

	Design			
	1	2	3	4
Mean time between failures (MTBF)	100	200	500	500
Mean downtime (MDT)	18	18	15	6
Availability*	0.847	0.917	0.971	0.988
Life-cycle cost (†)	51,000	49,000	50,000	52,000
Number of effective hours †	8470	9170	9710	9880
Cost/effective hour ($)	6.02	5.34	5.15	5.26

*Availability $= \dfrac{\text{MTBF}}{\text{MTBF} + \text{MDT}}$

†Number of effective hours = 10,000 h of life × availability.

TABLE 12.10
Value comparison of food-waste disposers

	Company design	Competitor designs	
		B	G
Grinding time	9/1.73 = 5.2	9/0.87 = 10.3	6/2.14 = 2.8
Fineness of grind	4/9.18 = 0.4	4/7.82 = 0.5	4/11.88 = 0.3
Frequency of jamming	9/2.25 = 4.0	9/1.98 = 4.6	9/2.46 = 3.7
Noise	4/0.40 = 10.0	4/0.45 = 8.9	4/0.52 = 7.7
Self-cleaning ability	4/0.62 = 6.5	2/0.49 = 4.1	4/0.58 = 6.9
Electrical safety	16/0.58 = 27.6	16/0.52 = 30.8	16/0.43 = 37.2
Particle protection	6/0.29 = 20.7	6/0.30 = 20.0	2/0.37 = 5.4
Ease of servicing	6/0.70 = 8.6	4/0.52 = 7.7	6/0.98 = 6.1
Cutter life	9/0.96 = 9.4	9/0.83 = 10.8	9/1.32 = 6.8
Ease of installation	9/0.54 = 16.7	9/0.33 = 27.3	9/0.70 = 11.8
Total	76/17.25 = 4.4	72/14.11 = 5.1	69/21.44 = 3.2

Note: Value = $\dfrac{\text{Design score}}{\text{Cost of achieving}}$.

→ 12.10 DESIGN REVIEW

Design review is a formal, documented, comprehensive, and systematic examination of a design to evaluate the design requirements and the capability of the design to meet these requirements and to identify problems and propose solutions.

Design review is not new. However, in the past, the term has referred to an informal evaluation of the design. Modern products often require a more formal program. A formal design review recognizes that many individual designers do not have specialized knowledge in reliability, maintainability, safety, producibility, and the other parameters that are important in achieving an optimum design. The design review aims to provide such knowledge.

For modern products, design reviews are based on the following concepts:

1. Design reviews become mandatory because of either customer demand or an upper management policy declaration.
2. Design reviews are conducted by a team consisting mainly of specialists who are not directly associated with the development of the design. These specialists must be highly experienced and must bring with them the reputation of objectivity. The combination of competence, experience, and objectivity is present in some people, but these people are in great demand. The success of design reviews largely depends on the degree to which management supports the program by insisting that the best specialists be made available for design review work. The program deteriorates into superficial activity if (1) inexperienced people are assigned to design reviews or (2) those on the design review team are not given sufficient time to study product information prior to design review meetings.

3. Design reviews are formal. They are planned and scheduled like any other legitimized activity. The meetings are built around prepared agendas and documentation sent out in advance. Minutes of meetings are prepared and circulated. Follow-ups for action are likewise formalized.
4. Design reviews cover all quality-related parameters and others as well. The parameters can include reliability, maintainability, safety, producibility, weight, packaging, appearance, cost, etc.
5. As much as possible, design reviews are based on defined criteria. Such criteria may include customer requirements, internal goals, and experience with previous products.
6. Design reviews are conducted at several phases in the progression of the design, such as design concept, prototype design and test, and final design. Reviews are done at several levels of the product hierarchy, such as system and subsystem.
7. The ultimate decision on inputs from the design review rests with the designer. The designer must listen to the inputs, but on matters of structural integrity and other creative aspects of the design, the designer retains a monopoly on decisions. The control and publication of the specification remain with the designer.

A universal obstacle to design review is the resistance of the design department. The common practice has been for this department to hold a virtual monopoly on design decisions, i.e., these decisions have historically been immune from challenge unless actual product trouble is encountered. With such a background, it is not surprising that designers have resisted the use of design reviews that challenge their designs. Designers have contended that such challenges are based purely on theory and analysis (at which they regard themselves as the top experts) rather than on the traditional grounds of "failed hardware." This resistance is further aggravated in companies that permit reliability engineers to propose competing designs. Designers have resisted the idea of having competitors even more than they have resisted the idea of design review.

➜ 12.11 SOFTWARE DEVELOPMENT

The software development life cycle can be divided into five phases:

1. Requirements
2. Build
3. Verify and test
4. Maintenance

For elaboration of these steps, see *JQH6*, Section 29. The approach to software development has evolved along with the technology, and not all methods follow exactly this formula. Newer approaches such as Agile development still incorporate these basic components while adding fast cycles to capture and act on user experience.

Assessments of software development have provided some surprising statistics (Table 12.11). Note that 67% of the total effort is spent on *maintenance* that involves

TABLE 12.11
Typical resource requirements for phases of a software development life cycle

	Total effort, %	Development effort, %
Requirements definition	3	9
Preliminary design	3	9
Detailed design	5	15
Coding	7	21
Testing and installation	15	46
Maintenance	67	—

Source: JQH4, p. 14.7.

modifications of the software due to (1) changes in the initial requirements and (2) errors not previously detected. Thus, we have an opportunity for improvement.

Also note that 46% of the basic development effort goes for testing and installation. Sometimes the extent of testing (and review) during software development is great.

EXAMPLE 12.5. In one software development organization, the number of errors per thousand lines of code (KLOC) measured at the *final* stage was 2/KLOC. At the *first* stage of review/inspection during development, the number was 50/KLOC. The intermediate stages included various forms of inspection, review, and testing to weed out errors. There was an opportunity for improvement.

According to the Pentagon and the Software Engineering Institute at Carnegie Mellon University, every 1000 lines of code have 5 to 15 bugs, costing up to $30,000 to correct (*BusinessWeek,* December 6, 1999, p. 108).

For a description of diagnosis involving software errors, see Levenson and Turner (1993). In testing software for use with medical equipment that delivers radiation to cancer patients, certain lines of code (used for the test only) were inadvertently left in the software. This error resulted in excessive radiation and led to 13 deaths.

But errors in software are only one part of the quality dimension. As with all products, quality must address both product features and freedom from deficiencies (errors). Prahalad and Krishnan (1999) explain the need to consider these several aspects of quality in information technology.

↛ 12.12 MEASUREMENT IN DESIGN

The management of quality-related activities in design—as in all functional activities—must include provision for measurement. A popular rallying cry is, "What gets measured, gets done." Readers are urged to review the 10 basic principles of quality measurement in Section 5.2.

Table 12.12 shows units of measure for various subject areas of product design.

Endres (1997), Chapters 3 and 4, provides extensive discussion of measuring quality in design, research, and development.

TABLE 12.12
Examples of quality measurement in design

Subject	Unit of measure
Overall design process	Cost of poor quality Number of months from first pilot unit to steady-state products
Design changes	Number of design changes (1) at design review, (2) at development testing, (3) after design release for steady-state production Number of design changes (1) to meet requirements, (2) to improve performance, (3) to facilitate manufacture Number of design changes requested by customer Number of waivers to specifications Number of drawing errors found by checkers on first check
Reliability, maintainability	Ratio of predicted reliability to actual reliability Ratio of actual reliability to reliability requirement Maintainability index compared to prior design
Software	Number of software errors per KLOC—first internal review Number of software errors per KLOC—final internal review Number of software errors per KLOC—discovered by customer Average score given by customers on overall quality of software
Ease of manufacture	Ratio of number of parts to theoretical minimum number Total assembly time Total number of operations

12.13 IMPROVING THE EFFECTIVENESS OF PRODUCT DEVELOPMENT

Chapter 3, "Organizationwide Assessment of Quality," presents four studies that provide a thorough picture of the current status of quality in an organization. These studies are cost of poor quality, standing in the marketplace, quality culture, and assessment of current quality activities (quality system). Such studies can also be applied to assess the current status of the product development process with respect to quality. Endres (1997), Chapters 4 and 5, presents examples of such studies in product development by companies, including Alcoa, Allied Signal, AT&T, Corning, Duracell, Eastman Chemical, IBM, and Olin. These assessments help define the actions needed to increase the effectiveness—pack more punch—of product development.

The trilogy of quality processes provides a useful model for product development. Thus the quality planning process (Chapter 5) provides the steps for effectively developing new products:

- Establish the project.
- Identify customers.
- Discover customer needs.
- Develop the product.
- Develop the process.
- Establish process controls; transfer to operations.

The quality control process (Chapter 6) measures the performance of the product development function. The steps are

- Choose control subjects.
- Establish measurement.
- Set standards.
- Measure performance.
- Compare to standards.
- Take action on the difference.

The quality improvement process (Chapter 4) identifies and solves chronic problems within product development. The steps are

- Prove the need.
- Identify projects.
- Organize project teams.
- Verify the project need and mission.
- Diagnose causes.
- Provide a remedy.
- Deal with resistance to change.
- Institute controls to hold the gains.

Applying the quality planning, quality control, and quality improvement processes to the product development process is a major undertaking that is likely to occur only if one or more of the four components of assessment clearly establish the need for action (see Section 12.1). Using the trilogy of quality processes, Bailey et al. (1999) examine research processes in the semiconductor industry.

A historical review of past product design changes can be a useful starting point for improvement. A study of 24 design changes revealed the following:

- Eleven of the changes were made to correct performance, reliability, or safety weaknesses; eight changes were made to correct administrative or paperwork errors; five changes were necessary to facilitate the manufacture of the product.
- Of the problems associated with these design changes, 23 were first found during production and one was found during field testing.
- In all 24 cases, notification of the change was given only to the original designer.

Two conclusions emerged from the study: The process of product development was finding problems too late, and feedback on problems was not shared with all designers.

The concept of self-control provides a framework for analyzing the work of designers. Universal in application, the concept holds that three criteria must be met before a person can be held responsible for controlling the quality of his or her activities. The designer is said to be in a state of self-control only if all three criteria are fully met. A weakness in any of the three criteria requires analyzing and correcting the *process of product development*, instead of looking to the individual designer for improvement. Problem 12.14 asks readers to identify questions for each of the criteria for self-control.

SUMMARY

- For complex products, errors during product development cause about 50% of fitness-for-use problems.
- Product development is a process with distinct phases that can incorporate forms of early warning of new-product troubles.
- Quality function deployment is a technique consisting of interlocking matrices that translate customer needs into product and process characteristics.
- Robust designs provide optimum performance regardless of variation in manufacturing and field conditions.
- Taguchi methods determine optimum values of product and process parameters that minimize variation while keeping the mean on target.
- Design for six sigma has five steps: define, measure, analyze, design, and verify.
- Reliability is the ability of an item to perform a required function under stated conditions for a stated period of time.
- Reliability quantification involves three phases: apportionment, prediction, and analysis. Maintainability quantification follows a familiar approach.
- Failure mode, effect, and criticality analysis and fault tree analysis are helpful qualitative tools for design assurance.
- The design-for-manufacturability technique aims to simplify a product design to facilitate manufacture.
- Value engineering and design-to-cost techniques analyze designs to achieve an optimum balance between performance and cost.
- Design review is a systematic examination of design requirements and the capability of the design in meeting the requirements.
- The process of product development can be examined by using the elements of quality planning, control, and improvement.
- In software development, 67% of the effort involves making changes in the initial software—thus there is an opportunity for improvement.
- The management of quality in design must include provision for measurement.

PROBLEMS

12.1. Make a failure mode, effect, and criticality analysis for one of the following products:
  (a) A product acceptable to the instructor.
  (b) A flashlight.
  (c) A toaster.
  (d) A vacuum cleaner.

12.2. Make a fault tree analysis for one of the products mentioned in problem 12.1.

12.3. Visit a local plant and determine whether any formal or informal numerical reliability and maintainability goals are issued to the design function for guidance in designing new products.

12.4. Obtain a schematic diagram of a product for which you can also obtain a list of components that fail most frequently. Show the diagram to a group of engineering students

most closely associated with the product (e.g., a mechanical product would be shown to mechanical engineering students). Have the students *independently* write their opinions of the components most likely to fail by ranking the top three.
(a) Summarize the results and comment on the agreement, or lack thereof, among the students.
(b) Comment on the students' opinions versus actual product history.

12.5. A reliability prediction can be made by the designer of a product or by an engineer in a staff reliability department that might be a part of the design function. One advantage for the designer making the prediction is that his or her knowledge of the design will likely make possible a faster and more thorough job.
(a) What is one other advantage in having the designer make the prediction?
(b) Are there any disadvantages to having the designer make the prediction?

12.6. Prepare a formal presentation to gain adoption of one of the following:
(a) Quantification of reliability goals, apportionment, and prediction.
(b) Formal design reviews.
(c) Failure mode, effect, and criticality analysis.
(d) Critical components program.

You will make the presentation to one or more people who will be invited into the classroom by the instructor. These people may be from industry or may be other students or faculty. (The instructor will announce time and other limitations on your presentation.)

12.7. Outline a reliability test for one of the following products:
(a) A product acceptable to the instructor.
(b) A household clothes dryer.
(c) A motor for a windshield wiper.
(d) An electric food mixer.
(e) An automobile spark plug.

The testing must cover performance, environmental, and time aspects.

12.8. Speak with some practicing design engineers and learn the extent of feedback of field information to them on their own design work.

12.9. You are the design engineering manager for a refrigerator. Most of your day is spent on administrative work. You do not have time to get into details on new or modified designs. However, you must approve (by sign-off) all new designs or changes. In reality, your sign-off consists of a brief review of the design, but you rely basically on the competence of your individual designers. You do not want to institute a formal reliability program for designers or set up a reliability group at this time. What action could you take to give yourself some assurance that a design presented to you has been adequately examined by the designer with respect to reliability? It is not possible to increase the testing, and any action you take must involve a minimum of additional costs.

12.10. You work for a public utility and your department employs outside contractors who design and build various equipment and building installations. You have just finished a value engineering seminar at a university, and you wonder whether your utility should consider establishing such a function to evaluate contractor designs. Someone in your management group has heard that value engineering can reduce costs but "by lowering the performance or reliability of a design." Comment.

12.11. The chapter discusses a number of concepts (e.g., reliability prediction, design review). Select several concepts and outline a potential application of the concept to an actual problem on a specific product. The outline should include the following items:
(a) The name of the concept.
(b) A brief statement of the problem.
(c) Application of the concept to the problem.
(d) Potential advantages.
(e) Obstacles to actual implementation.
(f) An approach to use to overcome each obstacle.

The outline for each concept should be about one page.

12.12. The section "Evaluating Designs by Testing" stated four risks. For each risk, recommend one or more preventive actions.

12.13. In designing the SX-70 camera system, the Polaroid Corporation first determined the top customer complaints on previous models. For example, customers had trouble getting the correct focus; customers forgot to change the batteries; customers did not like changing lenses; customers were not sure when to use the flash. Propose a design feature to prevent each complaint.

12.14. Review the concept of self-control in Chapter 6. Apply the concept to design engineering by creating three questions for each of the three criteria of self-control.

12.15. For an organization that has a system of documenting design changes, make a historical review of the documentation that initiated about 25 design changes. Draw some initial conclusions about the *process* of product development.

REFERENCES

Bailey, D. E., F. S. Settles, and D. Sanrow (1999). "Applying Continuous Quality Improvement Techniques to a Research Environment," *Quality Management Journal,* vol. 6, no. 2, pp. 62–77.

Beaton, G. N. (1959). "Putting the R&D Reliability Dollar to Work," *Proceedings of the Fifth National Symposium on Reliability and Quality Control,* Institute of Electrical and Electronics Engineers, New York, p. 65.

Brocato, R. C. and K. A. Potocki (1998). "A System Model for Quality Management Principles: Case Study Application to Technical Staff at an R&D Organization," *Annual Quality Congress Proceedings,* ASQ, Milwaukee, pp. 190–198.

Endres, A. (1997). *Improving R&D Performance the Juran Way,* John Wiley & Sons, New York.

Feitzinger, E. and H. L. Lee (1997). "Mass Customization at Hewlett-Packard: The Power of Postponement," *Harvard Business Review,* January–February, pp. 116–121.

Hammer, W. (1980). *Product Safety Management and Engineering,* Prentice-Hall, Englewood Cliffs, NJ.

Hatty, M. and N. Owens (1994). "Potential Failure Modes and Effects Analysis: A Business Perspective," *Quality Engineering,* vol. 7, no. 1, pp. 169–186.

Ireson, W. G., C. F. Coombs Jr., and R. Y. Moss (1996). *Handbook of Reliability Engineering and Management,* 2nd ed., McGraw-Hill, New York.

Montgomery, D. C. (1997). *Introduction to Statistical Quality Control,* 3rd ed., John Wiley & Sons, New York.

Park, R. (1999). *Value Engineering,* St. Lucie Press, Boca Raton, FL.

Perkins, D. (1996). "FMEA for a Real (Resource Limited) Design Community," *Annual Quality Congress Proceedings,* ASQ, Milwaukee, pp. 435–442.

Phadke, M. S., R. R. Kackar, D. V. Speeny, and M. J. Grieco (1983). "Off-Line Quality Control in Integrated Circuit Fabrication Using Experimental Design," *The Bell System Technical Journal,* vol. 62, no. 5, pp. 1273–1309.

Pignatiello, J. J. and J. S. Ramberg (1992). "Top Ten Triumphs and Tragedies of Genichi Taguchi," *Quality Engineering,* vol. 4, no. 2, pp. 211–225.

Prahalad, C. K. and M. S. Krishnan (1999). "The New Meaning of Quality in the Information Age," *Harvard Business Review,* September–October, pp. 109–118.

Ross, P. J. (1996). *Taguchi Techniques for Quality Engineering,* McGraw-Hill, New York.

Sobek, D. K. II, J. K. Liker, and A. C. Ward (1998). "Another Look at How Toyota Integrates Product Development," *Harvard Business Review,* July–August, pp. 36–49.

Turmel, J. and L. Gartz (1997). "Designing in Quality Improvement: A Systematic Approach to Designing for Six Sigma," *Annual Quality Congress Proceedings,* ASQ, Milwaukee, pp. 391–398.

CHAPTER 13

Managing Quality in Operations—Manufacturing

✦ 13.1 QUALITY IN MANUFACTURING IN THE 21ST CENTURY

Manufacturing operations is the nerve tactical center of an organization—where the action is. This chapter covers operations in the manufacturing sector; Chapter 14 covers operations in the service sector.

In manufacturing entities, operations are activities, typically carried out in a factory, that transform material into goods. Before we consider the planning, control, and improvement of manufacturing activities, we must recognize four important issues that will transform traditional manufacturing of the 20th century to a different manufacturing in the 21st century. As Dr. Juran clearly articulated "the 20th century was the century of productivity; the 21st century will be the century of quality." These issues are discussed next.

Customer Demands for Higher Quality, Reduced Inventories, and Faster Response Time

As products and processes become more complex, "world class" quality levels are now common and expected. To remain competitive, manufacturers must continually and intently focus on quality in the big Q framework. World-class enterprises can no longer view quality as being solely product focused. They now need to view it as business- and enterprise-focused (big Q). For many products, quality levels of 1% to 3% defective are being replaced by 1 to 10 defects per million parts (or 3.4 defects per million as in the six sigma approach). Customers demand reduced inventory levels based on the "just-in- time" (JIT) production system. Under JIT, the concept of large lot sizes is challenged by pull systems, reducing changeover times, redesigning processes, and standardizing jobs. The results are smaller lot sizes and lower inventory.

But JIT works only if product quality of supplied materials is high because little or no inventory exists to replace defective product. Finally, customers want faster response time from suppliers—to develop and manufacture new products (conception to commercialization). That faster response time puts pressure on the product development process and can result in inadequate review of new designs or enhancements to existing designs for product performance and for manufacturability. In some organizations, this results in a concept known as "tailoring," which is streamlining the integrated product development schedule often at the expense of quality. Collectively, these three parameters (quality, inventories, response time) place a heavy burden on operations.

Agile Competition

An agile organization is able to respond to constantly changing customer demands. This characteristic means agile supply chains, flexible manufacturing, mixed-model scheduling, single or pitch flow, changing over from one product to another quickly, manufacturing goods to customer order in small lot sizes, customizing goods for individual customers, and using the expertise of people and facilities within the company and also among groups of cooperating companies (partners). Goldman, Nagel, and Preiss (1995) describe the concept and include examples.

This concept includes the "virtual" organization or complete value stream—a group of companies linked by an electronic network to enable the partners to satisfy a common customer objective. The virtual organization may be partially created by transferring complete functions to a supplier—outsourcing. See Section 15.4 for a discussion of the pros and cons of outsourcing.

Impact of Technology

Technology (including computer information systems) is clearly improving quality by providing a wider variety of outputs and also more consistent output. The infusion of technology makes some jobs more complex, thereby requiring extensive job skills and quality planning; technology also makes other jobs less complex but may contribute to job monotony.

These four issues suggest that quality during operations can no longer focus on inspection and checking (reactive) but must respond to ever-increasing customer demands and changing competitive conditions (proactive).

We proceed now to examine specific methods for planning, controlling, and improving quality during manufacturing operations.

➔ 13.2 LEAN MANUFACTURING AND VALUE STREAM MANAGEMENT

Lean manufacturing is the process of optimizing operating systems by eliminating, or at least reducing, the waste within them. Anything that does not provide value to the customer is considered waste. The emphasis is on eliminating non-value-added

History of Manufacturing

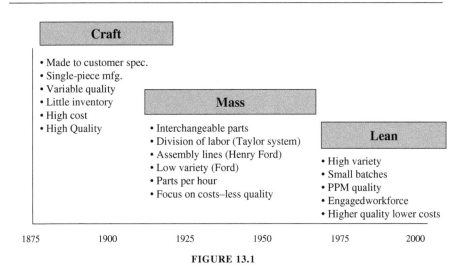

FIGURE 13.1

activities such as producing defective product, excess inventory charges due to work-in-process and finished goods inventory, excess internal and external transportation of product, excessive inspection, and idle time of equipment or workers due to poor balance of work steps in a sequential process. The goal of lean manufacturing has long been one of the goals of industrial engineering. Shuker (2000) provides a useful introduction to lean manufacturing based on the Toyota production system.

Lean Manufacturing

The history of manufacturing is summarized in Figure 13.1. The lean mission is to have the

- Shortest possible lead time
- Optimum level of strategic inventory
- Highest practical customer order service
- Highest possible quality (low defect rate)
- Lowest possible waste (low COPQ)

throughout the entire supply chain so as to win the marketplace. Lean is based on creating a "pull system." This is accomplished by synchronizing the flow of work (both internal and external to the company) to the "drumbeat" of the customer's requirements. All kinds of waste are driven out (time, material, labor, space, and motion). The overall intent is to reduce variation and drive out waste by letting customers pull value through the entire value stream (or supply chain).

The key principles of lean are

- Define value in the eyes of the customer
- Measure the value stream for each product

Lean Characteristics

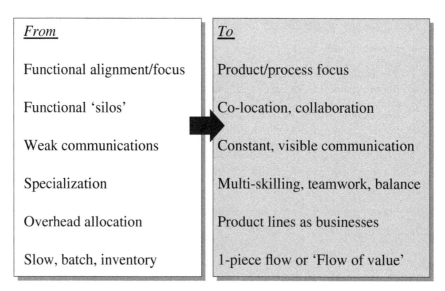

FIGURE 13.2
The characteristics of lean vs. traditional manufacturing practices.

- Analyze value flow without interruptions
- Improve process pull by reducing defects in products and deficiencies in processes
- Control processes by driving out variation (short and long term) and implementing sound controls to maintain improvements

Value is created by the customer. Lean starts by attempting to define value in terms of products and capabilities provided to the customer at the right time and appropriate price (see Figure 13.2).

The following are summaries of key concepts used in lean (in the context of both lean manufacturing and lean service).

The Eight Wastes

Taiischi Ohno (1988) identifies various types of waste. The following is an adapted list:

1. **Overproduction**—making or doing more than is required or earlier than needed
2. **Waiting**—for information, materials, people, maintenance, etc.
3. **Transport**—moving people or goods around or between sites
4. **Poor process design**—too many/too few steps, nonstandardization, inspection rather than prevention, etc.
5. **Inventory**—raw materials, work in progress, finished goods, papers, electronic files, etc.
6. **Motion**—inefficient layouts at workstations, in offices, poor ergonomics

7. **Defects**—errors, scrap, rework, nonconformance
8. **Underutilized personnel resources and creativity**—ideas that are not listened to, skills that are not used

Flow and Takt Time

Traditional operations have worked within a "push" system. A push system computes start times and then pushes product into operations based on artificial demand or based on overhead absorption. The concept of "flow" requires the rearrangement of mental thoughts regarding "typical" production processes. One must not think of just "functions" and "departments." We need to redefine how functions, departments, and organizations work to make a positive contribution to the value stream. Flow production requires that we produce at the customer's purchase rate and if necessary, make every product every day to meet customer's orders, i.e., to meet the pace or "drumbeat." The pace or drumbeat is determined by **takt time**. Takt time comes from the German word for meter, as in music, which establishes the pace, or beat, of the music. It is the time which reflects the rate at which customers buy one unit.

$$\text{Takt time} = \frac{\text{Available time (in a day)}}{\text{Average daily demand}}$$

For example, in Figure 13.3, the pace or takt time is calculated for the demand shown during a 10-day period.

Determine Pace

Takt Time Calculation Example:

Over 10 Days	Demand
1	30
2	40
3	50
4	60
5	10
6	30
7	40
8	20
9	60
10	40
10	380

Per Day:

$$\frac{\text{Time available in period (840 min.)}}{\text{Average demand (38)}} = 22.1 \text{ minutes}$$

Based on 2 shifts of 7 hours

FIGURE 13.3

Demand cannot be understated. A key component to satisfying the customer is understanding his or her demand (including seasonal or other particular demand). Demand variability can be mix driven, quantity driven, or both.

To be practical, takt time may need to be modified depending on the variability of the process. When modifying takt time beyond the simple equation, another name should be used, such as cell takt, practical takt, or machine takt. Although modifiers may be planned, they are still waste, or planned waste. Manpower staffing requirements can then be determined:

$$\text{Minimum staffing required} = \frac{\text{Total labor time in process}}{\text{Takt time}}$$

Once takt time has been calculated, constraints (such as long setup times) should be identified and managed (or eliminated) to enable smaller batches or ideally, single piece flow, to eliminate overproduction and excess inventory. Pull production scheduling techniques are used so that customer demand pulls demand through the value stream (from supplier to production to the customer). In pull production, materials are staged at the point of consumption. As they are consumed, a signal is sent back to previous steps in the production process to pull forward sufficient materials to replenish only what has been consumed.

The steps for improvement teams (or kaizen teams) to lean out an operation are

1. Determine pace (takt time and manpower)
2. Establish sequence and replenishment (product family turnover and set-up/changeover required)
3. Design the process (proximity, sequence, interdependence)
4. Feed the process (strategic inventory, standard WIP (SWIP), Murphy buffer)
5. Balance the process (load, standard work)
6. Stabilize and refine (6S [which includes safety], continuous improvement)

Value Stream Management

The value stream consists of all activities required to bring a product from conception to commercialization. It includes detailed design, sales, marketing, order taking, scheduling, production, and delivery. Understanding the value stream allows one to see value-added steps, non-value-added but needed steps, and non-value-added steps. Value-added activities **transform** or shape material or information to meet customer requirements. Non-valued-added activities take time or resources, but do not add value to the customer's requirement (but may meet company requirements).

The value stream improvement journey typically starts with training the team on key concepts in lean VSM, mapping the current state using value stream maps which document materials and information flow as well as any pertinent information on the process (such as wait times, processing times, and inventory levels). Improvements are identified. The desired future state is then documented as a future state value stream map, and the improvements and control features are implemented to drive toward the future state goal.

An example of a value stream map for a paint line for the current state and future state are shown in Figure 13.4a, b.

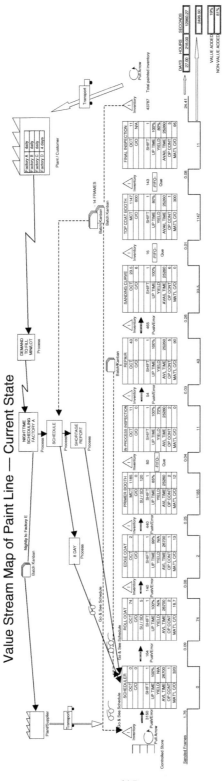

FIGURE 13.4a
(*Courtesy of Chris Arquette at a Juran Institute client.*)

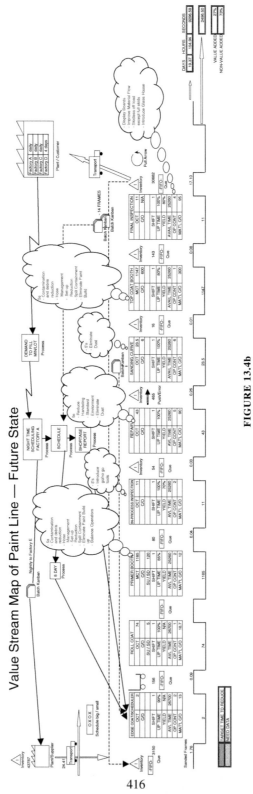

FIGURE 13.4b

(Courtesy of Chris Arquette at a Juran Institute client.)

13.3 INITIAL PLANNING FOR QUALITY

Planning starts with a review of product designs (see Section 12.10). Then we review the process designs to identify key product and process characteristics, determine the importance of product characteristics (typically can be classified as critical, major, minor, and incidental), analyze the process flow diagram, determine the vital process characteristics, determine process control and capability, error-proof the process, plan for a neat and clean workplace, validate the measurement systems, and plan for operator self-control. These elements are discussed next. The all important topic of process control and capability is covered with more detail in Section 20.10.

Review of Product Designs

There is a clear advantage to having a new product design reviewed by operations personnel before the design is finalized for the marketplace. In practice, the extent of such review varies greatly—from essentially nothing ("tossing it over the wall" to the operations people) to a structured review using formal criteria and follow-up on open issues (typically evidenced by sound-integrated product development schedules with provisions for robust design for manufacturability during various stages). Although a product design review often occurs during the design and development process (see Section 12.10), the emphasis is on the adequacy of field performance.

Review of product designs prior to release to operations must include an evaluation of manufacturability. This can only occur after there is a clear understanding of critical customer requirements (CCRs). This evaluation includes the following issues:

1. Identification of key product characteristics and key control characteristics (process features).
2. Relative importance of key product characteristics and key control characteristics.
3. Design for manufacturability (see Section 12.8).
4. Process robustness. A process is robust if it is flexible, easy to operate, error proof, and its performance will tolerate uncontrollable variations in factors internal and external to the process. Such an ideal is approached by careful planning of all process elements, e.g., cross-functional training of personnel to cover vacations. For a discussion of robustness, see Snee (1993).
5. Availability of capable manufacturing processes to meet product requirements, i.e., processes that not only meet specifications but do so with minimum variation.
6. Availability of capable measurement processes. This is discussed later and in Section 10.8.
7. Identification of special needs for the product, e.g., handling, transportation, and storage during manufacture.
8. Material control, e.g., identification, traceability, segregation, contamination control.
9. Special skills required of operations personnel.

This review of the product design must be supplemented by a review of the process design. The process review includes manufacturability issues initially raised in the product design review.

Identification of Key Product and Process Characteristics

Key *product* characteristics are outputs from a process that are measurable on, within, or about the product itself. They are the outputs perceived by the customer. Key *process* characteristics are inputs (process factors) that affect the outputs (KPCs). They are unseen by the customer and are measurable only when they occur. Product and process characteristics can be identified by using inputs from market research, quality function deployment, design review, and failure mode and effect analysis. Somerton and Mlinar (1996) describe the use of these and other tools to identify key characteristics.

Relative Importance of Product Characteristics

Planners are better able to allocate resources when they know the relative importance of the many product characteristics.

One technique for establishing the relative importance is the identification of critical items (see Section 12.5). Critical items are the product characteristics that require a high level of attention to ensure that all requirements are met. One company identifies "quality sensitive parts" by using criteria such as part complexity and high-failure-rate parts. For such parts, special planning includes supplier involvement before and during the contract, process capability studies, reliability verification, and other activities.

Another technique is the classification of characteristics as described earlier in the section. Under this system, the relative importance of characteristics is determined and indicated on drawings and other documents. The classification can be simply "functional" (or "critical to quality") or "nonfunctional." Another system uses several degrees of importance such as critical, major, minor, and incidental. The classification uses criteria that reflect safety, operating failure, performance, service, and manufacture. For elaboration, see Section 10.5 and *JQH5,* pages 22.6–22.7.

Analysis of the Process Flow Diagram

A process design can be reviewed by laying out the overall process in a flow diagram. Several types are useful. One type shows the paths followed by materials through their progression into a finished product. An example for a coating process at the James River Graphics Company is shown in Figure 13.5. Planners use such a diagram to divide the flow into logical sections called workstations. For each workstation, they prepare a formal document listing such items as operations to be performed, sequence of operations, facilities and instruments to be employed, and process conditions to be maintained. This formal document becomes the plan to be carried out by the production supervisors and workforce. The document is the basis for control activities by the inspectors. It also becomes the standard against which the process audits are conducted.

Correlation of process variables with product results

A critical aspect of planning during manufacture is to discover, by data analysis, the relationships between process features or variables and product features or results.

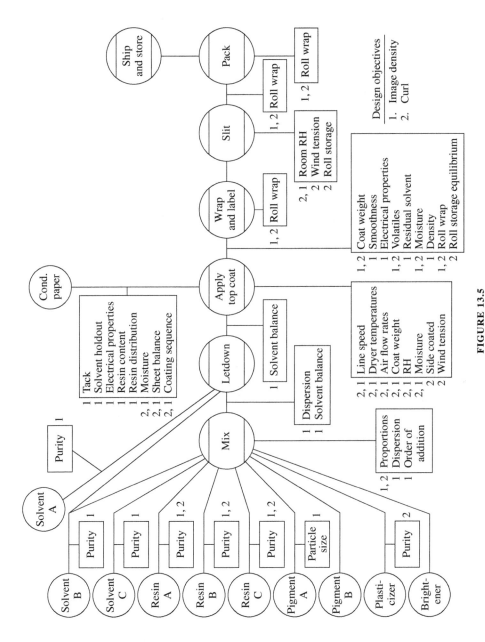

FIGURE 13.5
Product and process analysis chart. (*From Siff, 1984.*)

Such knowledge enables a planner to create process control features, including limits and regulating mechanisms on the variables, to keep the process in a steady state and achieve the specified product results. In Figure 13.5 each process feature is shown in a rectangle attached to the circle representing an operation; product results are listed in rectangles between operations at the point where conformance can be verified. Some characteristics (e.g., coat weight) are both process features and product results.

For each control station in a process, designers identify the numerous control subjects over which control is to be exercised. Each control subject requires a feedback loop made up of multiple process control features. A process control spreadsheet or control plan helps to summarize the detail. As an example, Figure 6.4 shows, for each control subject, the unit of measure, type of sensor, goal, frequency of measurement, sample size, and the criteria and responsibility for decision making. For elaboration, see *Juran's Quality Handbook, The Complete Guide to Performance Excellence,* 6th edition (2010, page 209).

Determining the optimal settings and tolerances for process features sometimes requires much data collection and analysis. Eibl et al. (1992) discuss such planning and analysis for a paint-coating process for which little information was available about the relationship between process features and product results.

Many companies have not studied the relationship between process features and product results. The consequences of this lack of knowledge can be severe. In the electronic component manufacturing industry, some yields are shockingly low and will likely remain that way until the process features are studied in depth. In all industries, the imposition of new quality demands such as six sigma requires much deeper understanding of product results and process features than in the past.

To understand fully the relationship between process features and product results, we often need to apply the concept of statistical design of experiments (see Section 18.11). See also the discussion of the Taguchi approach (Section 12.4). Under the six sigma approach, factorial experiments are becoming necessary to understand the interactions among several variables and product results. But upper management must supply the missing elements, i.e., the resources for full-time personnel to design and analyze the experiments and the training of process engineers to integrate these concepts in process planning.

Error-Proofing the Process

An important element of prevention is the concept of designing the process to be error free through "error-proofing" (the Japanese call it *poka-yoke*).

A widely used form of error-proofing is the design (or redesign of the machines and tools, the "hardware") to make human error improbable or even impossible. For example, components and tools may be designed with lugs and notches to achieve a lock-and-key effect, which makes it impossible to misassemble them. Tools may be designed to sense the presence and correctness of prior operations automatically or to stop the process on sensing depletion of the material supply. For example, in the textile industry a break in a thread releases a spring-loaded device that stops

TABLE 13.1
Summary of error-proofing principles

Principle	Objective	Example
Elimination	Eliminating the possibility of error	Redesigning the process or product so that the task is no longer necessary
Replacement	Substituting a more reliable process for the worker	Using robotics (e.g., in welding or painting)
Facilitation	Making the work easier to perform	Color-coding parts
Detection	Detecting the error before further processing	Developing computer software that notifies the worker when a wrong type of keyboard entry is made (e.g., alpha versus numeric)
Mitigation	Minimizing the effect of the error	Using fuses for overloaded circuits

the machine. Protective systems, e.g., fire detection, can be designed to be "fail-safe" and to sound alarms as well as all-clear signals.

In a classic study, Nakajo and Kume (1985) discuss five fundamental principles of error-proofing developed from an analysis of about 1000 examples collected mainly from assembly lines. These principles are elimination, replacement, facilitation, detection, and mitigation (see Table 13.1).

See *JQH6*, pages 11.351–11.353, for further examples of error-proofing.

An adjunct to error-proofing is the use of *poka-yoke* systems. These are control systems built into the process that stop the equipment when some type of irregularity occurs and signal the operator to address the problem (see Section 13.8).

Plan for Neat and Clean Workplaces

How obvious, but the reality is that many workplaces are dirty and disorganized. The benefits of a good workplace include the prevention of defects; prevention of accidents; and the elimination of time wasted searching for tools, documentation, and other ingredients of manufacture. A simple body of knowledge now provides us with a framework to create a neat and clean workplace. The approach is called 6S for sort, set in order, sweep and shine, standardize, self-discipline, and safety (see Figure 13.6). The steps are as follows:

1. *Sort.* Remove all items from the workplace that are not needed for current operations.
2. *Set in order.* Arrange workplace items so that they are easy to find, to use, and to put away.
3. *Sweep and shine.* Sweep, wipe, and keep the workplace clean.
4. *Standardize.* Make "shine" become a habit.
5. *Self-discipline.* Create the conditions (e.g., time, resources, rewards) to maintain a commitment to the 6S approach.
6. *Safety.* Implement behavioral-based safety processes and procedures that drive zero recordable injuries and zero lost time accidents.

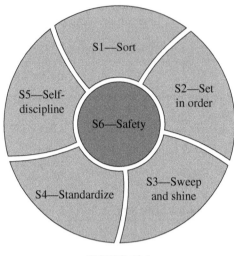

FIGURE 13.6
The 6S Techniques.

The Japanese developed the 6S technique to remove clutter from the work area, organize the workplace to help prevent defects from happening, and improve work flow. It is one of the basic elements of lean that an organization can implement. It is a great improvement tool that is very easy to install and brings "order out of chaos."

This technique should be put in place prior to the project solution being implemented because it can yield additional results on its own merits. Also, the final solution implementation will be easier to put in place if this technique is done first.

Decades ago, industries producing critical items (health care, aerospace) learned that clean and neat workplaces are essential in achieving extremely low levels of defects. The quality levels demanded by the six sigma approach now provide the same impetus.

Perhaps the significance of the 6S approach is its simplicity. The benefits are obvious; the tools are the simplest work-simplification tools; the tools are easy to understand and apply. Simple tools sometimes get dramatic results, and that's what has happened with 6S.

Validate the Measurement System

Particularly with the low defect levels demanded under the six sigma approach, it is important to understand the capability of the manufacturing process and also the capability of the measurement process. Thus planning and control of the measurement process become part of the six sigma approach. Previous studies assumed that variation of the measurement process was small compared to variation caused by the manufacturing process and thus could be ignored (in practice, the assumption was rarely tested in most industries). When variation due to the measurement process alone is even moderately large, the result will be mistakes in determining whether a product meets specifications—some "good" product will be incorrectly classified as defective, and some "bad" product will be incorrectly classified as good. To quantify the measurement process, see Section 10.8. Thus the time has come to evaluate measurement capability and to determine whether measuring equipment is accurately measuring process output.

Figure 13.7 provides perspective on the measurement issue. Note that the observed process variation (i.e., the variation of recorded measurements) is from two sources: variation in the process manufacturing the product and variation in the measurement process. Further, these two sources contain various components. The components shown in Figure 13.7 can be quantified and analyzed to determine measurement capability. The variation due to the process manufacturing the product can also be quantified—see Chapter 20 under "Process Capability."

Even in consumer product industries such as the manufacture of razor blades, tolerances can be on the order of the wavelength of visible light. Such tolerances are a long way from the days of tolerances in thousands or 10 thousands and using measuring instruments like micrometers and even supermicrometers. The measurement process must be capable of handling these conditions.

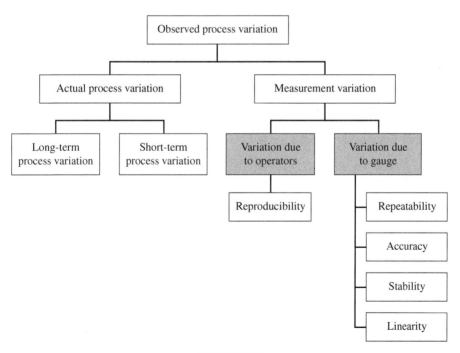

FIGURE 13.7
Possible sources of variation. (*Source: Juran Institute, Inc.*)

→ 13.4 CONCEPT OF CONTROLLABILITY: SELF-CONTROL

The concept of self-control was introduced in Section 6.3. An ideal objective for manufacturing planning is to place human beings in a state of self-control, i.e., to provide them with all they need to meet quality objectives. To do so, we must provide people with the following:

1. Knowledge of what they are supposed to do.
   - Clear and complete work procedures.
   - Clear and complete performance standards.
   - Adequate selection and training of personnel.
2. Knowledge of what they are actually doing (performance).
   - Adequate review of work.
   - Feedback of review results.
3. Ability and desire to regulate the process for minimum variation.
   - A process and job design capable of meeting quality objectives.
   - Process adjustments that will minimize variation.
   - Adequate worker training in adjusting the process.
   - Process maintenance to maintain the inherent process capability.
   - A strong quality culture and environment.

As we will see, most organizations do not adhere to the three elements and ten subelements of self-control.

The concept of self-control has objectives that are similar to those of the Toyota production system. For a perceptive dissection of the Toyota system into four basic rules, see Spear and Bowen (1999).

The three basic criteria for self-control make possible a separation of defects into categories of "controllability," of which the most important are

1. *Worker controllable.* A defect or nonconformity is worker controllable if all three criteria for self-control have been met.
2. *Management controllable.* A defect or nonconformity is management controllable if one or more of the criteria for self-control have not been met.

Only management can provide the means for meeting the criteria for self-control. Hence, any failure to meet these criteria is a failure of management, and the resulting defects are therefore beyond the control of the workers. This theory is not 100% sound. Workers commonly have a duty to call management's attention to deficiencies in the system of control, and sometimes they do not do so. (Sometimes they do, and management fails to act.) However, the theory is much more right than wrong.

Whether the defects or nonconformities in a plant are mainly management controllable or worker controllable is of the highest order of importance. Reducing the former requires a program in which the main contributions must come from managers, supervisors, and technical specialists. Reducing the latter requires a different kind of program in which much of the contribution comes from workers. The great difference between these two kinds of programs suggests that managers should quantify their knowledge of the state of controllability before embarking on major programs.

TABLE 13.2
Controllability study in a machine shop

Category	Percent
Management controllable	
Inadequate training	15
Machine inadequate	8
Machine maintenance inadequate	8
Other process problems	8
Material handling inadequate	7
Tool, gage, fixture (TGF) maintenance inadequate	6
TGF inadequate	5
Wrong material	3
Operation run out of sequence	3
Miscellaneous	5
Total	68
Worker controllable	
Failure to check work	11
Improper operation of machine	11
Other (e.g., piece mislocated)	10
Total	32

An example of a controllability study is given in Table 13.2. A diagnostic team was set up to study scrap and rework reports in six machine shop departments for 17 working days. The defect cause was entered on each report by a quality engineer who was assigned to collect the data. When the cause was not apparent, the team reviewed the defect and, when necessary, contacted other specialists (who had been alerted by management about the priority of the project) to identify the cause. The purpose of the study was to resolve a lack of agreement on the causes of chronically high scrap and rework. It did the job. The study was decisive in obtaining agreement on the focus of the improvement program. In less than one year, more than $2 million was saved, and important strides were made in reducing production backlogs.

Controllability can also be evaluated by posing specific questions for each of the three criteria of self-control. (Typical questions that can be posed are presented in this chapter.) Although this approach does not yield a quantitative evaluation of management and worker-controllable defects, it does show whether the defects are primarily management or worker controllable.

In the experience of the author, defects are about 85% management controllable. This figure does not vary much from industry to industry, but it does vary greatly among processes. Other investigators in Japan, Sweden, the Netherlands, and Czechoslovakia have reached similar conclusions.

Although the available quantitative studies make clear that defects are mainly management controllable, many industrial managers do not know this or are unable to accept the data. Their long-standing beliefs are that most defects are the result of worker carelessness, indifference, and even sabotage. Such managers are easily persuaded to embark on worker-motivation schemes that, under the usual state of affairs, aim at a small minority of the problems and hence are doomed to achieve minor results at best. The issue is not whether quality problems *in an industry* are management controllable.

The need is to determine the answer *in a given enterprise.* This answer requires solid facts, preferably through a controllability study of actual defects, as shown in Table 13.2.

We now discuss the three main criteria for self-control.

Criterion 1: Knowledge of "Supposed to Do"

This knowledge commonly consists of the following:

1. The product standard, which may be a written specification, a product sample, or other definition of the end result to be attained.
2. The process standard, which may be a written process specification, written process instructions, an oral instruction, or other definition of "means to an end."
3. A definition of responsibility, i.e., what decisions to make and what actions to take.

In developing product specifications, some essential precautions must be observed.

Unequivocal information must be provided

Specifications should be quantitative. If such specifications are not available, physical or photographic standards should be provided. But beyond the need for *clear* product specifications, there is also a need for *consistent* and *credible* specifications. In some organizations, production supervisors or even workers have a secret "black book" that contains the "real" specification limits used by inspectors for accepting product. The author has found this activity to be prevalent in many types of industries (electrical enclosures, automotive, construction equipment, flat glass, and even aircraft manufacturing to name a few). A further problem is communicating changes in specifications, especially when there is a constant parade of changes.

Information on seriousness must be provided

All specifications contain multiple characteristics, and these are not equally important. Production personnel must be guided and trained to meet all specification limits. But they must also be given information on the relative importance of each characteristic to focus on priorities. Section 10.5 explains methods of defining relative seriousness.

Reasons must be explained

Explanation of the purposes served by both the product and the specification helps workers to understand why both the nominal specification value and the limits must be met.

Process specifications must be provided

Work methods and process conditions (e.g., temperature, pressure, time cycles) must be unequivocally clear.

LTV, a steel manufacturer, uses a highly structured system of identifying key process variables, defining process control standards, communicating the information to the workforce, monitoring performance, and accomplishing diagnosis when problems arise. The process specification is a collection of process control standard procedures (Figure 13.8). A procedure is developed for controlling the key process variables

INTEGRATED PROCESS CONTROL
Standard Procedures
Process Control

LTV Steel

Plant: Indiana Harbor
Dept: No. 3 Sheet Mill

File No. 716-2.2.2
Date Orig Issue: _____
Revision No. 1
Date Revised: _____

Control Area	Control Point	Control Element	No.
Tandem Mill	Rolling	Rolling Solutions	2.2.2

Control Task
To maintain rolling solution characteristics at the proper levels.

Responsible for Control
Solution Attendant

Process Standard
- Oil concentration must be 2.5% to 3.5%
- Solution temperature must be 110°F–120°F
- SAP value must be above 120
- Iron fines must be less than 600 ppm

Reason for Control
- To provide the correct lubricity between work rolls and strip for reduced roll wear and control of strip temperature. This helps control strip flatness and avoids friction scratches.

Measurement	Routine Reporting of Data	Control Chart
Tools/Equipment – Standard chem. test set-up	Form No. Solution Attendant's Report	Type – X & Moving Range
Frequency – Twice/turn	By – Solution Attendant	By – Solution Att.
By – Solution Attendant		

Corrective Action
- Solution concentration approaching limits - add rolling oil or water.
- Solution temp. approaching limits - adjust temperature control.
- SAP reading between 100 and 120, skim solution tank and add new oil. SAP reading below 100, retest immediately and contact Operating Supervisor. If retest below 100, shut down mill and switch to alternative solution tank.
- Iron fines approaching limit, skim tank for 2 hours and add 100 gallons oil. Retest after 30 minutes second time; repeat procedure if still near or above limit.

Operating Procedure
See attached sheet

Disposition of Non-compliant Product
Identify coil(s) for special surface evaluation. Notify Metallurgical Supervisor.

Review Procedure
Once per turn the Operating Supervisor will
- Check Solution Attendant's Report
- Visually check temperature of solution

Developed By: _Robert Gorskie_ (IPC Coordinator)   _Richard H. Bump_

Approved: _D. J. Bird_ (Department Superintendent/Manager)   _Dale H. Dick Jr._ (Manager–Quality Control)   _R. Vatch_ (General Superintendent/Print Manager)

FIGURE 13.8
Process control standard procedure. (*From LTV Steel.*)

(variables that must be controlled to meet specification limits for the product). The procedure answers the following issues:

- What the process standards are.
- Why control is needed.
- Who is responsible for control.
- What and how to measure.
- When to measure.
- How to report routine data.
- Who is responsible for data reporting.
- How to audit.
- Who is responsible for audit.
- What to do with product that is out of compliance.
- Who developed the standard.

Often, detailed process instructions are not known until workers have become experienced with the process. Updating of process instructions based on job experience can be conveniently accomplished by posting a cause-and-effect diagram in the production department and attaching index cards to the diagram. Each card states additional process instructions based on recent experience.

A checklist must be created

The preceding discussion covers the first criterion of self-control: People must have the means of knowing what they are supposed to do. To evaluate adherence to this criterion, a checklist of questions can be created, including the following:

Adequate and complete work procedures
1. Are there written product specifications, process specifications, standard work, and work instructions? If written down in more than one place, do they all agree? Are they legible? Are they conveniently accessible to the worker? Are they under formal document control?
2. Does the worker receive specification changes automatically and promptly?
3. Does the worker know what to do with defective raw or work-in-progress material?
4. Have responsibilities for decisions and actions been clearly defined?

Adequate and complete performance standards
5. Do workers consider the standards attainable?
6. Does the specification define the relative importance of different quality characteristics? If control charts or other control techniques are to be used, is their relationship to product specifications clear, and do we also review for process control as well as process capability?
7. Are standards for visual defects displayed in the work area?
8. Are the written, physical, or pictorial specifications given to the worker the same as the criteria used by inspectors? Are deviations from the specification often allowed?
9. Does the worker know how the product is used?
10. Does the worker know the effect on future operations and product performance if the specification is not met?

Adequate selection and training

11. Does the personnel selection process adequately match worker skills with job requirements?
12. Has the worker been adequately trained to understand the specification and perform the steps needed to meet the specification?
13. Has the worker been evaluated by testing or other means to see whether he or she is qualified?

Criterion 2: Knowledge of Performance

For self-control, people must have the means of knowing whether their performance conforms to a standard. This conformance applies to

- The product in the form of specifications for product characteristics.
- The process in the form of specifications for process variables.

This knowledge is secured from three primary sources: measurements inherent in the process, measurements by production workers, and measurements by inspectors.

Criteria for good feedback to workers

The needs of production workers (as distinguished from supervisors or technical specialists) require that the data feedback can be read at a glance, deal only with the few important defects, deal only with worker-controllable defects, provide prompt information about symptom and cause, and provide enough information to guide corrective actions. Good feedback should

- *Be recognizable at a glance.* The pace of events on the factory floor is swift. Workers should be able to review the feedback while in motion. Where a worker needs information about process performance over time, charts can provide an excellent form of feedback, provided they are designed to be consistent with the assigned responsibility of the worker (Table 13.3). It is useful to use visual displays to highlight recurrent problems. A problem described as "outer hopper switch installed backwards" displayed on a wall chart in large block letters has much more impact than the same message buried away as a marginal note in a work folder.

**TABLE 13.3
Worker responsibility versus chart design**

Responsibility of the worker is to	Chart should be designed to show
Make individual units of product meet a product specification	Measurements of individual units of product compared to product specification limits
Hold process conditions to the requirements of a process specification	Measurements of the process conditions compared to the process specification limits
Hold averages and ranges to specified statistical control limits	Averages and ranges compared to the statistical control limits
Hold percentage nonconforming below some prescribed level	Actual percentage nonconforming compared to the limiting level

- *Deal only with the few important defects.* Overwhelming workers with data on all defects will result in diverting attention from the vital few.
- *Deal only with worker-controllable defects.* Any other course provides a basis for argument that will be unfruitful.
- *Provide prompt information about symptoms and causes.* Timeliness is a basic test of good feedback; the closer the system is to "real time" signaling, the better.
- *Provide enough information to guide corrective action.* The signal should be in terms that make it easy to decide on remedial action.

Feedback related to worker action

The worker needs to know what kind of *process* change to make to respond to a *product* deviation. Sources of this knowledge are

- The process specifications (see Figure 13.4 under "Corrective Action").
- Cut-and-try experience by the worker.
- The fact that the units of measure for both product and process are identical.

If they lack all of these, workers can only cut and try further or stop the process and sound the alarm.

Sometimes the data feedback can be converted into a form that makes the worker's decision easier about what action to take on the process.

For example, a copper cap had six critical dimensions. It was easy to measure the dimensions and to discover the nature of product deviation. However, it was difficult to translate the product data into process changes. To simplify this translation, use was made of a position–dimensions (P–D) diagram. The six measurements were first "corrected" (i.e., coded) by subtracting the thinnest from all the others. These corrected data were then plotted on a P–D diagram as shown in Figure 13.9. Such diagrams provided a way of analyzing the tool setup.

Feedback to supervisors

Beyond the need for feedback at workstations, there is a need to provide supervisors with short-term summaries. These take several forms.

Matrix summary. A common form of matrix is workers versus defects; the vertical columns are headed by worker names and the horizontal rows by the names of defect types. The matrix makes clear which defect types predominate, which workers have the most defects, and what the interaction is. Other matrices include machine number versus defect type and defect type versus calendar week. When the summary is published, it is usual to circle matrix cells to highlight the vital few situations that call for attention.

An elaboration of the matrix is to split the cell diagonally, thus permitting the entry of two numbers, e.g., number defective and number produced.

Pareto analysis. Some companies prefer to minimize detail and provide information on the total defects for each day plus a list of the top three (or so) defects encountered and how many of each there were. In some industries, a "chart room" displays performance against goals by product and by department.

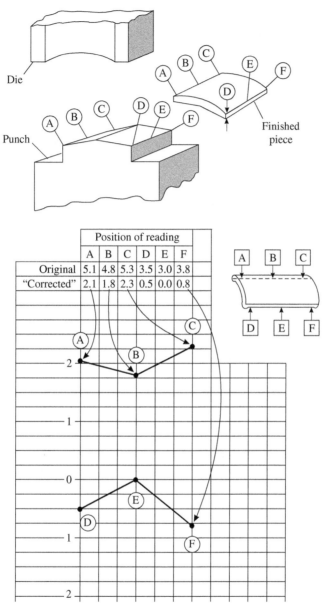

FIGURE 13.9
Method of drawing P–D diagram.

Automated quality information. Some situations justify the mechanization of both the recording and analysis of data. Entry of data into computer terminals on production floors is common now. Many varieties of software are available for analyzing, processing, and presenting quality information collected on the production floor.

The term *quality information equipment* (QIE) designates the physical apparatus that measures products and processes, summarizes the information, and feeds the information back for decision making. Sometimes such equipment has its own product development cycle to meet various product effectiveness parameters for the QIE. Remember QIE is also subject to sound measurement system analysis and gauge repeatability and reproducibility evaluation.

Checklist

A checklist for evaluating the second criterion of self-control includes questions such as these:

Adequate review of work
1. Are gauges provided to the worker? Do they provide numerical measurements rather than simply sorting good from bad? Are they precise enough? Are they regularly checked for accuracy?
2. Is the worker told how often to sample the work? Is sufficient time allowed?
3. Is the worker told how to evaluate measurements to decide when to adjust the process and when to leave it alone?
4. Is a checking procedure in place to ensure that the worker follows instructions on sampling work and making process adjustments?

Adequate feedback
5. Are inspection results provided to the worker, and are these results reviewed by the supervisor with the worker?
6. Is the feedback timely and in enough detail to correct problem areas? Have personnel been asked what detail is needed in the feedback?
7. Do personnel receive a detailed report of errors by specific type of error?
8. Does feedback include positive comments in addition to negative?
9. Is negative feedback given in private?
10. Are certain types of errors tracked with feedback from external customers? Could some of these be tracked with an internal early indicator?

Criterion 3: Ability and Desire to Regulate

Ability and desire to regulate is the third criterion for self-control. Regulating the process depends on various management-controllable factors, including these:

The process must be capable of meeting the tolerances. This factor is of paramount importance. In some organizations the credibility of specifications is a serious problem. Typically, a manufacturing process is created after release of the product design, a few trials are run, and full production commences. In cases where quality problems arise during full production, diagnosis sometimes reveals that the process is not capable of consistently meeting the design specifications. Costly delays in production then occur while the problem is resolved by changing the process or changing the specification. The capability of the manufacturing process should be verified during the product development cycle *before the product design is released for full production.* See Section 20.10 for a full discussion.

The process must be responsive to regulatory action in a predictable cause-and-effect relationship to minimize variation around a target value.

> **EXAMPLE 13.1.** In a process for making polyethylene film, workers were required to meet multiple product parameters. The equipment had various regulatory devices, each of which could vary performance with respect to one or more parameters. However, the workers could not "dial in" a predetermined list of settings that would meet all parameters. Instead, it was necessary to cut and try to meet all parameters simultaneously. During the period of cut and try, the machine produced nonconforming product to an extent that interfered with meeting standards for productivity and delivery. The workers were unable to predict how long the cut-and-try process would go on before full conformance was achieved. Consequently, it became the practice to stop cut and try after a reasonable amount of time and to let the process run, whether in conformance or not.

The worker must be trained on how to use the regulating mechanisms and procedures. This training should cover the entire spectrum of action—under what conditions to act, what kind and extent of changes to make, how to use the regulating devices, and why these actions need to be taken.

> **EXAMPLE 13.2.** Of three qualified workers on a food process, only one operated the process every week and became proficient. The other two workers were used only when the primary worker was on vacation or was ill, and thus they never became proficient. Continuous training of the relief people was considered uneconomical, and agreements with the union prohibited their use except under the situations cited earlier. This problem is management controllable, i.e., additional training or a change in union agreements is necessary.

The act of adjustment should not be personally distasteful to the worker, e.g., should not require undue physical exertion.

> **EXAMPLE 13.3.** In a plant making glass bottles, one adjustment mechanism was located next to a furnace area. During the summer months, this area was so hot that workers tended to keep out of it as much as possible. When the regulation consists of varying the human component of the operation, the question of process capability arises in a new form: Does the worker have the capability to regulate? This important question is discussed in Section 4.9, which includes some examples of discovering worker "knack."

The process must be maintained sufficiently to retain its inherent capability. Without adequate maintenance, equipment breaks down and requires frequent adjustments—often with an increase in both defects and variability around a nominal value. Clearly, such maintenance must be both preventive and corrective. The importance of maintenance has given rise to the concept of total productive maintenance (TPM) and reliability-centered maintenance (RCM). Under this approach, teams are formed to identify, analyze, and solve maintenance problems to maximize the uptime of process equipment. These teams consist of production line workers, maintenance personnel, process engineers, and others as needed. Problems are kept narrow in scope to encourage a steady stream of small improvements. Examples of improvements include a reduction in the number of tools lost, simplification of process adjustments, increased uptime, and gains in asset effectiveness as well as overall equipment effectiveness (OEE), particularly through improvements in unexpected losses.

Control systems and the concept of dominance

Specific systems for controlling characteristics can be related to the underlying factors that dominate a process. The main categories of dominance include the following:

- *Setup dominant.* Such processes have high reproducibility and stability for the entire length of the batch to be made. Hence, the control system emphasizes verification of the setup before production proceeds. Examples of such processes are drilling, labeling, heat sealing, printing, and presswork.
- *Time dominant.* Such a process is subject to progressive change with time (wearing of tools, depletion of a reagent, heating up of a machine). The associated control system will feature a schedule of process checks with feedback to enable the worker to make compensatory changes. Screw machining, volume filling, wool carding, and papermaking are examples of time-dominant processes.
- *Component dominant.* Here the quality of the input materials and components is the most influential. The control system is strongly oriented toward supplier relations along with incoming inspection and sorting of inferior lots. Many assembly operations and food formulation processes are component dominant.
- *Worker dominant.* For such processes, quality depends mainly on the skill and knack possessed by the production worker. The control system emphasizes such features as training courses and certification for workers, error-proofing, and rating of workers and quality. Workers are dominant in processes such as welding, painting, and order filling.
- *Information dominant.* In these processes, the job information usually undergoes frequent change. Hence, the control system emphasizes the accuracy and up-to-dateness of the information provided to the worker (and everyone else). Examples include order editing and "travelers" used in job shops.

The various types of dominance differ also in the tools used for process control. Table 13.4 lists the forms of process dominance along with the usual tools used for process control.

Checklist

A checklist for evaluating the third criterion of self-control typically includes questions such as these:

Process capable
1. Has the quality capability of the process been measured to include both inherent variability and variability due to time? Is the capability checked periodically?
2. Has the design of the job used the principles of error-proofing?
3. Has equipment, including any software, been designed to be compatible with the abilities and limitations of workers?

Process adjustments
4. Has the worker been told how often to reset the process or how to evaluate measurements to decide when the process should be reset?
5. Can the worker make a process adjustment to eliminate defects? Under what conditions should the worker adjust the process? When should the worker shut down the machine and seek more help? Whose help?

TABLE 13.4
Control tools for forms of process dominance

Setup dominant	Time dominant	Component dominant	Worker dominant	Information dominant
Inspection of process conditions	Periodic inspection	Supplier rating	Acceptance inspection	Computer-generated information
First-piece inspection	$\bar{X}$ chart	Incoming inspection	p chart	"Active" checking of documentation
Lot plot	Median chart	Prior operation control	c chart	Bar codes and electronic entry
Precontrol	$\bar{X}$ and R chart	Acceptance inspection	Operator scoring	Process audits
Narrow-limit gaging	Precontrol	Mockup evaluation	Recertification of workers	
Attribute visual inspection	Narrow-limit gaging		Process audits	
	p chart			
	Process variables check			
	Automatic recording			
	Process audits			

6. Have the worker actions that cause defects and the necessary preventive action been communicated to the worker, preferably in written form?
7. Can workers institute job changes that they show will provide benefits? Are workers encouraged to suggest changes?

Worker training in adjustments
8. Do some workers possess a hidden knack that needs to be discovered and transmitted to all workers?
9. Have workers been provided with the time and training to identify problems, analyze problems, and develop solutions? Does the training include diagnostic training to look for patterns of errors and determine sources and causes?

Process maintenance
10. Is there an adequate preventive maintenance program for the process?

Strong quality culture/environment
11. Is there sufficient effort to create and maintain awareness of quality?
12. Is there evidence of management leadership?
13. Have provisions been made for self-development and empowerment of personnel?
14. Have provisions been made for participation of personnel as a means of inspiring action?
15. Have provisions been made for recognition and rewards for personnel?

For a comparison of Russian and American practices with respect to worker control, see Pooley and Welsh (1994).

Use of Checklists on Self-Control

These checklists can help operations in the design (and redesign) of jobs to prevent errors; diagnose quality problems on individual jobs; identify common weaknesses in many jobs; and assist supervisors to function as coaches with personnel, prepare for process audits, and conduct training classes in quality. For elaboration, see *JQH6,* pages 20.692–20.69.

→ 13.5 AUTOMATED MANUFACTURING

The march to automation proceeds unabated. Several terms have become important:

- *Computer-integrated manufacturing (CIM).* CIM is the process of applying a computer in a planned fashion from design through manufacturing and shipping of the product.
- *Computer-aided manufacturing (CAM).* CAM is the process in which a computer is used to plan and control the work of specific equipment.
- *Computer-aided design (CAD).* CAD is the process by which a computer assists in the creation or modification of a design.

This trio of concepts is producing huge increases in factory productivity and efficiency. But automation, with proper planning, can also benefit product quality in several other ways:

- Automation can eliminate some of the monotony or fatiguing tasks that result in human errors. For example, when a manual seam-welding operation was turned over to a robot, the scrap rate plunged from 15% to zero, and manual grinding (a very dirty task) was virtually eliminated.
- Process variation can be reduced by the automatic monitoring and continuous adjustment of process variables.
- An important source of process troubles, i.e., the number of machine setups, can be reduced.
- Machines can automatically measure product and also record, summarize, and display the data for production line operators and staff personnel. Feedback to the worker can be immediate, thus providing early warning of impending troubles.
- With CAD, the quality engineer can provide inputs early in the design stage. When a design is placed into the computer, the quality engineer can review that design over and over again and thus keep abreast of design changes.

Collectively, these and other computer-based technologies for manufacturing are sometimes called "advanced manufacturing systems" (AMS).

With the continued enhancements of electronic information network provided by the Internet, a group of companies can operate as one virtual factory. This environment enables companies to exchange and act on information concerning inventory levels, delivery schedules, supplier lists, product specifications, and test data. It also means that CAD/CAM information and other manufacturing process information can be exchanged, data can be transferred to machines in a supplier's plant, and supplier software can be used to analyze producibility and to begin actual manufacturing.

Achieving the benefits of automated manufacturing requires a spectrum of concepts and techniques. Three of these are discussed here: the key functions of CIM, group technology, and flexible manufacturing systems.

Key Functions of Computer-Integrated Manufacturing

CIM integrates engineering and production with suppliers and customers (even globally) to interactively design, plan, and conduct the manufacturing activities. CIM activities include providing design tools to support innovation in remote sites, manufacturing planning, computer simulation models to evaluate and predict process performance, sensing and control tools to monitor processes, intelligence tools to gather and organize process data to share with other sites, information processing and transferring among geographically dispersed participants, and automated translation of text between different languages. For elaboration, see Lee (1995). This is tremendously important today with the continued emergence of manufacturing in developing countries around the world.

Group Technology

Group technology (or accelerated design replication) is the process of examining all items manufactured by a company to identify those with sufficient similarity that common design or manufacturing plans can be used. The aim is to reduce the number of new designs or new manufacturing plans. In addition to the savings in resources, group technology can improve both the quality of design and the quality of conformance by using proven designs and manufacturing plans. In many companies, only 20% of the parts initially thought to require a new design actually need it; of the remaining new parts, 40% could be built from an existing design, and the other 40% could be created by modifying an existing design. Accelerated replication of designs can reduce product design cycles from months to only weeks or even days.

Location of production machines can also benefit from the group technology concept. Machines are grouped according to value streams and can be sorted into cells of machines, each cell producing one or several part families (thus "cellular manufacturing"). For a discussion of shifting from traditional manufacturing organization to cellular manufacture and the impact on first-level supervisors, see *JQH6*, pages 11.332–11.349.

Flexible Manufacturing System

A flexible manufacturing system (FMS) is a group of several computer-controlled machine tools, linked by a materials handling system and a computer, to accommodate varying production requirements. The system can be reprogrammed to accommodate design changes or new parts. In contrast, in a fixed automation system, machinery, materials handling equipment, and controllers are organized and programmed for production of a single part or limited range of parts.

Typically, the individual machines are robots or other types of numerically controlled machine tools, each of which is run by a microcomputer. Several such tools are linked by a minicomputer, and then several of these minicomputers are tied into the mainframe computer.

From the one extreme of the (typically) mass production in automated industries, we can shift to the other extreme of the (typically) low-volume production in job shop industries. For more detail in quality planning for product based industries and continuous process based industries reference *JQH6*, Chapters 20 and 24.

Even traditional craft industries now benefit from technology. For example, the manufacture of fine pianos uses programmable logic controllers to make process adjustments; in wine making, aerial images of vineyards, taken by using digital sensors to collect data on the chlorophyll content of the vines, provide a strong indicator of taste (the aircraft will soon be replaced by satellites).

The potential benefits of the automated factory require significant time and resources for planning. Automation, however, will not be total—there will never be robot plumbers in the factory.

➜ 13.6 OVERALL REVIEW OF MANUFACTURING PLANNING

We incur great risk in going directly from a proposed manufacturing process plan into regular production. The delays and extra costs involved in quality failures require a review of the proposed process, including software used with the process. Such a review is most effectively accomplished through preproduction trials and runs.

Ideally, product lots should be put through the entire system, with deficiencies found and corrected before full-scale production begins. This should include a feasibility stage, prototype stage, preproduction verification and validation stage, and finally production verification and validation. In practice, companies usually make some compromises with this ideal approach. "Preproduction" may be merely the first runs of regular production but with special provision for prompt feedback and correction of errors as found. Alternatively, the preproduction run may be limited to features of product and process design that are so new that prior experience cannot reliably provide a basis for good risk taking. Although some companies do adhere to a strict rule of proving the product and process through preproduction lots, the more usual approach is one of flexibility, in which the use of preproduction lots depends on various factors:

- The extent to which the product embodies new or untested quality features.
- The extent to which the design of the manufacturing process embodies new or untried machines, tools, etc.
- The amount and value of product that will be out in the field before process, product, and use difficulties are fully known.

These trials sometimes include "production validation tests" to ensure that the full-scale process can meet the design intent.

The terms *process qualification* and *process certification* are also used to describe the review of manufacturing processes. Black (1993) describes how Caterpillar developed an internal certification program similar to its supplier certification program. The certification program is built around a 12-step manufacturing procedure to achieve process control and includes the identification of critical product characteristics and critical process parameters, the determination of process capability indexes, and actions for continuous quality improvement.

Preproduction trials and runs provide the ultimate evaluation—by manufacturing real product. Other techniques provide an even earlier warning before any product is made. For example, the failure mode, effect, and criticality analysis is useful in analyzing a proposed *product* design (see Section 12.5). The same technique can dissect potential failure modes and their effects on a proposed *process* design. Another technique uses highly detailed checklists to review proposed processes.

➔ 13.7 ORGANIZING FOR QUALITY IN MANUFACTURING OPERATIONS

The organization of today is influenced by the interaction of two systems: the technical system (design, equipment, procedures) and the social system (people, roles)—thus the name "sociotechnical systems" (STSs).

New ways of organizing work, particularly at the workforce level, have emerged. For example, supervisors are becoming "coaches"; they teach and empower rather than assign and direct. Operators are becoming "technicians"; they perform multiskilled jobs with broad decision making rather than narrow jobs with limited decision making. Team concepts play an important role in these new approaches.

In some organizations more than 40% of the people participate on teams; some organizations have a goal of 80% or greater. Permanent teams (e.g., process teams, self-managing teams) are responsible for all output parameters, including quality; ad hoc teams (e.g., quality project teams) are typically responsible for improvement in quality. A summary of the most common types of quality teams is given in Section 8.7.

Although these various types of quality teams are showing significant results, the reality is that, for most organizations, daily work in a department is managed by a supervisor who has a complement of workers performing various tasks. This configuration is the "natural work team" in operations. But team concepts can certainly be applied to daily work. One framework for a team in daily operations work is the control process from the trilogy of quality processes. As applied to daily work, the steps are choose control subjects, establish measurement, establish standards of performance, measure actual performance, compare to standards, and take action on the difference. For a detailed discussion of the steps, see Chapter 6, "Control of Quality to Maintain Superior Performance." When the natural work team of the department (self-directed work team) is trained in these concepts, the work team gains greater control over the key work processes so that it can meet customer needs while increasing employee involvement and empowerment.

13.8　PLANNING FOR EVALUATION OF PRODUCT

The planning must recognize the need for formal evaluation of product to determine its suitability for the marketplace. Three activities are involved:

1. Measuring the product for conformance to specifications
2. Taking action on nonconforming product
3. Communicating information on the disposition of nonconforming product.

These activities are discussed in Section 10.3. But the activities impinge on the manufacturing planning process. For example, several alternatives are possible for determining conformance, i.e., have the activity done by production workers, by an independent inspection force, or by a combination of both. Increasingly, the combination approach is being employed.

What has evolved is the concept of self-inspection combined with a product audit. Under this concept, all inspection and all conformance decisions, both on the process and on the product, are made by the production worker. (Decisions on the action to be taken on a nonconforming product are *not,* however, delegated to the worker.) However, an independent audit of these decisions is made. The quality department inspects a random sample periodically to ensure that the decision-making process used by workers to accept or reject a product is still valid. The audit verifies the decision process. Note that, under a pure audit concept, inspectors are not transferred to do inspection work in the production department. Except for those necessary to do audits, inspection positions are eliminated.

If an audit reveals that wrong decisions have been made by the workers, the product evaluated since the last audit is reinspected—often by the workers themselves.

Self-inspection has decided advantages over the traditional delegation of inspection to a separate department:

- Production workers are made to feel more responsible for the quality of their work.
- Feedback on performance is immediate, thereby facilitating process adjustments. Traditional inspection also has the psychological disadvantage of an "outsider" reporting the defects to a worker.
- The costs of a separate inspection department can be reduced.
- The job enlargement that takes place by adding inspection to the production activity of the worker helps to reduce the monotony and boredom inherent in many jobs.
- Elimination of a specific station for inspecting all products reduces the total manufacturing cycle time.

EXAMPLE 13.4. In a coning operation of textile yarn, the traditional method of inspection often resulted in finished cones sitting in the inspection department for several days, thereby delaying any feedback to production. Under self-inspection, workers received immediate feedback and could get machines repaired and setups improved more promptly. Overall, the program reduced nonconformities from 8 to 3%. An audit inspection of the products that were classified by the workers as "good" showed that virtually all of them were correctly classified. In this company, workers can also classify product as "doubtful." In one analysis, worker inspections classified 3% of the product as doubtful, after which an independent inspector reviewed the doubtful product and classified 2% as acceptable and 1% as nonconforming.

EXAMPLE 13.5. A pharmaceutical manufacturer employed a variety of tests and inspections before a capsule product was released for sale. These checks included chemical tests, weight checks, and visual inspections of the capsules. A 100% visual inspection had traditionally been conducted by an inspection department. Defects ranged from "critical" (e.g., an empty capsule) to "minor" (e.g., faulty print). This inspection was time-consuming and frequently caused delays in production flow. A trial experiment of self-inspection by machine operators was instituted. Operators visually inspected a sample of 500 capsules. If the sample was acceptable, the operator shipped the full container to the warehouse; if the sample was not acceptable, the full container was sent to the inspection department for 100% inspection. During the experiment, both the samples and the full containers were sent to the inspection department for 100% inspection with reinspection of the sample recorded separately. The experiment reached two conclusions: (1) the sample inspection by the operators gave results consistent with the sample inspection by the inspectors, and (2) the sample of 500 gave results consistent with the results of 100% inspection.

The experiment convinced all parties to switch to sample inspection by operators. Under the new system, good product was released to the warehouse sooner, and marginal product received a highly focused 100% inspection. In addition, the level of defects *decreased*. The improved quality level was attributed to the stronger sense of responsibility of operators (they themselves decided if product was ready for sale) and the immediate feedback received by operators from self-inspection. But there was another benefit—the inspection force was reduced by 50 people. These 50 people were shifted to other types of work, including experimentation and analysis on the various types of defects.

Criteria for Self-Inspection

For self-inspection, some criteria must be met:

- Quality must be the number one priority within an organization. If this requirement is not clear, a worker may succumb to schedule and cost pressures and classify products as acceptable that should be rejected.
- Mutual confidence is necessary. Managers must have sufficient confidence in the workforce to be willing to give workers the responsibility for deciding whether the product conforms to specification. In turn, workers must have enough confidence in management to be willing to accept this responsibility.
- The criteria for self-control must be met. Failure to eliminate the management-controllable causes of defects suggests that management does not view quality as a high priority, and this environment may bias the workers during inspections. Workers must be trained to understand the specifications and perform the inspection.
- Specifications must be unequivocally clear. Workers should understand the use that will be made of their products (internally and externally) to grasp the importance of a conformance decision.
- The process must permit assignment of clear responsibility for decision making. An easy case for application is a worker running one machine because there is clear responsibility for making both the product and the product-conformance decision. In contrast, a long assembly line or the numerous steps taken in a chemical process make it difficult to assign clear responsibility. Application of self-inspection to such multi-step processes is best deferred until experience is gained with some simple processes.

Self-inspection should apply only to products and processes that are stabilized and meet specifications and only to personnel who have demonstrated their competence. It is important to note again that the organization must demonstrate the process is not only capable, but in a state of statistical control.

Worker response to such delegation of authority is generally favorable; the concept of job enlargement is a significant factor. However, workers who do qualify for self-inspection commonly demand some form of compensation for this achievement, e.g., a higher grade, more pay. Companies invariably make a constructive response to these demands because the economics of delegating are favorable. In addition, the resulting differential tends to act as a stimulus to nonqualified workers to qualify themselves.

An adjunct to self-inspection is the use of *poka-yoke* devices as part of inspection. These devices are installed in a machine to inspect the process conditions and product results and provide immediate feedback to the operator. Devices such as limit switches and interference pins, used to ensure proper positioning of materials on machines, are *poka-yoke* devices for inspecting process conditions. Go/no go gauges are examples of *poka-yoke* devices for inspecting product. For elaboration, see The Productivity Press Development Team (1997). Grout and Downs (1998) compare the use of *poka-yoke* devices for controlling processes with statistical process control charts.

Another concept related to self-inspection is *jidoka*. Jidoka is one of the two main pillars of the Toyota Production System. It refers to the ability to stop production lines, by man or machine, in the event of problems such as equipment malfunction, quality issues, or late work. Jidoka helps prevent the passing of defects, helps identify and correct problem areas using localization and isolation, and makes it possible to "build" quality at the production process.

The three responsibilities of every operator are to:

- Check incoming work to ensure that it is defect free.
- Verify his or her work is free of defects.
- Never knowingly pass a defective product.

Start with your existing quality system and enhance by asking the following:

- What are the critical factors to the customer?
- How are the factors measured and recorded?
- Is the measurement system effective in identifying and preventing defects from reaching the customer?

For successful implementation of jidoka, an organization should:

1. Map and analyze its process.
2. Understand the customers' needs and translate them to its process.
3. Develop "in-station quality."
4. Mistake-proof the process and install automatic notification devices where possible.
5. Construct Andons signaling at points that are not 100% mistake-proofed.
6. Create a support structure—specific by issue.
7. Utilize root cause correction action when problems arise.
8. Implement visual control and Standard work throughout the process.
9. Develop a "stop-the-line" culture.

→ 13.9 CONDUCTING PROCESS QUALITY AUDITS

A quality audit is an independent review to compare some aspect of quality performance with a standard for that performance. Application to manufacturing has been extensive and includes both audits of activities (process audits) and audits of product (product audits). A full discussion of quality audits is given in Chapter 16, "Quality Assurance and Audits."

A *process quality audit* includes any activity that can affect final product quality. This onsite audit is usually done on a specific process by one or more persons and uses the process operating procedures. Adherence to existing procedures is emphasized, but audits often uncover situations of inadequate or nonexistent procedures. The checklists presented earlier in this chapter on the three criteria for self-control can suggest useful specific subjects for process audits. Audits must be based on a foundation of hard facts that are presented in the audit report in a way that will help those responsible to determine and execute the required corrective action.

Peña (1990) explains an audit approach for processes. Two types of audits are employed: engineering and monitor. The engineering process audit is conducted by a quality assurance engineer and entails an intense review of all process steps including

TABLE 13.5
Audit checklist

1. Is the specification accessible to production staff?
2. Is the current revision on file?
3. Is the copy on file in good condition and all pages accounted for?
4. If referenced documents are posted on equipment, do they match the specification?
5. If the log sheet is referenced in specifications, is a sample included in the specification?
6. Is the operator completing the log sheet according to specifications?
7. Are lots with out-of-specification readings authorized and taken care of in writing by the engineering department or the proper supervisor?
8. Are corrections to paperwork made according to specification?
9. Are equipment time settings according to specification?
10. Are equipment temperature settings according to specification?
11. Is the calibration sticker on equipment current?
12. Do chemicals or gases listed in the specification match actual usage?
13. Do quantities listed in the specification match the line setup?
14. Are changes of chemicals or gases made according to specification?
15. Is the production operator certified? If not, is this person authorized by the supervisor?
16. Is the production operating procedure according to specification?
17. Is the operator performing the written cleaning procedure according to specification?
18. If safety requirements are listed in the specification, are they being followed?
19. If process control procedures are written in the specification, are the actions performed by process control verifiable?
20. If equipment maintenance procedures are written in the specification, are the actions performed verifiable? according to specification?

From "Motorola's Secret to Total Quality Control" by E. Pena *Quality Progress,* October, 1990. Reprinted with permission form *Quality Progress,* © 1990 American Society for Quality.

process parameters, handling techniques, and statistical process control. Table 13.5 shows the audit checklist.

The monitor process audit covers a broad range of issues, e.g., whether specifications are correct and whether logs are filled in and maintained. Discrepancies (critical, major, or minor) are documented, and corrective action is required in writing. Critical defects must be corrected immediately; majors and minors must be resolved within five working days.

A product audit involves the reinspection of product to verify the adequacy of acceptance and rejection decisions. In theory, such product audits should not be needed. In practice, they can often be justified by field complaints. Such audits can take place at each inspection station for the product or after final assembly and packing. Sometimes an audit is required before a product may be moved to the next operation. For elaboration, see Section 16.12, "Product Audit."

➔ 13.10 QUALITY MEASUREMENT IN MANUFACTURING OPERATIONS

The management of key work processes must include provision for measurement. In developing units of measure, the reader should review the basics of quality measurement discussed in Section 6.2.

Managing Quality in Operations—Manufacturing 445

TABLE 13.6
Examples of quality measurement in manufacturing

Subject	Unit of measure
Quality of manufacturing output	Percentage of output meeting specifications at inspection ("first-time yield")
	Percentage of output meeting specifications at intermediate and final inspection
	Amount of scrap (quantity, cost, percentage, etc.), amount of rework (quantity, cost, percentage, etc.)
	Percentage of output shipped under waiver of specifications
	Number of defects found in product audit (after inspection)
	Warranty costs due to manufacturing defects
	Overall measure of product quality (defects in parts per million, weighted defects per unit, variability for critical characteristics, etc.)
	Amount of downgraded output
Quality of input to manufacturing	Percentage of critical operations with certified workers
	Amount of downtime of manufacturing equipment
	Percentage of product input meeting specifications
	Percentage of instruments meeting calibration schedules
	Percentage of specifications requiring changes after release

Table 13.6 shows examples for manufacturing activities. Note that many of the control subjects are forms of work output. In reviewing current units in use, a fruitful starting point is the measure of productivity. *Productivity* is usually defined as the amount of output related to input resources. Surprisingly, some organizations still mistakenly calculate only one measure of output, i.e., the total (acceptable *and* nonacceptable). Clearly, the pertinent output measure is that which is usable by customers (i.e., acceptable output).

The units in Table 13.6 become candidates for data analysis using statistical techniques such as control charts, discussed in Chapter 20. But there is a more basic point—the selection of the unit of measure and the periodic collection and reporting of data demonstrate to operating personnel that management regards the subject as having priority importance. This atmosphere sets the stage for improvement!

➔ 13.11 MAINTAINING A FOCUS ON CONTINUOUS IMPROVEMENT

Historically, the operations function has always been involved in troubleshooting sporadic problems. As chronic problems were identified, these were addressed using various approaches, such as quality improvement teams (see Chapter 4, "Improving

TABLE 13.7
Three types of action

Type of action to take	When to take action	Basic steps
Troubleshooting (part of quality control)	Performance indicator outside control limits	Identify problem
		Diagnose problem
	Performance indicator in clear trend toward control limits	Take remedial action
Quality improvement	The control limits are so wide that it is possible for the process to be in control and still miss the targets	Identify project
		Establish project
		Diagnose cause
	Performance indicator frequently misses its target	Remedy the cause
		Hold the gains
Quality planning	Many performance indicators for this process miss their targets frequently	Establish project
		Identify customers
	Customers have significant needs that the product does not meet	Discover customer needs
		Develop product
		Develop process
		Design controls

Source: Adapted from Juran Institute, Inc. (1995, pp. 5–7).

Quality While Decreasing Cost"). Often the remedies for improvement involve quality planning or replanning (see Chapter 5, "Quality by Design to Increase Sales"). These three types of action are summarized in Table 13.7.

Kannan et al. (1999) present the results of a survey of the application of 38 quality management practices (e.g., use of benchmark data to improve quality practices), 39 tools and techniques (e.g., statistical process control), 29 areas of documentation (e.g., quality assurance manual), and 12 quality measurements (e.g., customer complaints) at the operations level in manufacturing industries. The analysis includes the degree of usage and the impact on five organization performance measures.

Maintaining the focus on improvement clearly requires a positive quality culture in the organization. Therefore, we must first determine the present quality culture (see Section 3.7) and then take the steps to change the culture to one that will foster continuous improvement (see Chapter 9, "Creating a Quality Culture"). In addition, the operations function must be provided with the support to maintain the focus on improvement. A key source of that support should be the quality department. Thus the quality department should view operations as its key internal customer and provide the training, technical quality expertise, and other forms of support to enable operations to maintain the focus on improvement. Also, a quality department can urge upper management to set up cross-functional teams to address operations problems that may be caused by other functional departments such as engineering, purchasing, and information technology.

→ 13.12 CASE STUDY ON ERROR-PROOFING

Case study on error-proofing documentation in a biotech environment using six sigma and lean projects[1]

Background

In FDA-regulated manufacturing environments such as those in the life sciences, biotech, and pharmaceutical industries, complete and accurate documentation of processes and the completion of process steps are critical. However, such documentation is prone to a variety of human errors: slips, lapses, mistakes, and violations. This is a common problem across these industries, but the majority of document errors do not impact product quality.

Types of document errors caught during batch record review

Type of error	Description
Attentional lapse	• Forgot to record storage location and time and transfer after moving product from cold room • Forgot to make required entry on one of three equipment use logs tied to transfer lines and tanks • Forgot to enter information just entered on a ticket on a use log
Slip	• Transposed a lot number • Forgot to initial and date a page reviewed for completeness
Rule-based mistake	• Incorrectly exclude entry as "not applicable"

A decision by Genentech's senior management to error-proof such documentation resulted in the achievement of sustainable breakthrough improvements through the focused deployment of six sigma and lean project teams.

The Error-Proofing Project was chartered in September 2003 by senior management (GMP Core Team) at Genentech. The objective of the project was to reduce the inspection risk associated with document errors and lost tickets. This objective was expressed as the following dual goal:

Goal 1: Reduce the rate of document errors recorded in the discrepancy management system by 50% relative to the Q1/Q2 2003 baseline by September 2004. Maintain the document error rate within this new "zone of control": less than five document errors per 100 tickets (main batch record documents) for three months or more.

Goal 2: Reduce the overall volume of master ticket errors by 50% relative to the Q1/Q2 2003 baseline.

Six error-proofing teams were chartered in October 2003 to investigate the drivers for this pattern of error and variability. Consultants worked with each team. The teams

---

[1] Adapted from Bottome and Chua (2005) and with courtesy of Mr. Robert Bottome of Genentech.

were chartered such that their implementations would begin in March 2004 and deliver measurable document error rate reductions by June 2004.

The six projects were

1. Timely Feedback (a Six Sigma DMAIC project)
2. Change Volume and Timing (a Six Sigma DMAIC project)
3. Value-Stream Mapping in Growth Hormone and Recovery Operations (a Lean/Value-Stream Mapping project)
4. Document Rule Clarification (a Six Sigma DMAIC project)
5. Document Development (a Design for Six Sigma (DMADV) project)
6. Document Complexity (a Design for Six Sigma (DMADV) project)

Key Findings

1. Timely Feedback

Based on interviews across all shifts, the project team concluded that the timing and utility of feedback provided varied across production teams: in particular, the project team found that a mechanism for timely feedback was present only in certain instances. They set out to correlate the presence or absence of timely feedback to document error rates. As shown in the following hypothesis test (chi-square tables test of independence), the document error rate is dependent on the presence or absence of timely feedback.

```
              Chi-Square Test: <1wk FB S, 0 FB S
Expected counts are printed below observed counts
Chi-square contributions are printed below expected counts
                    <1wk
                    FB S    0 FB S    Total
          1          25        19       44
                   20.20     23.80
                   1.138     0.967
          2          20        34       54
                   24.80     29.20
                   0.928     0.788
      Total          45        53       98
         Chi-sq = 3.820, DF = 1, p-Value = .051
```
There is a statistical relationship of dependence between the lack of feedback and the groups we identified as high and low performers (on document errors).

Teams with timely feedback performed 50% better than those with incomplete or delayed feedback. As predicted by the team, the document error rate took its first significant shift toward a new zone of control in April 2004 immediately after the timely feedback mechanisms were introduced.

2. Change Timing

The Change Volume and Timing project team (a DMAIC project) set out to investigate the relationship between document effective and release dates with regard to

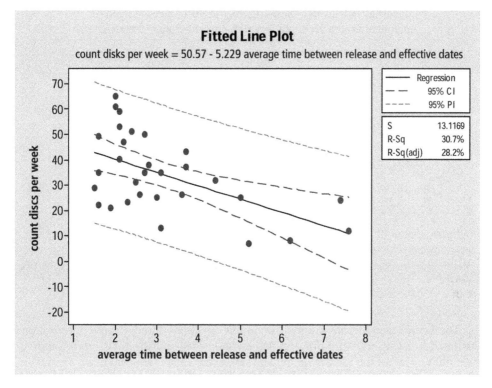

FIGURE 13.10
Relationship between the average time between release
and effective dates for documents and discrepancies.

document errors per document. By correlating the interval between released and effective dates to the document errors recorded per execution of the ticket, the team was able to show that the documents made effective within three days of their release dates carried a disproportionate amount of document error discrepancies (see Figure 13.10).

Because 45.7% of tickets had less than 2 days between release and effective dates, these documents represent a significant driver of document errors on the floor. As the team investigated further, they were able to confirm that 1–3 day intervals do not permit meaningful document training across shifts.

3. Change Volume

Another driver of document errors confirmed by the Change Volume and Timing team is the number of times a document gets changed in a year. More than 1440 ticket changes were initiated by SSF manufacturing in 2003. By comparing the discrepancy rates recorded for documents that were changed more than three times in a year to those that were changed less often, the team was able to confirm that frequent revisions drive document errors.

Kruskall-Wallis test shows significant differences in median document errors across different document change frequencies

Change frequency	N	Median	Ave rank	Z
0.08	215	0.1268	188.4	−3.06
0.17	127	0.1429	208.0	0.29
0.25	45	0.1875	243.8	2.30
0.33	20	0.5227	269.0	2.46
0.42	2	0.6483	317.0	1.33
0.58	1	0.6800	344.0	1.17
Overall	410		205.5	

$H = 18.09$ DF $= 5$ p $= 0.003$
$H = 18.28$ DF $= 5$ p $= 0.003$ (adjusted for ties)

Enforcement of a mechanism to pull documents with more than three changes per year, coupled with a system to make revisions more visible were proposed. The team remains convinced that these measures would yield further document error reduction benefits.

4. Ink Color

The Lean/Value Stream Mapping team in Growth Hormone Recovery operations generated a list of error-proofing implementation ideas during their rapid improvement phase. One of the ideas with evident universal utility grew out of an analysis of the factors that contribute to omission noted errors. The team noted that when entries are made in black ink on batch records, forms, and logs printed in black ink, it can be hard to see where an entry that was required may be missing. The team proposed switching to blue ink to enhance the contrast between entries and the printed text.

After negotiating the details with the Quality Assurance and Compliance group, the necessary SOP changes were made to permit the use of blue ink without requiring it. A memo was circulated explaining the rationale for the change and asking managers to enforce the consistent use of blue ink. The blue ink initiative was implemented in March 2004 just as the omission error rate dropped below 5 for the first time. Manufacturing Online Auditor (MOA) data collected in July and August 2004 suggest that omissions were detected twice as often now that blue ink was in routine use.

5. Equipment Log Location and Other Rapid Improvements

Another rapid improvement idea that was successfully implemented by the Growth Hormone Recovery Lean/VSM team looked at the centralization of equipment use logs. A survey tool was developed and spaghetti maps were used to test the hypothesis that having equipment use logs scattered throughout the work area increased technician wear and tear and incidentally the risk of document errors.

After centralizing the logs in one location, 95% of the survey respondents agreed that the system was much easier to use (less travel, less time, less difficult). Document error rates in the area continued to drop after the logs were centralized.

Rapid improvement efforts to identify single points of contact, provide radios, and structure communications for clarity, directness, and simplicity have contributed to a reduction in the impact of uncontrolled communications with the floor.

6. Documentation Rule Confusion

Interviews with operators who execute tickets confirmed that a high degree of confusion existed around the rules for cross outs, ruling out sections as not applicable, entering time correctly, and other fine points of batch record creation. This confusion was identified as a potentially significant driver of documentation errors.

Even though many of the errors that result from confusion are crossed out (see Figure 13.11) and then corrected during online review, cross-outs, as such, can be viewed as a concern by a FDA auditor because they reflect the operator's struggle to create the batch record correctly the first time.

The team systematically catalogued the points of confusion (sub-Y's) and then completed a cause and effect analysis for each one to identify the drivers [$f(x$'s)]. Examples of these validated sources of confusion (proven X's) include

Documents use ambiguous language (e.g., "approximately"), which causes confusion around doc rules and leads to doc errors.

Some documents include information only around alert limits; other documents include information around alert limits and action limits. This inconsistency causes confusion around doc rules, which leads to doc errors.

Documentation rules keep changing, which causes confusion around doc rules and leads to doc errors.

Operators are trained by different trainers and content is not consistent, which causes confusion around doc rules and leads to doc errors.

Justification behind the rules (criteria for the rules) is not communicated, which causes confusion around doc rules and leads to doc errors.

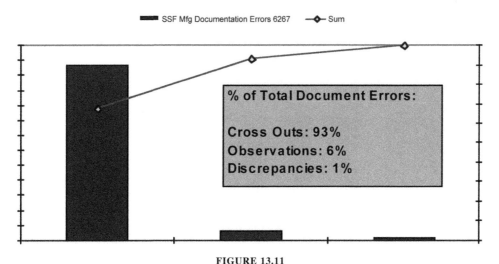

FIGURE 13.11
Baseline Document Error Performance based
on a review of documents completed before February 2004.

These validated causes of confusion were converted to customer needs and used as the basis for designing a clear, central SOP for documentation and associated training. These necessary tools were made effective and provided to operators in May 2004—just as the document error rate dropped below 5 per 100 tickets for the first time.

7. Low Accountability for Word Processing Errors and No Central Coordination for Document Changes

The Document Development and Change Process team (a DMADV team) defined its problem statement around the inability of the current process to reliably produce accurate/error-free documents within customer-defined timelines.

Data from Q1/Q2 of 2003 were analyzed by the team to identify master ticket errors and map them to process steps where the error could have originated or caught (individual errors were often mapped to multiple locations). This analysis identified the "Document Development" stage of the process (including word processing and SME redline steps) as the most significant generator of ticket errors (see Figure 13.12).

A Value-Stream Map of the process was created and used to show that only four of the 54 steps were value-add, 2 were non-value-add, and the remaining 48 were all classified as "non-value-add needed."

An FMEA of the process ranked the various identified failure modes, and root causes were assigned to the highest severity failure modes. The top 10 RPN scores

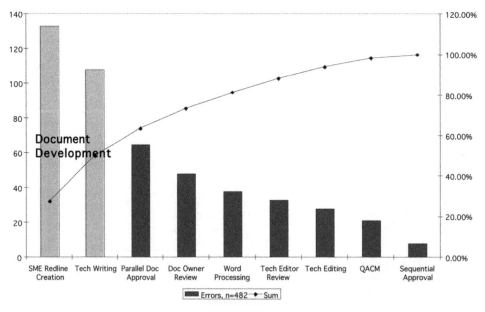

FIGURE 13.12

Pareto of performance gaps identified in Q1/Q2 2003 data, showing combined origin and "miss" steps where error could have been caught.

could be assigned to three root causes: communication gaps, business process inadequacies, and resource limitations.

An improved process was designed using DMADV methodology to address the identified issues; the three root causes were converted to customer needs. The improved process (which features a centralized coordinator, automated step references, resource management tools, and development checklists) was piloted in the fall of 2004. Compared to the baseline data, the campaign preparation effort delivered accurate documents in accordance with defined timelines.

8. Document Complexity

The problem statement devised by the Document Complexity Team (a DMADV project) emphasized the link between document errors and instructions to Manufacturing and QA staff that are often overly complicated or inconsistent.

The team interviewed customers (document users) across functions and prioritized their identified needs. A Pareto chart of these findings is provided in Figure 13.13.

The top three needs (identified most often) were fewer pieces of paper, no conflicting instructions, and fewer data entries.

Using these data, the team was able to devise a system for scoring instruction complexity and applied this to a set of standard instructions. These scores were then correlated to discrepancy rates per document. As shown in the table on the next page, for the "Combined" and "Ferm Only" data, the error rate correlated strongly with complexity score and 11 of 12 complexity elements (recovery data correlate strongly with only 9 of 12 elements).

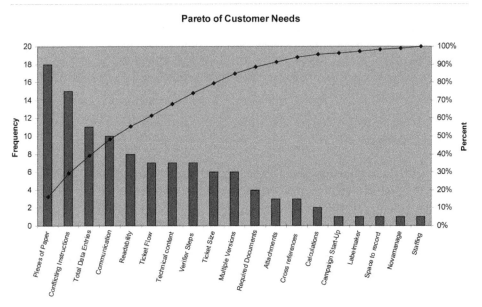

FIGURE 13.13
Pareto of customer needs for clear instructions.

Regression analyses were used to confirm that complexity scores could predict error rates and significant complexity elements.

Correlation results, complexity score, and error rate.

	Combined		Fermentation		Recovery	
	p	R	p	R	p	R
Complexity score	0.000	0.673	0.000	0.859	0.018	0.338
Total data entries	0.000	0.632	0.000	0.829	0.093	0.242
Total pages	0.000	0.618	0.000	0.925	0.006	0.385
SOP/SR refs	0.000	0.617	0.000	0.748	0.001	0.457
Operator steps	0.000	0.611	0.000	0.868	0.010	0.366
FNs to refer	0.000	0.600	0.000	0.773	0.018	0.336
Total SOPs/SRs	0.000	0.584	0.000	0.699	0.005	0.399
Other G-code	0.000	0.575	0.000	0.713	0.007	0.380
Verifier steps	0.000	0.491	0.000	0.877	0.016	0.343
Labeling steps	0.000	0.442	0.000	0.838	0.021	0.328
Calculations	0.000	0.395	0.019	0.466	0.134	0.217
Attachments	0.032	0.249	0.000	0.738	0.210	0.182
Ver/Op	0.470	−0.085	0.748	−0.068	0.832	0.031

The results of the analysis were used to identify two quick fixes which were successfully implemented: preprinted POMS labels and other attachments in a binder and a Departmental Controlled Document to list use logs. Both of these fixes were implemented by June 2004.

Breakthrough Results and New Zone of Control

As illustrated in Figure 13.14, the first goal was achieved in May 2004, and the document error has been held within the new zone of control ever since.

The second goal was achieved almost immediately, and the incidence of master ticket errors has not returned to Q1/Q2 2003 levels since the project began (see Figure 13.15).

Lessons Learned

Achieving and sustaining these breakthrough results demonstrated the power of applying project-by-project improvement using the right tools (in this case, six sigma and lean/value-stream management) coupled with the right management priority and commitment of the right resources and support. Lessons learned include

- Dedicated resources perform better than part-time teams; effective sponsors ensure that team members are not over booked.
- Active and engaged sponsorship is key, especially during the initial scope of work definition and chartering.
- Management must resist the impulse to direct teams to a solution prematurely while providing clear boundaries to the scope of the inquiry and the timing of the project.

Managing Quality in Operations—Manufacturing   455

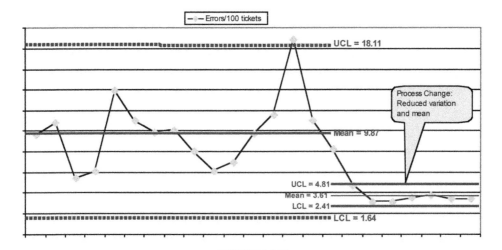

FIGURE 13.14
Document error rate: Jan 2003 through Nov 2004.
(Note from the author (Caldwell): This chart exemplifies the Juran Trilogy, demonstrating Quality Improvement's breakthrough results, followed by "holding the gains.")

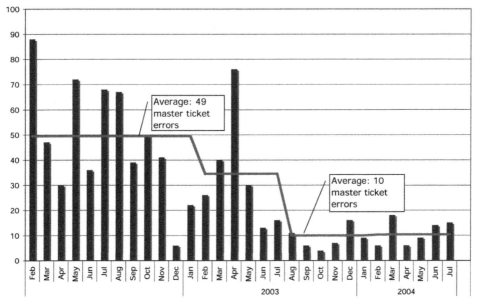

FIGURE 13.15
Master ticket errors per month.

- Take care to prepare for training events such that logistical issues, absent sponsors, and confusion about project goals do not distract participants.

In addition to these success factors, this effort taught management one truly fundamental lesson and that is, put your best people on the problem and make the investment to provide them with world class tools and support and they will deliver lasting results.

SUMMARY

- Activities to integrate quality in manufacturing planning have two objectives: to prevent defects and to minimize variability.
- By creating a flow diagram, we can dissect a manufacturing process and plan for quality at each workstation.
- To prevent defects and to minimize variability, we must discover the relationships between process features and product results.
- Error-proofing a process is an important element of prevention.
- For human beings to be in a state of self-control, they must have knowledge of what they are supposed to do, knowledge of what they are actually doing, and a process that is capable of meeting specifications and can be regulated.
- Failure to meet all three of these criteria means that the quality problem is management controllable. About 85% of quality problems are management controllable.
- The key factors of dominance in manufacturing are setup, time, component, worker, and information.
- Under the self-inspection concept, conformance decisions are made by the production worker, with an independent audit by the quality department (with the ultimate goal of operator self-control and self-directed work teams that eliminate independent audits).
- Computer-integrated manufacturing is the process of applying a computer in a planned fashion from design through manufacturing and shipping (the entire value stream); computer-aided manufacturing is the process in which a computer is used to plan and control the work of specific equipment; computer-aided design is the process by which a computer assists in the creation or modification of a design.
- In the group technology concept, all items are examined to identify those with sufficient similarity that a common design or manufacturing plan can be used.
- In a flexible manufacturing system, a group of several computer-controlled machine tools is linked by a materials handling system and a mainframe computer to process a group of parts completely.
- A process quality audit is an independent evaluation of any activity that can affect final product quality.

PROBLEMS

13.1. Visit a local manufacturing company and identify the departments that have the principal and collateral responsibilities for carrying out the planning activities for quality.

13.2. For a specific manufacturing operation with which you are familiar, describe how you would conduct a controllability study.

13.3. Your plant has just conducted a controllability study by analyzing the causes of a large sample of defective parts. The results showed that 45% of the causes were worker controllable and 55% management controllable. One program (diagnosis of specific causes, determination of remedies, etc.) is planned for the management-controllable problems. For the worker-controllable problems, a motivation program for the workers producing poor work is planned. You are asked to comment on this approach.

13.4. Study the system of performance feedback available to any of the following categories of people and report your conclusions on the adequacy of this feedback for controlling quality of performance.
(a) A motorist in city traffic.
(b) A student at school.
(c) A supermarket cashier.

13.5. For any process to which you can gain access, determine the major form of "dominance" affecting the attainment of quality.

13.6. A steel manufacturing plant has had a chronic problem of scrap and rework in its wire mill. The costs involved have reached a level where they are a major factor in the profits of the division. All levels of personnel in the wire mill are aware of the problem, and there is agreement on the vital few product lines that account for most of the problem. However, no reduction in scrap and rework costs has been achieved. What do you propose as a next step?

13.7. Select a manufacturing task regularly performed by a worker. Use the checklists on self-control to determine whether the worker can be held responsible for the quality of the output.

13.8. Why are the three criteria of self-control prerequisites for a successful motivation program?

REFERENCES

Black, S. P. (1993). "Internal Certification: The Key to Continuous Quality Success," *Quality Progress,* January, pp. 67–68.

Bottome, R. and R. C. H. Chua (2005). "Genentech Error-Proofs Its Batch Records," *Quality Progress,* July 2005, pp. 25–34.

Eibl, S., U. Kess, and F. Pukelsheim (1992). "Achieving a Target Value for a Manufacturing Process," *Journal of Quality Technology,* January, pp. 22–26.

Goldman, S. L., R. N. Nagel, and K. Preiss (1995). *Agile Competitors and Virtual Organizations,* Van Nostrand Reinhold, New York.

Grout, J. R. and B. T. Downs (1998). "Mistake-Proofing and Measurement Control Charts," *Quality Management Journal,* vol. 5, no. 2, pp. 67–75.

Juran, J. M. (1992). *Juran on Quality by Design,* Free Press, New York.

Lee, J. (1995). "Perspective and Overview of Manufacturing Initiatives in the United States," *International Journal of Reliability, Quality and Safety Engineering,* vol. 2, no. 3, pp. 227–233.

Nakajo, T. and H. Kume (1985). "The Principles of Foolproofing and Their Application in Manufacturing," *Reports of Statistical Application Research,* Union of Japanese Scientists and Engineers, Tokyo, vol. 32, no. 2, June, pp. 10–29.

Ohno, T. (1988). *The Toyota Production System—Beyond Large Scale Production,* Productivity Inc., Portland, Oregon.

Peña, E. (1990). "Motorola's Secret to Total Quality Control," *Quality Progress,* October, pp. 43–45.

Shuker, T. J. (2000). "The Leap to Lean," *Annual Quality Congress Proceedings,* ASQ, Milwaukee, pp. 105–112.

Siff, W. C. (1984). "The Strategic Plan of Control—A Tool for Participative Management," *ASQC Quality Congress Transactions,* Milwaukee, pp. 384–390.

Snee, R. D. (1993). "Creating Robust Work Processes," *Quality Progress,* February, pp. 37–41.

Somerton, D. G. and S. E. Mlinar (1996). "What's Key? Tool Approaches for Determining Key Characteristics," *Proceedings of the Annual Quality Congress,* ASQ, Milwaukee, pp. 364–369.

Spear, S. and H. K. Bowen (1999). "Decoding the DNA of the Toyota Production System," *Harvard Business Review,* September–October, pp. 96–106.

The Productivity Press Development Team (1997). *Mistake Proofing for Operators: The ZQC System,* Productivity Press, Portland, OR.

CHAPTER 14

Managing Quality in Operations—Service

➜ 14.1 THE SERVICE SECTOR

In many service organizations, the cost of poor quality ranges from 25 to 40% of operating expenses. For example, in one large bank, the cost of poor quality is 37% for automatic teller machines (ATMs), 26% for customer inquiry centers, and 50% for commercial loan operations (*JQH5,* page 33.11). Such numbers prove the need for addressing quality in the service sector.

As discussed in Chapter 1, "The Bottom Line: Quality and Business Performance," service is a product—i.e., the output of a process. As defined in *JQH6,* page 74, a *service* is work that is performed for someone else. A carpenter builds a home for a homeowner, the user; an automotive technician repairs cars for their owners; a nurse cares for patients; and a web browser provides information to meet the needs of its users. The overriding characteristic is that a service is intangible: It cannot be touched, gripped, handled, looked at, smelled, or tasted. Other characteristics are that the service is provided when the customer requests it, the service output is created as it is delivered, the service usually cannot be stored in an inventory, and completion time is critical. Services are often, but not necessarily, involved with the delivery of a tangible, physical good (a restaurant meal or plastic pellets).

Services account for more than two-thirds of the U.S. economy. The service sector consists of a wide spectrum of industries. The major categories of service industries are listed in Table 14.1. Additionally, almost all companies providing tangible goods also provide services to their customers.

As in the case of a manufactured product, quality means customer satisfaction and loyalty (see Section 1.3). Satisfaction and loyalty are achieved through two components: product features and freedom from deficiencies.

Research by Cronin and Taylor (1992) indicates that service quality is only one component in a consumer's buying preference—the total value of the offering

TABLE 14.1
Categories of service industries

- Transportation (railroads, airlines, bus lines, subways, common carrier trucking, pipelines)
- Public utilities (telephone communication, energy services, sanitation services)
- Marketing (retail food, apparel, automotive, wholesale trade, department stores)
- Finance (banks, insurance, sales finance, investment)
- Real estate
- Restaurants, hotels, and motels
- News media
- Business services (advertising, credit services, computer services)
- Health services (nursing, hospitals, medical laboratories)
- Personal services (amusements, laundry and cleaning, barber and beauty shops)
- Professional services (lawyers, doctors)
- Repair services (garages, television and home repairs)
- Government (defense, health, education, welfare, municipal services)

(service quality, price, convenience, and availability) must be considered. Additionally, they confirm that service quality is perceived and should be measured differently among service industries, specifically between high-involvement service providers (health care) and low-involvement providers (fast food). Finally, we note that because customer expectations of service continue to change rapidly, service design that incorporates the product features and a process design free of deficiencies must evolve rapidly.

➔ 14.2 INITIAL PLANNING FOR QUALITY

Section 5.11 presents a road map for planning the quality of a new product. The steps are establish the project, identify the customers, discover customer needs, develop the product, develop the process, and establish process controls/transfer to operations.

In service industries, the *service design* defines the features of the output provided to customers to meet their needs. Thus the concepts presented in Chapter 11, "Understanding Customer Needs," and Chapter 12, "New-Product Development's Role in Designing for Quality," apply to both the manufacturing and service sectors. The service design is turned into a reality by the *service process*, i.e., the process features such as the work activities, people, equipment, and physical environment to meet customer needs. The concept of quality function deployment (see Section 12.4 under "Quality Function Deployment") is helpful in designing both service products and service processes.

The service process can be reviewed for quality by several means: analyze the process flow diagram, reduce the cycle time, error-proof the process, provide for supplier quality, qualify the process by validating the process and measurement capability, and plan for personnel self-control. These topics are discussed here and in the next chapter.

Analysis of the Process Flow Diagram

Process flow diagrams (alias process maps or process blueprints) are also discussed in Chapters 4, 7, and 13. Figure 14.1 shows a flow diagram for handling a request for adjustments to customer bills (AT&T). The "line of interaction" is the boundary of activities where the customer and frontline employees ("online group") have discussions. The "line of invisibility" separates activities that are seen or not seen by customers. The "organization boundary" shows which activities occur in the three departments involved in the process. Note how this example illustrates both frontline direct customer contact and backroom or back-office operations. Also note that the time required is shown for some activities.

The symbol P denotes process points at which problems could occur that would cause customer dissatisfaction. To *prevent* problems, we must identify potential problems, usually based on past data or an analysis of the flow diagram. In the service sector, problems arise for recurring reasons (AT&T):

- Promises are not kept.
- Customers must contact several people to achieve problem resolution.
- Only partial service is provided, the service is performed incorrectly, the wrong service is performed, the wrong information is provided.
- Customers do not understand the service provided.
- Service is not provided when needed or takes too long.
- The customer is inconvenienced with paperwork or other matters.

These known or potential problems should be identified on the flow diagram and preventive actions put in place. If necessary, the problems can be regarded as process failures and analysis tracked using the approach of a failure mode, effect, and criticality analysis (see Section 12.5 under "Failure Mode, Effect, and Criticality Analysis").

Reduction of Process Cycle Time

One of the most widely used metrics used to measure service delivery is time. Customers continually demand shorter delivery times for services provided, and creating and analyzing a process flow diagram is an excellent tool to reduce cycle time in the context of a quality improvement project. See Chapter 4 for the universal, structured approach for diagnosing the causes and taking corrective action. This universal approach underpins the six sigma project approach. Lean efforts, focusing on the elimination of waste, are often used in used in service organizations to reduce cycle time. Among the common actions typically taken are:

- Eliminate rework loops to correct process errors.
- Eliminate or simplify steps of marginal value to the customer.
- Eliminate redundant steps such as inspections or reviews (but not until the causes of errors have been determined and eliminated).
- Combine steps and have them done by one worker, by several workers in a work cell, or by a multifunctional team.

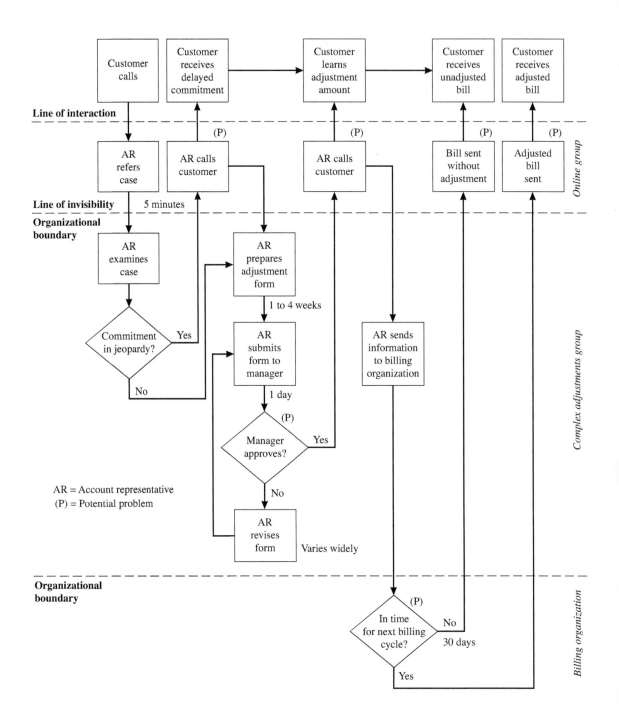

FIGURE 14.1
Service blueprint diagram. (*Reproduced with permission of AT&T.*)

- Transfer approval steps to lower levels.
- Change the sequence of activities from consecutive steps to simultaneous steps.
- Perform a step after serving a customer rather than before.
- Use technology to perform both routine and complex steps.

Marriott Hotels analyzed the elapsed time of the guest registration process. The analysis led to the integration of two subprocesses—the front-desk function and the bellperson function (to escort new guests to their room and carry their luggage). Under a revised process the front desk now preassigns a suitable guest room and assembles registration information and keys for the bellperson who greets and escorts the guests directly to their rooms (Hadley, 1995).

Error-Proofing the Process

An important element of prevention of errors is error-proofing the process. Five important principles of error-proofing a process are presented in Section 13.3 under "Error-Proofing the Process":

- Eliminate the error-prone activity.
- Substitute a more reliable process.
- Make the work easier for the worker.
- Detect errors earlier.
- Minimize the effect of errors.

In the service sector, examples of error-proofing include eliminating non-value-added activities, automating routine or unpleasant human tasks, using software to detect missing or grossly erroneous information, using bar codes at the supermarket checkout, using countdowns and checklists during surgical operations, entering medical prescriptions into a computer to replace handwritten prescriptions, and performing automatic built-in inspections. Obviously, technology is an important source of error-proofing aids.

Input from Information Technology as an Internal Supplier

Service industries need to plan, control, and improve the quality of inputs from external suppliers. The basic approach is provided in Chapter 15, "Managing the Supply Chain." The operations function within a service industry also has internal suppliers, and one of them, information technology (IT), deserves particular note.

Many of the significant advances in providing new service product features with amazing speed are due to the unending contributions of computers and related IT software and equipment.

These contributions have, unfortunately, been accompanied by significant problems in the services provided to operations people by the IT activity. Operations people report that these problems are sporadic and chronic quality-related problems that result in customer dissatisfaction and higher costs.

Research provides some understanding of the nature of these IT-related problems (Kittner et al., 1999). Operations managers from the financial services sector (but not from the IT or quality functions) were brought together to determine the perceived impact of the quality of IT services on the quality of the output of the operations function. The "operations" activities were process customer transactions, including both direct contact with external customers and backroom activities that did not involve direct contact.

First, the participants identified the outputs the IT department provided to operations as summarized in 11 categories in Table 14.2.

Next, participants were asked to identify problems involved in these services. The managers identified 115 issues—some were general, such as "downtime," and others were quite specific, such as "can't hire C++ programmers fast enough to keep up with user demand." These issues were further analyzed to identify 57 issues related to customer satisfaction. Participants were then asked to score each item on a scale of 1 to 10, where 10 was most important and 1 was least important. The mean scores of the top 10 issues are shown in Table 14.3.

In summary, Table 14.3 shows the perceptions of operations managers of problems with input received from IT and the relative importance of these problems to output of the operations function. Conclusion: Significant sporadic and chronic problems exist that cannot be solved by the traditional IT help desk. Chronic problems require the structured approach to quality improvement described in Chapter 4; sporadic problems require the troubleshooting approach described in Section 6.9 under "Troubleshooting."

TABLE 14.2
IT services provided to operations

Response category	Examples
Provide reports	Summary of sales activities, list of "at risk" accounts
Provide hardware and software support	Personal computer support services, guidance on hardware and purchasing
Record information	Posting to accounts, current rate information
Provide data communications support	Local area network connections and software access, data security
Provide for system availability and maintenance	Online availability of all systems, ensure e-mail systems are up and running
Provide online information	Automatic teller machine online, customer account information
Provide support for data processing	Download all accounts for customer information; download all workplace banking customers and their profiles
Develop new systems	Application development for new systems; systems development, testing, and design
Implement new systems	Assist in the development of testing plans and rollout, implementation of new systems
Provide programming support	Provide programming support
Provide training	Software training

TABLE 14.3
Top 10 quality issues

Issue	Mean
Accuracy of information	9.33
System downtime	8.92
System response time	8.58
Network reliability	8.42
System performance	8.33
Level of expertise	8.17
Hardware performance	8.09
Thorough testing—both IT and users need to be involved	8.08
Hardware problems	7.92
Lack of training	7.75

The demands continually made on the IT function for information services using current software and the development of new software make it almost impossible to address chronic problems effectively. But these problems often have a serious impact on external customer satisfaction. A quality department can help by taking a strong initiative to collect data to prove that action is necessary. Such proof can help to convince upper management to provide the IT and operations functions with the resources and a framework for improvement based on the project-by-project approach described in Chapter 4.

Input from Customers as an External Supplier

In service organizations, the customer is often a supplier. Customers may provide the detailed data for a service transaction, e.g., buy 100 shares of American Express stock at a price not to exceed $Y, order a dress of size 16 in the rose color, reserve a hotel room for March 12 to March 18. The customer input can be in error, resulting in delays to the customer and extra time and costs to the organization that must rectify the customer's error. *JQH5,* page 33.17, describes three approaches that financial service providers use to prevent customer errors: customer education, error-proofing, and monitoring and measuring customer input.

Measuring Process Capability

We need to ensure that the process can meet customer needs and goals under normal operating conditions. When customer needs can be quantified in particular parameters, then process capability indexes can be used to evaluate the process. In the six sigma approach, the process capability can be described in units of sigma, e.g., a process might be at a level of 4.8 sigma out of the ideal of 6 sigma. These matters are discussed in Section 20.15. An important quantitative parameter in the service sector is the time to complete a service transaction.

A preliminary measure of capability can be obtained by simply collecting a sample of data and comparing it to the specifications for a process. For example, consider the process for repairing ATM machines. An analysis concluded that the service

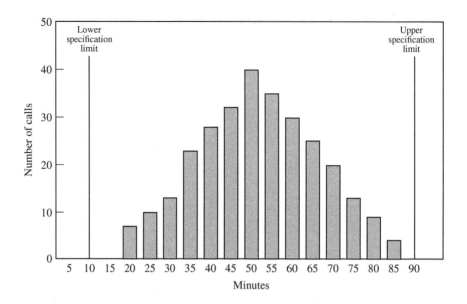

FIGURE 14.2
Time to repair machines. (*From Juran Institute, Inc.*)

organization should be able to respond in 10 minutes, and customers indicated that a machine should not be out of service for more than 90 minutes. A team looked at historical data measuring the repair time each day for a 6-month period. The results are shown in the histogram of Figure 14.2. Note that some observations exceeded the maximum time of 90 minutes. Thus the process is not capable. The process must be changed and tested again to verify that it is capable. This preliminary capability study is based on available data. A full study under controlled conditions should be conducted following the methods described in Section 20.10.

A broader approach to evaluating a service process measures four parameters: effectiveness (of output), efficiency, adaptability, and cycle time. For elaboration, see Section 7.6.

IBM defines five levels of process maturity. The highest level, level 1, designates a business process that operates at maximum effectiveness and efficiency and is a benchmark or leader; the lowest level, level 5, suggests a process that is ineffective and may have major deficiencies. Melan (1993) defines the specific criteria for each level. Criteria include organizational matters (e.g., a process owner) and technical matters (e.g., measurements of effectiveness and efficiency).

↬ 14.3 PLANNING FOR SELF-CONTROL

The concept of self-control was introduced in Section 6.3. Then, in Chapter 13, we applied the concept to manufacturing industries and provided checklists to evaluate manufacturing planning for quality. This chapter applies the concept to service industries and provides checklists to evaluate planning for quality in services.

For self-control, we must provide people with:

1. Knowledge of what they are supposed to do.
   - Clear and complete work procedures.
   - Clear and complete performance standards.
   - Adequate selection and training of personnel.
2. Knowledge of what they are actually doing (performance).
   - Adequate review of work.
   - Feedback of review results.
3. Ability and desire to regulate the process for minimum variation.
   - A process and job design capable of meeting quality objectives.
   - Process adjustments that will minimize variation.
   - Adequate worker training in adjusting the process.
   - Process maintenance to maintain the inherent process capability.
   - A strong quality culture and environment.

In designing for self-control, technology is of increasing help—for "supposed to do," information can be placed on line and kept up-to-date instantly; for "knowledge on how we are doing," we can now often provide instant feedback; for "regulating a process," mechanisms can be integrated into a process to provide adjustments when necessary.

Most service organizations do not adhere to the three elements and 10 subelements of self-control. Collins and Collins (1993) provide six examples from the manufacturing and service sectors illustrating problems that were originally blamed on people but were management controllable. Berry, Parasuraman, and Zeithaml (1994) identify 10 lessons learned in improving service quality. Three of these are service design (the real culprit is poor service system design), employee research (ask employees why service problems occur and what they need to do their jobs), and servant leadership (managers must serve by coaching, teaching, and listening to employees). Note that these three lessons are directly related to the concept of self-control.

The concept of self-control applies to the service sector—both to frontline direct customer contact operations and to backroom operations. Based on research with personnel in the financial services industry, Shirley and Gryna (1998) developed checklists for self-control that are presented here.

Criterion 1: Knowledge of "Supposed to Do"

Providing personnel in the service sector with the knowledge of what they are supposed to do is essential for self-control. The following checklist can help to evaluate this criterion.

Work procedures

1. Are job descriptions published, available, and up-to-date?
2. Do personnel know who their customers are? Have personnel and customers ever met?

3. Do personnel who perform the job have any impact on the formulation of the job procedure?
4. Are job techniques and terminology consistent with the background and training of personnel?
5. Do guides and aids (e.g., computer prompts) lead personnel to the next step in a job?
6. Do provisions exist to audit procedures periodically and make changes? Are changes communicated to all affected personnel?
7. Do provisions exist for deviations from corporate directives to meet local conditions?
8. Are procedures "reader friendly"?
9. Does supervision have a thorough knowledge of operations and provide assistance when problems arise?
10. Do procedures given to personnel fully apply to the job they do in practice?
11. Have personnel responsibilities been clearly defined in terms of decisions and actions?
12. Do personnel know what happens to their output in the next stage of operations and understand the consequences of not doing the job correctly?
13. If appropriate, is job rotation used?

Performance standards

14. Are formal job standards for quality and quantity needed? If yes, do they exist? Are they in written form?
15. Have personnel been told about the relative priority of quality versus quantity of output? Do personnel really understand the explanation?
16. Are job standards reviewed and changed when more tasks are added to a job?
17. Do personnel feel accountable for their output, or do they believe that shortcomings are not under their control?
18. Does information from a supervisor about doing a job always agree with information received from a higher level manager?

Training

19. Are personnel given an overview of the entire organization?
20. Is there regularly scheduled training to provide personnel with current information on customer needs and new technology?
21. Do personnel and their managers provide input to their training needs?
22. Does training include the why, not just the what?
23. Does the design of the training program consider the background of those to be trained?
24. Do the people doing the training provide enough detail? Do they know how to do the job?
25. Where appropriate, are personnel who are new to a job provided with mentors?

Criterion 2: Knowledge of Performance

For self-control, people must have the means of knowing whether their performance conforms to standard for product and process characteristics. The following checklist can help to evaluate this criterion.

Review of work

1. Are personnel provided with the time and instructions for performing self-review of their work?
2. Can errors be detected easily?
3. Are independent checks on quality needed? Are they performed? Are these checks performed by peer personnel or others?
4. Is a review of work performed at various checkpoints in process, not just when work is complete? Is the sample size sufficient?
5. Does an independent audit of an entire process ensure that individual work assignments are integrated to achieve process objectives?
6. Where appropriate, are detailed logs of customer contacts kept?

Feedback

7. Do upper management and supervision provide the same message and actions on the importance of quality versus quantity?
8. If needed, do standards exist for making corrections to output?
9. Where appropriate, is feedback provided to both individuals and a group of personnel? Is time provided for discussion with the supervisor, and does the discussion occur?
10. Is feedback provided to those who need it? Is it timely? Is it personnel specific?
11. Does feedback provide the level of detail needed particularly to correct problem areas? Have personnel been asked what detail is needed in the feedback?
12. Is feedback provided from customers (external or internal) to show the importance of the output and its quality?
13. Does feedback include information on both quality and quantity?
14. Are both positive and negative (corrective) feedback provided?
15. Is negative feedback given in private?
16. Do personnel receive a detailed report of errors by specific type of error?
17. Where appropriate, are reports prepared to describe trends in quality (in terms of specific errors)? Are such reports prepared for individual personnel and for an entire process performed by a group of people?
18. Are certain types of errors tracked with feedback from external customers? Could some of these be tracked with an internal early indicator?

A credit card provider has identified 18 key processes such as credit screening and payment processing. For the total of 18 processes, more than 100 internal and supplier process measures were identified. Daily and monthly performance results are available through video monitors and are also posted. Each morning, the head of operations meets with senior managers to discuss the latest results, identify problems, and propose solutions. Employees can access a summary of this meeting via telephone or electronic mail. The measurement system is linked to compensation by a daily bonus system that provides up to 12% of base salary for nonmanagers and 8 to 12% for managers (Davis et al., 1995).

Criterion 3: Ability to Regulate

The process given to personnel must be capable of meeting requirements, and the job design must include the necessary steps and authority for personnel to regulate the process. Here is a checklist to evaluate the ability to regulate.

Job design

1. Can the process (including procedures, equipment, software, etc.) given to personnel meet standards for quality and quantity of output? Has this capability been verified by trial under normal operating conditions?
2. Has the design of the job used the principles of error-proofing?
3. Does the job design minimize monotonous or unpleasant tasks?
4. Does the job design anticipate and minimize errors due to normal interruptions in the work cycle?
5. Can special checks be created (e.g., balancing of accounts) to detect errors?
6. Can steps be incorporated in data entry processes to reject incorrect entries?
7. Does the job design include provisions for action when wrong information is submitted or information is missing as an input to a job?
8. Is paperwork periodically examined, and are obsolete records destroyed to simplify working conditions?
9. When the volume of work changes significantly, do provisions exist to adjust individual responsibilities or add resources?
10. Do external factors (e.g., no account number on a check, cash received instead of a check) hinder the ability to perform a task?
11. Are enough personnel cross trained to provide an adequate supply of experienced personnel to fill in when needed?
12. If appropriate, is a "productive hour" scheduled each day in which phone calls and other interruptions are not allowed, thus providing time to be away from the work location to attend to other tasks?
13. Has equipment, including any software, been designed to be compatible with the abilities and limitations of personnel?
14. Does an adequate preventive maintenance program exist for computers and other equipment used by personnel?
15. Do some personnel possess a hidden knack that needs to be discovered and explained to all personnel?
16. For a job requiring special skills, have personnel been selected to ensure the best match of their skills and job requirements?

Changes in job design

17. Are proposed changes limited by technology (e.g., address fields on forms)?
18. Can personnel institute changes in a job when they show that the change will provide benefits? Are personnel encouraged to suggest changes?
19. What levels of management approval are required for instituting proposed changes? Could certain types of changes be identified that do not need any level of management approval?

20. Do management actions confirm that they are open to recommendations from all personnel?

Handling problems

21. Have personnel been provided with the time and training to identify problems, analyze problems, and develop solutions? Does this training include diagnostic training to look for patterns of errors and determine sources and causes?
22. Are personnel permitted to exceed process limits (e.g., maximum time on a customer phone call) if they believe it is necessary?
23. When personnel encounter an obstacle on a job, do they know where to seek assistance? Is the assistance conveniently available?

These checklists can help operations in the design (and redesign) of jobs to prevent errors, the diagnosis of quality problems on individual jobs, and the identification of common weaknesses in many jobs. The checklists can also help supervisors coach personnel, prepare for process audits, and develop training classes on quality. We proceed now from planning for quality in service operations to the control of quality in operations.

↦ 14.4 CONTROL OF QUALITY IN SERVICE OPERATIONS

The manufacturing sector has a long history of detailed procedures for controlling quality during operations. The formalization of such procedures in the service sector is still evolving. The basic steps for controlling quality are described in Chapter 6, "Control of Quality to Maintain Superior Performance." These steps are choose control subjects, establish measurement, establish standards of performance, measure actual performance, compare performance to standards, and take action on the difference. Applying these steps to service operations is discussed next.

Choose Control Subjects

To choose control subjects, we must first identify a major work process, identify the process objective, describe the process, identify customers, and discover customer needs. These steps, discussed in Chapter 6, lead to the selection of control subjects. Examples of control subjects in several service industries are given in Table 14.4.

Establish Measurement

In this step, we develop a unit of measure and the means of measurement (the sensor)—see Chapter 6. The unit of measure must be understandable to all, specific enough for decision making, and customer focused.

TABLE 14.4
Control subjects

Major work product	Major work process	Control subjects
Medical insurance	Claims processing	Accuracy of claim form
		Completeness of supporting documentation
Catering services	Food preparation	Freshness of ingredients
		Oven temperature
24-hour banking services	Maintenance of ATM machines	Availability of cash
		Number of service people available
Photo developing	Film processing	Maintenance of chemicals
		Accuracy of placement of film on spool

Source: Juran Institute, Inc.

Control subjects can be a mixture of features of the product, features of the process, and side effects of the process. Quantification of control subjects involves two kinds of indicators that must be explicit for those running the process:

1. *Performance indicators.* These indicators measure the output of the process and its conformance to customer needs, as defined by the unit of measure for the control subject.
2. *Process indicators.* These indicators measure activities or variation within the process that affect the performance indicators.

Table 14.5 shows 12 quality indicators at Federal Express along with the importance weights assigned to each indicator.

TABLE 14.5
Federal Express service quality indicators

Indicator	Weight
Abandoned calls	1
Complaints reopened	5
Damaged packages	10
International	1
Invoice adjustments requested	1
Lost packages	10
Missed pickups	10
Missing proofs of delivery	1
Overgoods (lost and found)	5
Right-day late deliveries	1
Traces	1
Wrong-day late deliveries	5

Source: American Management Association (1992).

These measures are tracked every day, both individually and in total.

Figure 14.3 shows how a division of AT&T related internal metrics to business processes and customer needs.

Specific measurements can be related to the underlying factors that dominate a process: setup, time, component, worker, and information. For elaboration, see Section 13.4 under "Control Systems and the Concept of Dominance."

Deyong and Case (1998) provide a methodology for linking customer satisfaction attributes with process metrics in service industries.

Establish Standards of Performance

This step involves setting target values for each performance indicator and each process indicator and also maximum and minimum limits.

	Business process	Customer need	Internal metric
Overall quality	Product (30%)	Reliability (40%)	% repair call
		Easy to use (20%)	% calls for help
		Features/functions (40%)	Function performance test
	Sales (30%)	Knowledge (30%)	Supervisor observations
		Response (25%)	% proposals made on time
		Follow-up (10%)	% follow-ups made
	Installation (10%)	Delivery interval (30%)	Average order interval
		Does not break (25%)	% repair reports
		Installed when promised (10%)	% installed on due date
	Repair (15%)	No repeat trouble (30%)	% repeat reports
		Fixed fast (25%)	Average speed of repair
		Kept informed (10%)	% customers informed
	Billing (15%)	Accuracy, no surprises (45%)	% billing inquiries
		Resolve on first call (35%)	% resolved first calls
		Easy to understand (10%)	% billing inquiries

FIGURE 14.3
Relating internal metrics to business processes and customer needs. From "Why Improving Quailty Doesn't Improve Quality (Or Whatever Happened to Marketing?)" by R. E. Kordupleski, et al., *California Management Review,* Spring, 1993, Vol. 35, No. 3. Copyright © 1993, by The Regents of the University of California. Reprinted from the *California Management Review,* Vol. 35, No. 3. By permission of The Regents.

Measure Actual Performance

Sometimes the measurement process can be automated. For example, the number of calls waiting to be answered in an insurance service center is clearly displayed; the elapsed time in filling a customer order at a fast-food franchise is clearly visible (the goal is 45 seconds).

In other cases, the measuring process is more complex. For example, to measure the quality of customer contact activities, banks and other organizations use "mystery shoppers"—researchers posing as customers to assess key dimensions of quality delivery.

Compare Performance to Standards

Sometimes the comparison must go beyond a simple analysis of average values. For example, in Figure 14.4, two banks are compared on the time to process a loan. Both banks average about 7 days. For bank A, almost all loans are processed in a 5- to 8-day window; bank B takes an average of 7 days, but some loans take about 21 days to process.

The comparison of actual to a target may use statistical process control techniques to distinguish between common and special causes. This approach is explained in Chapter 20, "Tools to Maintain Control of Quality."

Take Action on the Difference

Failure to meet standards may require one of three actions: troubleshooting, quality improvement, or quality planning. The steps for troubleshooting are similar to those for diagnosis in quality improvement. This process includes developing theories on

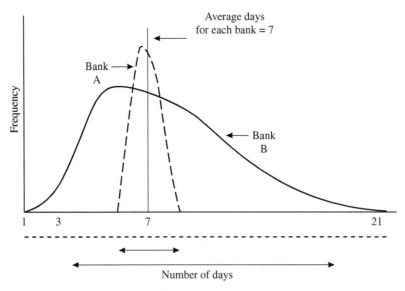

FIGURE 14.4
Time to process a loan. (*From Juran Institute, Inc.*)

possible causes and testing the theories (using data) to find the cause of the problem. The process for troubleshooting and diagnosing sporadic problems is usually simpler than the process for achieving improvement on chronic problems. For further discussion on the three types of action, see Section 13.11.

14.5 PROCESS QUALITY AUDITS

A *quality audit* is an independent review conducted to compare some aspect of quality performance with a standard for that performance. A discussion of various types of quality audits is presented in Chapter 16, "Quality Assurance and Audits."

A *process quality audit* includes any activity that can affect final product quality. This on-site audit is usually made on a specific process and uses process operating procedures. Adherence to existing procedures is emphasized, but audits often uncover situations of inadequate or nonexisting procedures. The self-control checklists presented earlier in this chapter can suggest useful specific subjects for process audits.

A major airline employs process audits to evaluate service in three areas: airport arrivals and departures, aircraft interior and exterior, and airport facilities. Forty-seven specific activities are audited periodically, and then performance measurements are made and compared with numerical goals. Two examples on the aircraft are the condition (appearance) of carpets inside the planes and the adhesion of paint on the planes.

McDonald's Corporation audits restaurants through a series of announced and unannounced visits. The audit includes quality, service, cleanliness, and sanitation. Highly detailed audit items include numerical standards on food-processing variables.

14.6 FRONTLINE CUSTOMER CONTACT

A basic activity in the service industries is the service encounter, i.e., the contact made with the client when meeting a customer's need. Typically, examples involve a bank teller processing deposits or withdrawals of money, a flight attendant providing services on an airplane, or a hotel clerk registering a guest. In all cases, the quality of the transaction involves both the technical adequacy of the result and the social skills of the "frontline" person who conducts the transaction. Three factors emerge as important: selection of the frontline employee, training of that employee, and "empowerment" of the employee to act to meet customer needs. (The reader should compare these factors to the three elements of self-control; see Section 6.3.)

In person-to-person type transactions, employee *selection* often has an immediate, direct, and lasting impact on customer perception. Some people have the necessary personal characteristics for frontline personnel; some people do *not* have these characteristics, even with training. Proper selection requires that we identify the personal characteristics required for a position (how about asking experienced employees?), use multiple interviews, train managers in interview procedures, identify nominees from among present employees, and ask for recommendations from present employees. As one example, Federal Express selects people using scientifically prepared profiles of successful performers.

Training, of course, is essential. Superior-quality service organizations devote from 1 to 5% of employee working hours to training. The content of the training depends on the job requirements, but often it stresses product knowledge. In addition, training involves activities such as role playing to handle situations when a transaction goes wrong and handling an irate customer who has a complaint. Such training takes time and special effort, e.g., at Lands' End, each mail-order telephone representative spends time in a warehouse viewing the products. Isn't it impressive when such a person is able to describe the color "medium gray" to a customer? Basically, however, the training must enable the employee to provide customers with dependable and regular service. Customers are impressed when they know that they can depend on uniformly good service from an organization. When an extraordinary situation arises and it is handled well, we have a delighted customer.

Empowerment, a key step beyond training, involves giving a new degree of authority to frontline employees. The term usually means encouraging employees to handle unusual situations that standard procedures don't cover. In the past, the employee would check with a superior—while the customer waited—and waited. The concept minimizes the use of the rule book and maximizes the use of the frontline employee's knowledge, initiative, and judgment to take the action necessary to meet the need of a customer standing at a service counter. For example, the policy manual at Nordstrom's department store states: "Use your own best judgment at all times." Risky? Some people would think so, but the opportunities for customer satisfaction and the employee attitude toward "ownership" of his or her job are strongly convincing. Section 9.7, under "Empowerment," provides further discussion.

Bank One Corp. uses an extensive shopper survey to measure the performance of bank tellers. The survey queries customers on specific contact items such as a friendly greeting, employee identification, eye contact, a smile, use of the customer's name during the transaction, undivided employee attention, accurate processing of the transaction, ability to give clear explanations, a professional appearance, and a neatly organized work area.

The investment in selection, training, and empowerment of employees can lead to the fulfillment of a company's dream—delighted customers. Table 14.6 provides some categories of action and examples of remarkable exploits to generate delighted customers.

The service encounter goes beyond "speed, content, and attitude." A research study of five service organizations and nine manufacturers identified 15 competencies that exemplify superior frontline service (Jeffrey, 1995). The top six competencies were considered the most important by the organizations and their customers:

1. *Build customer loyalty and confidence.* Meet customer needs and do what is sensible to maintain customer goodwill.
2. *Empathize with customers.* Be sensitive to customer feelings and show genuine concern and respect.
3. *Communicate effectively.* Be articulate and diplomatic.

TABLE 14.6
Some action taken to achieve delighted customers

Action	Example
Providing a service far beyond the scope of the company service	An airline flight attendant accompanied a sick passenger and her daughter to a hospital.
Providing a service beyond the call of a normal effort	To eliminate the sound of tinkling glass in a chandelier due to air conditioning, hotel employees removed every second piece of glass 30 minutes before the start of the meeting.
Providing extraordinary recognition of customer inconvenience	An automobile manufacturer paid a customer $985 for "lost time" incurred by the customer on excessive and improperly made repairs.
Providing a recognition of personal customer loss	A customer reported that she lost a pen, having sentimental value, in a grocery store. A clerk searched for the pen but with no success. The clerk presented the lady with three $20 gift certificates.

4. *Handle stress.* Stay organized and calm and show patience.
5. *Listen actively.* Interpret the meaning of the customer's words.
6. *Demonstrate mental alertness.* Process information quickly.

These competencies are not easy to achieve. Organizations must invest in careful selection of employees, training, and empowerment and then develop the employees into professionals to ensure retention of these key people.

Frontline personnel can serve as "listening posts" for an organization. When the listening posts are well designed to probe and ask specific questions, the information generated can help to develop new products and to sell existing products of which the customer is not aware. Some frontline personnel are uncomfortable asking certain questions that may lead to answers that the customer and the bank teller often prefer not to discuss, e.g., Why are you transferring your account to another bank? In some organizations, sales personnel and customer transaction personnel use every interaction as an opportunity to probe and ask specific questions. See Table 14.7. When the listening posts are well designed and the information they generate is well coordinated, their contribution can help to develop new products and to sell existing product that is already developed of which the customer is not aware.

USAA and American Express are companies that collect and analyze information generated in all types of customer contact. All written and telephone complaints,

TABLE 14.7
Questions to discover customer needs

- Is there anything else I can do for you?
- Are there any problems or needs today you have in managing your financial affairs?
- Are there any products or services you wish we offered?
- Are you aware of any products on the market that we don't offer?

compliments, and inquiries are collected, classified, and analyzed. The resulting information provides a clear indication of what is going well, what needs to be improved, and what new needs are emerging from customers.

Another way to gather customer information for future product development is to use customer management software to track every encounter with a customer.

Call centers and help desks are now a main mechanism for a service encounter. When the call concerns a customer problem, typically, resolution of the problem is not achieved with the first phone call reporting the problem. At one health maintenance organization, 36% of the calls were related to a lack of complete, accurate, and timely information between suppliers and customers. Cross (2000) discusses some of the issues involved.

14.7 ORGANIZING FOR QUALITY IN SERVICE OPERATIONS

As in all industries, top-down management understanding, appreciation, and support of any organizationwide quality program is mandatory. It all starts at the top. If only nodding approval or a "let's wait and see" attitude is perceived by the employees, a quality improvement program will fail. The employees watch what management does much more than what management says. A quality council made up of senior executives should be established to oversee the quality improvement program as soon as possible. The council should meet regularly—quarterly would be a desirable interval. Their interest and guidance to the program will speak loudly to the organization.

Teams are a common means of driving improvement in the service industries. For a summary of various types of teams, see Section 8.7.

Teams may be ad hoc to address a specific problem or may be permanent to be responsible for a specific activity. For example, the American Express Consumer Card Group uses semi-autonomous work teams. A team consists of 10 to 12 employees in the natural work group. Team members do customer service work, manage quality, inventory, and attendance; prepare work schedules; and prepare production reports and forecasts. Individual roles are defined to handle these team responsibilities. The team leader focuses on coaching, feedback, and special human resource issues.

One fast-food firm creates teams of crew members (workers at one location) who are trained to manage the site without a full-time manager (Harvard Business School, 1994). This approach means installing online sensor technology such as the time to prepare an order and providing crew members with the same operating and financial information provided to a restaurant general manager to run the site. Crew members make operating decisions such as ordering food materials. Thus knowledge that long separated "brain workers" from "hand workers" now resides in a computer on the operations floor.

The Ritz-Carlton Hotel company uses self-directed teams. These process teams are aligned with the way customers come in contact with the hotel: (1) prearrival team; (2) arrival, stayover, and departure team; (3) dining services team; (4) banquet services team; and (5) engineering and security team. In a self-directed work team, members may have specific individual roles, but the team shares accountability for meeting performance objectives.

TABLE 14.8
Examples of quality-in-daily-work projects

Department	Customer	Essential service	Service guarantee
Nursing unit 7-West	Ambulance emergency department	Pick up patients within specified time frame	The critical care unit staff will pick up the patient within 15 minutes of emergency room's call
Admitting	Elective surgery patients	Admission process including explanation of forms	Provide service within five minutes of entering department
Education	Diabetic patients	Outpatient diabetic education program	95% of all diabetic patients will be scheduled for diabetic education
Pediatrics	Member	Telephone access	98% of all calls will be handled within two minutes

Kaiser Permanente uses quality-in-daily-work teams. These frontline work teams focus on both work process improvements and the definition of a service guarantee to communicate service performance levels. The projects span across departments and include both clinical and support services. Table 14.8 shows a few examples (Centano et al., 1995).

For a discussion of research conducted on teams, see Katzenbach and Smith (1993). Mann (1994) explains how managers need to develop skills as coaches, developers, and managers of activities that reside in different departments ("boundary managers").

→ 14.8 SIX SIGMA PROJECTS IN SERVICE INDUSTRIES

The six sigma approach to improvement includes the phases of define, measure, analyze, improve, and control. These phases are explained in Section 4.5. Six sigma is increasingly being applied in the service sector. Table 14.9 shows examples of six sigma projects at American Express. Note the wide variety of projects in Table 14.9.

Hahn et al. (2000) describe how GE Capital applied six sigma (including a modified full-factorial design of experiment) to reduce losses due to delinquent credit card customers. Bott et al. (2000) explain six sigma steps for two projects at American Express. Bottome and Chua (2005) apply six sigma and lean to reduce documentation errors at Genentech (also see Section 13.12). Alonso, Gregorio, and Ansorena (2003) describe the implementation of six sigma at Telefonica de España (Telefonica), Spain's leading telecommunications company. Six sigma is deployed as part of a far-reaching corporate culture change program, focused on customer satisfaction and quality as key differentiators.

TABLE 14.9
Examples of six sigma projects at a financial services organization

Reduce cycle time of credit issuance
Reduce trading-call length
Reduce card issuance cycle time
Reduce encoding errors
Improve business travel contract-to-billing process
Eliminate nonreceived renewals
Eliminate incorrect fee adjustments
Improve XXX payment accuracy
Reduce YYY writeoffs

14.9 QUALITY MEASUREMENT IN SERVICE OPERATIONS

The management of key work processes must include provisions for measuring process control and also for monitoring overall operations. Readers should review the basics of quality measurement in Section 6.2.

Measuring quality in service-based industries is often perceived as more difficult than in manufacturing. Modern manufacturing is old, with many scholars tracing its roots to the 18th century. By its nature, manufacturing has always relied on measurements, quantities, weights, and so on. Measurements seem to be quantifiable and intrinsic to manufacturing. Service-based industries, however, have often relied on qualitative measurements such as good, better, best and economy, deluxe, and luxury.

One approach to measuring service quality is the SERVQUAL model (Zeithaml et al., 1990). This model identifies five dimensions of quality:

- *Tangibles:* appearance of facilities, equipment, personnel, materials.
- *Reliability:* ability to perform dependably and accurately.
- *Responsiveness:* provide timely service.
- *Assurance:* trust and confidence in employees.
- *Empathy:* individualized attention to customers.

Research by Cronin and Taylor (1992) confirms that service quality should be measured differently among service industries, specifically between high-involvement service providers (health care) and low-involvement providers (fast food).

Table 14.10 shows a few examples of quality measurements of overall operations in a credit card company. This organization can easily calculate the measures from basic raw data that are routinely collected in daily operations.

King and Dickinson (1996) present a framework (Figure 14.5) of three measures of process effectiveness that are particularly suited for service processes.

Results measures are primarily customer perceptions of outcomes. These measures drive priorities for improvement and monitor performance improvement. Overview in-process measures predict results and are lead indicators of process outcomes. Changes in an overview measure will thus lead to changes in the related results measure(s). Overview measures help to initiate the search for root causes of poor performance.

TABLE 14.10
Quality measures in a credit card company

Measure number	Quality measure
148	Number of abandoned calls/total calls
454	Average time to call pickup, seconds
458	Number of statement insertion errors
460	Number of applications not processed within standard
466	Number of hours credit card system down
467	Number of payments not posted

Source: JQH5, page 33.21.

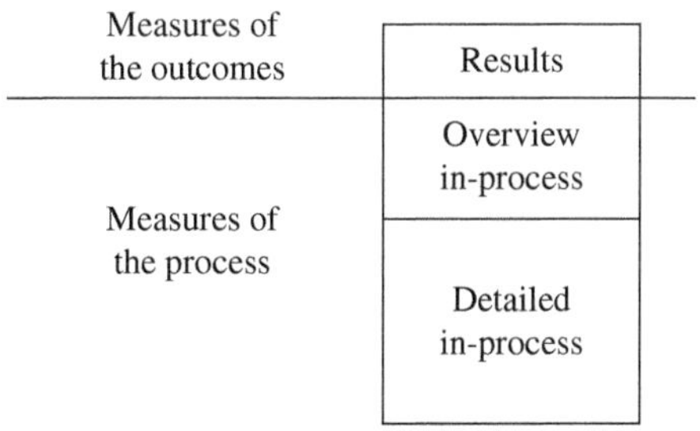

Measures of the outcomes	Results
Measures of the process	Overview in-process
	Detailed in-process

FIGURE 14.5
Framework of process effectiveness measures. From "A Framework for Process Effectiveness Measures" by M. N. King and T. Dickinson, *Quality Engineering,* Vol. 9, No. 1, September 1996, pp. 45–50. Reprinted with permission from *Quality Engineering,* © 1996 American Society for Quality.

Detailed in-process measures predict overview measures and are lead indicators of subprocess outcomes. Detailed in-process measures control the day-to-day operation of the process and provide early warnings and diagnostic information for improvement. Thus the approach forms a hierarchy of process control measures—exterior and interior to the organization and at different levels of a process.

As an example, consider the process of having a new phone installed at an arranged time. The results measure is the proportion of customers who report that the technician arrived on time. The overview in-process measure is also the proportion of appointments met, but this measure is logged by the technicians. Thus the results and the overview measures in this case are the same measure but collected both within and outside the process. The detailed in-process measures are many and might include proportion of customers who could not be given appointments at the time they requested or incidences and reasons for staff shortages.

Quality measurements are candidates for data analysis using statistical techniques such as the control charts discussed in Chapter 20. But the more basic point is that the reporting of data demonstrates to operating personnel that management regards quality as a high priority.

Figure 14.6 presents a flowchart for defining, collecting, and analyzing metrics in a service organization.

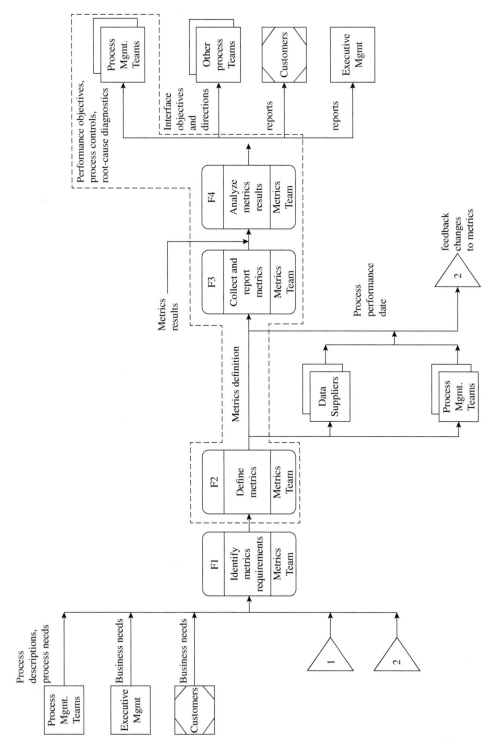

FIGURE 14.6
Flowchart for defining, collecting, and analyzing metrics in a service organization.

The Problem of "No Data!"

A frequent first lament of someone starting a quality program in a service industry is, "There are no data!" Although it is true in some instances that there are little useful or accurate data when starting out on a quality program, there often are more data than we think. Approach the information technology (IT) department of the organization early in a quality improvement program. It often has vast quantities of data that are routinely collected but not disseminated. A quick query of existing databases may yield a surprising quantity of data. The veracity of all data should be checked before using or relying on analysis of the data.

If there truly are little or no data available or if the data are corrupt and unreliable, then start gathering new data that will be useful. As with all change, suddenly starting to gather data where none had been gathered before can be alarming to employees or customers or both. There is a natural fear of change of being measured when one has not been measured in the past. Appropriate introduction of change methodologies should be used prior to gathering new data to allay such fears. If this step is skipped and misunderstanding of the real purpose of gathering the data exists, it can doom a quality program from the start. Too many individuals taking time out to just gather data or explain the purpose of gathering the data or the value of a quality system may seem a waste of time. The eagerness of the starting moment may be hard to resist. Keep in mind the old saying that applies here well. "Go slow to go fast." This saying is applicable to quality improvement efforts. This does not mean to imply that you should drag your feet or that implementation of a quality program is slow. It simply means to portray the obvious notion that to do things poorly fast is not improvement.

↛ 14.10 MAINTAINING A FOCUS ON CONTINUOUS IMPROVEMENT

Operations personnel in the service sector (and the manufacturing sector) are involved in addressing sporadic problems (fire drills) and also chronic problems. The actions required include troubleshooting, quality improvement, and quality planning. Maintaining the focus on improvement clearly requires a positive quality culture in an organization.

The broad approach to the three actions and the key elements of a positive quality culture are summarized in Section 13.11. A detailed discussion of each subject is provided in other chapters of this book.

Finally, the operations function must be provided with the support to maintain the focus on improvement. The quality department should regard operations as a key internal customer and provide training and technical quality expertise to support operations. In addition, operations managers must be guided in reviewing the mountain of reports they receive to identify and prioritize quality problems and set up the teams and other mechanisms to address those problems. Also, the quality department can urge upper management to set up cross-functional teams to address operations problems that may be caused by other functional departments such as IT (see Section 14.2), marketing, and purchasing.

SUMMARY

- Activities to integrate quality in service planning have two objectives: to incorporate product features and to prevent defects (minimize variability).
- By creating a flow diagram, we can dissect a service process and plan for quality at each workstation.
- Error-proofing a process is an important element of prevention.
- Information technology is an important internal supplier for the service operations process.
- Process capability can be measured for a service process.
- For self-control, we must provide personnel with the knowledge of what they are supposed to do, knowledge of what they are actually doing, and a process that can meet specifications and can be regulated.
- Failure to meet all three of these criteria means that the quality problem is management controllable. About 80% of quality problems are management controllable.
- The basic steps for controlling quality (Chapter 6) can be applied to service operations.
- Process quality audits apply to any activity that can affect final service quality.
- Three factors are important in frontline customer contact: selection, training, and empowerment of personnel.
- Various types of quality teams play a key role in service processes.
- Six sigma applies to service processes.
- Quality measures must be designed for each type of service organization.

PROBLEMS

14.1. For a specific service activity with which you are familiar, create a flow process diagram and indicate current or potential problem points.

14.2. For a specific service activity with which you are familiar, describe how you would conduct a controllability study to determine whether the quality problems are primarily management controllable or employee controllable.

14.3. Select a service task regularly performed by an employee. Use the checklists on self-control to determine whether the employee can be held responsible for the quality of the output.

14.4. For a service process with which you are familiar, define two control subjects and at least one measurement for each control subject.

14.5. Your work team performs blood tests that physicians request for their patients. Identify one control subject that will help manage the work process while meeting customer needs. Then define a unit of measure and a sensor for the control subject.

14.6. For a service process with which you are familiar, determine the major form of "dominance" affecting attainment of quality.

REFERENCES

Alonso, F., G. del Rey, and R. de Ansorena (2003). "How Telefonica Makes Its Management Connections," *European Quality,* vol. 8, no. 6, pp. 4–10.

American Management Association (1992). *Blueprints for Service Quality,* American Management Association, New York, pp. 51–64.

AT&T Quality Steering Committee (1990). *Achieving Customer Satisfaction,* AT&T Customer Information Center, Indianapolis, IN.

Berry, L. L., A. Parasuraman, and V. A. Zeithmal (1994). "Improving Service Quality in America: Lessons Learned," *Academy of Management Executive,* vol. 8, no. 2, pp. 32–52.

Bott, C., E. Keim, S. Kim, and L. Palser (2000). "Service Quality Six Sigma Case Studies," *Annual Quality Congress Proceedings,* ASQ, Milwaukee, pp. 225–231.

Bottome, R. and R. C. H. Chua (2005). "Genentech Error-Proofs Its Batch Records," *Quality Progress,* July 2005, pp. 25–34.

Collins, W. H. and C. B. Collins (1993). "Differentiating System and Execution Problems," *Quality Progress,* February, pp. 59–62.

Cronin, J. Joseph, Jr., and Taylor, Steven A. (1992). "Measuring Service Quality: A Reexamination and Extension," *Journal of Marketing,* July, pp. 55–68.

Cross, K. F. (2000). "Call Resolution: The Wrong Focus for Service Quality," *Quality Progress,* February, pp. 64–67.

Davis, R., S. Rosegrant, and M. Watkins (1995). "Managing the Link between Measurement and Compensation," *Quality Progress,* February, pp. 101–106.

Deyong, C. F. and K. E. Case (1998). "Linking Customer Satisfaction Attributes with Progress Metrics in Service Industries," *Quality Management Journal,* vol. 5, no. 2, pp. 76–90.

Hadley, H. (1995). Private communication to Patrick Mene, Marriott Hotels and Resorts, Washington, DC.

Hahn, G. J., N. Doganaksoy, and R. Hoerl (2000). "The Evolution of Six Sigma," *Quality Engineering,* vol. 12, no. 3, pp. 317–326.

Harvard Business School (1994). *Case 9-694-076, Taco Bell,* Boston.

Jeffrey, J. R. (1995). "Preparing the Front Line," *Quality Progress,* February, pp. 79–82.

Katzenbach, J. R. and D. K. Smith (1993). *Wisdom of Teams: Creating the High Performance Organization,* Harvard Business School Press, Boston.

King, M. N. and T. Dickinson (1996). "A Framework for Process Effectiveness Measures," *Quality Engineering,* vol. 9, no. 1, pp. 45–50.

Kittner, M., M. Jeffries, and F. M. Gryna (1999). "Operational Quality Issues in the Financial Sector: An Exploratory Study on Perception and Prescription for Information Technology," *Journal of Information Technology Management,* vol. X, nos. 1–2, pp. 29–39.

Kordupleski, R. E., R. T. Rust, and A. J. Zahorik (1993). "Why Improving Quality Doesn't Improve Quality (Or Whatever Happened to Marketing?)," *California Management Review,* Spring, vol. 35, no. 3, pp. 82–95.

Mann, D. W. (1994). "Re-engineering the Manager's Role," ASQC Quality Congress Transactions, Milwaukee, pp. 155–159.

Melan, E. H. (1993). *Process Management,* McGraw-Hill, New York.

Shirley, B. M. and F. M. Gryna (1998). "Work Design for Self-Control in Financial Services," *Quality Progress,* May, pp. 67–71.

Zeithaml, V. A., A. Parasuraman, and Leonard L. Berry (1990). *Delivering Service Quality,* Free Press, New York.

CHAPTER 15

Managing the Supply Chain

→ 15.1 SUPPLIER RELATIONS—A REVOLUTION

This step on the spiral of quality concerns the purchase of goods or services from suppliers, or vendors.

For many companies, purchases account for more than 60% of the sales dollar and are the source of over half of the quality problems. Poor quality of supplier items results in extra costs for the purchaser, e.g., for one appliance manufacturer, 75% of all warranty claims were traced to components purchased for the appliances.

Current emphasis on inventory reduction provides a further focus on quality. Under the just-in-time inventory concept, goods are received from suppliers only in the quantity and at the time that they are needed for production. The buyer stocks no inventories. If a portion of the purchased product is defective, production at the buyer's plant is disrupted because of the lack of backup inventory. With conventional purchasing, supplier quality problems can be hidden by excess inventory; with the just-in-time concept, purchased product must meet quality requirements.

The interdependence of buyers and suppliers has increased dramatically. Sometimes the interdependence takes the form of integrated facilities, e.g., a can manufacturer locates next door to a brewery. Sometimes technological skills are involved, e.g., an automobile manufacturer asks a supplier to propose a design for a purchased item. The supplier becomes an extension of the buyer's organization—a virtual department.

These circumstances have led to a revolution in the relationship between buyers and suppliers. In the past, the parties were often adversaries; some purchasers viewed suppliers as potential criminals who might try to sneak some defective product past the purchaser's incoming inspection. Today, the key phrase is *partnership alliance*, working closely together for the mutual benefit of both parties.

This new view of supplier relations requires changing the purchasing process from a traditional view to a strategic view. An overview of some changes is shown in Table 15.1. For elaboration, see *JQH5,* page 21.6.

Part of the revolution in supplier relations is the expansion of the traditional supplier concept to the broader supply chain concept (see Figure 15.1). Donovan and Maresca (in *JQH5*) define the supply chain as the tasks, activities, events, processes, and interactions undertaken by *all* suppliers and *all* end users in the development, procurement, production, delivery, and consumption of a specific good or service. Note that this definition includes end users, prime suppliers or distributors, and multiple tiers of suppliers to prime manufacturing or service organizations. Supply chain management is reserved for items that are of strategic importance to an organization.

The purchasing function has the primary role of managing the supply chain to achieve high quality and value throughout the supply chain. Admittedly, this ideal is lofty, but it highlights a new focus—*from managing purchasing transactions and troubleshooting to managing processes and supplier relationships.* Under supply chain management, mechanisms must be put in place to ensure adequate linkages among parties in the supply chain. Such mechanisms include clear contractual requirements and continuous feedback and communication. For further discussion of the supply chain concept, see *JQH5,* pages 21.4–21.9. Of course, managing supply chains is difficult. Fisher (1997) describes some of these difficulties and suggests a framework for supply chains based on the nature of the product demand and whether the products are primarily functional or primarily innovative. Supply chains also apply to the service sector. For an example involving physicians in a hospital and health care system, see Zimmerli (1996). This example also illustrates the use of internal customer surveys for the major processes of distribution, purchasing, and sterile processing.

TABLE 15.1
Traditional versus strategic view of the purchasing process

Aspect in the purchasing process	Traditional view	Strategic view
Supplier relationship	Adversarial, competitive, distrusting	Cooperative, partnership, based on trust
Length of relationship	Short term	Long term; indefinite
Quality assurance	Inspection upon receipt	No incoming inspection necessary
Supplier base	Many suppliers, managed in aggregate	Few suppliers, carefully selected and managed
Purchasing business plans	Independent of end-user organization business plans	Integrated with end-user organization business plans
Focus of purchasing decisions	Price	Total cost of ownership

Source: Adapted from *JQH5,* p. 21.6.

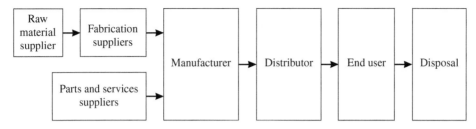

FIGURE 15.1
Elements of a supply chain. (*Source: JQH5, p. 21.4.*)

Much has been written about the importance of developing trust to replace the adversarial relationships of the past. Significant progress has been made, but in practice, suppliers (both large and small) still report pockets of arrogance exhibited by some purchasers.

✈ 15.2 SCOPE OF ACTIVITIES FOR SUPPLIER QUALITY

A purchasing system includes three key activities: specification of requirements, selection of a supplier, and supply chain management. The overall quality objective is to meet the needs of the purchaser (and the ultimate user) with a minimum of incoming inspection or later corrective action; this objective in turn leads to minimizing overall cost.

To achieve this quality objective, certain primary activities must be identified and responsibilities assigned. Table 15.2 shows a typical list of responsibilities assigned in one company. These activities are discussed in this chapter. Further elaboration is provided in *JQH5,* Section 21.

The responsibility matrix in Table 15.2 shows that the quality department has the principal responsibility for many supplier quality activities. Under an alternative policy, the purchasing department has the principal responsibility for quality, whereas others (e.g., product development and quality) have collateral responsibility. Such a shift in responsibility places a stronger focus on quality in setting priorities for delivery schedules, price, and quality. To meet this responsibility, most purchasing departments would need to supplement their technical capabilities. Some organizations have met that need by transferring technical specialists into the purchasing department.

To reflect a broad view of suppliers and the supply chain concept, some organizations are shifting from a function-based organization for purchasing transactions to a process-based organization for managing the supply chain (see *JQH5,* page 21.10). The process-based organization (see Chapter 7, "Business Process Management") uses a cross-functional team and process owner to focus on the total cost of ownership (rather than on the initial price of a purchased item), to identify opportunities for increased value, and to achieve a competitive advantage.

TABLE 15.2
Responsibility matrix—supplier relations

Activity	Participating departments		
	Product development	Purchasing	Quality
Defining product and program quality requirements	××		×
Evaluating alternative suppliers	×	×	××
Selecting suppliers		××	
Conducting joint quality planning	×		××
Cooperating with the supplier during the execution of the contract	×	×	××
Obtaining proof of conformance to requirements	×		××
Certifying qualified suppliers	×	×	××
Conducting quality improvement programs as required	×	×	××
Creating and utilizing supplier quality ratings		××	×

Note: ××, principal responsibility; ×, collateral responsibility.

Next we examine how quality relates to the three key activities of supply management: specification of requirements, selection of suppliers, and management of the supply chain.

➔ 15.3 SPECIFICATION OF QUALITY REQUIREMENTS FOR SUPPLIERS

Goals and requirements for suppliers must be aligned with those for each link in the supply chain, particularly the end user and the purchasing organization. These goals and requirements include quality parameters and general business issues.

For modern products, quality planning starts before a contract is signed. Such planning must recognize two issues:

1. The buyer must transmit to the supplier a full understanding of the use to be made of the product. Communicating usage requirements can be difficult even for a simple product.
2. The buyer must obtain information to be sure that the supplier can provide a product that meets all fitness-for-use requirements.

The complexity of many modern products makes it difficult to communicate usage needs to a supplier in a specification. Not only are the field usage conditions of a complex product sometimes poorly known, but the internal environments surrounding a particular component may not be known until the complete product is designed and tested. For example, specifying accurate temperature and vibration requirements to a supplier of an electric component may not be feasible until the complete system is developed. Such cases require, at the least, continuous cooperation between supplier and buyer. In special cases, it may be necessary to award separate development and production contracts to discover how to finalize requirements.

Circumstances may require two kinds of specifications:

1. Specifications defining the product requirements.
2. Specifications defining the quality-related activities expected of the supplier, i.e., the supplier's quality system.

Definition of Numerical Quality and Reliability Requirements for Lots

Beyond the quality and reliability requirements imposed on individual units or products, there is usually a need for added numerical criteria to judge conformance of *lots* of products.

These criteria are typically needed in acceptance sampling procedures, which makes it possible to accept or reject an entire lot of product based on the inspection and test result of a random sample from the lot. The application of sampling procedures is facilitated if lot quality requirements are defined in numerical terms. Examples of numerical indexes are shown in Table 15.3.

The selection of numerical values for these criteria depends on several factors and also on probability considerations. These criteria are also a means of indexing

TABLE 15.3
Forms of numerical sampling criteria

Quality index	Meaning	Typical values, %	Common misinterpretation
Parts per million (ppm)	Number of defects per million items	5–1000	—
Acceptable quality level (AQL)*	Percentage defective that has a high probability (say ≥ .90) of being accepted by the sampling plan	0.01–10.0	All accepted lots are at least as good as the AQL; all rejected lots are worse than the AQL
Lot tolerance percentage defective (LTPD)	Percentage defective that has a low probability (say ≤ .10) of being accepted by the sampling plan	0.5–10.0	All lots better than the LTPD will be accepted; all lots worse than the LTPD will be rejected
Average outgoing quality limit (AOQL)	Worse average percentage defective over many lots after sampling inspection has been performed and rejected lots 100% inspected	0.1–10.0	All accepted lots are at least as good as the AOQL; all rejected lots are worse than the AOQL

*Some sampling tables and other sources define AQL as the maximum percentage defective considered satisfactory as a process average.

sampling plans developed from statistical concepts. Unfortunately, many suppliers do not understand the statistical concepts and make incorrect interpretations of the quality-level requirement and also the results of sampling inspection (see Table 15.3). These criteria can also be a source of confusion in product liability discussions. Suppliers must understand that *all* product submitted is expected to meet specifications.

For complex and/or time-oriented products, numerical reliability requirements can be defined in supplier purchasing documents. Sometimes such requirements are stated in terms of mean time between failures. Numerical reliability requirements can help to clarify what a customer means by "high reliability."

> **EXAMPLE 15.1.** A capacitor manufacturer requested bids on a unit of manufacturing equipment that was to perform several manufacturing operations. Reliability of the equipment was important to maintaining production schedules, so a numerical requirement on "mean time between jams" (MTBJ) was specified to prospective bidders. (Previously, reliability had not been treated quantitatively. Equipment manufacturers had always promised high reliability, but results had been disappointing.) After several rounds of discussion with bidders, the manufacturer concluded that the desired level of reliability was unrealistic if the machine were to perform several operations. The capacitor manufacturer finally decided to revise the requirement for several operations and thereby reduce its complexity. The effort to specify a numerical requirement in the procurement document forced a clear understanding of reliability. Suppliers can also be required to demonstrate, by test, specified levels of reliability.

Definition of the Supplier Quality System

The second type of specification is a departure from the traditional practice of not telling a supplier how to run his or her plant. Defining required activities within a supplier's plant is sometimes necessary to ensure that a supplier has the expertise to conduct the full program needed for a satisfactory product. For some products, government regulations require that a buyer impose certain processing requirements (e.g., sanitary conditions for manufacturing pharmaceutical products) on suppliers. For other products, such as a complex mechanical or electronic subsystem, the overall system requirements may result in a need for a supplier to meet a numerical reliability or maintainability requirement and to conduct certain activities to ensure that such requirements are met (see Chapter 12). For still other products, suppliers are required to use statistical process control techniques on selected product characteristics or process parameters. Documents such as the ISO 9000 series and TS-16949 (formerly QS 9000), which define the elements of quality programs, can be cited as requirements in a contract with a supplier.

> **EXAMPLE 15.2.** Several suppliers were asked to submit bids on a battery needed in a space program. They were given a numerical reliability goal and asked to include in their bid proposal a description of the reliability activities that would be conducted to help meet the goal. Most of the prospective suppliers included a reliability program consisting of appropriate reliability activities for a battery. However, one supplier apparently had no expertise in formal reliability methodology and submitted a surprising write-up.

That supplier made a word-for-word copy of a reliability program write-up previously published for a missile system (the word *battery* was substituted for *missile*). This led to a suspicion, later confirmed, that the supplier knew little about reliability programs.

For complex products for which a supplier is asked to design and manufacture a product, the supplier can be required to include in the proposal a preliminary reliability prediction; a failure mode, effect, and criticality analysis; a reliability test plan; or other reliability analyses (see Chapter 12). The supplier's response provides some assurance on the design concept and also shows that the supplier has the reliability expertise to conduct the program and has included the funds and schedule time in the proposal.

We proceed next to the selection of suppliers.

➔ 15.4 SUPPLIER SELECTION; OUTSOURCING

Should we make or buy? This decision requires an analysis of factors such as the skills and facilities needed, available internal capacity, ability to meet delivery schedules, expected costs of making or buying, and other matters. This question brings us to the issue of outsourcing.

Outsourcing

Outsourcing is the process of subcontracting to a supplier external to the organization an activity that is currently conducted in house. Outsourcing is undertaken to reduce costs (the primary impetus), reduce cycle time, or improve quality. Estimates suggest that at least 85% of major corporations now outsource at least some activities. A trade association, the Outsourcing Institute, now exists.

But what activities should be outsourced? One principle holds that outsourcing should be confined to activities that are required but do not provide a competitive advantage, e.g., security, facility maintenance, administration of health benefits. Activities that are strategic and involve core competencies should not be outsourced. In practice, some organizations outsource significant (core) functional activities such as customer service, marketing, product design, and information technology.

Many important business issues enter into decisions about outsourcing. Bettis et al. (1992) offer important cautions about outsourcing core activities such as design and manufacturing. The more that outsourcing results in a supplier obtaining technical knowledge and market knowledge, the higher the risks to the company doing the outsourcing.

Outsourcing reduces internal costs by reducing personnel because the outsourcer (supplier) companies have the technology and knowledge to perform certain tasks more efficiently than some companies can internally. But there can be a serious impact on product quality if the supplier does not assign a high priority to quality. Outsourcing can also undermine employee morale and loyalty by creating fear that other activities will also be outsourced, resulting in a further loss of jobs. Some forward-looking companies like Eastman Chemical follow a policy that, if an activity is outsourced, no one will lose a job—people are retrained to take the positions of those who are normally retiring. Outsourcing also assumes that a capable supplier can be found and that adequate monitoring of the contract will ensure high quality. Sometimes these issues are glossed over in the zeal to reduce costs, and the result can be significant quality problems in the purchased items. Peterson (1998) analyzes some potential shortcomings when dealing with contract manufacturers. Bossert (1994) provides a checklist of 11 elements (e.g., inspection instructions, sufficient manufacturing controls) to compare contract manufacturing services.

Sharman (2002) discusses how the Internet is a driver of outsourcing and is accelerating supply chain trends. Ericson (2003) identifies six factors to help decide what and when to outsource. Meseck (2004) provides risk management strategies to help identify appropriate outsourcing opportunities and calculate the financial impact of global outsourcing. Weidenbaum (2004) discusses the high-level pros and cons of outsourcing, including when to and why to outsource and how it affects the United States. Hussey and Jenster (2003) provide an in-depth discussion of the types of outsourcing and the management issues associated with each from the supplier viewpoint. Ramachandran and Voleti (2004) discuss outsourcing from the perspective of suppliers in India, together with success factors and the need for suppliers to manage both growth and consolidation. Both Soliman (2003) and Weerakkody et al. (2003) discuss global outsourcing of Application Service Providers (ASPs). Karmaker (2004) argues that the offshoring issue should be one of competitiveness, not job loss. He examines current changes in the service industry and recommends strategies to realign, redesign, and restructure. Clott (2004) provides background facts and figures on outsourcing, its impact on work with emphasis on IT. He also discusses proponent and dissenter views and briefly discusses labor and ethics factors. Lee (2004) argues that the best supply chains are not just fast and cost-effective; they are agile and adaptable, and they ensure that all their companies' interests stay aligned.

Once skills are lost through outsourcing, it is difficult to reverse the process if later events require that the activity be brought back into an organization. This situation could be devastating for activities such as product design and selected operations.

Clearly, outsourcing can be a sensible, viable business decision—after all, the subcontracting of selected manufacturing activities has been a part of manufacturing history. But the desire for cost reduction may be taking outsourcing too far. Perhaps we should first study the entire activity to be outsourced as a quality improvement project using the road map provided in Chapter 4, "Improving Quality While Decreasing Cost." Thus, suppose customer service is a candidate for outsourcing. The *process* of customer service would be studied for both effectiveness and efficiency, and the necessary internal changes would be made. The result (quality and costs) could then be compared to those of outside suppliers. Some organizations even set up the activity under question as a separate profit center (to compete against outside suppliers) to spur internal improvement—the threat of job loss is a powerful spur.

To sum up on a sensitive issue: Outsourcing can provide superior quality and lower costs for an activity that a company cannot easily develop and maintain on its own, e.g., information technology. Outsourcing can also enable a company to focus resources on the core competencies that are important for competitive advantage, e.g., product design, operations, marketing. But these core activities vary by organization. The authors believe that the core competencies must be carefully identified within each organization, and once identified they should be performed internally and not be outsourced.

Multiple Suppliers versus Single Source

Multiple sources of supply have advantages. Competition can result in better quality, lower costs, better service, and minimum disruption of supply from strikes or other catastrophes.

A single source of supply also has advantages. The size of the contract given to a single source will be larger than that with multiple sources, and the supplier will attach more significance to the contract. With a single source, communications are simplified and more time is available for working closely with the supplier. The most dramatic examples of single sources are multidivisional companies in which some divisions are suppliers to others.

A clear trend has emerged. Organizations are significantly reducing the number of multiple suppliers. Since about 1980, reductions of 50 to 70% in the supplier base have become common. This trend does *not* necessarily mean that businesses are going to single sources for all purchases; it *does* mean a single source for some purchases and fewer multiple suppliers for other purchases. Working with a smaller number of suppliers helps to achieve useful partnerships by providing the time and skills necessary to facilitate in-depth cooperation. The forms of cooperation are discussed later in this chapter.

Whether a single source or multiple suppliers, selection must be based on the reputation of the supplier, qualification tests of the supplier's design, survey of the supplier's manufacturing facility, and information from data banks and other sources on supplier quality.

✥ 15.5 ASSESSMENT OF SUPPLIER CAPABILITY

Evaluating supplier quality capability involves one or both of the following actions:

1. Qualifying the supplier's design through the evaluation of product samples.
2. Qualifying the supplier's capability to meet quality requirements on production lots, i.e., the supplier's quality system.

Qualifying the Supplier's Design

In some cases, the supplier is asked to create a new design to meet the functions desired by the purchaser. In these cases, the supplier makes samples based on the proposed design. (Such samples are often made in an engineering model shop because a manufacturing process for the new design has not yet been created.) The samples are tested (the "qualification test") either by the purchaser or by the supplier, who then submits the results to the purchaser. Qualification test results are often rejected. Two reasons are common: (1) The test results show that the design does not provide the product functions desired, or (2) the test procedure is not adequate to evaluate the performance of the product. Such rejections (and ensuing delays in shipments) can be prevented by starting with a rigorous definition of product requirements and by requiring an approval of the test procedure before the tests commence.

Qualification test results do show whether the supplier has created a design that meets the performance requirements; such test results do *not* show whether the supplier is capable of manufacturing the item under production conditions.

A supplier may be required to submit a failure mode, effects, and criticality analysis as evidence of analyses to prevent product or process failures. Increasingly, this requirement is part of the six sigma approach to quality during design.

Qualifying the Supplier's Manufacturing Process

Evaluation of the supplier's manufacturing capability can be done by reviewing past data on similar products, performing process capability analysis, or evaluating the supplier's quality system through a quality survey.

Data showing the supplier's past performance on the same or similar products may be available within the local buyer's organization, other divisions of the same corporation, government data banks, or industry data banks.

With the process capability analysis approach, data on key product characteristics are collected from the process and evaluated by using statistical indexes for process capability (see Chapter 20). All evaluation occurs before the supplier is authorized to proceed with full production. Typically, process capability analysis of a supplier's process is reserved for significant product characteristics, safety-related items, or products requiring compliance with government regulations.

The third approach, a quality survey, is explained next.

When all three approaches can be used, the information collected can provide a sound prediction of supplier capability.

Supplier Quality Survey (Supplier Quality Evaluation)

A supplier quality survey is an evaluation of the ability of a supplier's quality system to meet quality requirements on production lots, i.e., to prevent, identify, and remove any product that does not meet requirements. The results of the survey are used in the supplier selection process, or, if the supplier has already been chosen, the survey alerts the purchaser to areas where the supplier may need help in meeting requirements. The survey can vary from a simple questionnaire mailed to the supplier to a visit to the supplier's facility.

The questionnaire poses explicit questions such as these submitted to suppliers of a manufacturer of medical devices:

- Has your company received the quality requirements on the product and agreed that they can be fully met?
- Are your final inspection results documented?
- Do you agree to provide the purchaser with advance notice of any changes in your product design?
- What protective garments do your employees wear to reduce product contamination?
- Describe the air-filtration system in your manufacturing areas.

The more formal quality survey consists of a visit to the supplier's facility by a team of observers from departments such as quality, engineering, manufacturing, and purchasing. Such a visit may be part of a broader survey of the supplier covering financial, managerial, and technological competence. Depending on the product involved, the activities included in the quality portion of the survey can be chosen from the following list:

- *Management:* philosophy, quality policies, organization structure, indoctrination, commitment to quality.
- *Design:* organization, systems in use, caliber of specifications, orientation to modern techniques, attention to reliability, engineering change control, development laboratories.
- *Manufacture:* physical facilities, maintenance, special processes, process capability, production capacity, caliber of planning, lot identification and traceability.
- *Purchasing:* specifications, supplier relations, procedures.
- *Quality:* organizational structure, availability of quality and reliability engineers, quality planning (materials, in-process, finished goods, packing, storage, shipping, usage, field service), audit of adherence to plan.
- *Inspection and test:* laboratories, special tests, instruments, measurement control.
- *Quality coordination:* organization for coordination, order analysis, control over subcontractors, quality cost analysis, corrective action loop, disposition of nonconforming product.
- *Data systems:* facilities, procedures, effective use reports.
- *Personnel:* indoctrination, training motivation.
- *Quality results:* performance attained, self-use of product, prestigious customers, prestigious subcontractors.

Following the survey, the team reports its findings. These consist of (1) some objective findings as to the supplier's facilities (or lack of facilities), (2) subjective judgments on the effectiveness of the supplier's operations, (3) a further judgment on the extent of assistance needed by the supplier, and (4) a highly subjective prediction whether the supplier will deliver a good product if awarded a contract.

The quality survey is a technique for evaluating the supplier's ability to meet quality requirements on production lots. The evaluation of various quality activities can be quantified by a scoring system.

A scoring system that includes importance weights for activities is illustrated in Table 15.4. This system is used by a manufacturer of electronic assemblies. In this case,

TABLE 15.4
Scoring of a supplier quality survey

Activity	Receiving inspection			Manufacturing			Final inspection		
	R	W	R × W	R	W	R × W	R	W	R × W
Quality management	8	3	24	8	3	24	8	3	24
Quality planning	8	4	32	8	4	32	10	4	40
Inspection equipment	10	3	30	10	3	30	10	3	30
Calibration	0	3	0	10	3	30	0	3	0
Drawing control	0	3	0	10	2	20	10	2	20
Corrective action	10	3	30	8	3	24	8	3	24
Handling rejects	10	2	20	8	2	16	10	3	30
Storage and shipping	10	1	10	10	1	10	10	1	10
Environment	8	1	8	8	1	8	8	1	8
Personnel experience	10	2	20	10	3	30	10	2	20
Area total			174			224			206

Note: R, rating; W, weight.

Interpretation of area totals:
  Fully approved: Each of the three area totals is 250.
  Approved: None of the three area totals is less than 200.
  Conditionally approved: No single total is less than 180.
  Unapproved: One or more of the area totals is less than 180.

the importance weights (W) vary from 1 to 4 and must total 25 for each of the three areas surveyed. The weights show the relative importance of the various activities in the overall index. The actual ratings (R) of the activities observed are assigned as follows:

10: The specific activity is satisfactory in every respect (or does not apply).
8: The activity meets minimum requirements but improvements could be made.
0: The activity is unsatisfactory.

    Supplier quality surveys have both merits and limitations. On the positive side, such surveys can identify important weaknesses such as a lack of special test equipment or an absence of essential training programs. Further, the survey opens up lines of communication and can stimulate action on quality by the supplier's upper management. On the negative side, surveys that emphasize the supplier's organization, procedures, and documentation have had only limited success in predicting future performance of the product.

    Suppliers in some industries have been burdened with quality surveys from many purchasers. These repeat surveys (called "multiple assessment") are time-consuming for suppliers. In another approach, a standard specification of the elements of a quality system (e.g., the ISO 9000 series) is created and assessors are trained to use the specification to evaluate supplier capability. A list of suppliers that have passed the assessment is published, and other purchasers are encouraged to use these results instead of making their own assessment of a supplier. The assessors are independent of the supplier or purchasing organization—thus the term *third-party assessment.* In some countries, a national standards organization acts in this role.

Bossert (1998) describes how supplier evaluation can start with an ISO 9000 assessment and then be supplemented by a quality survey including a supplier visit. The supplier visit covers contract and specification review, process audit, process risk analysis (using a failure mode and effects analysis), and statistical techniques (including the measurement process).

On to the third phase of supplier relations—management of the supply chain through quality planning, quality control, and quality improvement.

15.6 SUPPLY CHAIN QUALITY PLANNING

Donovan and Maresca (in *JQH5*) suggest the following steps for a purchasing process that involves the purchasing organization, suppliers, and end users. This approach is sometimes called a "sourcing process."

1. Document the organization's historic, current, and future procurement activity.
2. Identify a commodity from the procurement activity that represents both high expenditure and high criticality to the business.
3. For this commodity, assemble a cross-functional team.
4. Determine the sourcing needs of the customer through data collection, survey, and other activities.
5. Analyze the supply industry's structure, capabilities, and trends.
6. Analyze the cost components of the commodity's total cost of ownership.
7. Translate the customer needs into a sourcing process that will satisfy the customer and provide the opportunity to manage and optimize the total cost of ownership.
8. Obtain management endorsement to transfer the sourcing strategy into operation. Implement it. For elaboration, see *JQH5,* pages 21.18–21.20.

In doing the detailed quality planning with suppliers, three approaches emerge:

- *Inspection.* The focus is on various forms of product inspection.
- *Prevention.* The premise is that quality must be built in by the supplier with the purchaser's help. But there is still an arm's-length relationship between purchaser and supplier.
- *Partnership.* Suppliers are offered the financial security of a long-term relationship in exchange for a supplier's commitment to quality that includes a strong teamwork relationship with the buyer.

Partnership—involving not just quality but also other business issues—is clearly the wave of the future. Teamwork actions vary greatly, e.g., training a supplier's staff in quality techniques, including suppliers in a design review meeting to gain ideas on how supplier parts can best be used, sharing confidential sales projections with suppliers to assist in supplier production scheduling. Such partnerships often lead to formation of supplier quality councils, which help provide new approaches for the benefits of both the buyer and suppliers. Various opportunities for teamwork are discussed later. But such teamwork depends on truly open communication between buyers and suppliers.

Such cooperation can best be achieved by setting up multiple channels of communication: Designers must communicate directly with designers, quality specialists with quality specialists, etc. These multiple channels are a drastic departure from the single channel, which is the method in common use for purchase of traditional products. In the single-channel approach, a specialist in the buyer's organization must work through the purchasing agent, who in turn speaks with the salesperson in the supplier's organization, to obtain information. Of course, the concept of multiple channels seems sensible, but wouldn't it be useful to determine whether multiple channels yield better results in quality? Carter and Miller (1989) did just that.

> **EXAMPLE 15.3.** In an innovative research study, they compared quality levels for two communication structures: serial (single channel) and parallel (multiple channel). At a manufacturer of mechanical seals, one section of a plant followed the serial communication concept while a second area used parallel communication. Over a 19-month period, the section using parallel communication improved the average percentage of items rejected from 30.3 to 15.0%, a statistically significant difference; the section with serial communication had no such improvement—in fact, its rejection percentage increased slightly.

Next we address how partnership can be achieved through joint economic planning, joint technological planning, and cooperation during contract execution.

Joint Economic Planning

The economic aspects of joint quality planning concentrate on two major approaches:

- *Value rather than conformance to specification.* The technique used is to analyze the value of what is being bought and to try to effect an improvement. The organized approach is known as value engineering. Applied to supplier quality relations, value engineering looks for excessive costs due to (1) overspecification for the use to which the product will be put, e.g., a special product ordered when a standard product would do; (2) emphasis on original price rather than on cost of use over the life of the product; and (3) emphasis on conformance to specification, not fitness for use. Suppliers are encouraged to make recommendations on design or other requirements that will improve or maintain quality at a lower cost.
- *Total cost of ownership.* The buyer must add a whole array of quality-related costs to the purchase price: incoming inspection, materials review, production delays, downtime, extra inventories, etc. However, the supplier also has a set of costs it is trying to optimize. The buyer should put together the data needed to understand the life-cycle costs or the cost of use and then press for a result that will optimize them.

> **EXAMPLE 15.4.** A heavy-equipment manufacturer bought 11,000 castings per year from several suppliers. It was decided to calculate the total cost of the purchased casting as the original purchase price plus incoming inspection costs plus the costs of rejections detected later in assembly. The unit purchase price on a contract given to the lowest bidder was $19.

The inspection and rejection costs amounted to an additional $2.11. The variation among bid prices was $2. Thus the lowest bid does not always result in the lowest total cost.

Joint Technological Planning

The standard elements of such planning include

1. Agreement on the meaning of performance requirements in the specifications.
2. Quantification of quality, reliability, and maintainability requirements.

   EXAMPLE 15.5. A supplier was given a contract to provide an air-conditioning system with a mean time between failures of at least 2000 hours. As part of joint planning, the supplier was required to submit a detailed reliability program early in the design phase. The program write-up was submitted and included a provision to impose the same 2000-hour requirement on each supplier of parts for the system. This revealed a complete lack of understanding by the supplier of the multiplication rule (see Section 19.4, "The Relationship between Part and System Reliability").

3. Definition of reliability and maintainability tasks to be conducted by the supplier.
4. Preparation of a process control plan for the manufacturing process. The supplier can be asked to submit a plan summarizing the specific activities which will be conducted during the manufacture of the product. Typically, the plan must include statistical process control techniques to prevent defects by detecting problems early.
5. Definition of special tasks required of the supplier. These may include activities to ensure that good manufacturing practices are met, special analyses are prepared for critical items, etc.
6. Seriousness classification of defects to help the supplier understand where to concentrate efforts.
7. Establishment of sensory standards for qualities that require use of the human being as an instrument.

   EXAMPLE 15.6. The federal government was faced with the problem of defining the limits of color on a military uniform. It was finally decided to prepare physical samples of the lightest and darkest acceptable colors. Such standards were then sent out with the provision that the standards would be replaced periodically because of color fading.

8. Standardization of test methods and test conditions between supplier and buyer to ensure their compatibility.

   EXAMPLE 15.7. A carpet manufacturer repeatedly complained to a yarn supplier about yarn weight. The supplier visited the customer to verify the test methods. Their test methods were alike. Next an impartial testing lab verified the tests at the carpet plant. Finally, the mystery was solved. The supplier was spinning (and measuring) the yarn in bone-dry conditions, but the carpet manufacturer measured at standard conditions. During this period, $62,000 more was spent for yarn than if it had been purchased at standard weight.

9. Establishment of sampling plans and other criteria relative to inspection and test activity. From the supplier's viewpoint, the plan should accept lots having the usual process average. For the buyer, the critical factor is the amount of damage caused

by one defect getting through the sampling screen. Balancing the cost of sorting versus sampling can be a useful input in designing a sampling plan (see Chapter 10). In addition to sampling criteria, error of measurement can also be a problem.

10. Establishment of quality levels. In the past, suppliers were often given "acceptable quality levels" (AQL). The AQL value was just one point on the "operating characteristic" curve that described the risks of sampling plans. A typical AQL value might be 2.0%. Many suppliers interpreted this to mean that product which included 2% defective was acceptable. It is best to make clear to the supplier through the contract that *all* product submitted is expected to meet specifications and that any nonconforming product may be returned for replacement. In many industries, the unit of measurement is defects per million (DPM).

11. Establishment of a system of lot identification and traceability. This concept has always been present in some degree, e.g., heat numbers of steel, lot numbers of pharmaceutical products. More recently, with intensified attention to product reliability, this procedure is more acutely needed to simplify the localization of trouble, to reduce the volume of product recall, and to fix responsibility. These traceability systems, though demanding some extra effort to preserve the order of manufacture and identify the product, make greater precision in sampling possible.

12. Establishment of a system of timely response to alarm signals resulting from defects. Under many contracts, the buyer and supplier are yoked to a common timetable for completion of the final product. Usually, a separate department (e.g., materials management) presides over major aspects of scheduling. However, upper management properly looks to the people associated with the quality function to set up alarm signals to detect quality failures and to act positively on these signals to avoid deterioration, whether in quality, cost, or delivery.

Such depth of joint technological planning bears no resemblance to the old approach of sending a supplier a blueprint with a fixed design and a schedule.

→ 15.7 SUPPLY CHAIN QUALITY CONTROL

Donovan and Maresca (*JQH5*, Section 21) suggest these steps for successful supplier control:

1. Create a cross-functional team.
2. Determine critical performance metrics.
3. Determine minimum standards of performance.
4. Reduce the supplier base to those able to meet minimum performance requirements.
5. Assess supplier performance:
    a. Supplier quality systems assessment.
    b. Supplier business management.
    c. Supplier product fitness for use.

For elaboration, see *JQH5,* pages 21.20–21.23.

The detailed quality control activities focus on cooperation during contract execution, supplier certification, supplier rating, and quality measurement for supplier relations. These activities emphasize continuous feedback to suppliers.

Cooperation during Contract Execution

This cooperation usually concentrates on the following activities.

Evaluation of initial samples of product

Under many circumstances, the supplier must submit test results of a small initial sample produced from production tooling and a sample from the first production shipment before the full shipment is made. The latter evaluation can be accomplished by having a buyer's representative visit the supplier's plant and observe the inspection of a random sample selected from the first production lot. A review can also be made of process capability of process control type data from that lot.

Design information and changes

Design changes may take place at the initiative of either the buyer or the supplier. Either way, the supplier should be treated like an in-house department when developing procedures for processing design changes. This need is especially acute for modern products, for which design changes can affect products, processes, tools, instruments, stored materials, procedures, etc. Some of these effects are obvious, but others are subtle, requiring a complete analysis to identify the effects. Failure to provide adequate design change information to suppliers has been a distinct obstacle to good supplier relations.

Surveillance of supplier quality

Quality surveillance is the continuing monitoring and verification of the status of procedures, methods, conditions, processes, products, services, and analysis of records in relation to stated references to ensure that specified requirements for quality are being met (ISO 8402). Surveillance by the buyer can take several forms: inspecting the product, meeting with suppliers to review quality status, auditing elements of the supplier quality program, monitoring of the manufacturing practices of the supplier, reviewing statistical process control data, and witnessing specific operations or tests. Major or critical contracts require on-site presence or repeat visits.

Evaluating delivered product

Evaluation of supplier product can be achieved by using one of the methods listed in Table 15.5.

In previous decades, incoming inspection often consumed a large amount of time and effort. With the advent of modern complex products, many companies have found that they do not have the necessary inspection skills or equipment. This situation has forced them to rely more on the supplier's quality system or inspection and test data, as discussed later in this chapter.

TABLE 15.5
Methods of evaluating supplier product

Method	Approach	Application
100% inspection	Every item in a lot is evaluated for all or some of the characteristics in the specification	Critical items where the cost of inspection is justified by the cost of risk of defectives; also used to establish the quality level of new suppliers
Sampling inspection	A sample of each lot is evaluated by a predefined sampling plan and a decision is made to accept or reject the lot	Important items where the supplier has established an adequate quality record by the prior history of lots submitted
Identifying inspection	The product is examined to ensure that the supplier sent the correct product; no inspection of characteristics is made	Items of less importance where the reliability of the supplier laboratory has been established in addition to the quality level of the product
No inspection	The lot is sent directly to a storeroom or processing department	For purchase of standard materials or goods not used in the product, e.g., office supplies
Using supplier data (supplier certification)	Data of the supplier inspection is used in place of incoming inspection	Items for which a supplier has established a strong quality record

The choice of evaluation method depends on a variety of factors:

- Prior quality history on the part and supplier.
- Criticality of the part on overall system performance.
- Criticality on later manufacturing operations.
- Warranty or use history.
- Supplier process capability information.
- The nature of the manufacturing process. For example, a press operation depends primarily on the adequacy of the setup. Information on the first few pieces and last few pieces in a production run is usually sufficient to draw conclusions about the entire run.
- Product homogeneity. For example, fluid products are homogeneous, and the need for large sample sizes is thus less.
- Availability of required inspection skills and equipment.

A useful tool for learning about a supplier's process and comparing several suppliers' manufacturing product to the same specification is the histogram. A random sample is selected from a lot, and measurements are made on selected quality characteristics. The data are charted as frequency histograms. The analysis consists of comparing the histograms to the specification limits.

An application of histograms to evaluating the hardenability of a particular grade of steel from three suppliers is shown in Figure 15.2. The specification was a maximum Rockwell C reading of 43 measured at Jominy position J8.

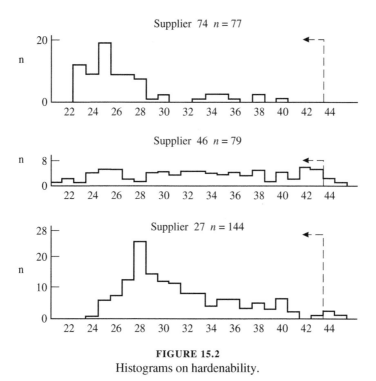

FIGURE 15.2
Histograms on hardenability.

Histograms were also prepared for carbon, manganese, nickel, and chromium content. Analysis revealed:

- Supplier 46 had a process without any strong central tendency. The histogram on nickel for this supplier was also rectangular in shape, indicating a lack of control of the nickel content and resulting in several heats of steel with excessively high Rockwell values.
- Supplier 27 had several heats above the maximum, although the process had a central value of about 28. The histograms for manganese, nickel, and chromium showed several values above and apart from the main histogram.
- Supplier 74 showed much less variability than the others. Analysis of other histograms for this supplier suggested that about half of the original heats of steel had been screened out and used for other applications.

Note how these analyses can be made without visiting the supplier plants, i.e., "the product tells on the process." Histograms have limitations (see Chapter 17), but they are an effective tool for incoming inspection.

Action on nonconforming product

During the performance of the contract, there will arise instances of nonconformance. These may be on the product itself or on process requirements or procedural requirements. Priority effort should go to cases where a product is unfit for use.

Communications to the supplier on nonconformance must include a precise description of the symptoms of the defects. The best description is in the form of samples, but if this is not possible, the supplier should have the opportunity to visit the site of the trouble. There are numerous related questions: What disposition is to be made of the defective items? Who will sort or repair them? Who will pay the costs? What were the causes? What steps are needed to avoid a recurrence? These questions are outside the scope of pure defect detection; they require discussion among departments within each company and further discussions between buyer and supplier.

Supplier Certification

A "certified" supplier is one whose quality data record establishes that it is not necessary to perform routine inspection and test on each lot or batch received. A "preferred" supplier produces quality better than the minimum. An "approved" supplier meets minimum requirements. Some organizations use different terms and even different rankings, but usually certified suppliers are the ideal. Unfortunately, they are in the minority. Spooner and Collins (1995) describe how criteria were developed at Walker Manufacturing to define these categories, e.g., a certified supplier has performed "at an overall 90% compliance level and met the individual rating component requirements for four consecutive quarters within two years."

ASQ recommends eight criteria for certification. These are summarized in Table 15.6.

Supplier certification provides a model for the low DPM levels necessary for just-in-time manufacture, drastically reduces buyer inspection costs, and identifies suppliers for partnerships. Certified suppliers receive preference in competitive bidding and achieve industry recognition by their certified status.

Schneider et al. (1995) explain how process capability indexes are used as part of the certification process at Dow Chemical. The concept of supplier certification

TABLE 15.6
Criteria for supplier certification

Criteria	Examples
No product-related lot rejections for at least one year	An alternative is volume related, e.g., no rejects in 20 consecutive lots
No non-product-related rejections for at least six months	The marking on a container or the timeliness of an analysis document
No production-related negative incidents for at least six months	Ease with which the supplier's product can be used in the buyer's process or product
Passed a recent on-site quality system evaluation	A supplier survey on defined criteria
Has a totally agreed on specification	No ambiguous phrases like "characteristic odor" or "clear of contamination"
Fully documented process and quality system	The system must include plans for continuous improvement
Timely copies of inspection and test data	Real-time availability of data
Process is stable and in control	Statistical control and process capability studies

Source: Adapted from Maass et al. (1990).

applies equally to the service sector. Brown (1998) describes the approach used by a telecommunications company for suppliers of building leasing, maintenance, security, and food service.

Supplier Quality Rating

Supplier quality rating provides a quantitative summary of supplier quality over a period of time. This type of rating is useful in deciding how to allocate purchases among suppliers. Rating furnishes both buyer and supplier with common factual information that becomes a key input for identification and tracking of improvement efforts and for allocating future purchases among suppliers.

To create a single numerical quality score is difficult because there are several units of measure, such as

- The quality of multiple lots expressed as lots rejected versus lots inspected.
- The quality of multiple parts expressed as percentage nonconforming.
- The quality of specific characteristics expressed in numerous natural units, e.g., ohmic resistance, percentage of active ingredient, mean time between failures.
- The economic consequences of bad quality, expressed in dollars.

Because these units of measure vary in importance among companies, published rating schemes differ markedly in emphasis.

Measures in use

Supplier quality rating plans are based on one or more of the following measures:

Product percentage nonconforming. This measure is a ratio of the amount of defective items received to the total number of items received. On a lot-by-lot basis, the formula is the number of lots rejected divided by the number of lots received; on an individual piece basis, the formula is the number of individual pieces rejected divided by the number of individual pieces received.

Overall product quality. This plan summarizes supplier performance at incoming inspection and later phases of product application. Points are assigned for each phase; the maximum number of points is given when no problems are encountered. Table 15.7 shows an example from AT&T. Note that the phases are incoming inspection, production failures, vendor response to problems, and AT&T customer complaints. Each rating element has further detailed criteria that are used to assign points for the element, e.g., if 3% of the lots are rejected for "visual/mechanical" reasons in a rating period, then one point is deducted from the maximum of five for that element. Note that the overall rating evaluates the supplier response to problems, whereas detailed criteria include both timeliness and adequacy of the response.

Economic analysis. This type of plan compares suppliers on the total dollar cost for specific purchases. The total dollar cost includes the quoted price plus quality costs for defect prevention, detection, and correction.

Composite plan. Supplier performance is not limited to quality. It includes delivery against schedule, price, and other performance categories. These multiple needs

TABLE 15.7
AT&T quality performance rating

Rating element	Maximum points
Incoming inspection	
Visual mechanical PPM	10
Visual/mechanical—percentage of lot rejections	5
Testing PPM	10
Testing—lot rejections	5
Ship-to-stock credit	—
Production failures	
Shop complaints	20
Quality appraisal	10
Vendor response	
Response to problems	10
Failure analysis response	20
AT&T customer complaints	10
Total	100

Source: Nocera et al. (1989).

suggest that the supplier rating should include overall supplier performance rather than just supplied quality performance. The purchasing department is a strong advocate of this principle and has valid grounds for this advocacy. Table 15.8 illustrates this approach with an example from Tecumseh Products Company. The overall rating of 92.46 is calculated by combining the four ratings using weights of 40% for quality, 30% for delivery, 20% for cost, and 10% for responsiveness to problems. Walker Manufacturing uses categories and weights of 35% for quality, 35% for delivery, 20% for price, and 10% for supplier support (Spooner and Collins, 1995).

Some organizations use a periodic supplier rating to determine the share of future purchases given to each supplier. The rating system and its effect on market share are fully explained to all suppliers. The approach has been used successfully by both automotive and appliance manufacturers to highlight the importance of quality to their suppliers.

TABLE 15.8
Supplier rating report

Overall combined rating	92.46
Total quality rating	99.05
Total delivery rating	95.58
Total cost rating	79.22
Total response rating	83.30
Total lots received	18
Total parts received	398,351
Total parts rejected	3,804

Source: Wind (1991).

TABLE 15.9
Examples of quality measurement in supplier relations

Subject	Units of measure
Quality of submitted lots	Percentage of lots rejected Cost of poor quality Percentage of lots accepted on waiver Number of rejected lots classified "use as is"
Supplier relations program	Percentage of suppliers certified Percentage of suppliers classified acceptable as a result of a supplier survey Percentage of qualification test *procedures* approved on first submission Percentage of qualification test *results* approved on first submission Percentage of initial product samples approved on first submission Percentage of first production shipments approved on first submission Percentage of suppliers submitting data Average time to resolve problems
Business relationships	Average number of multiple suppliers per item Percentages of purchases as single source Percentage of purchases to lowest bidder Average time to secure bids Average time to secure answers to technical inquiries
Adequacy of inventory	Percentage of stockouts
Service to suppliers	Average number of days to pay supplier invoice Number of accounts payable beyond X days

Quality Measurement in Supplier Relations

The management of quality-related activities in supplier relations must include provision for measurement. Readers are urged to review the 10 basic principles of quality measurement in Section 6.2, "The Importance of Information and Measurement."

Table 15.9 shows units of measure for various subject areas of supplier relations. Klenz (2000) discusses the use of a data warehouse for supplier quality analysis.

➔ 15.8 SUPPLY CHAIN QUALITY IMPROVEMENT

Donovan and Maresca (*JQH5,* Section 21) propose a sequence of five tiers of progression for improvement:

1. Create a joint team of the end user and supplier to align goals, analyze the supply chain business process, and work on chronic problems.
2. Focus on cost reduction, including the cost of poor quality.

3. Evaluate the value added by each link in the supply chain.
4. Exchange information and ideas routinely throughout the chain.
5. Have the supply chain work as a single process with all parties routinely collaborating on improvement opportunities to generate value for customers as well as suppliers.

For elaboration, see *JQH5,* pp. 21.23–21.25.

The general approach to handling chronic supplier problems follows the step-by-step approach to improvement explained in Chapter 4. This process includes the early steps of establishing the proof of the need for the supplier to take action and the application of Pareto analysis to identify the vital few problems. The following section on Pareto analysis of suppliers explains the form of such analyses of suppliers' problems.

Cooperation often requires providing technical assistance to suppliers. Miller and Kegaris (1986) describe how businesses may need to share proprietary information on a "need to know" basis. This often represents a major breakthrough in communications.

Sometimes upper management must provide the leadership in obtaining action from suppliers. Amazing results can be achieved when the initial step in an improvement program is a meeting of both the buyer's and supplier's upper management teams, who plan the action steps for improvement together. Such discussions have much more impact than a meeting of the two quality managers.

> **EXAMPLE 15.8.** For an appliance manufacturer, 75% of the warranty costs were due to suppliers' items. The president and his staff met individually with the counterpart team from each of 10 key suppliers. Warranty data were presented to establish the "proof of the need." A goal was set for a 50% reduction in warranty costs over a five-year period. Each supplier was asked to develop a quality improvement program. The purchaser provided an eight-hour training session for the president and staff members of the key suppliers. Follow-up meetings were held. A system of supplier recognition awards was set up and purchasing practices were changed to transfer business to the best suppliers. The result: a decline in service calls from 41 to 13 calls per 100 products and a saving of $16 per unit in warranty costs.

Pareto Analysis of Suppliers

Supplier improvement programs can fail because the vital few problems are not identified and attacked. Instead, the programs consist of broad attempts to tighten up all procedures. The Pareto analysis (see Section 4.7 under "The Pareto Principle") can be used to identify the problem in a number of forms:

1. *Analysis of losses (defects, lot rejections, etc.) by material number* or *part number.* Such analysis serves a useful purpose applied to catalog numbers involving substantial or frequent purchases.
2. *Analysis of losses by product family.* This process identifies the vital few product families present in small but numerous purchases of common product families, e.g., fasteners, paints.
3. *Analysis of losses by process,* i.e., classification of the defects or lot rejections in terms of the processes to which they relate, e.g., plating, swaging, coil winding.

4. *Analysis by supplier across the entire spectrum of purchases.* This process can help to identify weaknesses in the supplier's managerial approach, as contrasted with the technological, which is usually correlated with products and processes. One company had 222 suppliers on the active list. Of these, 38 (or 17%) accounted for 53% of the lot rejections and 45% of the bad parts.
5. *Analysis by total cost of the parts.* In one company, 37% of the part numbers purchased accounted for only 5% of the total dollar volume of purchases but for a much higher percentage of the total incoming inspection cost. The conclusion was that these "useful many" parts should be purchased from the best suppliers, even at top prices. The alternative of relying on incoming inspection would be even more costly.
6. *Analysis by failure mode.* This technique is used to discover major defects in the management system. For example, suppose that studies disclose multiple instances of working to the wrong issue of the specification. In such cases, the system used for specification revision should be reexamined. If value analysis discovers multiple instances of overspecification, the design procedures for choosing components should be reexamined. These analyses by failure mode can reveal how the buyer is contributing to his or her own problems.

The cross-functional team approach for quality improvement described in Chapter 4 also applies to supplier quality. This means that there should be joint customer-supplier teams and also that suppliers must be encouraged to set up an infrastructure (quality council, formation of teams, identification of projects, execution of projects) internally to address quality. Chen and Batson (1996) describe how Johnson & Johnson Consumer Products use 17 steps in this approach to supplier quality improvement. *JQH5,* Section 29, presents quality improvement in the automotive industry. In the service sector, Sun Health Alliance employs an innovative approach to stimulate improvement (Nussman, 1993). Sun provides grant funding to partner hospitals and corporate partners to support quality improvement demonstration projects. The projects include patient care topics (e.g., establishing clinical pathways for specific diagnoses), nonclinical topics (e.g., reducing turnaround time for lab results), and employee-specific topics (e.g., reducing turnover or employee "needle sticks").

Handfield et al. (2000) discuss the results of research to identify "pitfalls" in supplier development. The research involved 84 companies in the fields of telecommunications, automobiles, electronics, computers, services, chemicals, consumer nondurable goods, and aerospace. The pitfalls were mainly concerned with identifying key projects, defining the details of the agreement between the buyer and supplier organizations, and monitoring the status and modifying strategies when necessary.

Some of the pitfalls were supplier specific; some were buyer specific; some were specific to the supplier-buyer relationship.

The supplier-specific pitfalls stemmed chiefly from the suppliers' lack of commitment and lack of technical or human resources. To avoid these pitfalls, companies took these actions:

1. Show suppliers where they stand.
2. Tie business relationships to performance improvement.
3. Illustrate supplier benefits clearly.

4. Ensure follow-up through a supplier champion (a supplier employee).
5. Keep initial improvements simple.
6. Draw on buyer's resources.
7. Offer personnel support
8. Build training centers.

The buyer-specific pitfalls occur when buyers see no obvious potential benefits from working on supplier development. To avoid this situation, companies found these tactics helpful:

1. Consolidate to fewer suppliers.
2. Keep a long-term focus.
3. Determine the total cost of ownership.
4. Set small goals.
5. Make executive commitment in the buyer organization a priority.

The supplier-buyer relationship pitfalls involved lack of trust between the organizations, poor alignment of cultures, and insufficient inducements to suppliers. Constructive solutions were to

1. Delegate an ombudsman from the buyer organization.
2. Make provisions for handling confidential information.
3. Spell out clearly a cooperative purchasing relationship with well-defined objectives beyond the purchase price.
4. Minimize legal involvement.
5. Adapt to local cultures.
6. Create a road map that defines responsibilities and expectations for both organizations.
7. Offer financial incentives.
8. Show suppliers how they can become "designed in" to buyer products and thus have greater potential for future business.
9. Offer repeat business as an incentive.

Supplier quality improvement needs upper management at all links in the supply chain to provide a structured approach (see Chapter 4) to improvement. Cheerleading and flag waving will not work.

SUMMARY

- A revolution in the relationship between buyers and suppliers has emerged in the form of supplier partnerships and the supply chain.
- Quality specifications often define requirements for both the product and the quality system.
- Organizations are significantly reducing the number of multiple suppliers.
- Outsourcing has both benefits and risks.
- Core competencies must be identified and performed internally.

- Evaluating supplier quality capability involves qualifying the supplier's design and the manufacturing process.
- Supplier partnerships require joint economic planning, joint technological planning, and cooperation during contract execution.
- A certified supplier is one that, after extensive investigation, supplies material of such quality that routine testing on each lot received is not necessary.
- Measurements for supplier relations should be based on input from customers; provide for both evaluation and feedback; and include early, concurrent, and lagging indicators of performance.
- Quality and reliability requirements should be stated in quantitative terms.
- Suppliers must understand that *all* product submitted is expected to meet specifications.
- The results of supplier surveys can be stated in quantitative terms.
- Histogram analyses of supplier data can reveal much information about the supplier's process.
- Pareto analysis of supplier data helps to establish priorities for improvement efforts.
- Supplier quality rating provides a quantitative summary of supplier quality over a period of time.
- Supplier quality improvement needs upper management involvement at all links in the supply chain.

PROBLEMS

15.1. Visit the purchasing agent of some local institution to learn the overall approach to supplier selection and the role of supplier quality performance in this selection process. Report your findings.

15.2. Visit a sampling of local suppliers (printer, merchant, repair shop, etc.) to learn the role of quality performance in their relationship with their clients. Report your findings.

15.3. A government agency contracted with a company to design and build a satellite system. Months after the contract was signed, the company discovered that the design would not be immune to certain types of radar interference. The agency claimed that it had described the performance desired for the satellites. The company disagreed (with respect to the radar interference). If this need had been realized at the start of the project, creating an appropriate design would have been relatively simple. The satellites are in an advanced stage of design and construction, and the necessary changes would cost $100 million. There was further confusion. The company had chosen a supplier to manufacture the satellites. This supplier had previous experience with such products, and some people claimed that the supplier should have been aware of the radar interference problem. Comment on the actions that should be taken by three such organizations to prevent such a situation on a future project (*BusinessWeek*, 1978).

15.4. During World War II, many manufacturers made products that were totally new to them. For example, the Ford Motor Company was asked to produce fuselage sections for B-24 aircraft. To do this, Ford had to work closely with the Consolidated Company, which was responsible for the manufacture of the entire aircraft. Thus Ford was a supplier for Consolidated. There was much friction between the companies. Lindbergh (1970, pages 644–676) describes the background of this classic case:

> In short, if the Consolidated men were carrying a chip on one shoulder, the Ford men arrived with a chip on each shoulder. Instead of taking the attitude that they had come to San Diego to learn how to build Consolidated bombers from the company that had developed those bombers, they took the attitude that they were there only as a preliminary to showing Consolidated how to build Consolidated bombers better and on mass production. The inevitable result was a deep-rooted antagonism which still exists.

The first article delivered by Ford was "not only as bad but considerably worse than the aviation people said it would be—rivets missing . . . badly formed skin . . . cracks *already* started . . . etc." However, this article had been passed both by Ford inspection and by the Army inspector stationed at Ford. Lindbergh concluded:

> What has happened is clear enough: under pressure, and encouraged by the desire to get production under way at Willow Run, and more than a little due to lack of experience, both Army and Ford inspection passed material that should have been rejected (and which was rejected by the more experienced and impartial inspectors at Tulsa).

Describe the *specific* actions that you would recommend to correct the immediate problem and prevent a recurrence in the future.

15.5. Apply the composite plan of supplier rating (Section 15.7) to compare three suppliers for one of the following: (a) any product or service acceptable to the instructor; (b) an automatic washing machine; (c) a new automobile; (d) a lawn mower.

15.6. Can you think of a situation other than 100% inspection that would result in the histogram shown in Figure 15.3?

15.7. Can you describe what caused the unusual histogram plots in Figure 15.4?

15.8. You have been asked to propose a specific quality rating procedure for use in one of the following types of organizations: (a) a company acceptable to the instructor; (b) a large municipal government; (c) a manufacturer of plastic toys; (d) a bank; (e) a manufacturer of whisky. Research the literature for specific procedures and select (or create) a procedure for the organization.

15.9. Visit a local organization and learn how it determines the quality of purchased items. Define the specific procedures used and what use is made of the information compiled.

15.10. Outline a potential application of several concepts in this chapter. Follow the instructions given in problem 15.11.

15.11. In an organization with which you are familiar, identify one activity as a candidate for outsourcing. Analyze the potential benefits and disadvantages of outsourcing that activity. Also, discuss how you might address the disadvantages.

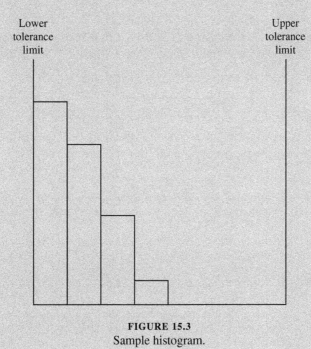

FIGURE 15.3
Sample histogram.

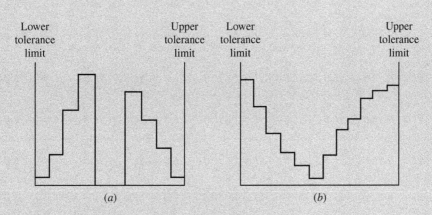

FIGURE 15.4
Sample histograms.

REFERENCES

Bettis, R. A., S. P. Bradley, and G. Hamel (1992). "Outsourcing and Industrial Decline," *Academy of Management Executive,* vol. 6, no. 1, pp. 7–22.

Bossert, J. L., ed. (1994). *Supplier Management Handbook,* ASQ Quality Press, Milwaukee, p. 212.

Bossert, J. L. (1998). "Considerations for Global Supplier Quality," *Quality Progress*, January, pp. 29–32.

Brown, J. O. (1998). "A Practical Approach to Service-Supplier Certification," *Quality Progress*, January, pp. 35–39.

BusinessWeek (1978). "A $100 Million Satellite Error," August 7, p. 52.

Carter, J. R. and J. G. Miller (1989). "The Impact of Alternative Vendor/Buyer Communication Structures on the Quality of Purchased Materials," *Decision Sciences*, Fall, pp. 759–776.

Chen, B. A. and R. G. Batson (1996). "A Team Based Supplier Quality Improvement Process," *Annual Quality Congress Proceedings*, ASQ, Milwaukee, pp. 537–544.

Clott, C. B. (2004). "Perspectives on Global Outsourcing and the Changing Nature of Work," *Business and Society Review*, 109(2): 153–170.

Ericson, C. (2003). "A Global Look at e-Sourcing," *Supply Chain Management Review* 7(6):13.

Fisher, M. L. (1997). "What Is the Right Supply Chain for Your Product?" *Harvard Business Review*, March–April, pp. 105–116.

Handfield, R. B., D. R. Krause, T. V. Scannell, and R. M. Monczka (2000). "Avoid the Pitfalls in Supplier Development," *Sloan Management Review*, vol. 41, no. 2, pp. 37–49. By permission of the publisher. All rights reserved.

Hussey, D. and P. Jenster (2003). "Outsourcing: The Supplier Viewpoint," *Strategic Change* 12(1): 7–20.

Klenz, B. W. (2000). "Leveraging the Data Warehouse for Supplier Quality Analysis," *Annual Quality Congress Proceedings*, ASQ, Milwaukee, pp. 519–528.

Lee, H. L. (2004). "The Triple-A Supply Chain," *Harvard Business Review*, October, pp. 102–112.

Lindbergh, C. A. (1970). *The Wartime Journals of Charles A. Lindbergh*, Harcourt Brace Jovanovich, New York.

Maass, R. A., J. O. Brown, and J. L. Bossert (1990). *Supplier Certification—A Continuous Improvement Strategy*, ASQ Quality Press, Milwaukee.

Meseck, G. (2004). "Risky Business: How to Calculate the Financial Impact of Global Outsourcing," *Logistics Today*, 45(8): 34.

Miller, G. D. and R. J. Kegaris (1986). "An Alcoa-Kodak Joint Team," *Juran Report Number Six*, Juran Institute, Inc., Wilton, CT, pp. 29–34.

Nocera, C. D., M. K. Foliano, and R. E. Blalock (1989). "Vendor Rating and Certification," *Impro Conference Proceedings*, Juran Institute, Inc., Wilton, CT, pp. 9A-29 to 9A-38.

Nussman, H. B. (1993). "The Sun Health Alliance for Quality—A Unique Customer-Supplier Partnership," *Impro Conference Proceedings*, Juran Institute, Inc., Wilton, CT, pp. 3A.1-1 to 3A.1-4.

Peterson, Y. S. (1998). "Outsourcing: Opportunity or Burden," *Quality Progress*, June, pp. 63–64.

Ramachandran, K. and S. Voleti (2004). "Business Process Outsourcing (BPO): Emerging Scenario and Strategic Options for IT-Enabled Services," *Vikalpa*, 29(1): 49–62.

Schneider, H., J. Pruett, and C. Lagrange (1995). "Uses of Process Capability Indices in the Supplier Certification Process," *Quality Engineering*, vol. 8, no. 1, pp. 225–235.

Sharman, G. (2002). "How the Internet Is Accelerating Supply Chain Trends," *Supply Chain Management Review*, 6(2):18.

Soliman, K. S. (2003). "A Framework for Global IS Outsourcing by Application Service Providers," *Business Process Management Journal*, 9(60): 375.

Spooner, G. R. and D. W. Collins (1995). "A Cross Functional Approach to Supplier Evaluation," *Proceedings of the Annual Quality Congress*, ASQ, Milwaukee, pp. 825–832.

Weerakkody, V., W. L. Currie, and Y. Ekanayake (2003). "Re-Engineering Business Processes Through Application Service Providers: Challenges, Issues and Complexities," *Business Process Management Journal,* 9(6) 776.

Weidenbaum, M. (2004). "Outsourcing; Pros and Cons," *Executive Speeches* 19(1): 31–35.

Wind, J. F. (1991). "Revolutionize Supplier Rating by Computerization," *Quality Congress Transactions,* ASQ, Milwaukee, pp. 556–564.

Zimmerli, B. (1996). "Re-Engineering the Supply Chain," *Impro Conference Proceedings,* Juran Institute, Inc., Wilton, CT, pp. 4F-1 to 4F-18.

CHAPTER 16

Quality Assurance and Audits

→ 16.1 DEFINITION AND CONCEPT OF ASSURANCE

In this book, *quality assurance* is the activity of providing evidence to establish confidence that quality requirements will be met. ISO defines quality assurance as all the planned and systematic activities implemented within the quality system, and demonstrated as needed, to provide adequate confidence that an entity will fulfill requirements for quality. Readers are warned that other meanings are common. For example, "quality assurance" is sometimes the name of a department concerned with many quality management activities such as quality planning, quality control, quality improvement, quality audit, and reliability.

Many quality assurance activities provide protection against quality problems through early warnings of trouble ahead. The assurance comes from evidence—a set of facts. For simple products, the evidence is usually some form of inspection or testing of the product. For complex products, the evidence is inspection and test data and also reviews of plans and audits of the execution of plans. A family of assurance techniques is available to cover a wide variety of needs.

Quality assurance is similar to the concept of the financial audit, which provides assurance of financial integrity by establishing, through "independent" audit, that the plan of accounting is (1) such that, if followed, it will correctly reflect the financial condition of the company and (2) is actually being followed. Today, independent financial auditors (certified public accountants) have become an influential force in the field of finance.

Many forms of assurance previously discussed in this book are performed within functional departments (Table 16.1). This chapter discusses three forms of company-wide quality assurance: quality audits, quality assessments, and product audits.

TABLE 16.1
Examples of departmental assurance activities

Department	Assurance activity
Marketing	Product evaluation by a test market Controlled use of product Product monitoring Captive service activity Special surveys Competitive evaluations
Product development	Design review Reliability analysis Maintainability analysis Safety analysis Human factors analysis Manufacturing, inspection, and transportation analysis Value engineering Self-control analysis
Supplier relations	Qualification of supplier design Qualification of supplier process Evaluation of initial samples Evaluation of first shipments
Production	Design review Process capability analysis Preproduction trials Preproduction runs Failure mode, effect, and criticality analysis for processes Review of manufacturing planning (checklist) Evaluation of proposed process control tools Self-control analysis Audit of production quality
Inspection and test	Interlaboratory tests Measuring inspector accuracy
Customer service	Audit of packaging, transportation, and storage Evaluation of maintenance services

Source: JQH4, p. 9.3.

➔ 16.2 CONCEPT OF QUALITY AUDITS

A *quality audit* is an independent review conducted to compare some aspect of quality performance with a standard for that performance. It is a process conducted by either an internal or external auditor that helps to ensure that an organization's systems are in place and are being followed. The objective of the audit is to draw attention to needed improvements and ensure legal and regulatory requirements are being followed in order to bring consistent quality goods and services to the consumer. A successful quality audit concentrates on the needs of the organization. The term *independent* is critical and is used in the sense that the reviewer (called the "auditor") is neither the person responsible for the performance under review nor the immediate supervisor of that person. An independent audit provides an unbiased picture of performance. The terms *quality assessment* (or *quality evaluation*) and *quality audit* have similar meanings, but in common usage, *assessment* refers to a total spectrum of quality activities often including managerial matters such as the cost of poor quality, standing in the market place, and quality culture (see Chapter 3).

The ISO 19011:2002 First Edition definition spells out some additional aspects: A quality audit is a systematic and independent examination to determine whether quality activities and related results comply with planned arrangements and whether these arrangements are implemented effectively and are suitable for achieving objectives. (A product audit, discussed later in this chapter, is a review of *physical product*; a quality audit is a review of an *activity*.)

Internal audits, sometimes called *first-party audits*, are conducted by, or on behalf of, the organization itself for management review and other internal purposes and may form the basis for an organization's self-declaration of conformity. In many cases, particularly in smaller organizations, independence can be demonstrated by the freedom from responsibility for the activity being audited. External audits include those generally termed *second-* and *third-party audits*. Second-party audits are conducted by parties having an interest in the organization, such as customers, or by other persons on their behalf. Third-party audits are conducted by external, independent auditing organizations, such as those providing registration or certification of conformity to the requirements of ISO 9001 or ISO 14001.

Companies use quality audits to evaluate their own quality performance and the performance of their suppliers, licensees, agents, and others; regulatory agencies use quality audits to evaluate the performance of organizations they regulate.

The specific purpose of quality audits is to provide independent assurance that:

- Plans for attaining quality are such that, if followed, the intended quality will be attained.
- Products are fit for use and safe for the user.
- Standards and regulations defined by government agencies, industry associations, and professional societies are being followed.
- There is conformance to specifications.
- Procedures are adequate and are being followed.
- The data system provides accurate and adequate information on quality to all concerned.
- Deficiencies are identified, and corrective action is taken.
- Opportunities for improvement are identified, and the appropriate personnel are alerted.

A key question in establishing an audit program is whether the audits should be compliance oriented or effectiveness oriented or both. In practice, many quality audits are compliance oriented; the audits compare quality-related activities to some standard or requirement for those activities (e.g., do written work instructions exist for production and service operations?). The emphasis is on determining conformance to the requirement of written work instructions and maintenance of procedures for those instructions as evidence of that conformance. These audits have matured to "process-centered" audits, whereby the auditor will follow a product or particular service through its process life cycle, checking for conformance to requirements. Effectiveness audits evaluate whether the requirement is achieving the desired result (for external and internal customers) and whether the activity is making efficient use of resources (for elaboration, see Russell and Regel, 1996).

At first glance, it seems that audits should be both compliance oriented and effectiveness oriented—and sometimes they can be both. When audits are conducted internally, they can and should be both compliance and effectiveness oriented. But when audits are conducted by external parties, the companies audited can have serious and reasonable issues if an audit concerns matters of effectiveness of operations including use of resources. Effectiveness evaluations of internal operations by external auditors open a broad range of considerations about customer satisfaction and requirements and internal management processes that make it difficult to conduct such audits in a fair and useful way.

→ 16.3 PRINCIPLES OF A QUALITY AUDIT PROGRAM

Five principles are essential to a successful quality audit program:

1. An uncompromising emphasis on conclusions based on facts. Any conclusions lacking a factual base must be so labeled.
2. An attitude on the part of auditors that the audits provide assurance to management and also a useful *service* to line managers in managing their departments. Thus audit reports must provide sufficient detail on deficiencies to facilitate analysis and action by line managers.
3. An attitude on the part of auditors to identify opportunities for improvement (this is not usually the case with third-party audits because independence must be respected, but it is important for first- and second-party audits). Such opportunities include highlighting good ideas used in practice that are not part of formal procedures. Sometimes an audit can help to overcome deficiencies by communicating through the hierarchy the reasons for deficiencies that have a source in another department.
4. Addressing the human relations issues discussed.
5. Competence of auditors. The basic education and experience of the auditors should be sufficient to enable them to learn in short order the technological aspects of the operations they are to audit. Lacking this background, they will be unable to earn the respect of the operations personnel. In addition, they should receive special training in the human relations aspects of auditing. The American Society for Quality provides a program for the certification of quality auditors.

These five essentials for a successful quality audit activity were responsible for a dramatic tribute to an audit activity within one company. Line managers voluntarily give up part of their own budget each year to provide funds for a quality audit group.

➔ 16.4 SUBJECT MATTER OF AUDITS

For simple products, the range of audits is also simple and is dominated by product audits (discussed later). For complex products, the audit is far more complex. In large companies, even the division of the subject matter is a perplexing problem. For such companies, the programs of audit use one or more of the following approaches to divide up the subject matter:

- *Organizational units.* Large companies comprise several layers of organization, each with specific assigned missions: corporate office, operating divisions, plants, etc. Such companies commonly use multiple teams of quality auditors; each reviews its specialized subject matter and reports the results to its own "clientele."
- *Product lines.* Here the audits evaluate the quality aspects of specific product lines (e.g., printed circuit boards, hydraulic pumps) all the way from design through field performance.
- *Quality systems.* Here the audits are directed at the quality aspects of various segments of the overall systematic approach to quality such as design, manufacturing, supplier quality, and other processes. A system-oriented audit reviews any such system over a whole range of products. Table 16.2 provides an example from Mallinckrodt Inc., a medical products manufacturer.
- *Product and process control systems.* These type audits evaluate whether current product and process controls (including measurement) are in place and able to meet the needs of customers with regard to quality.
- *Specific activities.* Audits may also be designed to single out specific procedures that have special significance to the quality mission: disposition of nonconforming products, documentation, instrument calibration, software (see Table 16.3).

Audits of quality systems as well as specific activities may take the form of (1) audit of the plans or (2) audit of the execution versus the plans. Further, the subject matter may include internal activities or external activities such as those conducted by suppliers.

Identifying Opportunities

Experienced auditors often can discover opportunities for improvement as a by-product of their search for nonconformance to stated needs. These opportunities may even be known to the operations personnel so that the auditor is only making a rediscovery. However, these personnel may have been unable to act due to any of a variety of handicaps: preoccupation with day-to-day control, inability to communicate through the layers of the hierarchy, and lack of diagnostic support.

The auditor's relatively independent status and lack of preoccupation with day-to-day control may enable him or her to prevail over these handicaps. In addition, the auditor's reports go to multiple layers of the hierarchy and thereby have a greater likelihood of reaching the ear of someone who has the power to act on the opportunity. For example, the auditor may find that the quality cost reports

TABLE 16.2
Quality systems evaluation—components and elements

A. Organizational design
   1. Management responsibility
   2. Job descriptions
B. Customer management practices
   1. Corrective action
   2. Servicing
   3. Complaint and inquiry handling
   4. Recall and field correction
C. Organizational and individual development practices
   1. Training
   2. Personnel hygiene
D. Product development practices
   1. Device design control
   2. Concept generation
   3. Device development
   4. Transfer to operations
   5. Life-cycle maintenance/postmarket surveillance
E. Product and process control practices
   1. Process control
   2. Special processes
   3. Process capability
   4. Facilities and equipment
   5. Control of contamination
   6. Recovered material
F. Procurement practices
   1. Purchasing
   2. Contract review
G. Warehousing and distribution practices
   1. Handling, storage, distribution, and installation
H. Quality assurance practices
   1. Product identification and traceability
   2. Acceptance activities
   3. Nonconforming goods
   4. Labeling
   5. Internal quality audits
   6. Electronic data processing
I. Information analysis practices
   1. Inspection, measuring, and test equipment
   2. Statistical techniques
   3. Analytical methods and laboratories
J. Document management practices
   1. Document control
   2. Quality records
   3. Product registration and approval dossiers

are seriously delayed owing to backlogs of work in the accounting department. His or her recommendation to expedite the reports may reach the person who can act, whereas the same proposal by the operations personnel may never reach that level.

ISO 9001:2008 has a clear and strong emphasis on continuous improvement.

TABLE 16.3
Examples of audits of tasks

Scope or activity	Examples of specific tasks audited
Engineering documentation	Use of latest issue of specifications by operators; time required for design changes to reach shop
Job instructions	Existence and adequacy of written job instructions
Machines and tools	Use of specified machines and tools; adequacy of preventive maintenance
Calibration of measuring equipment	Existence of calibration procedures and degree to which calibration intervals are met
Production and inspection	Adequacy of certification program for critical skills; adequacy of training
Production facilities	General cleanliness and control of critical environmental conditions
Inspection instructions	Existence and adequacy of written instructions
Documentation of inspection results	Adequacy of detail; feedback and use by production personnel
Material status	Identification of inspection status and product configuration; segregation of defective product
Materials handling and storage	Procedure for handling critical materials; protection from damage during handling; control of in-process storage environments

✧ 16.5 STRUCTURING THE AUDIT PROGRAM

Audits of individual tasks or systems of tasks are usually structured. For example, they are designed to carry out agreed upon purposes and are conducted under agreed rules of conduct. Reaching agreement on these rules and purposes requires collaboration among three essential participating groups:

- The heads of the activities which are to be the subject of audit.
- The heads of the auditing department(s).
- The upper management, which presides over both.

Without such collective agreements, the audit program may fail. The usual failure modes are (1) an abrasive relationship between auditors and line managers or (2) a failure of line managers to heed the audit reports.

Table 16.4 depicts the typical flow of events through which audit programs are agreed on and audits are carried out. A published statement of purposes, policies, and methods becomes the charter that legitimizes audits and provides continuing guidelines for all concerned.

Audits are often done by full-time auditors who are skilled in both technical and human relations aspects. Audit teams of upper managers, middle managers, and specialists can also be effective. See *JQH5*, pages 41.16–41.19, for a discussion.

TABLE 16.4
Steps in structuring an audit program

	Audit department	Line department	Upper management
Discussion of purposes to be achieved by audits and general approach for conducting audits	X	X	X
Draft of policies, procedures, and other rules to be followed	X	X	
Final approval			X
Scheduling of audits	X	X	
Conduct of audits	X		
Verification of factual findings		X	
Publication of report with facts and recommendations	X		
Discussion of reports	X	X	X
Decisions on action to be taken		X	
Subsequent follow-up	X		

16.6 PLANNING AUDITS OF ACTIVITIES

The main steps in performing an audit are planning, opening meeting, performing, reporting, follow-up on corrective action, and closure. The flowchart in Figure 16.1 describes these steps in some detail. An excellent reference for the auditing process is ASQ Quality Audit Division (2013).

Behind the steps in Figure 16.1 are a number of important policy issues:

Legitimacy

The basic right to conduct audits is derived from the "charter" that has been approved by upper management, following participation by all concerned. Beyond this basic right are other questions of legitimacy: What is the scope and objective? What shall be the subject matter for audit? Should the auditor be accompanied during the tour? Whom may the auditor interview? The bulk of auditing practice provides for legitimacy—the auditor acts within the provisions of the charter plus supplemental agreements reached after discussion with all concerned.

Scheduled versus unannounced

Most auditing is done on a scheduled basis. "No surprises, no secrets." This practice enables all concerned to organize workloads, assign personnel, etc., in an orderly manner. It also minimizes the irritations that are inevitable when audits are unannounced.

Customer

The *customer* of the audit is anyone who is affected by the audit (see Section 1.3). The key customer is the person responsible for the activity being audited. Other customers include upper management and functions affected by the activity. Each customer has needs that should be recognized during the planning of the audit (see Chapter 11). Note that this orientation of service to the activity audited means that the audit must go beyond compliance with a requirement. Such an orientation is *not* practiced (or even accepted) by all auditors, but the author believes that the concept is basic to useful audits.

Audit team

Audits are conducted by individuals or by a team. A team usually has a lead auditor who plans the audit, creates the audit schedule, assembles or creates checklists or aid memoirs, conducts the meetings, reviews the findings and comments of the auditors, prepares the audit report, evaluates corrective action, and presents the audit report.

Clearly, auditors must be open-minded and possess sound judgment, have the trust and respect of line management, and be knowledgeable in the area audited. ISO 19011:2002 First Edition recommends other qualifications for auditors, including education, training, experience, personal attributes, and management capabilities.

Auditing is a sensitive task. A survey of auditors and auditees in the financial services industry investigated five attributes of the audit and auditor: professionalism, business knowledge, risk perspective, audit planning and conduct, and reporting audit results. Several surprises: Auditees viewed professionalism (objectivity, knowledge of the area being audited) as three times more important than auditors did; auditors thought risk perspective (coverage of key risk areas, audit in sufficient detail) as three times more important than auditees did. Notice how this disparity illustrates the importance of understanding the needs of audit customers.

Use of reference standards and checklists

As far as possible, the auditor is expected to compare activities as they are with some objective standard of what they should be. Where such standards are available, there is less need for the auditor to make a subjective judgment and thereby less opportunity for wide differences of opinion. However, provision should be made for challenge of the standard itself. The reference standards normally available include

- Written policies of the company as they apply to quality.
- Stated objectives in the budgets, programs, contracts, etc.
- Customer and company quality specifications.
- Pertinent government specifications and handbooks.
- Company, industry, and other pertinent quality standards on products, processes, and computer software.
- Published guides for conduct of quality audits.
- Pertinent quality departmental instructions.
- General literature on auditing.

528 Quality Management and Analysis

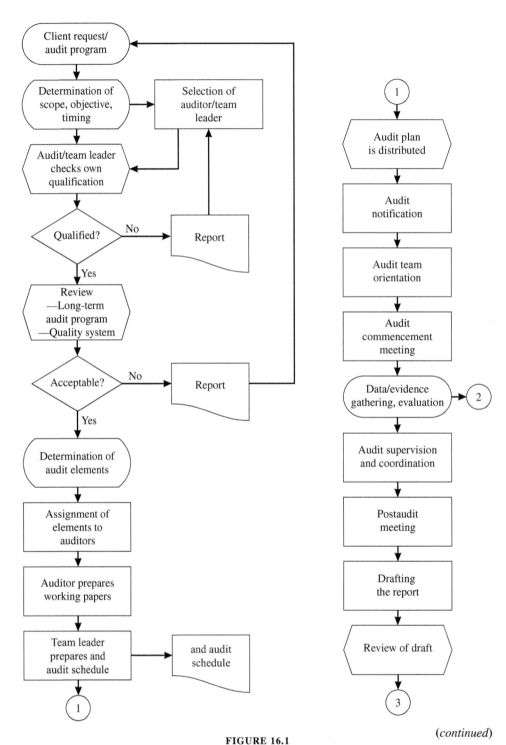

FIGURE 16.1
Flowchart for quality audit. (*Adapted from ANSI/ASQC, 1986, pp. 9–13.*)

(*continued*)

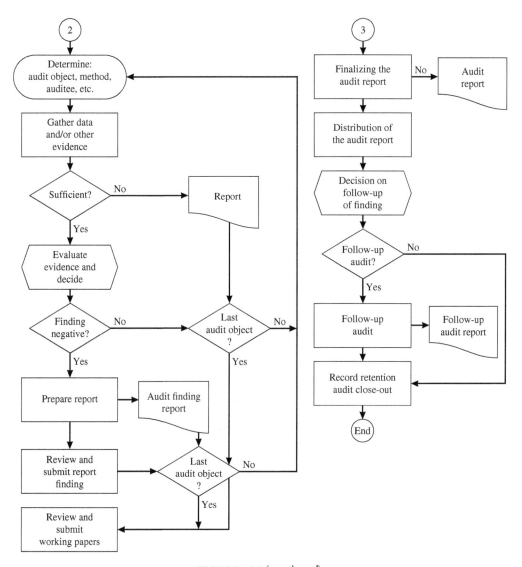

FIGURE 16.1 (*continued*)

One type of checklist identifies areas of subject matter that are to be checked, leaving it to the auditor to supply the detailed checklist. Typical examples of such areas are maintenance of machines and tools or control of engineering change orders. Having a standard for comparison in audits is an important matter. Steven Ehrhardt of Mallinckrodt Inc. uses a perceptive principle. When his audit group is asked to make an audit, he first determines whether the company has a standard or clear job specification for the activity. If not, then *before* making the audit, he asks those responsible for the activity to define a standard. Without a standard, Ehrhardt believes, people do not know what they are supposed to do (the first element of self-control, see

TABLE 16.5
Potential stages of product auditing

Stage at which product auditing is conducted	Pros and cons of using this stage
After acceptance by inspectors	Most economical, but does not reflect effects of packing, shipping, storage, or usage
After packing but before shipment to field	Requires unpacking and repacking, but evaluates effect of original packing
Upon receipt by dealers	Difficult to administer at such multiple locations, but reflects effects of shipping, storage
Upon receipt by users	Even more difficult to administer, but evaluates the added effects of dealer handling and storage plus effects of shipment to user and unpacking
Performance in service	The ideal, but also the most difficult to administer because of the number and variety of usages; can be simplified through sampling

Section 6.3), and thus an audit is not appropriate. Clarification of standards may be more important than the audit itself. Imagine how this initiative to provide service by an auditor helps to build a trusting relationship with operations people.

Some checklists go into great detail, requiring the auditor to check numerous items of operational performance (and to record the fact that such items were checked). For example, an auditor checking a test performed by an inspector might be required to check the work of the inspector for the correctness of the specification issue number used, the list of characteristics checked, the type of instruments employed, the sample size, data entries, etc. In a hospital, an audit checklist could include questions such as, "Are applicable intravenous solutions stored under refrigeration prior to delivery?" and "Are all drugs, chemicals, and biologicals clearly, accurately, and appropriately labeled?" For an example, see Table 16.5.

Section 13.9 discusses process quality audits in manufacturing operations; Section 14.5 explains applications of process quality audits in service operations.

→ 16.7 AUDIT PERFORMANCE

Several policy issues affect audit performance.

Verification of facts

Auditors are universally expected to review with the line supervision the facts (outward symptoms) of any deficiencies discovered during the audit. The facts should be agreed on before the item enters a report that will go to higher management.

Discovery of causes

Many companies expect the auditor to investigate major deficiencies to determine their causes. This investigation then becomes the basis of the auditor's recommendation.

Other companies expect the auditor to leave such investigations to the line people; audit recommendations will then include proposals for such investigations. As mentioned earlier, typically recommendations are not given for third-party audits.

Recommendations and remedies

Auditors are invariably expected to make recommendations to reduce deficiencies and improve performance. In contrast, auditors are commonly told to avoid becoming involved in designing remedies and making them effective. However, auditors are expected to follow up recommendations to ensure that something specific is done, i.e., that the recommendation is accepted or else considered and rejected.

Policy issues are often incorporated into a "quality audit manual." Such a manual also includes details on the subject matter to be covered in audits; checklists of items to be checked and questions to be asked; classification of the seriousness of deficiencies observed; use of software for entry, processing, storage, and retrieval of audit data; and guidelines for audit reports.

Status of the audit

The key customer should be kept informed about progress of the audit—what has been covered and what remains to be done. The status of lengthy audits can be reported through debriefing meetings, informal discussions, and electronic mail. Status reporting includes explaining what deficiencies or problems have been detected—even before preparing a draft of the audit report. This is usually accomplished during the report close-out meeting at the conclusion of the audit or could be a daily report close-out meeting. Status reports enable the company to check the accuracy of the auditors' observations and give the people responsible for the activity audited a chance to explain their plan to correct the deficiency.

→ 16.8 AUDIT REPORTING

Audit results should be documented in a report, and a draft should be reviewed (preferably at the postaudit meeting) with the management of the activity that was audited. Auditors and the activity audited should agree in advance on the distribution of the audit report. If desired, the report may be issued by the auditor and the auditee. All members of an audit team and the auditee should sign the report.

The report should include the following items:

- Executive summary.
- Purpose and scope of the audit.
- Details of the audit plan, including audit personnel, dates, the activity that was audited (personnel contacted, material reviewed, number of observations made, etc.). Details should be placed in an appendix.
- Standards, checklist, or other reference documents that were used during the audit.
- Audit observations, including supporting evidence, conclusions, and recommendations—using the audit customer's terminology.

- Recommendations, if applicable, for improvement opportunities.
- Recommendations, if applicable, for follow-up on the corrective action that is to be proposed and implemented by line management, along with subsequent audits if necessary.
- Distribution list for the audit report.

Summarizing audit data

In an audit most elements of performance are found adequate, whereas some are found in a state of nonconformance to established standards. Reporting of these findings requires two levels of communication:

1. *Reports of nonconformance to secure corrective action.* These reports are made promptly to the responsible operating personnel, with copies to some of the managerial levels.
2. *A report of the overall status of the subject matter under review.* To meet these requirements, the report should
    - Evaluate overall quality performance in ways that provide answers to the major questions raised by upper managers, e.g., Is the product safe? Are we complying with legal requirements? Is the product fit for use? Is the product marketable? Is the performance of the department under review adequate?
    - Provide evaluations of the status of the major subdivisions of the overall performance—the quality systems and subsystems, the divisions, the plants, the procedures, etc.
    - Provide some estimate of the frequency of inadequacies in relation to the number of opportunities for inadequacies (see "Units of Measure").
    - Provide some estimate of the trend of this ratio (of inadequacies found to inadequacies possible) and of the effectiveness of programs to control the frequency of occurrence of inadequacies.

Seriousness classification

Some audit programs use seriousness classification of inadequacies. This approach is quite common in product audits, where defects are classified in terms such as critical, major, and minor, each with some "weight" in the form of demerits. These systems of seriousness classification are highly standardized (see Section 10.5).

Some audit programs also apply seriousness classification to discrepancies found in planning, in procedures, in decision making, in data recording, and so on. The approach parallels that used for product audits. Definitions are established for such terms as "serious," "major," and "minor"; demerit values are assigned; and total demerits are computed.

Units of measure

For audits of plans, procedures, documentation, etc., it is desirable to compare the inadequacies found against some estimate of the opportunities for inadequacies. Some companies provide an actual count of the opportunities, such as the number of criteria or check points called out by the plans and procedures. Another form is to count the inadequacies per audit with a correction factor based on the length of time consumed by the audit. The obvious reason is that more time spent in auditing means more ground covered and more inadequacies found.

Distribution of audit report

Traditionally, copies of the audit report are sent to upper management for notification, review, and possible follow-up. Clearly, managers of audited activities are not happy with audit reports listing various nonconformances that are sent to their superiors. To promote harmony and a constructive viewpoint on audits, some organizations have adopted a different policy. The audit report is sent only to the manager whose activity is audited, and a follow-up audit is scheduled. If the nonconformances are corrected in time for the follow-up audit, the audit file is closed; otherwise, a copy of both audit reports is sent to upper management.

Another approach is to establish prior to the audit the single point of accountability (SPA) for the audited function. This individual is recognized as the point of contact for the lead auditor and the individual singularly responsible in the audited function for report receipt and internal distribution.

In the spirit of ongoing improvement, after the report is issued, the auditees should be asked about the value they received from the audit and the report.

Regel (2000) reports on a survey of quality auditors. Of their seven primary concerns in achieving closure on their audits, the one ranked first in importance was "report issues" that accounted for 25.3% of the total concerns. Examples of reasons included "insufficient detail," "not linked to business goals," "not stated clearly," and "not stated in management language."

✦ 16.9 CORRECTIVE ACTION FOLLOW-UP

The final phase of the audit is follow-up to confirm that corrective action has been taken by the audited activity and that the corrective action is effective. The steps are shown in Figure 16.2 (from Russell and Regel, 1996; reprinted by permission of the ASQ). ASQ Quality Audit Division (2013) provides details on this process.

It is important to remember the key purpose of an audit: improvement. If corrective action is not implemented for some reason, the auditor should first verify that the conclusions in the report are correct and agreed upon by the audited area. If this is not the case, then the lack of agreement must be resolved. If there is agreement with the report but the audited area has not been able to implement corrective action because of lack of resources or other reasons, the auditor should determine whether he or she can somehow help the audited area. One possible approach is to review how the seriousness of the deficiency is presented to management. Restating the deficiency in monetary or other terms that will have an impact on management can help to obtain the necessary resources or remove obstacles to implementation of the corrective action. For an example, see Section 16.13.

534   Quality Management and Analysis

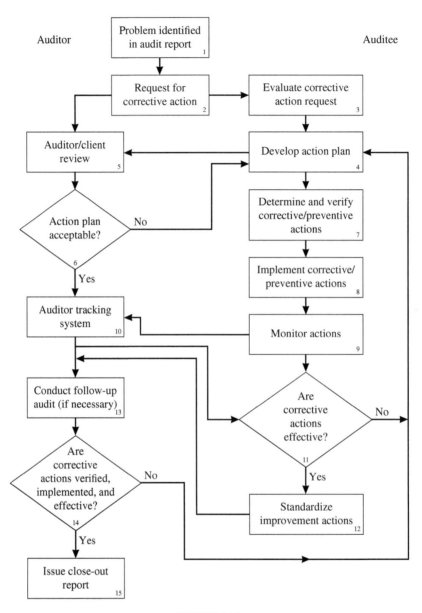

FIGURE 16.2
The audit function improvement process.

→ 16.10   HUMAN RELATIONS IN AUDITING

In theory, the audit is an instrument plugged into operations to secure an independent source of information. Where it is a physical instrument, e.g., the propeller speed indicator on the bridge of a ship, there is no clash of personalities. However, auditors are human beings, and in practice, their relationships with those whose work is being audited can become quite strained. Deficiencies turned up in the audit may be resented because of the implied criticism.

Recommendations in the audit may be resented as an invasion of responsibilities. In the reverse direction, auditors may regard slow responses to requests for information as a form of grudging cooperation. These and other human relations problems are sufficiently important to warrant extensive discussion plus indoctrination of both auditing personnel and operations personnel with respect to the following issues:

- *The reasons behind the audits.* These reasons may have been well discussed during the basic formulation of the audit program. However, that discussion was held among the managers. There is also a need to explain to both supervisors and nonsupervisors the "why" of the audits. (It is not enough to explain that upper management wants audits done.) Obviously, all employees are also customers, consumers, and concerned citizens, so it is easy to point out the benefits they derive from audits conducted in other companies. In addition, it can be made clear that the managers, customers, regulators, etc., of this company likewise require added assurance.
- *Avoiding an atmosphere of blame.* A sure way to cause a deterioration in human relations is to look for someone to blame rather than how to achieve improvement. Line managers as well as auditors can fall into this trap. An atmosphere of blame breeds resentment and also dries up the sources of information. Audit reports and recommendations should be problem oriented rather than person oriented.
- *Balance in reporting.* An audit that reports only deficiencies may be factual as far as it goes. Yet it will be resented because nothing is said about the far greater number of elements of performance that are done well. ("Even a broken clock is correct twice a day"—Anonymous.) Some companies require that the auditors start their reports with "commendable observations." Others have evolved overall summaries or ratings that consider deficiencies and also the opportunities for deficiencies.
- *Depersonalizing the report.* In many companies, auditors derive much influence from the fact that their reports are reviewed by upper management. Auditing departments should be careful to avoid misusing this influence. The ideal is to depersonalize the reports and recommendations. The real basis of the recommendations should be the facts rather than the opinion of the auditor. Where there is room for a difference of opinion, auditors have a right and a duty to give their opinions as input to the decision-making process. However, any position of undue advocacy should be avoided because this tends to reduce the auditor's credibility as an objective observer. (The ultimate responsibility for results rests on the line managers, not on the auditors.)
- *Postaudit meeting.* An important part of the implementation phase is the postaudit meeting that is held with the manager and his or her team of the audited activity. At this meeting, the audit observations are presented so that the manager and the team can plan for corrective action. In addition, the manager and team can point out to the auditor any mistakes with respect to the facts that have been collected.

A self-audit and an independent audit can be combined to provide a two-tier audit. Each audit has an audit plan, execution, and report. The advantages include using the expertise of the person responsible for the function, assuring objectivity with an independent auditor, and minimizing some of the human relationship issues.

The aim of both the self-audit and the independent audit is to build an atmosphere of trust based on the prior reputation of the auditors, the approach used during the audit, and an emphasis on being helpful to the activity audited. Even such small matters as the title of the audit process should be carefully considered. Occasionally, people try to avoid the use of the term *audit* when what will be done *is* observation and evaluation. Also, audits may be hidden in a company education program. Such subterfuges detract from the trust that must be developed for audits to be effective and useful.

Auditors must function within the culture of an organization. Anthropology is the study of the origin; behavior; and physical, social, and cultural development of human beings. For the astute comments of a quality manager (who is a certified quality auditor) with a formal background in anthropology, see Hunt (1997).

Quality auditors must, of course, adhere to the highest standards of ethics and professional conduct. For a discussion of the specifics including the American Society for Quality Code of Ethics and the Institute of Internal Auditors Code of Ethics, see ASQ Quality Audit Division (2013).

→ 16.11 PRODUCT AUDIT

A product audit is an independent evaluation of a product's quality to determine its fitness for use and conformance to specification. Product auditing takes place after inspections have been completed. The purposes of product auditing include:

1. Estimating the quality level delivered to customers.
2. Evaluating the effectiveness of the inspection decisions in determining conformance to specifications.
3. Providing information useful in improving the outgoing product quality level and improving the effectiveness of inspection.
4. Providing additional assurance beyond routine inspection activities.

There is a good deal of logic behind such product audits. In many cases, the inspection and testing department is subordinate to a manager who is also responsible for meeting other standards (schedules, costs, etc.). In addition, there is a value in reviewing the performance of the entire quality control function, which includes inspection and test planning as well as the conduct of the tests themselves. Finally, the more critical the product, the greater the need for some redundancy as a form of assurance.

Stage of evaluation

Ideally, the product audit should compare actual service performance with users' service needs. This ideal is so difficult and costly to administer that most product auditing consists of an approximation (see Table 16.5).

For many simple, stable products, the approximation of test results versus specifications is a useful, economical way of conducting a product audit. Even for products not so simple, most quality characteristics identifiable by the user are also completely identifiable while the product is still at the factory. Thus product characteristics that are essential to use are properly evaluated at some appropriate stage, whether in the factory or at some more advanced stage.

As products become increasingly complex, product auditing is increasingly conducted at several of the stages shown in Table 16.5. The bulk of the characteristics may be evaluated at the most economical stage, i.e., shortly after factory inspection. However, the remaining (and usually more sophisticated) characteristics may be evaluated at other stages.

Scope of the product audit

The scope of some product audits completely misses the mark in measuring customer reaction.

As one example, the plant manager of an electronics manufacturing firm received a rating of 98% on a product audit from the plant. For this rating, the plant received an award for quality. When the mean time between failures of that same product was measured in the field, the value was only 200 hours. This problem was a known reason for customer complaints, but such matters had not been evaluated by the product audit.

In another case, a vehicle manufacturer had a system of taking a weekly product audit sample from production. A comparison of separate market research results with the internal product audit was devastating. Only 18% of the characteristics that customers claimed were important to them were being checked in the product audit.

For simple products, a representative sample of finished goods may be bought on the open market. These samples are then checked for fitness for use and conformance to specification. Some companies conduct such audits annually as part of broad annual planning for the product line. Such audits may include a review of competitive product as well.

For complex consumer products, e.g., household appliances, it is feasible to secure product audit data at multiple stages of the product progression shown in Table 16.5. The most extensive product audit takes place immediately following factory inspection and testing. Additional audit data are then secured from selected distributors and dealers under a special joint "open and test" audit. Similar arrangements are made to secure data from selected servicing dealers. In addition, the data from consumer "arrival cards" are used. When properly arranged with due regard to time lags, all of these data sources can be charted to show trends as well as levels.

Audit plans must spell out, or give guidance on, the selection of detailed product dimensions or properties that are to be checked. Provision should be made for two types of audit—random and focused. The former is based on a random selection of product characteristics to yield an unbiased picture of quality status. A focused audit, on the other hand, concentrates on a specific area of the product that experience suggests needs to be studied. Many companies use audit manuals to spell out the design of the audit for the auditor, almost to the last level of detail. For example, the manual may specify particular categories of dimensions to be audited (i.e., length) but may rely on the auditor to select which length dimension to audit.

→ 16.12 SAMPLING FOR PRODUCT AUDIT

For products manufactured by mass production, sample sizes for product audit can often be determined by using conventional statistical methods. These methods determine the sample size required for stated degrees of risk (see Section 18.4). Sample sizes for product audit determined by these methods when applied to mass production still represent a small fraction of the product that needs to be sampled. In contrast, for products manufactured as large units or in small quantities, the conventional concepts of statistical sampling are prohibitively costly. In such cases, sample sizes are often arbitrary, and they seem small from the viewpoint of probability. For example, a vehicle manufacturer uses a product audit sample consisting of 2% of production per shift with a minimum of five vehicles—whichever number is larger. Even though the number of vehicles sampled may be small, the total number of characteristics that is sampled may be quite large. Auditors check 380 items on each vehicle, and the product audit test includes a 17-mile road test. In some cases of highly homogeneous production, a sample of one unit taken from batch production can be adequate for a product audit.

→ 16.13 REPORTING THE RESULTS OF PRODUCT AUDIT

The results of a product audit appear in the form of the presence or absence of defects, failures, etc. A continuing score or "rating" of quality is then prepared based on the audit results.

Product audit programs often use a seriousness classification of defects. Defects are classified in terms such as critical, major, minor A, minor B, and incidental at times, each with some "weight" in the form of demerits. In product audits, the usual unit of measure is *demerits per unit of product.*

> **EXAMPLE 16.1.** A product audit system uses four classes of seriousness of defects. During one month, the product auditors inspected 1200 finished units of product with the following results:
>
> | Type of defects | Number found | Demerits per defect | Total demerits |
> |---|---|---|---|
> | Critical | 1 | 100 | 100 |
> | Major | 5 | 25 | 125 |
> | Minor A | 21 | 5 | 105 |
> | Minor B | 64 | 1 | 64 |
> | Total | 91 | | 394 |

Although the 91 defects found represented many defect types and four classes of seriousness, the total of 394 demerits, when divided by the 1200 units inspected, gives a single number, i.e., 0.33 demerit per unit.

The actual number of demerits per unit for the current month is often compared against historical data to observe trends. (Sometimes it is compared with competitors'

TABLE 16.6
Audit data of 50 units

Class of defect	Probability	Number of defects revealed by audit	Cost per service call, $	Expected costs, $	Expected number of service calls
V1	1.00	1	15.00	15.00	1.00
V2	0.60	3	15.00	27.00	1.80
E1	1.00	3	30.00	90.00	3.00
E2	0.60	4	30.00	72.00	2.40
P1	1.00	1	25.00	25.00	1.00
P2	0.60	2	25.00	30.00	1.20
P3	0.20	2	25.00	10.00	0.40
Totals				269.00	10.80

products to judge the company's quality versus market quality.) A major value of a measure such as demerits per unit is that it compares discrepancies found with the opportunity for discrepancies. Such an index appeals to operating personnel as eminently fair.

A scoreboard in demerits per unit is by no means universally accepted. Managers in some industries want ready access to the figures on critical and major defects. These managers believe that such figures represent the real problems regardless of demerits per unit.

It is often useful to summarize the product results in other languages. A manufacturer of consumer products classifies defects in a product audit as visual (V), electrical (E), and performance (P) and then predicts service costs for products in the field. This is done by first establishing classes for each type of defect in terms of the probability of receiving a field complaint (e.g., a class 2 visual defect has a 60% probability). Service call costs are then combined with audit data. For example, Table 16.6 shows the results of an audit of 50 units. The expected cost is the product of the probability, the number of defects, and the cost per service call. The expected service cost per unit is then estimated as $269/50 = $5.38. Alternatively, as indicated in Table 16.6, the expected number of service calls is the product of the probability and the number of defects. The expected number of service calls per unit is then estimated as 10.8/50 = 0.22, or about 22 of every 100 products delivered to the field can be expected to have service calls.

In addition to summarizing the defects found (in both number and relative seriousness), the audit results can be tallied by functional responsibility (i.e., design, purchasing, production).

Audit results can also be summarized to show the effectiveness of the previous inspection activities. Typically, a simple ratio is used, such as the percentage of total defects detected by inspection. For example, if the previous inspection revealed a total of 45 defects in a sample of N pieces and if the product audit inspection revealed five additional defects, the inspection effectiveness would be (45/50)(100), or 90%.

Stravinskas (1989), Lane (1989), and Williams (1989) describe how AT&T Microelectronics changed from a traditional audit approach of reinspection of product

(product audit) to an approach that uses system audits and process audits combined with a reduced amount of product audit. During one period, inspection costs were reduced by 12%, and the savings were then used to provide additional prevention activity, which resulted in a $2 saving in failure costs for each $1 added in prevention cost.

SUMMARY

- Quality assurance is the activity of providing evidence to establish confidence that quality requirements will be met.
- A quality audit is an independent review conducted to compare some aspect of quality performance with a standard for that performance. We conduct quality audits on activities that have an impact on product quality.
- Five ingredients are essential for successful audits: emphasis on facts, attitude of service on the part of auditors, identification of opportunities for improvement, addressing human relations issues, and the competence of auditors.
- Quality surveys provide a broader review of quality activities than audits of specific activities.
- A product audit is an independent evaluation of product quality to determine fitness for use and conformance to specifications.

PROBLEMS

16.1. Visit a retail establishment that sells consumer electronics and observe the extent to which customers use their senses in securing quality assurance of the products they buy. Report your findings.

16.2. Visit an apartment building and discuss with the superintendent the various means used to obtain early warning of potential dangers, e.g., burglary, fire. Report your findings.

16.3. List the early-warning devices in use in a private home. Report your findings.

16.4. You are a quality manager. On one of your company's product lines, a report shows that power consumption has risen from the usual level of 35.5 W to a level of 35.9 W. This difference is, without a doubt, statistically significant. However, the line manager has taken no action to investigate the reason for the change on the grounds that (1) the product still conforms to the specification limit of maximum 36.4 W, and (2) he must give priority to several other products in which there are failures to comply with specification. What action do you take?

16.5. Visit a nearby facility that is a part of a chain of such facilities, e.g., food market, restaurant, motel, gasoline station. You will most likely find that it is subject to periodic quality audits from some headquarters. Obtain a copy of the auditor's checklist, study it, and report on its contents with respect to the various aspects of quality audits discussed in this chapter.

16.6. For any manufacturing company to which you have access, secure a copy of a quality audit. Study it and report on its contents with respect to the various aspects of quality audits discussed in this chapter.

REFERENCES

ASQ Quality Audit Division (2013). *The Quality Audit Handbook,* 2nd ed., ASQ, Milwaukee.
Hunt, J. R. (1997). "The Quality Auditor: Helping Beans Take Root," *Quality Progress,* December, ISO 19011:2002: *A New Auditing Standard for QMS and EMS,* pp. 27–33.
Lane, P. A. (1989). "Continuous Improvement—AT&T QA Audits," *ASQC Quality Congress Transactions,* Milwaukee, pp. 772–775.
Regel, T. (2000). "Management Audit and Compliance Audit Compatibility," *Annual Quality Congress Proceedings,* ASQ, pp. 606–609.
Russell, J. P. and T. Regel (1996). *After the Quality Audit: Closing the Loop on the Audit Process,* Quality Press, ASQ, Milwaukee.
Stravinskas, J. M. (1989). "Manufacturing System and Process Audits," *ASQC Quality Congress Transactions,* Milwaukee, pp. 91–94.
Williams, C. A. (1989). "Improving Your Quality Auditing Systems," *ASQC Quality Congress Transactions,* Milwaukee, pp. 797–799.

CHAPTER 17

The Role of Statistics and Probability

17.1 STATISTICAL TOOLS IN QUALITY

Statistics or *statistical inference* is the science of making decisions under uncertainty. We must confront uncertainty in our decisions because the real world is subject to random variation (discussed later). Statistical methods help us make good decisions—and avoid bad ones—in the face of that variation-induced uncertainty. At a deeper level, quality, in the sense of freedom from deficiencies (see Chapter 1), requires statistical tools because most deficiencies arise as the result of random variation.

The term *statistics* is also used in everyday speech to mean any systematic collection, tabulation, analysis, or interpretation of data. It can even mean just data. Of course, in this more mundane usage, we also apply statistics for managing and improving quality, but the topic of the next four chapters relates to the application of science of statistics for dealing with quality issues.

Probability is a measure that describes the chance that an event will occur based on the underlying variation in the phenomenon being analyzed. Measures of probability allow us to make quantitative statements about the risks associated with the decisions we must make based on the data available to us.

Of course, statistical tools are only a part of the capabilities one must master to achieve outstanding quality, but they are a critical part.

17.2 THE CONCEPT OF VARIATION

The concept of *variation* states that no two items will be perfectly identical. Variation is a fact of nature and a fact of industrial life. For example, even "identical" twins vary slightly in height and weight at birth.

544 Quality Management and Analysis

The dimensions of a large-scale, integrated chip vary from chip to chip; the proportions of the various ingredients in cans of tomato soup vary slightly from can to can; the time required to check in at an airline check-in counter varies from passenger to passenger. To disregard the existence of variation (or to rationalize falsely that it is small) can lead to seriously incorrect decisions on major problems. Statistics helps to analyze data properly and draw conclusions, taking into account the existence of variation.

Statistical variation—variation due to random causes—is much greater than most people think. Often we decide what action to take based on the most recent data point, and we forget that the data point is part of a history of data. Malcolm Roberts once said: "Many managers run their systems by the last data point."

Data summarization can take several forms: tabular, graphical, and numerical. Sometimes one form will provide a useful, complete summarization. In other cases, two or even three forms are needed for complete clarity.

→ 17.3 TABULAR SUMMARIZATION OF DATA: FREQUENCY DISTRIBUTION

A *frequency distribution* is a tabulation of data arranged according to size. The raw data of the electrical resistance of 100 coils are given in Table 17.1. Table 17.2 shows the frequency distribution of these data with all measurements tabulated at their actual values. For example, there were 14 coils each of which had a resistance of 3.35 ohms (Ω); there were 5 coils each of which had a resistance of 3.30 Ω. The frequency distribution spotlights where most of the data are grouped (the data are centered about a resistance of 3.35) and how much variation there is in the data (resistance runs from 3.27 to 3.44 Ω). Table 17.2 shows the conventional frequency distribution and the cumulative frequency distribution in which the frequency values are accumulated to show the number of coils with resistances equal to or less than a specific value. The particular problem determines whether the conventional, cumulative, or both distributions are required. Note, of course, that resistance is measured here only to the nearest hundredth of an ohm. The five 3.30 values are not likely equal if measured to additional significant figures.

TABLE 17.1
Resistances of 100 coils, Ω

3.37	3.34	3.38	3.32	3.33	3.28	3.34	3.31	3.33	3.34
3.29	3.36	3.30	3.31	3.33	3.34	3.34	3.36	3.39	3.34
3.35	3.36	3.30	3.32	3.33	3.35	3.35	3.34	3.32	3.38
3.32	3.37	3.34	3.38	3.36	3.37	3.36	3.31	3.33	3.30
3.35	3.33	3.38	3.37	3.44	3.32	3.36	3.32	3.29	3.35
3.38	3.39	3.34	3.32	3.30	3.39	3.36	3.40	3.32	3.33
3.29	3.41	3.27	3.36	3.41	3.37	3.36	3.37	3.33	3.36
3.31	3.33	3.35	3.34	3.35	3.34	3.31	3.36	3.37	3.35
3.40	3.35	3.37	3.35	3.32	3.36	3.38	3.35	3.31	3.34
3.35	3.36	3.39	3.31	3.31	3.30	3.35	3.33	3.35	3.31

TABLE 17.2
Tally of resistance values of 100 coils

Resistance, Ω	Tabulation	Frequency	Cumulative frequency
3.45			
3.44	|	1	1
3.43			
3.42			
3.41	||	2	3
3.40	||	2	5
3.39	||||	4	9
3.38	⧕ |	6	15
3.37	⧕ |||	8	23
3.36	⧕ ⧕ |||	13	36
3.35	⧕ ⧕ ||||	14	50
3.34	⧕ ⧕ ||	12	62
3.33	⧕ ⧕	10	72
3.32	⧕ ||||	9	81
3.31	⧕ ||||	9	90
3.30	⧕	5	95
3.29	|||	3	98
3.28	|	1	99
3.27	|	1	100
3.26			
Total		100	

When there are a large number of highly variable data, the frequency distribution can become too large to serve as a summary of the original data. The data may be grouped into *cells* to provide a better summary. Table 17.3 shows the frequency distribution for these data grouped into six cells, each 0.03 Ω wide. Grouping the data into cells condenses the original data, and therefore some detail is lost.

The following is a common procedure for constructing a frequency distribution:

1. Decide on the number of cells. Table 17.4 provides a guide.
2. Calculate the approximate cell interval i. The cell interval equals the largest observation minus the smallest observation divided by the number of cells. Round this

TABLE 17.3
Frequency table of resistance values

Resistance, Ω	Frequency
3.415–3.445	1
3.385–3.415	8
3.355–3.385	27
3.325–3.355	36
3.295–3.325	23
3.265–3.295	5
Total	100

**TABLE 17.4
Number of cells in
frequency distribution**

Number of observations	Recommended number of cells
20–50	6
51–100	7
101–200	8
201–500	9
501–1000	10
Over 1000	11–20

result to some convenient number (preferably the nearest *uneven* number with the same number of significant digits as the actual data).
3. Construct the cells by listing cell boundaries.
   a. Each cell boundary should be to one more significant digit than the actual data and should end in a 5.
   b. The cell interval should be constant throughout the entire frequency distribution.
4. Tally each observation in the appropriate cell and then list the total frequency f for each cell.

This procedure should be adjusted when necessary to provide a clear summary of the data and to reveal the underlying pattern of variation.

➔ 17.4 GRAPHICAL SUMMARIZATION OF DATA: THE HISTOGRAM

A *histogram* is a vertical bar chart of a frequency distribution. Figure 17.1 shows the histogram for the electrical resistance data. Note that as in the frequency distribution, the histogram highlights the center and amount of variation in the sample of data. The simplicity of construction and interpretation of the histogram makes it an effective tool in the elementary analysis of data.

Graphical methods are essential to effective data analysis and clear presentation of results. Many of these methods are used throughout this book. More are available. Experience dictates that the first step in data analysis is to plot the data using an appropriate graphical tool. While most graphical analysis will also require statistical support, simply calculating the various statistics without examining the data graphically can lead to faulty analysis.

One variation of the histogram is the stem-and-leaf plot. Heyes (1985) presents data on wire break strength in grams (see Table 17.5) for supplier A. The corresponding stem-and-leaf plot is shown in Figure 17.2. Note that the stem is the first digit(s) of each value and the leaf is the remaining digits, e.g., for a value of 216, the stem is 2 and the leaf is 16. Note that this plot reveals the shape of the histogram and also makes it possible to regain the original values of the data.

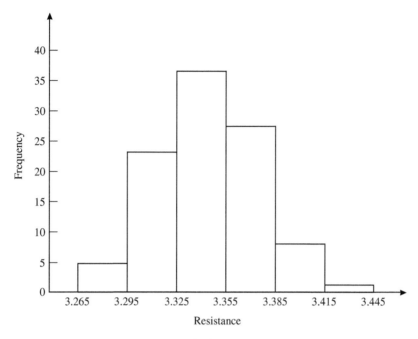

FIGURE 17.1
Histogram of resistance.

TABLE 17.5
Original data on wire break strength

1. 346	6. 402	11. 368
2. 338	7. 635	12. 376
3. 323	8. 281	13. 311
4. 438	9. 431	14. 379
5. 398	10. 390	15. 216

Stem	Leaf
2	16, 81
3	11, 23, 38, 46, 68, 76, 79, 90, 98
4	02, 31, 38
5	–
6	35

FIGURE 17.2
Stem-and-leaf plot. (*From Heyes, 1985.*)

→ 17.5 BOXPLOTS

A simple, clever, and effective way to summarize data is a boxplot. The *boxplot* is a graphical five-number summary of the data. In the basic boxplot, the five values are the median, maximum value, minimum value, first quartile, and third quartile. The quartiles are the values below which one-fourth and three-fourths of the observations lie.

Using the wire break strength data, the data are first arranged in rank order (see Table 17.6). The median is the middle value (the eighth rank, or 376). The maximum and minimum values are 216 and 635. The quartiles are 323 and 402 because those values divide the data into quarters. Figure 17.3 shows the resulting boxplot. The box, bounded by the two quartiles with the median inside the box, summarizes the middle part of the data. The lines extending out to the extreme values are the "whiskers." The longer whisker on the right suggests that the data include some values that are much larger than the other values. Also, the location of the median indicates that the values above the median are, as a group, closer to the median than the values below the median.

Innovative methods of graphical analysis and display of data are discussed in a now classic text by Tukey (1977). A good summary of graphical methods including the boxplot is presented in Wadsworth et al. (2001).

In another example, a large service organization had a problem of poor reliability of copy machines. The copiers were repaired by two different firms (A and B). An improvement team conducted a study to see which contractor provided the faster and more consistent response to a repair call. The response time, in minutes, was recorded for a sample of 10 repair calls for each contractor. The resulting boxplots are shown in Figure 17.4. The team concluded that contractor B was superior because B was usually faster and more consistent than A.

TABLE 17.6
Ordered data on wire break strength

1. 216	6. 346	11. 398
2. 281	7. 368	12. 402
3. 311	8. 376	13. 431
4. 323	9. 379	14. 438
5. 338	10. 390	15. 635

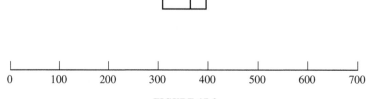

FIGURE 17.3
Boxplot. (*From Heyes, 1985.*)

The Role of Statistics and Probability   549

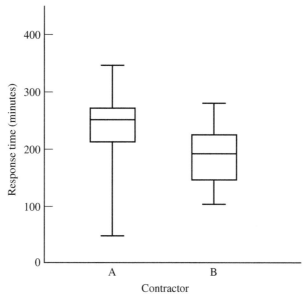

FIGURE 17.4
Basic boxplots of copier repair response time.
(*Source: Juran Institute, Inc.*)

✦ 17.6 GRAPHICAL SUMMARIZATION OF TIME-ORIENTED DATA: THE RUN CHART

The histogram is a simple and effective way to summarize data variation according to size. The run chart is a plot of the data versus time. Such plots can reveal trends, cycles, and other changes over time. An example is presented in Figure 17.5.

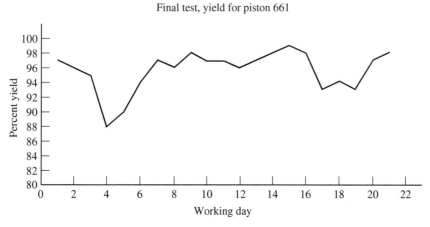

FIGURE 17.5
Run chart. (*Source: Juran Institute, Inc.*)

More powerful plots of data over time are presented in Chapter 20, "Tools to Maintain the Control of Quality."

Readers should be cautious in employing certain graphical methods of the popular process. Some of these (e.g., three-dimensional bar graphs) sacrifice clarity for glitz. A summary of graphical methods for quality is presented in Wadsworth et al. (2000). Harris (1996) provides a comprehensive reference on "information graphics." For unusually creative ideas on displaying all types of information, see Tufte (1997).

→ 17.7 METHODS OF SUMMARIZING DATA: NUMERICAL

Data can also be summarized by computing (1) a measure of central tendency to indicate where most of the data are centered and (2) the measure of dispersion to indicate the amount of scatter in the data. Often these two measures provide an adequate summary.

The key measure of the central tendency is the *arithmetic mean,* or average. The definition of the *mean* is

$$\overline{X} = \frac{\sum_{i=1}^{n} X_i}{n}$$

where $\overline{X}$ = sample mean
X_i = individual observations
n = number of observations
$\sum_{i=1}^{n} X_i$ = summation of the X_i

Another measure of central tendency is the *median*—the middle value when the data are arranged according to size. The median is useful for asymmetric distributions or for data that can be ranked but are not measured on a continuous metric, such as color or visual appearance.

Two measures of dispersion are commonly calculated.

In general, the *standard deviation* is the most useful measure of dispersion. Like the mean, the definition of the standard deviation is a formula:

$$s = \sqrt{\frac{\sum_{i=1}^{n}(X_i - \overline{X})^2}{n-1}}$$

where s is the sample standard deviation. The square of the standard deviation, s^2, is called the *variance.*

The standard deviation is the square root of the average of squared deviations of the observations from their mean. There is usually difficulty in understanding the "meaning" of the standard deviation. The only definition is the formula. There is no hidden meaning to the standard deviation, and it is best viewed as an index that shows the amount of variation in a set of data. Later, applications of the standard deviation to predictions will help clarify its meaning.

One useful technique is to calculate a relative measure of variation as the standard deviation divided by the mean (the coefficient of variation).

When the amount of data is small (10 or fewer observations), the *range* is a more useful measure of dispersion. The range is the difference between the maximum value and the minimum value in the data. Because the range is based on only two values, it contains less information about the data. For data sets of more than 10, the standard deviation is usually the best choice.

A problem that sometimes arises in the summarization of data is that one or more extreme values are far from the rest of the data. A simple (but not necessarily correct) solution is available: drop such values. The reasoning is that a measurement error or some other unknown factor makes the values "unrepresentative." Unfortunately, this approach may be rationalizing to eliminate an annoying problem of data analysis and may seriously mislead the investigation. Extreme values should be excluded only if evidence can be developed that, in fact, the outlier arises from some special cause such as a measurement error. Remember that we are using statistics to help us make better decisions. If we change the data to fit our priors, then we have defeated the whole purpose of using data.

→ 17.8 PROBABILITY DISTRIBUTIONS: GENERAL

A *sample* is a limited number of items taken from a larger source. A *population* is a large source of items from which the sample is taken. We take measurements from a sample and calculate a sample statistic, e.g., the mean. A *statistic* is a quantity computed from a sample to estimate a population parameter. It is usually assumed that the sample is random, i.e., each possible sample of n items has an equal chance of being selected.

A *probability distribution function* is a mathematical formula that relates the values of the characteristic with their probability of occurrence in the *population*. The collection of these probabilities is called a probability distribution. The mean (μ) of a probability distribution is often called the expected value. Some distributions and their functions are summarized in Figure 17.6. Distributions are of two types:

1. *Continuous* (for "variable" data). When the characteristic being measured can take on any value (subject to the fineness of the measuring process), its probability distribution is called a "continuous probability distribution." For example, the probability distribution of the resistance data of Table 17.1 is an example of a continuous probability distribution because the resistance could have any value, limited only by the fineness of the measuring instrument. Most continuous characteristics follow one of several common probability distributions: the normal distribution, the exponential distribution, and the Weibull distribution. These distributions find the probabilities associated with occurrences of the *actual values* of the characteristic. Other continuous distributions (e.g., t, F, and chi square) are important in data analysis but are not helpful in directly predicting the probability of occurrence of actual values.
2. *Discrete* (for "attribute" data). When the characteristic being measured can take on only certain specific values (e.g., integers 0, 1, 2, 3), its probability distribution

Distribution	Form	Probability function	Comments on application
Normal	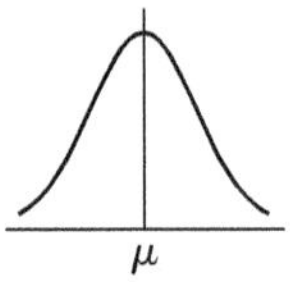 (bell curve centered at μ)	$y = \dfrac{1}{\sigma\sqrt{2\pi}} e^{-\dfrac{(X-\mu)^2}{2\sigma^2}}$ μ = Mean σ = Standard deviation	Applicable when there is a concentration of observations about the mean and it is equally likely that observations will occur above and below the mean. X can take any real positive or negative number.
Exponential	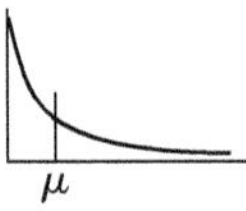 (decaying curve from μ)	$y = \dfrac{1}{\mu} e^{-\dfrac{x}{\mu}}$	Applicable when it is likely that more observations will occur below the average than above. $X > 0$.
Weibull	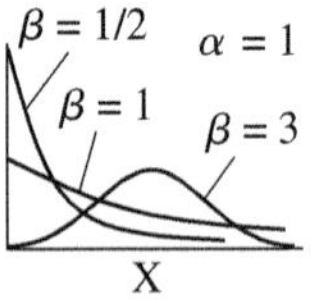 $\beta = 1/2$, $\alpha = 1$ $\beta = 1$ $\beta = 3$ X	$y = \alpha\beta(X - \gamma)^{\beta-1} e^{-\alpha(X-\gamma)^\beta}$ α = Scale parameter β = Shape parameter γ = Location parameter	Applicable in describing a wide variety of patterns in variation. $X > \gamma$.
Poisson*	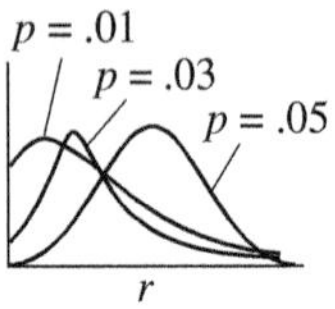 $p = .01$ $p = .03$ $p = .05$ r	$y = \dfrac{(np)^r e^{-np}}{r!}$ n = Number of trials r = Number of occurrences p = Probability of occurrence	Applicable in defining the probability of occurrence per unit time or unit space. np = average occurrences per unit and can be less than 1.
Binomial*	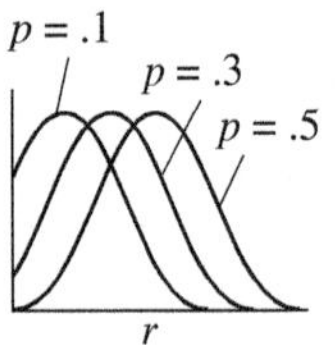 $p = .1$ $p = .3$ $p = .5$ r	$y = \dfrac{n!}{r!(n-r)!} p^r q^{n-r}$ n = Number of trials r = Number of occurrences p = Probability of occurrence $q = 1 - p$	Applicable in defining the probability of r occurrences in n trials of an event that has a constant probability of occurrence on each independent trial.

FIGURE 17.6
Summary of common probability distributions. (Asterisks indicate that these are discrete distributions, but the curves are shown as continuous for ease of comparison with the continuous distributions. Strictly, the actual plots are discrete, not continuous.)

is called a "discrete probability distribution." For example, the distribution of the number of defects r in a sample of five items is a discrete probability distribution because r can be only 0, 1, 2, 3, 4, or 5. The most common discrete distributions are the Poisson and binomial (see Figure 17.6).

The following paragraphs explain how probability distributions can be used with a sample of observations to make predictions about the larger population. Such predictions assume that the data come from a process that is stable over time, which is not always the case. Plotting the data points in order of production provides a rough test for stability; plotting the data on a statistical control chart provides a rigorous test (see Chapter 20, "Tools to Maintain the Control of Quality").

→ 17.9 THE NORMAL PROBABILITY DISTRIBUTION

Many quality characteristics can be approximated by the *normal density function:*

$$y = \frac{1}{\sigma\sqrt{2\pi}} e^{-(X-\mu)^2/2\sigma^2}$$

where $e = 2.7183$ (approximate)
$\pi = 3.1416$ (approx.)
$\mu =$ population mean
$\sigma =$ population standard deviation

Problems are solved with a table, but note that the distribution requires only the mean μ and standard deviation σ of the population.[1] The curve for the normal probability distribution is related to a frequency distribution and its histogram. As the sample becomes larger and larger and the width of each cell becomes smaller and smaller, the histogram approaches a smooth curve. If the entire population were measured and if it were normally distributed, the result would be as shown in Figure 17.6. Thus the *shape* of a histogram of sample data provides some indication of the probability distribution for the population. If the histogram resembles[2] the "bell" shape shown in Figure 17.6, this is a basis for suspecting that the population follows a normal probability distribution. There are statistical tests to support determining the underlying distribution when that is necessary.

Using the Normal Probability Distribution for Predictions

Predictions require just two estimates and a table. The estimates are

$$\text{Estimate of } \mu \text{ is } \overline{X} \quad \text{Estimate of } \sigma \text{ is } s$$

The calculations of the sample $\overline{X}$ and s are made by the methods previously discussed.

For example, from past experience, a manufacturer concludes that the burnout time of a particular light bulb follows a normal distribution. A sample of 50 bulbs has been tested, and the average life is 60 days with a standard deviation of 20 days. How many bulbs in the entire population of light bulbs can be expected to be still working after 100 days of life?

The problem is to find the area under the curve beyond 100 days (see Figure 17.7). The area under a distribution curve between two stated limits represents the probability of an occurrence between those limits. Therefore, the area under the curve is

---

[1] Unless otherwise indicated, Greek symbols will be used for population values and Roman symbols for sample values.
[2] The sample histogram may not look as if it came from a normal population. Small deviations from exact normality are expected in random samples. Statistical tests are available to determine whether the deviations are significant.

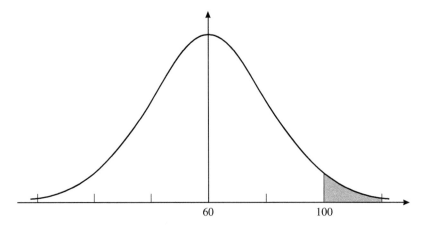

FIGURE 17.7
Distribution of light bulb life.

the probability that a bulb will last more than 100 days. To find the area, calculate the difference Z between a particular value and the average of the curve in units of standard deviation. Every conceivable combination of means and standard deviation defines a unique normal curve, but these unique curves are all related to each other in a fairly simple way. Every set of data (Xs) can be transformed into Zs using the following:

$$Z = \frac{X - \mu}{\sigma}$$

This Z statistic has a mean of 0 and a standard deviation of 1.0. So, a table to describe the areas under this standard normal distribution is all we need to compute the probabilities for any normal distribution.

In this problem $Z = (100 - 60) \div 20 = +2.0$. Table A in Appendix I shows a probability of .9773 for $Z = 2$. The statistical distribution tables provide probabilities that cover the span from $-\infty$ up to and including the value of Z included in the formula (i.e., cumulative probabilities). Thus .9773 is the probability that a bulb will last 100 days or less. The normal curve is symmetrical about the average, and the total area is 1.000. The probability of a bulb lasting more than 100 days then is $1.0000 - .9773$, or .0227, or 2.27% of the bulbs in the population will still be working after 100 days.

Similarly, if a characteristic is normally distributed and if estimates of the average and standard deviation of the population are obtained, this method can estimate the total percentage of production that will fall within engineering specification limits.

Figure 17.8 shows representative areas under the normal distribution curve. Thus 68.26% of the *population* will fall between the average of the population plus or minus 1 standard deviation of the population, 95.46% of the population will fall between the average and $\pm 2\sigma$, and finally, $\pm 3\sigma$ will include 99.73% of the population.

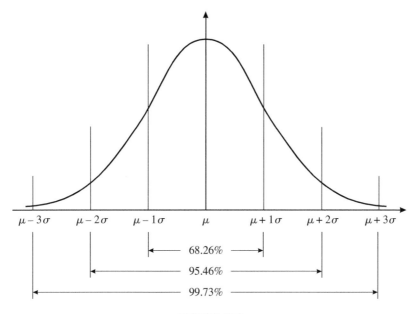

FIGURE 17.8
Areas of the normal curve (derived from Appendix I, Table A).

✦ 17.10 PROBABILITY CURVES AND HISTOGRAM ANALYSIS

We can combine the histogram concept and the probability curve concept to yield a practical working tool known as *histogram analysis.*

A random sample is selected from the process, and measurements are made for the selected quality characteristics. A histogram is prepared and specification limits are added. Knowledge of the production process is then combined with insights provided by the histogram to draw conclusions about the ability of the process to meet the specifications.

Figure 17.9 shows 16 typical histograms. Readers are encouraged to interpret each of these pictures by asking two questions:

1. Can the process meet the specification limits? This is the question of process capability.
2. What action on the process, if any, is appropriate?

These questions can be answered by analyzing the following characteristics:

1. *The centering of the histogram.* This defines the aim of the process.
2. *The width of the histogram.* This defines the variability about the aim.
3. *The shape of the histogram.* When a normal or bell-shaped curve is expected, then any significant deviation or other aberration is an alert that either normality does not apply to the process or there are quality problems. For example, histograms with two or more peaks may reveal that several "populations" have been mixed together or that several processes are at work.

556 Quality Management and Analysis

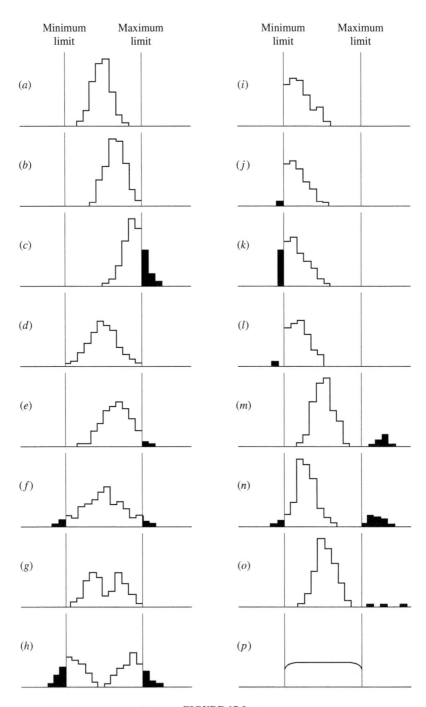

FIGURE 17.9
Distribution patterns related to tolerances. (*Adapted from Armstrong and Clarke, 1946.*)

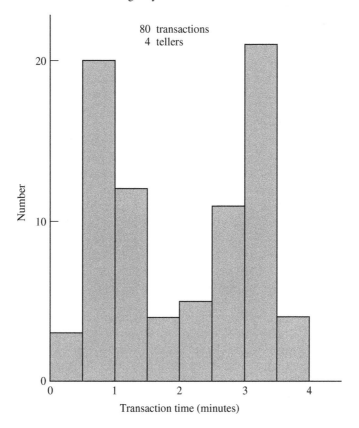

FIGURE 17.10
Histogram of bank teller transaction times. (*Juran Institute, Inc.*)

An example of a histogram from the service industry is presented in Figure 17.10. Data were collected from four bank tellers on the time to conduct a transaction during the busy lunch-hour period. Note the similarity of the histogram to Figure 17.9g. The two peaks suggest a combination of two bell-shaped distributions, which in turn suggests two different work processes. Even when separate histograms by teller were plotted, the same double-peaked histograms resulted. Then the types of transactions were examined. Developing histograms by type of transaction revealed the reason for the double-peaked histograms. Simple transactions such as withdrawals and deposits resulted in short transaction times; complex transactions such as opening retirement accounts and obtaining certificates of deposit resulted in longer transaction times. Originally, management believed that two inexperienced tellers were the primary reason for delays in serving customers. Histogram analysis revealed the true causes as the extra steps required for complex transactions.

Histograms illustrate how variable (or continuous) data provide much more information than attribute (or categorical) data. For example, Figures 17.9b, d, g, and i warn of potential trouble, even though all units in the sample are within the specification limits. With attribute measurement, all units would simply be classified as acceptable, and the inspection report would have stated "50 inspected, 0 defective"—therefore no problem. One customer had a dramatic experience based on a lot that yielded a sample histogram similar to Figure 17.9i. Although the sample indicated that the lot met quality requirements, the customer realized that the supplier must have made much scrap and screened it out before delivery. A rough calculation indicated that full production must have been about 25% defective. The histogram enabled the customer to deduce this *without ever having been inside the supplier's plant.* Note how the "product tells on the process." As the customer would eventually pay for this scrap (in the selling price), he wanted the situation corrected. The supplier was contacted, and advice was offered in a constructive manner.

As a general rule, at least 40 measurements are needed for the histogram to reveal the basic pattern of variation. Histograms based on too few measurements can lead to incorrect conclusions because the shape of the histogram may be incomplete without the observer realizing it.

Histograms have limitations. Because histograms do not display the order of production, the time-to-time process variations during production are not disclosed. Hence the seeming central tendency of a histogram may be illusory—the process may have drifted substantially.

The histogram is an important analytical tool. The key to its usefulness is its simplicity. It speaks a language that everyone understands—comparison of product measurements against specification limits. To draw useful conclusions from this comparison requires little experience in interpreting frequency distributions and no formal training in statistics. The experience soon expands to include applications in development, manufacturing, supplier relations, and field data.

➔ 17.11 THE EXPONENTIAL PROBABILITY DISTRIBUTION

The *exponential probability function* is

$$y = \frac{1}{\mu} e^{-X/\mu} \quad (x > 0)$$

Figure 17.6 shows the shape of an exponential distribution curve. Note that the normal and exponential distributions have distinctly different shapes. An examination of the tables of areas shows that 50% of a normally distributed population occurs above the mean value and 50% below. In an exponential population, 36.8% is above the mean and 63.2% below the mean. This refutes the intuitive idea that the mean is always associated with a 50% probability. The exponential curve can be used to describe the loading pattern for some structural members because smaller loads are more numerous than larger loads. The exponential curve is also useful in describing the distribution of failure times of complex equipment.

A fascinating property of the exponential distribution is that the standard deviation equals the mean.

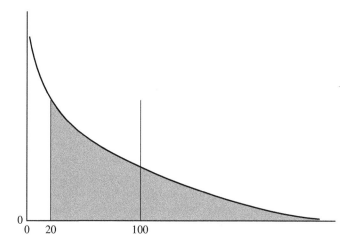

FIGURE 17.11
Distribution of time between failures.

Using the Exponential Probability Distribution for Predictions

Predictions based on an exponentially distributed population require only an estimate of the population mean. For example, the time between successive failures of a complex piece of repairable equipment is measured, and the resulting histogram is found to resemble the exponential probability curve. For the measurement made, the *mean time between failures* (commonly called MTBF) is 100 hours. What is the probability that the time between two successive failures of this equipment will be at least 20 hours?

The problem is finding the area under the curve beyond 20 hours (Figure 17.11). Table B in Appendix I gives the area under the curve beyond any particular value X that is substituted in the ratio X/μ. In this problem,

$$\frac{X}{\mu} = \frac{20}{100} = 0.20$$

From Table B, the area under the curve beyond 20 hours is 0.8187. The probability that the time between two successive failures is greater than 20 hours is .8187, i.e., there is about an 82% chance that the equipment will operate continuously without failure for 20 or more hours. Similar calculations would give a probability of .9048 for 10 or more hours.

→ 17.12 THE WEIBULL PROBABILITY DISTRIBUTION

The *Weibull distribution* is a family of distributions having the general function

$$y = \alpha\beta(X - \gamma)^{\beta-1} e^{-\alpha(X - \gamma)^\beta} \quad \text{(for } x > 0\text{)}$$

where α = scale parameter
β = shape parameter
γ = location parameter

The curve of the function (Figure 17.6) varies greatly depending on the numerical values of the parameters. Most important is the shape parameter β, which reflects the pattern of the curve. Note that when β is 1.0, the Weibull function reduces to the exponential. When β is about 3.5 (and $\alpha = 1$ and $\gamma = 0$), the Weibull closely approximates the normal distribution. In most practical applications, β varies from about 1/3 to 5. The scale parameter α is related to the peakedness of the curve, i.e., as α becomes larger the curve becomes flatter. The location parameter γ is the smallest possible value of X. It is often 0, thereby simplifying the equation. It is often unnecessary to determine the values of these parameters because predictions are made directly from Weibull probability paper, but King (1981) gives procedures for graphically finding α, β, and γ. Table J in Appendix I provides a sample of Weibull paper. With this sample paper, β can be estimated by drawing a line parallel to the line of best.

The Weibull covers many shapes of distributions. This feature makes the Weibull popular in practice because it reduces the problems of examining a set of data and deciding which of the common distributions (e.g., normal or exponential) fits best. Computer software such as Excel or MINITAB is useful in performing Weibull analyses.

Using the Weibull Probability Distribution for Predictions

Consider the case of seven heat-treated shafts that were stress tested until each of them failed. The fatigue life (in terms of number of cycles to failure) was as follows:

11,251	40,122
17,786	46,638
26,432	52,374
28,811	

The problem is to predict the percentage of failure of the population for various values of fatigue life. The solution is to plot the data on a probability plot grid, observe whether the points fall approximately in a straight line, and if so, read the probability predictions (percentage of failure) from the graph or compute them from the fitted parameters. Before the ready availability of laptop computing power, a paper and pencil were required for the application of a Weibull probability paper.

In a Weibull plot, the original data are usually[3] plotted against *mean ranks*. (Thus the mean rank for the ith value in a sample of n ranked observations refers to the mean value of the percentage of the population that would be less than the ith value in repeated experiments of size n.) The mean rank is calculated as $i/(n + 1)$. The mean ranks necessary for this example are based on a sample size of seven failures and are as shown in Table 17.7. The cycles to failure are now plotted on the probability graph against the corresponding values of the mean rank (see Figure 17.12). Note that the horizontal scale is logarithmic. The vertical scale reflects the cumulative percentages of the data expected in the fitted Wiebull distribution. The vertical axis gives the cumulative percentage of failures in the population corresponding to the fatigue life

---

[3]There are other plotting positions such as $(i - 5)/n$ and $(i - 3)(n + 4)$. The latter is used in most software.

TABLE 17.7
Table of mean ranks

Failure number (i)	Mean rank
1	0.125
2	0.250
3	0.375
4	0.500
5	0.625
6	0.750
7	0.875

shown on the horizontal axis. For example, about 50% of the population of shafts will fail in fewer than 31,000 cycles. About 80% of the population will fail in fewer than 44,000 cycles. By appropriate subtractions, predictions can be made of the percentage of failures between any two fatigue life limits. The Minitab software permits reading the exact values from the graph.

It is tempting to extrapolate particularly to predict life. For example, suppose that the minimum fatigue life were specified as 8000 cycles and the seven measurements displayed earlier were from tests conducted to evaluate the ability of the design to meet 8000 cycles. Because all seven tests exceeded 8000 cycles, the design seems adequate and might therefore be released for production. However, extrapolation on the fitted distribution predicts that about 2.3% of the *population* of shafts would fail

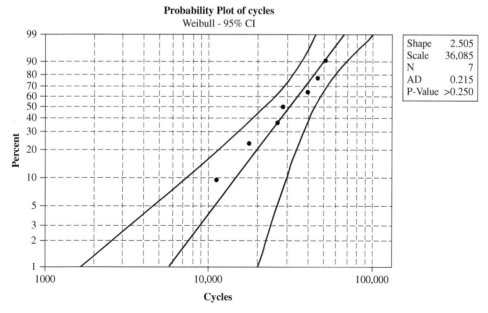

FIGURE 17.12a
Probability plot of fatigue life in cycles.

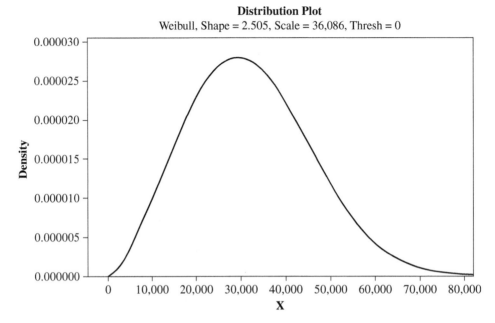

FIGURE 17.12b
Distribution of fatigue life in cycles.

in less than 8000 cycles. The available confidence intervals around the fitted distribution suggest that at a 95% level of confidence, the 8000 cycle failure rate could be as high as 11%. This information suggests a review of the design before release to production. Thus the small *sample* (all *within* specifications) gives a deceptive result, but the Weibull plot acts as an alarm signal by highlighting a potential problem and the risks associated with it.

Extrapolation can go in the other direction. Note that a probability plot of life-test data does *not* require completing all tests before the plotting starts. As each unit fails, the failure time can be plotted against the mean rank. If the early points appear to be following a straight line, it is tempting to draw in the line *before* all tests are finished. The line can then be extrapolated beyond the actual test data, and the life predictions can be made without accumulating a large amount of test time. The approach has been applied to predicting, *early* in a warranty period, the "vital few" components of a complex product that will be most troublesome. However, extrapolation has dangers. It requires the judicious melding of statistical theory, process experience, and judgment.

To make a valid Weibull plot, at least seven points are needed. Such small samples can have significant sampling variability, and it is wise to calculate and evaluate the confidence limits. Any fewer data points casts doubt on the ability of the plot to reveal the underlying pattern of variation. The Anderson-Darling (A-D) statistic on the graph is a measure of how well the actual data fit the proposed distribution. If the *P*-value is less than .05, then the fit is too poor to be used. Other distributions should be evaluated or outlying values evaluated for indications of special cause variation.

Probability distributions for normal, exponential, Weibull, and other probability functions can also be tested as appropriate against life data. Although the mathematical functions and tables provide the same information, the graph reveals *relationships* between probabilities and values of X that are not as readily apparent from the calculations. For example, the reduction in percentage defective in a population as a function of wider and wider specification limits can be easily portrayed by the graph.

Other continuous distributions include the continuous uniform (roughly speaking, all values having equal probabilities), log normal (logarithms of the original values are normally distributed), and the multinormal (e.g., a product with two measurement parameters, each normally distributed, is called a bivariate normal).

→ 17.13 THE POISSON PROBABILITY DISTRIBUTION

The fundamental Poisson distribution provides the probability of a specific number of occurrences for some unit of time or space, as follows: If the probability of occurrence p of an event is constant in each of n independent trials of the event, the probability of r occurrences in n trials is

$$\frac{(np)^r e^{-np}}{r!}$$

where n = number of trials
p = probability of occurrence
r = number of occurrences

Using the Poisson Distribution for Predictions

Poisson distributions are widely used in estimating and predicting defects per unit of time or for defects per unit (DPU) calculations, especially for complex electronic components. The Poisson is also useful as an approximation in calculating probabilities associated with sampling procedures. Table C in Appendix I gives cumulative Poisson probabilities directly, i.e., the probability of r or fewer occurrences in n trials of an event having probability p. For example, suppose that a lot of 300 units of product is submitted by a vendor whose past quality has been about 2% defective. A random sample of 40 units is selected from the lot. Table C in Appendix I provides the probability of r or fewer defectives in a sample of n units. Entering the table with a value of np equal to 40(0.02), or 0.8, for various values of r results in Table 17.8. Individual probabilities can be found by subtracting cumulative probabilities. Thus the probability of exactly two defectives is .953 − .809, or .144. Of course, the probabilities in Table 17.8 could also be found by substituting the formula six times (r = 0, 1, 2, 3, 4, 5).

The Poisson, in this case, is an approximation of the more exact, but complex, binomial distribution and applies when the sample size is at least 16, the population size is at least 10 times the sample size, and the probability of occurrence p in each trial is less than .1. These conditions are often met.

TABLE 17.8
Table of Poisson probabilities

r	Probability of r or fewer in sample with 0.8 expected per sample.
0	.449
1	.809
2	.953
3	.991
4	.999
5	1.000

➔ 17.14 THE BINOMIAL PROBABILITY DISTRIBUTION

If the conditions of the Poisson distribution are not met, the binomial distribution may be applicable. If the probability of occurrence p of an event is constant in each of n independent trials of the event, then the probability of r occurrences in n trials is

$$\frac{n!}{r!(n-r)!} p^r q^{n-r}$$

where $q = 1 - p$.

In practice, the assumption of constant probability of occurrence is considered reasonable when the population size is at least 10 times the sample size.[4]

Tables for the binomial are available, but today it is usually easiest to use the calculation capability of software like Minitab or Excel.

Using the Binomial Probability Distribution for Predictions

A lot of 100 units of product is submitted by a vendor whose past quality has been about 5% defective. A random sample of six units is selected from the lot. The probabilities of various sample results are given in Table 17.9.

TABLE 17.9
Table of binomial probabilities

r	P (Exactly r defectives in 6) $= [6!/r! \, (6-r)!](0.05)^r (0.95)^{6-r}$
0	.7351
1	.2321
2	.0306
3	.0021
4	.0001
5	.000
6	.000

---

[4] Under this condition, the change in probability from one trial to the next is negligible. If this condition is not met, the hypergeometric distribution should be used (*JQH5*, page 44.24, 44.25).

In using the formula, note that $0! = 1$.

Other discrete distributions include the hypergeometric (used when the assumptions of the Poisson and binomial cannot be met), discrete uniform (all values have equal probabilities), and the multinomial (when two or more parameters are observed in a sample).

➔ 17.15 BASIC THEOREMS OF PROBABILITY

Probability is expressed as a number that lies between 1.0 (certainty that an event will occur) and .0 (impossibility of occurrence).

A convenient definition of probability is one based on a frequency interpretation: If an event A can occur in s cases out of a total of n possible and equally probable cases, the probability that the event will occur is

$$P(A) = \frac{s}{n} = \frac{\text{Number of successful cases}}{\text{Total number of possible cases}}$$

EXAMPLE 17.1 A lot consists of 100 parts. A single part is selected at random, and thus each of the 100 parts has an equal chance of being selected. Suppose that a lot contains a total of eight defectives. Then the probability of drawing a single part that is defective is 8/100, or .08.

The following theorems are useful in solving problems:

THEOREM 17.1 If $P(A)$ is the probability that an event A will occur, then the probability that A will not occur is $1 - P(A)$. This theorem is sometimes called the complementary rule of probability.

THEOREM 17.2 If A and B are two events, then the probability that either A or B or both will occur is

$$P(A \text{ or } B) = P(A) + P(B) - P(A \text{ and } B)$$

A special case of this theorem occurs when A and B cannot occur simultaneously (i.e., A and B are *mutually exclusive*). Then the probability that either A or B will occur is

$$P(A \text{ or } B) = P(A) + P(B)$$

EXAMPLE 17.2 The probability of r defectives in a sample of six units from a 5% defective lot was previously found by the binomial (sampling is with replacement). The probability of zero defectives was .7351; the probability of one defective was .2321. The probability of zero or one defective is then .7351 + .2321, or .9672.

THEOREM 17.3 If A or B are two events, then the probability that events A and B occur together is

$$P(A \text{ and } B) = P(A) \times P(B \mid A)$$

where $P(B \mid A)$ = probability that B will occur assuming that A has already occurred.

A special case of this theorem occurs when the two events are independent, i.e., when the occurrence of one event has no influence on the probability of the other event. If A and B are independent, then the probability of both A and B occurring is

$$P(A \text{ and } B) = P(A) \times P(B)$$

EXAMPLE 17.3 A complex system consists of two major subsystems that operate independently. The probability of successful performance of the first subsystem is .95; the corresponding probability for the second subsystem is .90. Both subsystems must operate successfully to achieve total system success. The probability of the successful operation of the total system is therefore $.95 \times .90 = .855$.

The theorems preceding have been stated in terms of two events but can be expanded for any number of events.

→ 17.16 COMPUTER SOFTWARE FOR STATISTICAL ANALYSIS

With the advent of statistical software packages, the practitioner can now use many statistical techniques that were not previously considered because of the difficulty in doing the calculations. Currently, most software packages provide a basic explanation of a technique, define the inputs required, and then present the results. But such accessibility has a danger. The practitioner must understand the assumptions behind the methods and what the final results do and do not mean. In the haste to obtain an answer and avoid tedious detail, there is a danger of applying a technique incorrectly or misunderstanding a result. Beware of these serious consequences.

SUMMARY

- Statistical methods are essential in the modern approach to quality.
- Variation is a fact of nature and a fact of industrial life.
- In summarizing data, useful tabular and graphical tools include the frequency distribution, histogram, boxplot, and probability plots.
- In summarizing data, useful numerical indexes include the average, median, range, and standard deviation.
- A sample is a limited number of items taken from a larger source called the population.
- A probability distribution function relates the values of a characteristic to their probability of occurrence in the population.
- The important continuous probability distributions are the normal, exponential, and Weibull; important discrete distributions are the Poisson and binomial.
- Three theorems of probability are basic in analyzing the probability of specific events.

PROBLEMS

Note: Many of the statistical problems in the book have intentionally been stated in business language. Thus the specific statistical technique required will often *not* be specified. With this approach the student should gain some experience in translating the business problem into a statistical formulation and then choosing the appropriate statistical technique.

17.1. The following data consist of 80 potency measurements of the drug streptomycin.

4.1	5.0	2.0	2.6	4.5	8.1	5.7	2.5
3.5	6.3	5.5	1.6	6.1	5.9	9.3	4.2
4.9	5.6	3.8	4.4	7.1	4.6	7.4	3.5
4.9	5.1	4.6	6.3	8.3	6.3	8.8	1.0
5.3	5.4	4.4	2.9	7.5	5.7	5.3	3.0
4.2	5.2	7.0	3.7	6.7	5.8	6.9	2.8
6.0	8.2	6.1	7.3	8.2	6.2	4.3	2.2
5.2	5.5	3.5	7.1	7.9	5.6	5.4	3.9
6.8	8.2	4.2	4.2	5.5	6.2	3.5	3.4
6.8	4.7	4.6	4.1	4.7	5.0	3.4	7.1

(a) Summarize the data in tabular form.
(b) Summarize the data in graphical form.

17.2. Compute a measure of central tendency and two measures of variation for the data given in problem 17.1. Calculate the following three sets of limits: $\overline{X} \pm 1s, \overline{X} \pm 2s, \overline{X} \pm 3s$. For each set, calculate the percentage of data values that fall within the limits. Compare these percentages to the theoretical percentages based on the normal distribution.

17.3. Examine the histograms in Figure 17.7. For each histogram, comment on (a) the ability of the process to meet specification limits and (b) what action, if any, on the process is appropriate.

17.4. Heyes (1985) presents the following data on the wire break strength for supplier B:

470	425	438	620	452
573	382	486	526	300
520	450	389	371	598

Prepare a boxplot for this data.

17.5. A company has a filling machine for low-pressure oxygen shells. Data collected over the past two months show an average weight after filling of 1.433 g with a standard deviation of 0.033 g. The specification for weight is 1.460 ± 0.085 g. Weight is normally distributed.
(a) What percentage will not meet the weight specification?
(b) Would you suggest a shift in the aim of the filling machine? Why or why not?

17.6. A company that makes fasteners has government specifications for a self-locking nut. The locking torque has both a maximum and a minimum specified. The offsetting machine used to make these nuts has been producing nuts with an average locking torque of 8.62 in-lb and a variance σ^2 of 4.49-squared in-lb. Torque is normally distributed.

568   Quality Management and Analysis

(a) If the upper specification is 13.0 in-lb and the lower specification is 2.25 in-lb, what percent of these nuts will meet the specification limits?
(b) Another machine in the offset department can turn out the nuts with an average of 8.91 in-lb and a standard deviation of 2.33 in-lb. In a lot of 1000 nuts, how many would have too high a torque?
Answer: (a) 97.95%.   (b) 40 nuts.

17.7. A power company defines service continuity as providing electric power within specified frequency and voltage limits to the customer's service entrance. Interruption of this service may be caused by equipment malfunctions or line outages due to planned maintenance or to unscheduled reasons. Records for the entire city indicate that there were 416 unscheduled interruptions in 1997 and 503 in 1996.
(a) Calculate the mean time between unscheduled interruptions, assuming that power is to be supplied continuously.
(b) What is the chance that power will be supplied to all users without interruption for at least 24 hours? For at least 48 hours? Assume an exponential distribution.

17.8. An analysis was made of repair time for an electrohydraulic servovalve used in fatigue test equipment. Discussions concluded that about 90% of all repairs could be made within 6 hours.
(a) Assuming an exponential distribution of repair time, calculate the average repair time.
(b) What is the probability that a repair would take between 3 and 6 hours?
Answer: (a) 2.6 hours.   (b) 0.217.

17.9. Three designs of a certain shaft are to be compared. The information on the designs is summarized as follows:

	Design I	Design II	Design III
Material	Medium-carbon alloy steel	Medium-carbon unalloyed steel	Low-carbon special analysis steel
Process	Fully machined before heat treatment, then furnace-heated, oil-quenched, and tempered	Fully machined before heat treatment, then induction-scan-heated, water-quenched, and tempered	Fully machined before heat treatment, then furnace-heated, water-quenched, and tempered

	Design I	Design II	Design III
Equipment cost	Already available	$125,000	$500
Cost of finished shaft	$57	53	55

Fatigue tests were run on six shafts of each design with the following results (in units of thousands of cycles to failure):

I	II	III
180	210	900
240	360	1400
100	575	1500
50	330	340
220	130	850
110	575	600

(a) Rearrange the data in ascending order and make a Weibull plot for each design.
(b) For each design, estimate the number of cycles at which 10% of the population will fail. (This value is called the B_{10} life.) Do the same for 50% of the population.
(b) Calculate the average life of each design based on the test results. Then estimate the percentage of the population that will fail within this average life. Note that it is not 50%.
(b) Comment on replacing the current design I with II or III.

17.10. Life tests on a sample of five units were conducted to evaluate a component design before release to production. The units failed at the following times:

Unit number	Failure time, hours
1	1200
2	1900
3	2800
4	3500
5	4500

Suppose that the component was guaranteed to last 1000 hours. Any failures during this period must be replaced by the manufacturer at a cost of $200 for each component. Although the number of test data is small, management wants an estimate of the cost of replacements. If 4000 of these components are sold, provide a dollar estimate of the replacement cost.

17.11. The following data consist of 24 measurements (in seconds) of the time to answer a hot-line telephone inquiry at an AT&T call center (AT&T, 1990).

24	24	21	24	24	24
27	26	25	23	23	23
26	22	22	28	21	25
25	27	25	26	22	20

(a) Summarize the data in tabular form.
(b) Summarize the data in graphical form.

17.12. For the data of problem 17.11, perform the analyses stated in problem 17.2.

REFERENCES

Armstrong, G. R. and P. C. Clarke (1946). "Frequency Distribution vs Acceptance Table," *Industrial Quality Control,* vol. 3, no. 2, pp. 22–27.
AT&T (1990). *Analyzing Business Process Data: The Looking Glass,* AT&T Bell Laboratories, Indianapolis, IN.
Harris, R. L. (1996). *Information Graphics,* Management Graphics, Atlanta, GA.
Heyes, G. B. (1985). "The Box Plot," *Quality Progress,* December, pp. 12–17.
Juran Institute, Inc. (1989). *Quality Improvement Tools—Box Plots,* Wilton, CT, p. 7.
King, J. R. (1981). *Probability Charts for Decision Making,* rev. ed., TEAM, Tamworth, NH.
Tufte, E. R. (1997). *Visual Explanations,* Graphics Press, Cheshire, CT.
Tukey, J. W. (1977). *Exploratory Data Analysis,* Addison-Wesley, Reading, MA.
Wadsworth, H. M., K. S. Stephens, and A. B. Godfrey (2001). *Modern Methods for Quality Control and Improvement,* 2nd ed., John Wiley and Sons, New York.

CHAPTER 18

Tools for Analyzing Data

18.1 SCOPE OF DATA PLANNING AND ANALYSIS

Here are some types of problems that can benefit from statistical analysis:

1. Determining the usefulness of a limited number of test results in predicting the true value of a product characteristic.
2. Determining the number of tests required to provide adequate data for evaluation.
3. Comparing test data between two alternative designs.
4. Predicting the amount of product that will fall within specification limits.
5. Predicting system performance.
6. Controlling process quality by early detection of process changes.
7. Planning experiments to discover the factors that influence a characteristic of a product or process, i.e., exploratory experimentation.
8. Determining the quantitative relationship between two or more variables.

The central issue in all of these problems is *prediction* of population parameters on the basis of sample results. Statistics ties in with the scientific method—for a perceptive discussion, see Box and Liu (1999) and Box (1999).

These statistical methods involve both simple and complex techniques. Fortunately, computer software has eliminated the drudgery of doing detailed calculations (and making errors in those calculations). Understanding the basic concepts is essential to interpreting computer outputs correctly, and this chapter focuses on both the concepts and the tools.

Providing training in statistical tools within an organization requires careful planning to achieve actual implementation. Finn (1995) identifies common problems and provides recommendations for successful implementation.

18.2 STATISTICAL INFERENCE

Drawing conclusions from limited data is notoriously unreliable. The "gossip" of a small sample can be dangerous. Examine the following problems concerned with the evaluation of test data. For each, give a yes or no answer based on your intuitive analysis of the problem. (Write your answers on a piece of paper *now* and then check for the correct answers at the end of this chapter.) Some of the problems are solved in the chapter.

Examples of Engineering Problems That Can Be Solved by Using the Concepts of Statistical Inference

1. A single-cavity molding process has been producing insulators with an average impact strength of 5.15 ft-lb [6.9834 Newton-meters (N-m)]. A group of 12 insulators from a new lot shows an average of 4.952 ft-lb (6.7149 N-m). Is this enough evidence to conclude that the new lot is lower in average strength?
2. Past data show that the average hardness of brass parts is 49.95. A new design claims to have higher hardness. A sample of 61 parts of the new design shows an average of 54.62. Does the new design actually have a different hardness?
3. Two types of spark plugs were tested for wear. A sample of 10 of design 1 showed an average wear of 0.0049 in. (0.0124 cm). A sample of eight of design 2 showed an average wear of 0.0064 in. (0.0163 cm). Are these enough data to conclude that design 1 is better than design 2?
4. Only 11.7% of the 60 new alloy blades on a turbine rotor failed on test in a gas turbine where 20% have shown failures in a series of similar tests in the past. Are the new blades better than the old ones?
5. Of 1050 resistors supplied by one manufacturer, 3.71% were defective. Of 1690 similar resistors from another manufacturer, 1.95% were defective. Can one reasonably assert that the product of one plant is inferior to that of the other?

You probably had some incorrect answers. Truth be told, in some of the cases, we didn't actually give you enough information to answer the questions accurately. The statistical methods used to analyze these problems are called *statistical inference*. We start the chapter with the concept of sampling variation and sampling distributions.

18.3 SAMPLING VARIATION AND SAMPLING DISTRIBUTIONS

Suppose that a battery is to be evaluated to ensure that life requirements are met. A mean life of 30 hours is desired. Preliminary data indicate that the life follows a normal distribution and that the standard deviation is equal to 10 hours. A sample of four batteries is selected at random from the process and tested. If the mean of the four is close to 30 hours, it is concluded that the battery meets the specification. Figure 18.1 plots the distribution of *individual* batteries from the population, assuming that the true *mean* of the population is exactly 30 hours.

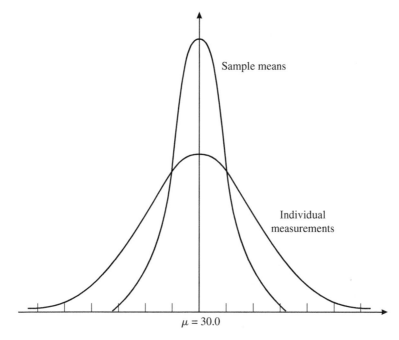

FIGURE 18.1
Distributions of individual measurements and sample means.

If a sample of four is life-tested, the following lifetimes might result: 34, 28, 38, and 24, giving a mean of 31.0. However, this random sample is selected from the many batteries made by the same process. Suppose that another sample of four is taken. The second sample of four is likely to be different from the first sample. Perhaps the results would be 40, 32, 18, and 29, giving a mean of 29.8. If the process of drawing many samples (with four in each sample) is repeated over and over, different results would be obtained in most samples. The fact that samples drawn from the *same* process can yield different sample results illustrates the concept of sampling variation.

Returning to the problem of evaluating the battery, a dilemma exists. In the actual evaluation, only one sample of four can be drawn (because of time and cost limitations). Yet the experiment of drawing many samples indicates that samples vary. The question is, How reliable is the sample of four that will be the basis of the decision? The final decision can be influenced by the luck of which sample is chosen. The key point is that the existence of sampling variation means that any one sample cannot always be relied upon to give an adequate decision. The statistical approach analyzes the results of the sample, *taking into account the possible sampling variation that could occur.* Formulas have been developed to define the expected amount of sampling variation. Knowing this, a valid decision can be reached by evaluating one sample of data.

The problem, then, is to define how means of samples vary. If sampling is continued and for each sample of four the mean is calculated, these means could be compiled into a histogram. Figure 18.1 shows the resulting probability curve superimposed on the curve for the population. The narrow curve represents the distribution of life for the sample *means* (where each average includes four individual batteries). This is called the *sampling distribution of means*. The curve for means is narrower than the curve for individuals because in calculating means, extreme individual values are offset. The mathematical properties of the curve for averages have been studied and the following relationship was developed:

$$\sigma_{\bar{x}} = \frac{\sigma}{\sqrt{n}}$$

where $\sigma_{\bar{x}}$ = standard deviation of means of samples (sometimes called the standard error of the mean)
σ = standard deviation of individual items
n = number of items in *each* sample

The mathematical justification for the relationship is the central limit theorem. This limit states that if $x_1, x_2, \ldots x_n$ are outcomes of a sample of n independent observations of a random variable X, then the mean of the samples of n will approximately follow a normal distribution with mean μ and standard deviation $\sigma_{\bar{x}} = \sigma/\sqrt{n}$ when n is large ($n > 30$).

The relationship is significant because if an estimate of the standard deviation of *individual* items can be obtained, then the standard deviation of sample means can be calculated from the foregoing relationship instead of running an experiment to generate sample averages. The problems of evaluating the battery can now be portrayed graphically (Figure 18.2). Using the calculated mean and standard deviation of the mean from

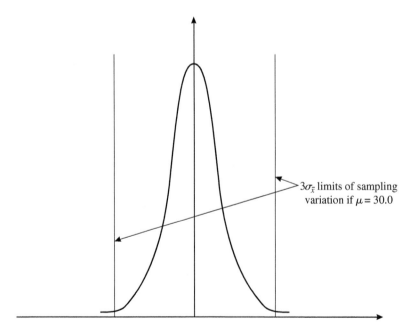

FIGURE 18.2
Distribution of sample means.

the sample, this figure sets the limits on the range of values for the means that will occur 99.3% of the time if the sampling experiment were repeated many times.

This concept of a sampling distribution is basic to the two major areas of statistical inference, estimation and tests of hypotheses, which are discussed next.

18.4 STATISTICAL ESTIMATION: CONFIDENCE LIMITS

Estimation is the process of analyzing a sample result to predict the corresponding value of the population parameter. In other words, the process is to estimate a desired population parameter by an appropriate measure calculated from the sample values. For example, the sample of four batteries previously mentioned had a mean life of 31.0 hours. If this is a representative sample from the process, what estimate can be made of the true average life of the entire population of batteries?

The estimation statement has two parts:

1. The *point estimate* is a single value used to estimate the population parameter. For example, 31.0 hours is the point estimate of the average life of the population. The sample mean has some important statistical properties: On average, sample means will equal the population mean, and the sample mean is usually the best point estimate of the population mean.
2. The *confidence interval* is a range of values that include (with a preassigned probability called a *confidence level*) the true value of a population parameter. *Confidence limits* are the upper and lower boundaries of the confidence interval. A confidence level is the probability that an assertion about the value of a population parameter is correct.[1]

Confidence limits should not be confused with other limits, e.g., control limits, statistical tolerance limits (see Chapter 19, "Statistical Tools to Design for Quality," for distinctions among several types of limits).

Table 18.1 summarizes confidence limit formulas for common parameters. The following examples illustrate some of these formulas.

> **EXAMPLE 18.1.** Mean of a normal population. Twenty-five specimens of brass have a mean hardness of 54.62 and an estimated standard deviation of 5.34. Determine the 95% confidence limits on the mean. The standard deviation of the population is unknown.
>
> **Solution.** Note that when the standard deviation is unknown and is estimated from the sample, the *t* distribution (Table D in Appendix I) must be used. The *t* value for 95% confidence is found by entering the table at 0.975 and 25 − 1, or 24, degrees of freedom[2] and reading a *t* value of 2.064.

---

[1] Confidence levels of 90, 95, or 99% are usually used in practice.

[2] A mathematical derivation of degrees of freedom is beyond the scope of this book, but the underlying concept can be stated. *Degrees of freedom* (DF) is the parameter involved when, e.g., a sample standard deviation is used to estimate the true standard deviation of a universe. DF equals the number of measurements in the sample minus some number of constraints estimated from the data to compute the standard deviation. In this example, it was necessary to estimate only one constant (the population mean) to compute the standard deviation. Therefore, DF = 25 − 1 = 24.

576 Quality Management and Analysis

TABLE 18.1
Summary of confidence limit formulas
$(1 - \alpha)$ = confidence level

Parameters	Formulas
Mean of a normal population (standard deviation known)	$\bar{X} \pm Z_{\alpha/2} \dfrac{\sigma}{\sqrt{n}}$ where $\bar{X}$ = sample average Z = normal distribution coefficient σ = standard deviation of population n = sample size
Mean of a normal population (standard deviation unknown)	$\bar{X} \pm t_{\alpha/2} \dfrac{s}{\sqrt{n}}$ where t = distribution coefficient (with $n - 1$ degrees of freedom) s = estimated σ (s is the sample standard deviation)
Standard deviation of a normal population	Upper confidence limit = $s\sqrt{\dfrac{n-1}{\chi^2_{\alpha/2}}}$ Lower confidence limit = $s\sqrt{\dfrac{n-1}{\chi^2_{1-\alpha/2}}}$ where χ^2 = chi-square distribution coefficient with $n - 1$ degrees of freedom $1 - \alpha$ = confidence level
Population fraction defective	See Table F in Appendix I.
Difference between the means of two normal populations (standard deviations σ_1 and σ_2 known)	$(\bar{X}_1 - \bar{X}_2) \pm Z_{\alpha/2} \sqrt{\dfrac{\sigma_1^2}{n_1} + \dfrac{\sigma_2^2}{n_2}}$
Difference between the means of two normal populations ($\sigma_1 = \sigma_2$ but unknown)	$(\bar{X}_1 - \bar{X}_2) \pm t_{\alpha/2} \sqrt{\dfrac{1}{n_1} + \dfrac{1}{n_2}}$ $\times \sqrt{\dfrac{\Sigma(X - \bar{X}_1)^2 + \Sigma(X - \bar{X}_2)^2}{n_1 + n_2 - 2}}$
Mean time between failures based on an exponential population of time between failures	Upper confidence limit = $\dfrac{2rm}{\chi^2_{\alpha/2}}$ Lower confidence limit = $\dfrac{2rm}{\chi^2_{1-\alpha/2}}$ where r = number of occurrences in the sample (i.e., number of failures) m = sample mean time between failures DF = $2r$

$$\text{Confidence limits} = \bar{X} \pm t\frac{s}{\sqrt{n}}$$

$$= 54.62 \pm (2.064)\frac{5.34}{\sqrt{25}}$$

$$= 52.42 \text{ and } 56.82$$

There is 95% confidence that the true mean hardness of the brass is between 52.42 and 56.82.

EXAMPLE 18.2. Mean of an exponential population. A repairable radar system has been operated for 1200 hours, during which time eight failures occurred. What are the 90% confidence limits on the mean time between failures for the system?

Solution

$$\text{Sample MTBF} = \frac{1200}{8} = 150 \text{ h between failures}$$

Upper confidence limit = 2(1200)/7.962 = 301.4

Lower confidence limit = 2(1200)/26.296 = 91.3

The values 7.962 and 26.296 are obtained from the chi-square table (Table E in Appendix I). There is 90% confidence that the true mean time between failures is between 91.3 and 301.4 h.

Confusion has arisen in the application of the term *confidence level* to a reliability index such as MTBF. Using a different example, suppose that the numerical portion of a reliability requirement reads as follows:

"The MTBF shall be at least 100 hours at the 90% confidence level." This statement means that

1. The minimum MTBF must be 100 hours.
2. Actual tests shall be conducted on the product to demonstrate with 90% confidence that the 100-hour MTBF has been met.
3. The test data shall be analyzed by calculating the observed MTBF and the lower one-sided 90% confidence limit on MTBF. The true MTBF lies above this limit with 90% of confidence.
4. The lower one-sided confidence limit must be ≥100 hours.

The term *confidence level* from a statistical viewpoint has great implications for a test program. The observed MTBF must be *greater* than 100 if the lower confidence limit is to be ≥100. Confidence level means that sufficient tests must be conducted to demonstrate, with statistical validity, that a requirement has been met. Confidence level does *not* refer to the qualitative opinion about meeting a requirement. Also, confidence level does *not* lower a requirement, i.e., a 100-hour MTBF at a 90% confidence level does *not* mean that 0.90 × 100, or 90 hours, is acceptable. Such serious misunderstandings have occurred. When the term *confidence level* is used, a clear understanding should be verified and not assumed.

18.5 IMPORTANCE OF CONFIDENCE LIMITS IN PLANNING TEST PROGRAMS

Additional tests will increase the accuracy of estimates. Accuracy here refers to the agreement between an estimate and the true value of the population parameter. The increase in accuracy does not vary linearly with the number of tests—doubling the number of tests usually does *not* double the accuracy. Examine the graph (Figure 18.3) of the confidence interval for the mean against sample size (a standard deviation of 50.0 was assumed): When the sample size is small, an increase has a great effect on the width of the confidence interval; after about 30 units, an increase has a much smaller effect. The effect diminishes because of the square root in the formula for confidence limits. Doubling the accuracy requires a sample size four times larger. The inclusion of the cost parameter is vital here. The cost of additional tests must be evaluated against the value of additional accuracy.

Further, if the sample is selected randomly and if the sample size is less than 10% of the population, accuracy depends primarily on the absolute size of the sample rather than the sample size expressed as a percentage of the population size. Thus a sample size that is 1% of the population of 100,000 may be better than a 10% sample from a population of 1000.

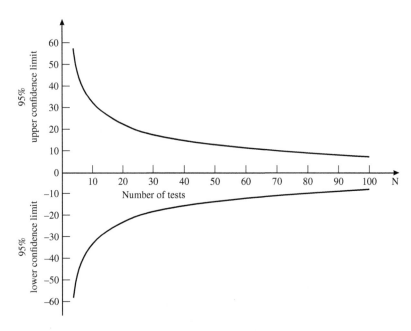

FIGURE 18.3
Width of confidence interval versus number of tests.

18.6 DETERMINATION OF THE SAMPLE SIZE REQUIRED FOR A SPECIFIED ACCURACY IN AN ESTIMATE

Confidence limits can help to determine the size of a test program required to estimate the mean of a product characteristic within a specified accuracy. It is desired to estimate the true mean of the battery previously cited where $\sigma = 10$. The estimate must be within 2.0 hours of the true mean if the estimate is to be of any value. A 95% confidence level is desired on the confidence statement. The desired confidence interval is ± 2.0 hours, or

$$2.0 = \frac{(1.96)(10)}{\sqrt{n}} \quad n = 96$$

A sample of 96 batteries will provide an average that is within 2.0 hours of the true mean (with 95% confidence). Notice the type of information required for estimating the mean of a normal population: (1) the desired width of the confidence interval (the accuracy desired in the estimate), (2) the desired confidence level, and (3) the variability of the characteristic under investigation. The number of tests required cannot be determined until the engineer furnishes these items of information. Past information may also have a major role in designing a test program (see Section 18.11).

18.7 TEST OF HYPOTHESIS

Basic Concepts

A hypothesis, as used here, is an assertion about population. Typically, the hypothesis is stated as a pair of hypotheses, as follows: the *null hypothesis* H_0 and an *alternate hypothesis*, H_a. The null hypothesis, H_0, is a statement of no change or no difference. Hence the term "null." The alternate hypothesis is the statement of change or difference, that is, if we reject the null hypothesis, then the alternate is true by default.

For example, to test the hypothesis that the mean life of a population of batteries equals 30 hours, we state:

$$H_0: \mu = 30.0 \text{ hours}$$
$$H_a: \mu \text{ not equal to } 30.0 \text{ hours}$$

A hypothesis test is a test of the validity of the assertion and is carried out by analyzing a sample of data.

Sample results must be carefully evaluated for two reasons. First, there are many other samples that, by chance alone, could be drawn from the population. Second, the numerical results in the sample actually selected can easily be compatible with several different hypotheses. These points are handled by recognizing the two types of sampling errors.

TABLE 18.2a
Type I (α) error and type II (β) error

	Suppose the H_0 is	
Suppose decision of analysis is	True	False
Fail to Reject H_0	Correct decision $Pr = 1 - \alpha$	Wrong decision $Pr = \beta$
Reject H_0	Wrong decision $Pr = \alpha$	Correct decision $Pr = 1 - \beta$

The Two Types of Sampling Errors.

In evaluating a hypothesis, two errors can be made:

1. *Reject* the null hypothesis when it is *true*. This is called *type I error* or the *level of significance*. The maximum probability of type I error is denoted by α.
2. *Fail to reject* the null hypothesis when it is *false*. This is called *type II error* and the probability is denoted by β.

These errors are defined in terms of probability numbers and can be controlled to desired values. The results possible in testing a hypothesis are summarized in Table 18.2a. Definitions are found in Table 18.2b.

The type I error is shown graphically in Figure 18.4 for the hypothesis $H_0 : \mu_0 = 30.0$, a known population standard deviation of 10, and $\alpha = 0.05$. The interval on the

TABLE 18.2b
Hypothesis testing definitions

- **Null Hypothesis (H_0):** statement of no change or no difference. This statement is assumed true until sufficient evidence is presented to reject it.
- **Alternative Hypothesis (H_a):** statement of change or difference. This statement is considered true if H_0 is rejected.
- **Type I error:** the error in rejecting H_0 when it is true, or in saying there is a difference when there is no difference.
- **Alpha Risk:** the maximum risk or maximum probability of making a type I error. This probability is preset, based on how much risk the researcher is willing to take in committing a type I error (rejecting H_0 wrongly), and it is usually established at 5% (or .05). If the *p*-value is less than alpha, then reject H_0.
- **Significance level:** the risk of committing a type I error.
- **Type II error**: the error in failing to reject H_0 when it is false, or in saying there is no difference when there really is a difference.
- **Beta risk**: the risk or probability of making a type II error, or overlooking an effective treatment or solution to the problem.
- **Significant difference:** the term used to describe the results of a statistical hypothesis test where a difference is too large to be reasonably attributed to chance.
- ***p*-value:** the probability of obtaining samples as different, when there is really no difference in the population(s) i.e., the actual probability of committing a type I error. The *p*-value is the actual probability of incorrectly rejecting the null hypothesis (H_0) (i.e., the chance of rejecting the null when it is true). When the *p*-value is less than alpha, reject H_0. If the *p*-value is greater than alpha, fail to reject H_0.
- **Power:** the ability of a statistical test to detect a real difference when there really is one, or the probability of rejecting H_0, when it is, in fact, false. Commonly used to determine if sample sizes are sufficient to detect a difference in treatments if one exists. Power = $(1 - \beta)$, or 1 minus the probability of making a type II error.

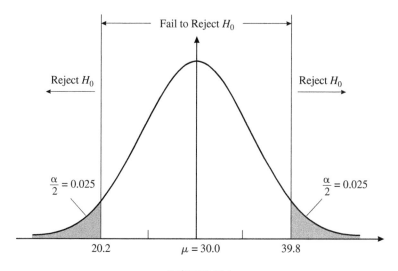

FIGURE 18.4
Rejection regions for $H_0 : \mu = 30.0$ and $H_a : \mu \neq 30.0$.

horizontal axis between the vertical lines represents the *fail to reject region* for the test of a null hypothesis. If the sample result (e.g., the mean) falls within this region, the null hypothesis is not rejected. Otherwise, it is rejected. The terms *fail to reject* and *reject* require careful interpretation. The meanings are explained in Section 18.9 of this chapter. Notice that a small portion of the curve falls outside this region. This area (α) represents the probability of obtaining a sample result outside this region, even though the null hypothesis is correct.

Suppose it has been decided that the type I error must not exceed 5%. This value is the maximum probability of rejecting the hypothesis when the true mean life is 30.0. The rejection region can be obtained by locating values of mean life that have only a 5% chance of being exceeded when the true mean life is 30.0. Further, suppose a sample n of four measurements is taken and $\sigma = 10.0$.

Remember that the curve represents a population of sample means because the decision will be made on the basis of a sample mean. Sample means vary less than individual measurements according to the relationship $\sigma_{\bar{x}} = \sigma/\sqrt{n}$ (see Section 18.3, "Sampling Variation and Sampling Distributions").

Further, the distribution of sample means is approximately normal even if the distribution of the individual measurements (going into the means) is not normal. The approximation holds best for large values of n but is adequate for n as low as 4.

Table A in Appendix I shows that a 2.5% area in each tail is at a limit that is 1.96 standard deviations from 30.0. Then, under the hypothesis that $\mu_0 = 30.0$, 95% of sample means will fall within $\pm 1.96\sigma_{\bar{x}}$ of 30.0, or

$$\text{Upper limit} = 30.0 + 1.96 \frac{10}{\sqrt{4}} = 39.8$$

$$\text{Lower limit} = 30.0 - 1.96 \frac{10}{\sqrt{4}} = 20.2$$

The *do not reject* H_0 region is thereby defined as 20.2 to 39.8. If the mean of a random sample of four batteries is within this region, the hypothesis is not rejected. If the mean falls outside the region, the hypothesis is rejected. This decision rule provides a type I error of .05.

The type II, or β, error, the probability of *failing to reject* a null hypothesis when it is false, is shown in Figure 18.5 as the shaded area. Notice that it is possible to obtain a sample result within the *do not reject* H_0 region, even though the population has a true mean that is *not* equal to the mean stated in the hypothesis. The numerical value of β depends on the true value of the population mean (and also on n, σ, and α). The various probabilities are depicted by an *operating characteristic* (OC) *curve*.

The problem now is to construct an OC curve to define the magnitude of the type II (β) error. Because β is the probability of *failing to reject* the original hypothesis ($\mu_0 = 30.0$) when it is *false,* the probability that a sample mean will fall between 20.2 and 39.8 must be found when the true mean of the population is something other than 30.0. This is done by finding the area under the normal curve for all possible values of the true mean of the population.

The results form the OC curve shown in Figure 18.6. The OC curve is a plot of the probability of *failing to reject* the original hypothesis as a function of the true value of the population parameter (and the given values of n, σ, and α). Note that for a mean equal to the hypothesis (30.0), the probability of *failing to reject* is $1 - \alpha$. This curve should not be confused with that of a normal distribution of measurements. In some cases, the shape is similar, but the meanings of an OC curve and a distribution curve are entirely different.

The Use of the Operating Characteristic Curve in Selecting an Acceptance Region

The *reject* H_0 region was determined by dividing the 5% allowable α error into two equal parts (see Figure 18.4). This process is called a *two-tail test.* The entire 5% error could also be placed at either the left or the right tail of the distribution curve. These are *one-tail tests.*

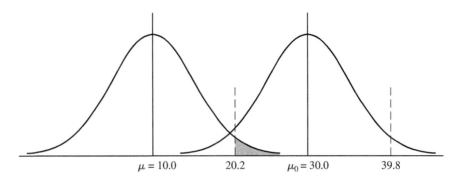

FIGURE 18.5
Type II, or β, error.

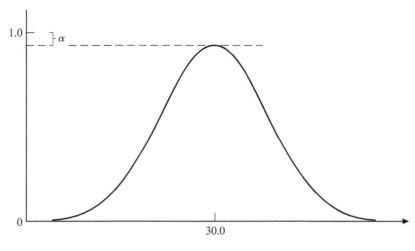

FIGURE 18.6
Operating characteristic curve.

Operating characteristic curves for tests having these one-tail acceptance regions can be developed following the approach used for the two-tail region. Although the α error is the same, the β error depends on whether a one-tail or two-tail test is used.

In some problems, knowledge is available to indicate that if the true mean of the population is *not* equal to the hypothesis value, then it is on one side of the hypothesis value. For example, a new material of supposedly higher mean strength will have a mean equal to or *greater than* that of the present material. Such information will help select a one-tail or two-tail test to make the β error as small as possible. The following guidelines are based on the analysis of OC curves:

Use a one-tail test with the entire α risk on the right tail if (1) it is suspected that (if μ_0 is not true) the true mean is $>\mu_0$ or (2) values of the population mean $<\mu_0$ are acceptable and we want to detect only a population mean $>\mu_0$.

Use a one-tail test with the entire α risk on the left tail if (1) it is suspected that (if μ_0 is not true) the true mean is $<\mu_0$ or (2) values of the population mean $>\mu_0$ are acceptable and we want to detect only a population mean $<\mu_0$.

Use a two-tail test if (1) there is no prior knowledge of the location of the true population mean or (2) we want to detect a true population mean less than or more than the μ_0 stated in the original hypothesis.[3]

The selection of a one- or two-tail test will be illustrated by some examples in Section 18.8. Every test of hypothesis has an OC curve. (An alternative to the OC is the "power" curve, but power is simply one minus the probability of acceptance, or $1 - \beta$. Minitab uses power curves.)

---

[3]With a two-tail test, the hypothesis is sometimes stated as the original hypothesis $H_0 : \mu_0 = 30.0$ against the alternative hypothesis $H_a : \mu_0 \neq 30.0$. With a one-tail test, $H_0 : \mu_0 = 30.0$ is used against the alternative $H_a : \mu_1 < 30.0$ if α is placed in the left tail, or $H_a : \mu_1 > 30.0$ is used if α is placed in the right tail.

With this background, the discussion now proceeds to the steps for testing a hypothesis.

→ 18.8 TESTING A HYPOTHESIS WHEN THE SAMPLE SIZE IS FIXED IN ADVANCE

Ideally, desired values for type I and type II errors are defined in advance, and the required sample size is determined (see Section 18.10). If the sample size is fixed in advance because of cost or time limitations, then usually the desired type I error is defined and this procedure is followed:

1. State the practical problem.
2. State the null hypothesis and alternate hypothesis.
3. Choose a value for α (alpha). Common values are .01, .05, and .10.
4. Choose the test statistic for testing the hypothesis.
5. Obtain a sample of observations, compute the test statistic, and confirm that assumptions required by the test are true.
6. Determine the rejection region for the test (i.e., the range of values of the test statistic that results in a decision to reject the null hypothesis), and compare the value to the rejection region to decide whether to reject the hypothesis. (Or compute the p-value for the test statistic and compare that to the α that has been set.)
7. Draw the practical conclusion.

Table 18.3 summarizes some common tests of hypotheses. The procedure is illustrated through the following examples. In these examples a type I error of .05 will be assumed.

1. Test for a population mean, μ. (Standard deviation of the population is known.)

   EXAMPLE 18.3. A single-cavity molding press has been producing insulators with a mean impact strength of 5.15 ft-lb (6.98 N-m) and with a standard deviation of 0.25 ft-lb (0.34 N-m). A new lot shows the following data from 12 specimens:

Specimen	Strength
1	5.02
2	4.87
3	4.95
4	4.88
5	5.01
6	4.93
7	4.91
8	5.09
9	4.96
10	4.89
11	5.06
12	4.85
	$\overline{X} = 4.95$

TABLE 18.3
Summary of formulas on tests of hypotheses

Hypothesis	Test statistic and distribution
$H_0: \mu = \mu_0$ (the mean of a normal population is equal to a specified value μ_0; σ is known)	$Z = \dfrac{\overline{X} - \mu_0}{\sigma/\sqrt{n}}$ Standard Normal distribution
$H_0: \mu = \mu_0$ (the mean of a normal population is equal to a specified value μ_0; σ is estimated by s)	$t = \dfrac{\overline{X} - \mu_0}{s/\sqrt{n}}$ t distribution with $n - 1$ degrees of freedom (DF)
$H_0: \mu_1 = \mu_2$ (the mean of population 1 is equal to the mean of population 2; assume that $\sigma_1 = \sigma_2$ and that both populations are normal)	$t = \dfrac{\overline{X}_1 - \overline{X}_2}{\sqrt{1/n_1 + 1/n_2}\sqrt{[(n_1 - 1)s_1^2 + (n_2 - 1)s_2^2]/(n_1 + n_2 - 2)}}$ t distribution with DF $= n_1 + n_2 - 2$
$H_0: \sigma = \sigma_0$ (the standard deviation of a normal population is equal to a specified value σ_0)	$\chi^2 = \dfrac{(n-1)s^2}{\sigma_0^2}$ Chi-square distribution with DF $= n - 1$
$H_0: \sigma_1 = \sigma_2$ (the standard deviation of population 1 is equal to the standard deviation of population 2; assume that both populations are normal)	$F = \dfrac{s_1^2}{s_2^2}$ F distribution with $DF_1 = n_1 - 1$ and $DF_2 = n_2 - 1$
$H_0: \hat{p} = p_0$ (the fraction defective in a population is equal to a specified value p_0; assume that $np_0 \geq 5$) $\hat{p}$ = sample proportion	$Z = \dfrac{\hat{p} - p_0}{\sqrt{p_0(1 - p_0)/n}}$ Standard Normal distribution approximation good for very large samples and $.23 < p > .75$
$H_0: p_1 = p_2$ (the fraction defective in population 1 is equal to the fraction defective in population 2; assume that $n_1 p_1$ and $n_2 p_2$ are each ≥ 5)	$Z = \dfrac{X_1/n_1 - X_2/n_2}{\sqrt{\hat{p}(1 - \hat{p})(1/n_1 + 1/n_2)}} \qquad \hat{p} = \dfrac{X_1 + X_2}{n_1 + n_2}$ Standard Normal distribution approximation good for very large samples and $.23 < p < .75$
To test for independence in a J × K contingency table that cross-classifies the variables A and B H_0: A is independent of B H_a: A is dependent on B	$\chi^2 = \sum_{j=1}^{J}\sum_{k=1}^{K} \dfrac{(f_{jk} - e_{jk})^2}{e_{jk}}$ Chi-square distribution with DF $= (J - 1)(K - 1)$ where: f_{jk} = the observed frequency of data for category j of variable A and to category k of variable B e_{jk} = the expected frequency $= f_{j0} f_{0k} / f_{00}$ f_{j0} = frequency total for category j for variable A f_{0k} = frequency total for category k of variable B f_{00} = frequency total for J × K table

For nonnormal data, nonparametric tests are used. As a result, medians (instead of means) are compared. For comparing the median(s)
- to a target: One-Sample Wilcoxon test or sign test for data with many outliers
- of two populations: Mann-Whitney test or Mood's Median for data with many outliers
- of two or more populations: Kruskal-Wallis test or Mood's Median for data with many outliers

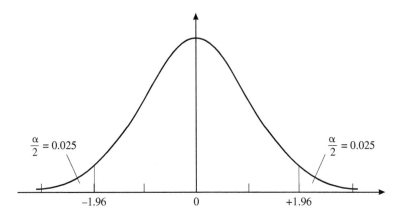

FIGURE 18.7
Distribution of Z (two-tail test).

Is the new lot from which the sample of 12 was taken different in mean impact strength from the past performance of the process? Assume $\alpha = .05$.

Solution. $H_0 : \mu_0 = 5.15$ ft-lb (6.98 N-m). (The mean of the population from which the sample was taken is the same as the past process average.) $H_a : \mu_0 \neq 5.15$ ft-lb (6.98 N-m)

Test statistic

$$Z = \frac{\bar{X} - \mu_0}{\sigma/\sqrt{n}}$$

Rejection region. Assuming no prior information and that a deviation on either side of the hypothesis average is important to detect, a two-tail test (Figure 18.7) is applicable. From Table A in Appendix I, the *do not reject* H_0 region is Z between -1.96 and $+1.96$.

Analysis of sample data

$$Z = \frac{4.95 - 5.15}{0.25/\sqrt{12}} = -2.75$$

Conclusion. Because Z is in the rejection region, the null hypothesis is rejected. Therefore, sufficient evidence is present to conclude that the mean impact strength of the new process is significantly different from the mean of the past process. The answer to the first question in Section 18.2 is yes.

2. Test for two population means, μ_1 and μ_2, when the standard deviation is unknown but believed to be the same for the two populations. (This assumption can be evaluated by test 4.)

   EXAMPLE 18.4. Two makes of spark plugs were operated in the alternate cylinders of an aircraft engine for 100 hours and the following data were obtained:

	Make 1	Make 2
Number of spark plugs tested	10	8
Average wear per 100 h($\bar{X}$), in.	0.0049	0.0064
Variability (s), in.	0.0005	0.0004

Can it be said that make 1 wears less than make 2?

Solution. $H_0: \mu_1 = \mu_2$.
$H_a: \mu_1 < \mu_2$.

Test statistic

$$t = \frac{\bar{X}_1 - \bar{X}_2}{\sqrt{1/n_1 + 1/n_2} \sqrt{[(n_1 - 1)s_1^2 + (n_2 - 1)s_2^2]/(n_1 + n_2 - 2)}}$$

with degrees of freedom $= n_1 + n_2 - 2$.

Rejection region. We are concerned only with the possibility that make 1 wears less than make 2; therefore, use a one-tail test (Figure 18.8) with the entire α risk in the left tail. From Table D in Appendix I, the *do not reject H_0* region is $t > -1.746$.

Analysis of sample data

$$t = \frac{0.0049 - 0.0064}{\sqrt{1/10 + 1/8} \sqrt{[(10 - 1)(0.0005)^2 + (8 - 1)(0.0004)^2]/(10 + 8 - 2)}} = -6.9$$

Conclusion. Because t is in the rejection region, the null hypothesis is rejected. Therefore, sufficient evidence is present to conclude that make 1 wears less than make 2. The answer to the third question in Section 18.2 is yes.

The paired-comparison test is a special case of the test for two population means. Here the two samples represent two different treatments of the same sets of objects or people. The data are taken in pairs for the same object or person and the differences, d, between pairs, are analyzed with a t-test against a target mean of zero.

3. Test for a population standard deviation, σ.

   EXAMPLE 18.5. For the insulator strengths tabulated in the first example, the sample standard deviation is 0.036 ft-lb (0.049 N-m). The previous variability, recorded over a period, has been established as a standard deviation of 0.25 ft-lb (0.34 N-m). Does the low value of 0.036 indicate that the new lot is significantly more uniform (i.e., standard deviation is less than 0.25)?

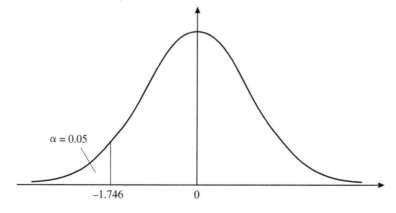

FIGURE 18.8
Distribution of t with α on left tail.

Solution. $H_0: \sigma_0 = 0.25$ ft-lb (0.34 N-m).
$H_a: \sigma < 0.25$ ft-lb (0.34 N-m).

Test statistic

$$\chi^2 = \frac{(n-1)s^2}{\sigma_0^2}$$

with degrees of freedom $= n - 1$.

Rejection region. We believe that the standard deviation is smaller; therefore, we will use a one-tail test (Figure 18.9) with the entire risk on the left tail. From Table E in Appendix I the *do not reject* H_0 region is $\chi^2 \geq 4.57$.

Analysis of sample data

$$\chi^2 = \frac{(12-1)(0.078)^2}{(0.25)^2} = 1.08$$

Conclusion. Because χ^2 is in the rejection region, the null hypothesis is rejected. Therefore, sufficient evidence is present to conclude that the new lot is more uniform than the old.

When the chi-square distribution is used to test a sample variance, the original data must come from a normally distributed population.

4. Test for the difference in variability (s_1 versus s_2) in two samples.

EXAMPLE 18.6. A materials laboratory was studying the effect of aging on a metal alloy. Researchers wanted to know whether the parts were more consistent in strength after aging than before. The following data were obtained (assume $\alpha = .05$):

	At start (1)	After 1 year (2)
Number of specimens (n)	9	7
Average strength ($\bar{X}$), psi	41,350	40,920
Standard deviation (s), psi	934	659

Solution. $H_0: \sigma_1 = \sigma_2$.
$H_a: \sigma_1 > \sigma_2$.

Test statistic

$$F = \frac{s_1^2}{s_2^2} \text{ with } DF_1 = n_1 - 1, \quad DF_2 = n_2 - 1$$

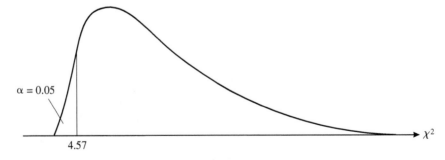

FIGURE 18.9
Distribution of χ^2 with α on the left tail.

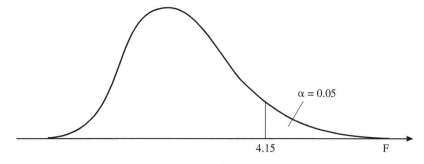

FIGURE 18.10
Distribution of F with α on right tail.

Rejection region. We are concerned with an improvement in variation; therefore, we will use a one-tail test (Figure 18.10) with the entire α risk in the right tail.
From Table G in Appendix I, the *do not reject* H_0 region is $F \leq 4.15$.

Analysis of sample data

$$F = \frac{(934)^2}{(659)^2} = 2.01$$

Conclusion. Because F is inside the *do not reject* H_0 region, we fail to reject the null hypothesis. Therefore, there is not sufficient evidence to conclude that the parts were more consistent in strength after aging.

In this test and other tests that compare two samples, the samples must be independent to ensure valid conclusions.

5. Testing the difference between two proportions (using MINITAB)

   EXAMPLE 18.7 The tabulated data in worksheet Control-Test is from a drug clinical trial. The trial consisted of 1200 specimens divided into two 600 specimen groups. A "1" indicates an instance of the disease, a "0" indicates no disease detected. One group (Test Group) received a new formula drug while the other received no medication. In the Control Group there were 141 instances of the disease under study. In the Test Group, there were 117 instances of the disease.

   a. Can the manufacturer claim, with 95% confidence, that the new formula is more effective in combating the disease?
   b. What is the most error the manufacturer would have in making that claim?
   c. What is the confidence interval for the effect of the new formula (p1-p2)? (MINITAB filename Control-Test.mpj)

   Answer: a) No. See proportions test later.

   Tally for Discrete Variables: Test Group, Control Group

   ```
   Test                    Control
   Group      Count        Group      Count
     0         483           0         459
     1         117           1         141
    N=         600          N=         600
   ```

Test and CI for Two Proportions: Test Group, Control Group
```
Event = 1
Variable              X     N    Sample p
Test Group          117   600    0.195000
Control Group       141   600    0.235000


Difference = p (Test Group) - p (Control Group)
Estimate for difference: -0.04
95% CI for difference: (-0.0864329, 0.00643295)
Test for difference = 0 (vs not = 0): Z = -1.69  P-Value = 0.091
```

b) The manufacturer would have a 9.1% chance of being incorrect in claiming the new product is more effective.

c) confidence interval is from −8.6% (less effective) to .64% more effective).

But there are two issues still to be addressed. We are interested really only in the alternative hypothesis that the test group is less than the control group, and the preceding test is simply for "not equal." We should conduct a one-tail test. If we do, the result is as follows:

```
Difference = p (1) - p (2)
Estimate for difference:  -0.04
95% upper bound for difference:  -0.00103224
Test for difference = 0 (vs < 0):   Z = -1.69   P-Value = 0.046
```

This result would lead us to reject the null hypothesis in favor of the alternative that the test group infection rate is lower.

But there is one further issue. For a binomial test, this sample is not very large and the proportions, not to mention their difference, are also relatively small. Modern software like MINITAB lets us do an exact test based on the actual binomial distributions called Fishers exact test. The results of that test do not let us reject the null:

```
Fisher's exact test: P-Value = 0.053
```

While we are now back at the original conclusion, it is only by accident because now we are using the more appropriate alternative hypothesis and using the most accurate statistical test available.

6. Testing the difference between two means (using MINITAB)

EXAMPLE 18.8 To investigate a problem with warping of wood panels, a theory was proposed that warping was due to differing moisture content in the layers of the laminated product before drying. The data shown was taken between layers 1-2 and 2-3. Is there a significant difference in the moisture content?

Layer 1-2		Layer 2-3	
4.43	4.40	3.74	5.14
6.01	5.99	4.30	5.19
5.87	5.72	5.27	4.16
4.64	5.25	4.94	5.18
3.50	5.83	4.89	4.78
5.24	5.44	4.34	5.42
5.34	6.15	5.30	4.05
5.99	5.14	4.55	3.92
5.75	5.72	5.17	4.07
5.48	5.00	5.09	4.54
5.64	5.01	4.74	4.23
5.15	5.42	4.96	5.07
5.64		4.21	

(Filename: Moisture Content.mpj)

Answer: (MINITAB: Graph>Prob Plot; Stat>Basic Stats>Two Variances; Stat>Basic Stats> 2 Sample t with equal variances)

A probability plot and Anderson-Darling test indicate that the assumption of normality in the data sets is appropriate. Equal variance test concludes variances are equal. While equality of variances is not required, if they are equal, the test uses a pooled estimate of the standard deviation, which provides a more sensitive test than the weighted average of unequal standard deviations.

The following two sample *t* tests conclude that there is a significant difference in the moisture content in the two areas tested. Boxplots are shown in Figure 18.11.

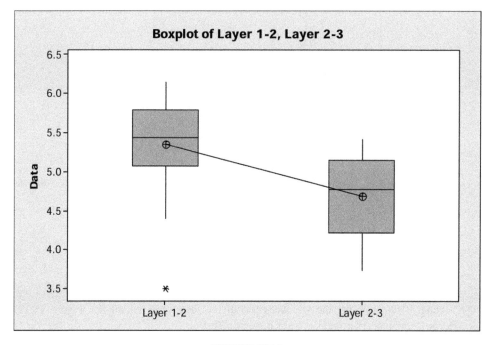

FIGURE 18.11

Two-Sample T-Test and CI: Layer 1-2, Layer 2-3

```
Two-sample T for Layer 1-2 vs Layer 2-3

            N      Mean     StDev    SE Mean
Layer 1-2   25     5.350    0.613    0.12
Layer 2-3   25     4.689    0.499    0.10

Difference = mu (Layer 1-2) 2 mu (Layer 2-3)
Estimate for difference: 0.660901
95% CI for difference: (0.343158, 0.978644)
T-Test of difference = 0 (vs not =): T-Value = 4.18 P-
Value = 0.000 DF = 48
  Both use Pooled StDev = 0.5587
****************************************************
```

7. Testing the difference in means across three groups (using MINITAB)

    EXAMPLE 18.9 The production supervisor thought that a circuit response time had to do something with type of circuit. It is just a "hunch." He wants to know for sure if that is true.

Circuit Type	Response_time
1	19
1	22
1	20
1	18
1	25
2	20
2	21
2	33
2	27
2	40
3	16
3	15
3	18
3	26
3	17

    To compare the means of three populations, we would use an ANOVA. The hypothesis being tested is:

    H_0: There is no difference in mean response times of the 3 circuit types.

    H_a: There is a difference in mean response times of the 3 circuit types.

    First ANOVA requires that the data for each factor be normal. These data meet that requirement.

    Second, it requires that the variances for all the factors, be equal. See Figure 18.12. Because the data are normal, we use Bartlett's Test.

    The p-values > 0.05 means that we may accept the variances as equal and can use ANOVA.

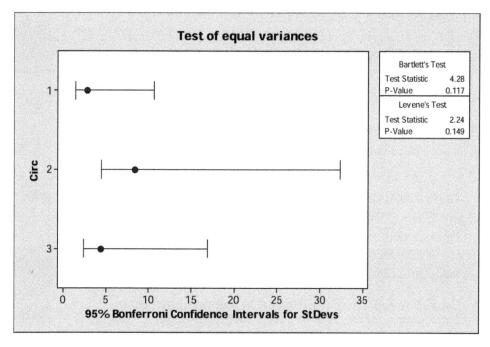

FIGURE 18.12

Our results are:

One-way ANOVA: Resp_t versus Circ

```
Source   DF      SS      MS       F      P
Circ      2   260.9   130.5    4.01  0.046
Error    12   390.8    32.6
Total    14   651.7

S = 5.707     R-Sq = 40.04%   R-Sq(adj) = 30.04%

                          Individual 95% CIs For Mean Based on
                          Pooled StDev
Level    N     Mean    StDev    -----+---------+---------+---------+
1        5   20.800    2.775         (------*-----)
2        5   28.200    8.408                       (-----*-----)
3        5   18.400    4.393    (------*-----)
                                -----+---------+---------+---------+
                                    18.0      24.0      30.0      36.0

Pooled StDev = 5.707
```

Here the *p*-value is lower than our acceptable risk of 0.05 (alpha), so we can reject the null hypothesis.

Conclusion: The production supervisor is correct. There is a difference in response times among the circuits. The confidence intervals show that the response time for circuit 2 is longer than for circuit 3. The time for circuit 1, however, cannot be differentiated from either 2 or 3.

8. Testing for the independence of factors

EXAMPLE 18.10 A six sigma project team is trying to determine if treatment outcomes for a particular illness depend on the hospital where patients underwent treatment. The following table shows the number of patients who underwent the treatment in each hospital last year and the corresponding patient outcome—full, partial, or no recovery.

	Hospital			
Outcome of Treatment	A	B	C	D
Full Recovery	98	113	82	32
Partial Recovery	15	15	17	15
No Recovery	10	12	11	14

To test for the independence of two factors, the chi-square contingency tables test of independence is used. The hypothesis being tested is:

H_0: Outcome of treatment is independent of hospital
H_a: Outcome of treatment is dependent on hospital

Running the chi-square tables test using MINITAB, our results are:

Chi-Square Test: Hospital A, Hospital B, Hospital C, Hospital D

```
Expected counts are printed below observed counts
Chi-Square contributions are printed below expected counts

            Hospital A   Hospital B   Hospital C   Hospital D   Total
Full              98         113           82           32        325
               92.11      104.84        82.37        45.68
                0.377       0.635        0.002        4.097
Partial           15          15           17           15         62
               17.57       20.00        15.71         8.71
                0.376       1.250        0.105        4.534
None              10          12           11           14         47
               13.32       15.16        11.91         6.61
                0.828       0.659        0.070        8.276
Total            123         140          110           61        434

Chi-Sq 21.209,   DF = 6,   P-Value = 0.002
```

The calculated Chi-Sq = 21.209. This is greater than the Critical Chi-Sq = 12.59 (DF = 6 and assume $\alpha = 0.05$) in Table E of Appendix I. Hence, we reject the null hypothesis of independence.

Because the software computes p-value of 0.002, we can reject the null hypothesis without using the table because the p-value is lower than our acceptable risk of $\alpha = 0.05$.

The chi-square contributions show that Hospital D has substantially fewer no-recovery cases than expected. There are also fewer-than-expected partial-recovery cases. Full-recovery cases are more frequent than expected.

Conclusion: Outcome of patient treatment is dependent on hospital where treatment was performed. Hospital D has a lower rate of no recovery and partial recovery than the other hospitals, and it has a higher rate of full recovery.

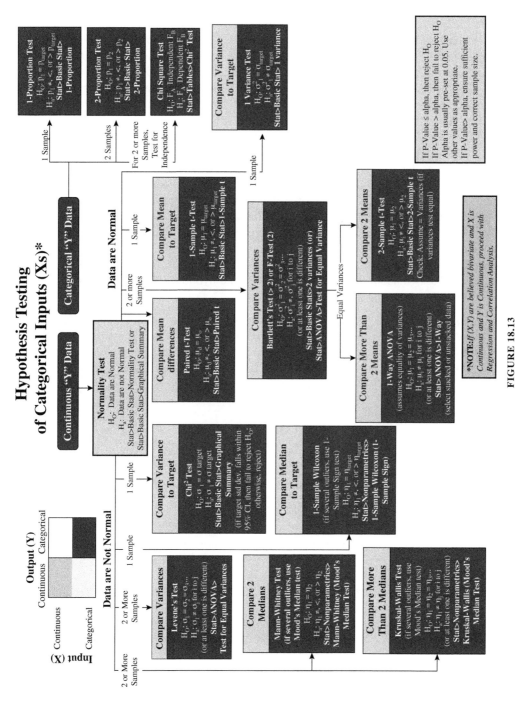

FIGURE 18.13
Hypothesis Testing Road Map (with MINITAB). Courtesy of Juran Institute Inc. All rights reserved.

→ 18.9 DRAWING CONCLUSIONS FROM TESTS OF HYPOTHESES

The payoff for these tests of hypotheses comes from reaching useful conclusions. The meaning of *reject the null hypothesis* or *fail to reject the null hypothesis* is shown in Table 18.4, together with some analogies to explain the subtleties of the meanings.

When a null hypothesis is rejected, the practical conclusion is that the parameter value specified in the hypothesis is wrong. The conclusion is drawn with strong conviction—at a confidence level of $(1 - \alpha)$ percent. A related question is: "What is a good estimate of the value of the parameter for the population?" Help can be provided on this question by calculating the confidence limits for the parameter (see Section 18.4 and Figure 18.13 for an additional outline).

When a null hypothesis is not rejected, the numerical value of the parameter stated in the null hypothesis has not been proved, but it has not been disproved. It is not correct to say that the null hypothesis has been proved as correct at the $(1 - \alpha)$ percent confidence level. Many other hypotheses could be accepted for the given sample of observations, and yet only one hypothesis can be true. Therefore, a failure to reject does not mean a high probability of proof that a specific hypothesis is correct. (All other factors being equal, the smaller the sample size, the more likely it is that the null hypothesis will not be rejected. Less evidence certainly does not imply proof.)

With a failure to reject a hypothesis, a key question then is, what conclusion, if any, can be drawn about the parameter value in the hypothesis? Two approaches are suggested:

1. *Calculate confidence limits on the sample result.* These confidence limits define an interval within which the true population parameter lies. If this interval is small, a *fail to reject* decision on the test of hypothesis means that the true population value is either equal to or close to the value stated in the hypothesis. Then it is reasonable

TABLE 18.4
The meaning of a conclusion from tests of hypotheses

	If null hypothesis is rejected	If null hypothesis is not rejected
Adequacy of evidence in the sample of observations	Sufficient to conclude that null hypothesis is false, and we can adopt the alternative	Not sufficient to conclude that hypothesis is false; without further evidence, we must assume it is true
Difference between sample result (e.g., $\bar{X}$) and hypothesis value (e.g., μ_0)	Unlikely that difference was due to chance (sampling) variation	Difference likely due to chance (sampling) variation
Analogy of guilt in a court of law	Guilt has been established beyond a reasonable doubt	Have not established guilt beyond a reasonable doubt
Analogy of a batting average in baseball	If player got 300 base hits out of 1000 times at bat, this is sufficient to conclude that his overall batting average is about .300	If player got 3 hits in 10 times, this is not sufficient to conclude that his overall average is about .300

to act as if the parameter value specified in the hypothesis is correct. A relatively wide confidence interval is a stern warning that the true value of the population might be far different from that specified in the hypothesis. For example, the confidence limits of 21.2 and 40.8 on battery life in Section 18.4 would lead to not rejecting the hypothesis of $\mu = 30.0$, but note that the confidence interval is relatively wide.

2. *Construct and review the operating characteristic curve (or power curve) for the test of the hypothesis.* This curve defines the probability that other possible values of the population parameter could have been accepted by the test. Knowing these probabilities for values relatively close to the original hypothesis can help draw further implications about not rejecting the original hypothesis. For example, Figure 18.6 shows the OC curve for a hypothesis that specified that the population mean is 30.0. Note that the probability of not rejecting the hypothesis when the population mean is 30.0 is .95 (or $1 - \alpha$). But also note that if μ really is 35.0, then the probability of not rejecting $\mu = 30.0$ is still high (about .83). If μ really is 42.0, the probability of not rejecting $\mu = 30.0$ is only about .33.

Care must be taken in drawing business conclusions from statistical conclusions, particularly when the null hypothesis is not rejected.

A hypothesis test determines whether a statistically significant difference exists between the sample result and the value of the population parameter stated in the null hypothesis. A decision to reject the null hypothesis means that a statistically significant difference is present. However, the difference does not necessarily have practical importance. Large sample sizes, although not generally available, can detect small differences that may not have practical effects. Conversely, *failing to reject the null hypothesis* means that a statistically significant difference was not found, but this outcome may have been due to a small sample size. A larger sample size could result in a "reject null hypothesis" and thus detect a significant difference.

Instead of determining a rejection region for a test of hypothesis and seeing whether the test statistic falls within the rejection region, another approach is to determine the significance probability (*p*-value). In this approach we calculate the test statistic and then look in the appropriate probability distribution table to determine the probability value of the test statistic by chance (i.e., the smallest statistical significance level at which we can reject the hypothesis). Computer software often provides the *p*-value as part of the data analysis. This *p*-value is then compared to the significance level (α) originally chosen: If *p* is equal to or less than the significance level (α), the hypothesis is rejected; if *p* is greater than α, the hypothesis is not rejected.

➔ 18.10 DETERMINING THE SAMPLE SIZE REQUIRED FOR TESTING A HYPOTHESIS

The preceding sections assume that the sample size is fixed by nonstatistical conditions and that only the type I error is predefined for the test. The ideal procedure is to predefine the desired type I and II errors and calculate the sample size required to cover both types of errors.

The sample size required will depend on (1) the sampling risks desired (α and β), (2) the size of the smallest true difference that is to be detected, and (3) the variation in the characteristic being measured. The sample size can be determined by using the OC curve for the test. Most software provides the needed calculations and curves. While the following example is relatively simple, sample size calculations for some types of tests are more complex.

The sample size can also be calculated directly:

$$n = \left[\frac{(Z_{\alpha/2} + Z_\beta)\sigma}{\mu - \mu_0}\right]^2$$

Suppose that it was important to detect the fact that the mean life of the battery cited previously was 35.0 hours. Specifically, we want to be 80% sure of detecting this change ($\beta = .2$). Further, if the true mean was 30.0 hours (as stated in the hypothesis), we want to have only a 5% risk of wrongly rejecting the hypothesis ($\alpha = .05$). Then,

$$n = \left[\frac{(1.96 + 0.84)10}{35 - 30}\right]^2 = 31.4$$

The required sample size is 32.

If, as is usually the case, the population standard deviation is not known, the Z-statistics will be replaced with the corresponding t-statistics and α with s. For comparing two means, the standard deviation will be for the difference of two random variables.

❖ 18.11 THE DESIGN OF EXPERIMENTS

The body of knowledge called design of experiments (DOE) had early roots in the field of agriculture with the pioneering work of Sir Ronald A. Fisher. Next, the application of DOE initially focused in the manufacturing sector, particularly the chemical industry.

Two developments have promoted broader and more robust application of the DOE discipline. First, the widespread availability of accessible and inexpensive software has made it easier to perform needed mathematical operations. Second, as organizations pursue ever higher levels of quality, the problems that must be solved often demand greater capabilities for problem solving, such as DOE.

As a matter of fact, most of us experiment. We try different approaches to see if we can get better results. Unstructured approaches to experimentation are at best inefficient; at worst, they can actually give the wrong answers. The basic principles of designing experiments are not terribly complex and should be observed in all experimentation. The complexities in experimental design generally arise either from the inherent complexity of a particular problem or from the work to extract the most useful information from the experiment.

Experimental design in manufacturing has now expanded to everything from basic metals manufacturing and fabrication to high-tech electronic circuit designs. Outside manufacturing, experimental designs have optimized performance in call centers, back-office financial services processing, and a range of logistics operations.

When conducting experiments to improve quality, we usually want to know two things: (1) which variables have the most impact on our quality outcome and (2) a mathematical relationship that allows us to predict the outcome from the variables.

The plan for conducting such experiments is the *design of the experiment.* We will first cover an example that presents several alternative designs and defines the basic terminology and concepts.

Suppose that three detergents are to be compared for their ability to clean clothes in a washing machine. The "whiteness" readings obtained by a special measuring procedure are the *dependent,* or *response, variable.* The independent variable under investigation (detergent) is a *factor,* and each value of the factor is called a *level,* i.e., there are three levels. A treatment is a single level assigned to a single factor, detergent A. A treatment combination is the set of levels for all factors in a given experimental run. A factor may be discrete (different detergents) or continuous (water temperature).

Figure 18.14 illustrates six designs of experiments starting with the simple, basic design in Figure 18.14*a*.

Each "x" in the chart represents one run of the experiment using the detergent at the top of the column. This design is a beginning for systematic experimentation in two respects. First, it makes three measurements on each of the experimental treatments. In DOE, we call these three *replicates.* As we have seen in hypothesis testing, sampling error is always present. With a larger sample (more replicates), we get smaller sampling error. Second, although we cannot tell it from the diagram, this design also held as many other factors constant as possible. The same washing machine, same water temperature, same washing time, and so on, were used for all nine tests. These other variables were not of interest for the experiment, but the experimenters did not want variation in them to affect the measurements, so they held them constant for all runs. When a variable that is not one of the factors being tested is held constant for all or a portion of the runs, we call this *blocking*—in the same way as a paired t-test blocks on the individual cases.

One drawback of this design is that the conclusions about detergent brands apply only to the specific conditions of the experiment. We cannot determine whether the results would be the same for all machines, temperatures, etc.

Figure 18.14*b* introduces a second factor at three levels, i.e., washing machines brands I, II, and III. The entry in each cell represents a run using the machine at the top of the column and the detergent in the cell. This is a faulty design. It would not be possible to determine whether an observed difference was the result of the detergents or the washing machines.

In Figure 18.14*c*, nine conditions are assigned at random (although one can see a pattern), thus the name *completely randomized design.* However, detergent A is not used with machine brand III, and detergent *B* is not used with machine brand I. Randomization has removed the failure we noted in case *b*, but we cannot compute the effects of the machines in the result.

Randomization is a very important feature of a good experimental design. In particular, we almost always randomize the sequence in which the various runs are executed. While we may be able to isolate and control or block known variables that can be regulated, there are also unknown variables and variables we cannot regulate readily. Examples include ambient temperature, machine temperature, wear, minor fluctuations in fluid or granular raw material composition, and a whole host of others that we may never know.

As example *c* shows, pure randomization, however, can eliminate what might otherwise be useful information, so one of the foundational rules in experimental design is:

Block or control what you can and randomize what you cannot.

600 Quality Management and Analysis

(a)

A	B	C
x	x	x
x	x	x
x	x	x

(b)

I	II	III
A	B	C
A	B	C
A	B	C

(c)

I	II	III
C	B	B
A	C	B
A	A	C

(d)

I	II	III
B	A	C
C	C	A
A	B	B

(e)

	I	II	III
1	C	A	B
2	B	C	A
3	A	B	C

(f)

	I ABC	II ABC	III ABC
1	xxx	xxx	xxx
2	xxx	xxx	xxx
3	xxx	xxx	xxx

FIGURE 18.14
Some experimental designs.

Figure 18.14d shows a *randomized block design* that implements this principle. Here each block is a machine brand, and the detergents are run in random order within each block. This design guards against any possible bias from the order in which the detergents are used and has advantages in the subsequent data analysis and conclusions. First, a test of hypothesis can be run to compare detergents and a separate test of hypothesis run to compare machines; all nine observations are used in both tests. Second, the conclusions concerning detergents apply for the three machines on average, and vice versa, thus providing conclusions over a wider range of conditions.

Now suppose that a third factor such as water temperature is also to be studied. One can use the *Latin square design* shown in Figure 18.14e. Note that this design requires using each detergent only once with each machine and only once with each temperature. Thus, all three factors can be evaluated (by three separate tests of hypothesis) with only nine observations. Not shown in the graphic is the fact that the temporal order for running each of the nine should be randomized to mitigate the effects of unspecified variables. Even this design, however, has a weakness. This design assumes no interaction among the factors. No interaction between detergent and machine means that the effect of changing from detergent A to B to C does not depend on which machine is used, and similarly for the other combinations of factors. The concept of interaction is shown in Figure 18.15.

Finally, the main factors and possible interactions could be investigated by the *factorial design* in Figure 18.14f. *Factorial* means that at least one test is run for every

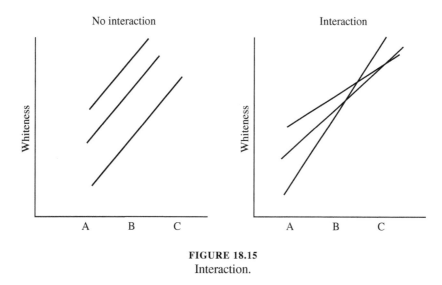

FIGURE 18.15
Interaction.

combination of main factors, in this case 3 × 3 × 3 or 27 combinations. Separate tests of hypothesis can be run to evaluate the main factors and also test each of the possible interactions. Again, all the observations contribute to each comparison.

In addition to calculating the main effects of the detergent, machine, and temperature, a factorial design also permits calculation of *interaction* effects. In this case, we can calculate three different two-way interactions (detergent x machine, detergent x temperature, and machine x temperature). We can also calculate a three-way interaction among all three factors.

Figure 18.15 illustrates interactions graphically. Whiteness (the dependent or response variable) is plotted against temperature, with a separate line drawn for each detergent. In the left-hand diagram, the three lines are approximately parallel. This is a graphic indication that temperature and detergent have no interaction. Increasing the temperature will increase whiteness by about the same amount for each of the detergents. And, at each temperature, detergent C creates more whiteness than B, which creates more than A.

The right-hand diagram illustrates interaction. While detergent B produces higher whiteness than A at all temperatures, the difference is much smaller. The magnitude of the difference is reduced by the interaction of the detergent with the temperature. For detergent C, the interaction actually reverses the direction. Detergent C performs better at lower temperatures, and C is better than A and B at lower temperatures while being worse at higher temperatures. Graphically, interactions show up as non-parallel lines.

Factorial designs are very powerful, and their broad adoption has greatly improved experimental design and analysis. Their chief drawback is that the number of required experimental runs grows very quickly as more variables are added. Two strategies help the experimenter work with this challenge. First, initial experiments are confined to only two levels (so-called 2^k designs). Second, *fractional* factorial can be used for initial screening to reduce the number of variables. These fractional designs introduce some limitations, but they can reduce the cost of experimentation while still avoiding the weaknesses and flaws of designs not based on factorials. (See Section 18.15.)

The analyses of designs discussed and illustrated in this chapter assume that any underlying relationships among the variables are linear. Of course, many relationships are linear, or at least approximate linearity over the relatively small region of interest for an experiment. But far too many relationships are not linear to ignore the possibility of nonlinearity. Factorial designs can be enhanced to detect the presence of nonlinearity by including one or more *center points* in the experiment. These treatments are run using the exact mean of the high and low values in a 2^k experiment. The center point runs use only the means. There are no interaction terms with other values. If these center points turn out to be significantly different from what would be forecasted by the high–low combinations, then curvature has been demonstrated and further research is required because simple 2^k designs with center points can only detect the presence of curvature. The functionality therefore cannot be estimated.

One of the common tools for experimenting and modeling nonlinear relationships is response surface modeling (RSM), which can both detect the presence of curvature and estimate all linear, interaction, and quadratic terms. Box et al. (2005) offer a number of creative ways to deal with nonlinearity.

For a general accessible and comprehensive treatment of the topic of experimental design see Box et al. (2005). A sequential approach to experimentation can often be helpful. For a series of four papers, see Carter (1996). Emanuel and Palanisamy (2000) discuss sequential experimentation at two levels and a maximum of seven factors.

Increasingly, experimental design is focusing on exploratory experimentation as a means of discovering new knowledge (Box, 1999). This emphasis may require the use of extensive designs such as full factorials, response surface methodology, and other designs.

→ 18.12   AN ILLUSTRATIVE CASE OF PAINT BLISTERS

Experiment Objective

Design of Experiment (DOE) example for a paint line to eliminate or reduce blisters on painted product[4]

"Key" Factors for Blisters

- Primer gram weight
- Top coat gram weight
- Primer oven temperature
- Primer flash air flow

A mathematical model needs to be developed to determine the best operational settings.

The following are slides from the DOE report (Figures 18.16 to 18.22).

---

[4] *Courtesy of a Juran Institute client Merillat Industries.*

Blisters D.O.E.:

- 2^4 Factorial, Half Fractional, 8 Runs
- Develop method to Apply Forced Air through Flash Tunnel

D.O.E. Run #	Primer Flash Wind Setting (Fans)	Primer Gram Weight Setting	Top Coat Gram Weight Setting	Turn (Frame) Temp	Data To Be Collected					Remarks
					Description/Data Type	Sample Size, Number of Samples	Where To Collect Data	Who Will Collect Data	How Will Data Be Recorded	
1	On	12	13.8	90	% Blisters Found	One Day Production	Paint Line	QC Inspectors	Data Sheet	Set Primer Oven Temp to 135
2	On	13.8	12	90	% Blisters Found	One Day Production	Paint Line	QC Inspectors	Data Sheet	Set Primer Oven Temp to 135
3	Off	13.8	12	105	% Blisters Found	One Day Production	Paint Line	QC Inspectors	Data Sheet	Set Primer Oven Temp to 145
4	Off	12	13.8	105	% Blisters Found	One Day Production	Paint Line	QC Inspectors	Data Sheet	Set Primer Oven Temp to 145
5	Off	12	12	90	% Blisters Found	One Day Production	Paint Line	QC Inspectors	Data Sheet	Set Primer Oven Temp to 135
6	On	13.8	13.8	105	% Blisters Found	One Day Production	Paint Line	QC Inspectors	Data Sheet	Set Primer Oven Temp to 145
7	Off	13.8	13.8	90	% Blisters Found	One Day Production	Paint Line	QC Inspectors /Green Belts	Data Sheet	Set Primer Oven Temp to 135
8	On	12	12	105	% Blisters Found	One Day Production	Paint Line	QC Inspectors	Data Sheet	Set Primer Oven Temp to 145

FIGURE 18.16
Data collection plan.

Blisters D.O.E.: Interactions Plots

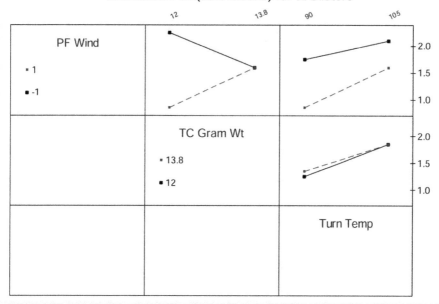

FIGURE 18.17

Blisters D.O.E.: Main Effects Plots

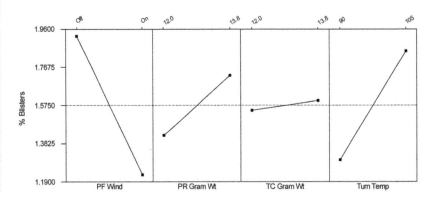

FIGURE 18.18

Blisters D.O.E.: Cube Plot

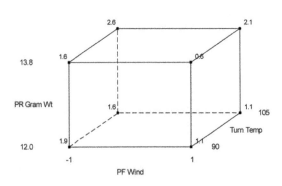

FIGURE 18.19

Blisters D.O.E.: Session Window Results

Fractional Factorial Fit: % Blisters versus PF Wind, PR Gram Weight, TC Gram Weight, and Turn Temp.

Estimated Effects and Coefficients for % (coded units)

```
Term            Effect      Coef     SE Coef      T        P
Constant                   1.5750   0.07289    21.61    0.002
PF Wind         -0.7000   -0.3500   0.07289    -4.80    0.041  ⬅
PR Gram          0.3000    0.1500   0.07289     2.06    0.176  *
TC Gram          0.0500    0.0250   0.07289     0.34    0.764
Turn Tem         0.5500    0.2750   0.07289     3.77    0.064  ⬅
PF Wind*TC Gram  0.7000    0.3500   0.07289     4.80    0.041  ⬅
```

Conclusion: Significant Factors

Analysis of Variance for % (coded units)

P is <.10, Reject Ho.

```
Source              DF   Seq SS    Adj SS    Adj MS     F       P
Main Effects         4   1.77000   1.77000   0.44250  10.41   0.090
2-Way Interactions   1   0.98000   0.98000   0.98000  23.06   0.041
Residual Error       2   0.08500   0.08500   0.04250
Total                7   2.83500
```

Estimated Coefficients for % using data in uncoded units

```
Term              Coef
Constant         -3.40833
PF Wind          -5.36667
PR Gram           0.166667
TC Gram           0.0277778
Turn Tem          0.0275000
PF Wind*TC Gram   0.388889
```

Ho: Factors tested have no effect on the occurrence of blisters.

*We want to include primer gram weight because we feel it has a significant enough effect to warrant control.

FIGURE 18.20

Blisters D.O.E.: Model Equation in Uncoded Units

% Blisters = -3.408 - 5.367(PF Wind) +.167(PR Gram) +.028(Turn Temp) +.389(PF Wind*TC Gram)

Lets look at fits and Residuals:

FITS1	RESI1
1.975	-0.075
1.125	-0.025
2.525	0.075
0.575	0.025
1.875	-0.275
1.325	-0.225
1.325	0.275
1.875	0.225

Ho: The data is Normal

Ha: The data is not normal

P>.05, Fail to Reject, Data is normal

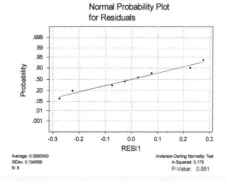

FIGURE 18.21

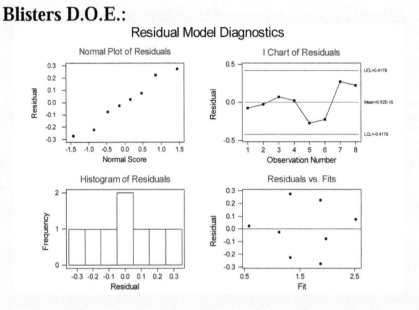

FIGURE 18.22
Everything looks Good!
Residuals satisfy assumptions of normality, time, independence, and homoscedasticity.

➔ 18.13 CONTRAST BETWEEN THE CLASSICAL AND MODERN METHODS OF EXPERIMENTATION

The contrast between the classical method of experimentation (varying one factor at a time, holding everything else constant) and the modern approach is striking. Table 18.5 compares these two approaches for an experiment in which there are two factors (or variables) whose effects on a characteristic are being investigated. (The same conclusions hold for an experiment with more than two factors.)

This discussion has been restricted to the design or planning of the experiment. After the data are collected, the analysis phase beings. For simple experiments, some of the basic tests of hypotheses and confidence limits (previously discussed) provide the tools of analysis. For more complex experiments, we use additional tools such as the analytical analysis of variance and the graphical analysis of means (see Wadsworth, 1998).

For a perceptive discussion of the relation of design of experiments to the scientific method, see Box and Liu (1999).

TABLE 18.5
Comparison of classical and modern methods of experimentation

Criteria	Classical	Modern
Basic procedure	Hold everything constant except the factor under investigation. Vary that factor and note the effect on the characteristic of concern. To investigate a second factor, conduct a separate experiment in the same manner.	Plan the experiment to evaluate both factors in one main experiment. Include, in the design, measurements to evaluate the interaction effect of varying both factors simultaneously.
Experimental conditions	Care taken to have material, workers, and machine constant throughout the entire experiment.	Realizes difficulty of holding conditions reasonably constant throughout an entire experiment. Instead, experiment is divided into several groups or blocks of measurements. Within each block, conditions must be reasonably constant (except for deliberate variation to investigate a factor).
Experimental error	Recognized but not stated in quantitative terms.	Stated in quantitative terms that support decision-making under uncertainty.
Basis of evaluation	Effect due to a factor is evaluated with only a vague knowledge of the amount of experimental error.	Effect due to a factor is evaluated by comparing variation due to that factor with the quantitative measure of experimental error.
Possible bias due to sequence of measurements	Often assumed that sequence has no effect.	Guarded against by randomization.
Effect of varying both factors simultaneously ("interaction")	Not adequately planned into experiment. Frequently assumed that the effect of varying factor 1 (when factor 2 is held constant at some value) would be the same for any value of factor 2.	Experiment can be planned to include an investigation for interaction between factors.
Validity of results	Misleading and erroneous if interaction exists and is not realized.	Even if interaction exists, a valid evaluation of the main factors can be made.
Number of measurements	For a given amount of useful and valid information, more measurements needed than in the modern approach.	Fewer measurements needed for useful and valid information.
Definition of problem	Objective of experiment frequently not defined as necessary.	Designing the experiment requires defining the objective in detail (how large an effect do we want to determine, what numerical risks can be taken, etc.).
Application of conclusions	Sometimes disputed as applicable only to the controlled conditions under which the experiment was conducted.	Broad conditions can be planned in the experiment, thereby making conclusions applicable to a wider range of actual conditions.

18.14 ANALYSIS OF VARIANCE

Analysis of variance (ANOVA) is an important tool employed to analyze the results of an experiment, as well as many other analytical situations in which the effects of multiple factors on a continuous variable are to be analyzed. In this approach, the total variation of all measurements around the overall mean is divided into sources of variation that are then analyzed for statistical significance. The following example explains this technique.

Four types of diets (1, 2, 3, 4) were fed to 24 animals. Blood coagulation time (the response variable) was recorded and is shown in Table 18.6.

The issue is whether diet has an effect on coagulation time. The null hypothesis is $\mu 1 = \mu 2 = \mu 3 = \mu 4$, i.e., the average coagulation time is the same for both diets.

The ANOVA analysis (using MINITAB®) is shown in Figure 18.23. The Total variation is divided into variation due to DietNum and Error (the residual when variation due to diet is subtracted from the total). The sum of the squares (SS) for the total variation is the sum of the squared deviations of the 24 measurements around the overall average of the 24. The SS for diet is the sum of the squared deviations of the individual measurements for each diet around the averages for each diet. The SS for the error is found by subtracting the SS for diet from the SS for the total variation. The mean square (MS) is the sum of the squares divided by the degrees of freedom (DF). The mean square value is essentially a variance (square of the standard deviation). The hypothesis is tested by comparing the MS for diet to the MS for error. This step is done by using the F test and calculating F as 76.00/5.60 or 13.57. If a significance level of .05 is used, the acceptance region on F is equal to or less than 3.10. The calculated F is much greater than 3.10, and thus the hypothesis is rejected; therefore, we conclude that diet does have an effect on coagulation time. Note that the MINITAB output shows the significance probability (p-value) as zero, meaning that it is highly unlikely that an F-value of 13.57 could have occurred by chance if the means for the different diets were equal. This outcome is equivalent to rejecting the hypothesis. Also note that confidence intervals are provided for the diet means. Confidence intervals that do not overlap suggest that the diet means are not equal, and we reject the hypothesis. Note how the confidence intervals provide useful information on exactly which diets are different from each other beyond the basic F test result that indicates at least one is different from another.

TABLE 18.6
Diet and coagulation time

Diet 1	Diet 2	Diet 3	Diet 4
62	63	68	56
60	67	66	62
63	71	71	60
59	64	67	61
	65	68	63
	66	68	64
			63
			59

Source: Juran Institute, Inc.

Source	DF	SS	MS	F	P
DietNum	3	228.00	76.00	13.57	0.000
Error	20	112.00	5.60		
Total	23	340.00			

Individual 95% confidence intervals for mean based on pooled StDev

Level	N	Mean	StDev
1	4	61.000	1.826
2	6	66.000	2.828
3	6	68.000	1.673
4	8	61.000	2.619

Pooled StDev = 2.366

```
                    (-------*-------)
                                   (-------*------)
                                           (-----*-----)
   (-----*-----)
---+---------+---------+---------+---
  59.5      63.0      66.5      70.0
```

FIGURE 18.23
One-way analysis of variance and confidence intervals for CoagTime.
(*From Juran Institute, Inc.*)

This example introduces only the basic concept of ANOVA. ANOVA requires that the data satisfy certain assumptions. The responses for each level must be independent and normally distributed, and the population variances for levels of the factor under study must be equal.

The analysis of variance determines which sources of variation (factors) contribute a significant amount of the total variation of the response variable, y. Knowing which factors are significant can lead to a mathematical model for y as a function of the dependent variables, x, found to be significant; i.e., $y = f(x_1, x_2, \ldots x_n)$. For elaboration, see DeVor et al. (1992), Section 18.

→ 18.15 FRACTIONAL FACTORIAL EXPERIMENTAL DESIGN

Even with just two levels, the number of experimental runs required for a full factorial design grows very rapidly, e.g., requiring 128 runs for seven factors at two levels. For designs with four or more factors, it is possible to conduct fewer runs in a way that preserves most of the benefits of a factorial design. The reduced number must be selected in very special ways that ensure (1) *balance,* i.e., each factor is represented with an equal number of low-value and high-value runs, and (2) *orthogonality,* i.e., for each factor, its low value must be paired with an equal number of low and high values for every other factor and similarly for each high value.

The price to be paid for the reduced sample sizes is the loss of estimates for higher levels of interaction and some mixing of effects in interactions. In our seven-factor case, the following are the effects of adopting:

- Half fraction (64 runs): no four-way, five-way, six-way or seven-way interactions.
- Quarter fraction (32 runs): also no three-way interactions.

- Smaller fractions suffer from having significant confounding of the main effects estimates with interaction terms.

The good news is that with many factors, it is not unusual for one or more to show up as insignificant. When the insignificant variable or variables are removed, there are enough observations for a proper estimation of the model. If the results do not support this *reduced model* estimation, then the experimenter has learned that all the variables are important and can make an informed decision on whether to expand the sample.

Some practitioners have urged that even smaller fractions can be used in particular cases. For example, Genichi Taguchi argues for their use in robust design.

A sixteenth-fraction factorial design with seven factors (A, B, C, D, E, F, and G) each at two levels (1 and 2) is shown in Figure 18.24. This design requires only eight trials of the 128 combinations (the eight are indicated in black). The matrix shows the level (1 or 2) to be used for each of the seven factors. The column numbers (1 through 7) represent the factors (A through G). For example, trial number 1 runs each factor at level 1.

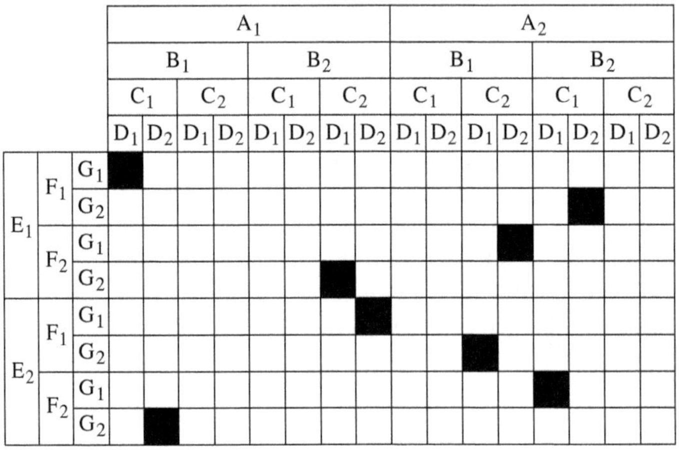

Trial no.	Column no.						
	1	2	3	4	5	6	7
1	1	1	1	1	1	1	1
2	1	1	1	2	2	2	2
3	1	2	2	1	1	2	2
4	1	2	2	2	2	1	1
5	2	1	2	1	2	1	2
6	2	1	2	2	1	2	1
7	2	2	1	1	2	2	1
8	2	2	1	2	1	1	2

FIGURE 18.24
Fractional factorial design with matrix of levels. (*From Ross, 1996.*)

For elaboration of the Taguchi approach and tables of orthogonal array designs, see Ross (1996).

Although Taguchi's approach has been used to good effects in some cases, many practitioners remain concerned that such small fractions run real risks of incorrect results for the main effects as well as missing important interactions. In the example cited here, the main effects are each confounded by three different two-way effects.

18.16 PLANNING FOR EXPERIMENTAL DESIGN

Experienced statisticians have long known that formal design of experiments involves much more than selecting a design and making statistical calculations. A successful experiment requires careful planning of the activities before, during, and after the experiment.

Bisgaard (1999) describes 11 points to be covered in a "preexperiment proposal" (reprinted by courtesy of Marcel Dekker, Inc.).

1. Statement of objective(s) of the experiment.
2. List of factors and their levels.
3. Description of how to measure the responses.
4. A table showing the design and possible confounding.
5. Layout of data collection sheets.
6. Description of the experimental procedure.
7. Schedule for the experiment.
8. An outline of how to analyze the data.
9. Budget for time, money, and other resources.
10. Anticipated problems and how to deal with them.
11. List of team members and responsibilities.

This list particularly applies to factorial experiments conducted in a sequential mode.

Barton (1997) describes some graphical methods useful in preexperiment planning. These methods show how graphical plots can be prepared based on preexperiment discussions with engineers who provide information on the product and process. These plots help to define the goals of the experiment; identify dependent, independent, and "nuisance" variables; and choose a mathematical relationship (a regression) between the independent and dependent variables. See also Coleman and Montgomery (1993) for further discussion of planning for experiments.

18.17 REGRESSION ANALYSIS

Quality problems sometimes require a study of the mathematical relationship between two or more continuous variables, or *regression analysis*. The uses of regression analysis include forecasting and prediction, determining the important variables that influence some result, and locating optimum operating conditions.

TABLE 18.7
Cutting speed (X, in feet per minute) versus tool life (Y, in minutes)

X	Y	X	Y	X	Y	X	Y
90	41	100	22	105	21	110	15
90	43	100	35	105	13	110	11
90	35	100	29	105	18	110	6
90	32	100	18	105	20	110	10

The steps in a regression study are

1. Clearly define the objectives of the study. This step must include a definition of the dependent or response variable and the independent variables that are thought to be related to the dependent variable.
2. Collect data on the independent and dependent variables.
3. Prepare scatter diagrams (plots of one variable versus another).
4. Calculate the regression equation.
5. Study the equation to see how well it fits the data.
6. Provide measures of the precision of the equation.

The following example illustrates these steps.

Suppose that the life of a tool varies with the cutting speed of the tool and we want to predict life based on cutting speed. Thus life is the dependent variable (Y) and cutting speed is the independent variable (X). Data are collected at four cutting speeds (Table 18.7).

The plot of the data is a *scatter diagram* (Figure 18.25). This plot should always be prepared before making any further analysis. The graph alone may provide sufficient information on the relationship between the variables to draw conclusions about the immediate problem, but the graph is also useful for suggesting possible forms of

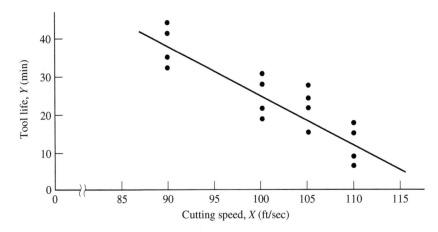

FIGURE 18.25
Tool life Y versus cutting speed X.

an estimating equation. Figure 18.25 suggests that life varies with cutting speed (i.e., life decreases with an increase in speed) and also in a linear manner (i.e., increases in speed result in a certain decrease in life that is the same over the range of the data). Note that the relationship is not perfect—the points scatter about the line.

Usually it is valuable to determine a regression equation. For linear relationships, this can be done approximately by drawing a straight line by eye and then graphically estimating the Y intercept and slope. The linear regression model is

$$Y = \beta_0 + \beta_1 X + \epsilon$$

where β_0 and β_1 are the unknown population intercept and slope and ϵ is a random-error term that may be due to measurement errors and/or the effects of other independent variables. This model is estimated from sample data by the form

$$\hat{Y} = b_0 + b_1 X$$

where $\hat{Y}$ is the predicted value of Y for a given value of X and b_0 and b_1 are the sample estimates of β_0 and β_1.

These estimates are usually found by the least-squares method, so named because it minimizes the sum of the squared deviations between the observed and predicted values of Y. The least-squares estimates are

$$b_1 = \frac{\Sigma(X_m - \bar{X})(Y_m - \bar{Y})}{\Sigma(X_m - \bar{X})^2} = \frac{\Sigma X_m Y_m - (\Sigma X_m \Sigma Y_m)/N}{\Sigma X_m^2 - (\Sigma X_m)^2/N}$$

$$b_0 = \bar{Y} - b_1 \bar{X}$$

The summations range from $m = 1$ to $m = N$, where N is the total number of sets of values of X and Y.

The detailed calculations are easily handled by statistical software, most spreadsheet applications, and even handheld calculators. For these data,

$$b_1 = \frac{-1191.25}{875} = -1.3614$$

$$b_0 = 23.06 - (-1.3614)(101.25) = 160.9018$$

and hence the prediction equation is

$$\hat{Y} = 106.90 - 1.3614 X$$

After estimating the coefficients of the prediction equation, the equation should be plotted over the data to check for gross calculation errors. Roughly half the data points should be above the line and half below it, with this approximately equal distribution applying throughout the range of the data. In addition, the equation should pass exactly through the point $\bar{X}, \bar{Y}$.

A number of criteria exist for judging the adequacy of the prediction equation. One common measure of the adequacy of the prediction equation is R^2, the proportion of variation explained by the prediction equation. R^2, or the *coefficient of determination*, is the ratio of the variation due to the regression, $\Sigma(\hat{Y}_m - \bar{Y})$, to the total variation, $\Sigma(Y_m - \bar{Y})^2$. $\hat{Y}_m$ is the predicted Y value for X_m. The calculation formula is

$$R^2 = \frac{b_1 \Sigma(X_m - \bar{X})(Y_m - \bar{Y})}{\Sigma(Y_m - \bar{Y})^2}$$

$$= \frac{(-1.3614)(-1191.25)}{1958.94} = 0.828$$

Thus for this example, the prediction equation explains 82.8% of the variation in tool life. The coefficient of determination and *all* other measures of the precision of a regression relationship must be interpreted with great care. Regression analysis is not an area for the amateur.

An example of a MINITAB output for a regression analysis is shown in Figure 18.26. Note that the regression plot provides two sets of confidence limits. The inner set shows 95% confidence limits for the mean value (i.e., for predictions using the relationship represented by the line), and the outer set shows 95% prediction limits for individual values. Also note that the coefficient of determination ("R-sq") is 69.5% and the ANOVA table shows that the regression is significant with an F-value of 43.29 (comparing the regression variation to the experimental error variation). The p-value of 0.000 shows that the F-value lies in the rejection region and the equation as a whole is significant.

The preceding discussion covers simple *linear regression*—the prediction of a dependent variable, Y, from a single predictor variable, X. *Multiple regression* cases involve two or more predictor variables. For two predictor variables, the multiple linear regression model is

$$Y = \beta_0 + \beta_1 X_1 + \beta_2 X_2 + \epsilon$$

Another measure of the degree of association between two variables is the *simple linear correlation coefficient, r*. This coefficient varies between -1.0 and $+1.0$. Calculations for r are explained in *JQH5,* Section 44. Many software programs provide r as an output. Positive values indicate that as one variable increases, so does the other; negative values mean that as one variable decreases, so does the other. A coefficient of zero means that there is no association between the two variables; values of -1.0 or $+1.0$ indicate perfect correlation.

Interpretation of the coefficient of correlation requires great care. First, r indicates the strength of a *linear* relationship. Thus, an r may be close to zero when there is actually a strong *nonlinear* relationship. Also, a high value of r does *not* necessarily imply that one variable *causes* the other. Variable X may contribute to variable Y but not be the sole cause, both X and Y may result from some common (but unknown) cause, or other reasons may explain the apparent correlation.

This brief treatment of regression is just an introduction to a complex subject. Further topics include confidence intervals and other measures of precision, multiple regression, and nonlinear regression.

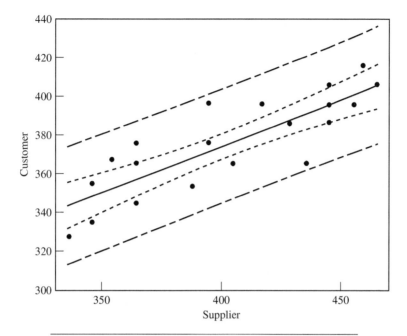

FIGURE 18.26
MINITAB output for regression analysis. (*From Juran Institute, Inc.*)

18.18 ADVANCED TOOLS OF DATA ANALYSIS

The growth of computer databases is resulting in large amounts of data that are easily accessible for analysis—*data mining* is the new term. This environment is leading to the use of various data analysis tools—some are new, some are old, some were developed by statisticians, some by other practitioners. The area of multivariate analysis includes tools such as principal component analysis, factor analysis, cluster analysis, discriminant analysis, multiple regression, neural networks, and time series analysis.

616   Quality Management and Analysis

Other tools include Bayesian analysis, fuzzy logic, image recognition, voice recognition, and artificial intelligence.

Fortunately, computer software is available for many of these tools, and the time saved in making calculations can be invested in thoroughly understanding the meaning of the results.

Answers to questions at the beginning of the chapter:

1. Yes
2. Yes
3. Yes
4. No
5. Yes

WORKED EXAMPLES USING MINITAB

Statistical Tools for Analyzing Data

1. A medical group formed a six sigma team to reduce the time it took for patients to see a physician. The mean days to appointment was reduced from 8.6 to 6.2 days. Analyze the data provided. Does this represent a statistically significant reduction in time?

Time data collected	Days to Appointment
Before improvement	11, 9, 11, 13, 8, 8, 13, 10, 8, 8, 7, 10, 6, 10, 6, 7, 4, 8, 8, 7
After improvement	6, 3, 11, 6, 5, 8, 7, 8, 9, 3, 7, 2, 6, 5, 6, 6, 7, 7, 5,7

MINITAB Analysis

Data are normally distributed

$H_0: \mu_{before} = \mu_{after}$
$H_a: \mu_{before} > \mu_{after}$

Two-sample t for before vs after (one-sided)

```
           N     Mean    StDev   SE Mean
Before    20     8.60    2.30    0.52
After     20     6.20    2.09    0.47

Difference = mu (Before) - mu (After)
Estimate for difference: 2.40000
95% lower bound for difference: 1.22603
t-test of difference = 0 (vs >): t-value = 3.45 p-value = .001 DF = 37
```

Conclusions. Reject the null hypothesis and conclude there is a significant difference (reduction) in days to appointment.

2. A cabinet maker was experiencing keying errors in the order placement process. It was hypothesized that certain errors were a function of the condition of orders from builders and/or dealers sent by facsimile (FAX), that is, typed or handwritten.

   Questions. (a) Is the error rate for builders independent of the format (typed or handwritten)?
   (b) Is the error rate for dealers independent of the format (typed or handwritten)?

   Data

   Builders

Input method	Good	Bad
Handwritten	3698	12
Typewritten	7261	45

   Dealers

Input method	Good	Bad
Handwritten	700	9
Typewritten	1310	18

   H_0: error rate is independent of format
   H_a: error rate is dependent on format

   Conclusions. (a) No. The null hypothesis of independence cannot be rejected based on a chi-square value of 4.089 (degrees of freedom = 1) and corresponding p-value of .43.
   (b) The null hypothesis of independence cannot be rejected given the chi-square value of 0.026 (degrees of freedom = 1) and corresponding p-value of .872.

3. In a facility manufacturing pharmaceuticals in three fermentation vessels, a project team is trying to determine if the quality of production runs is dependent on the fermentation vessel in which the product is made. The following table shows the number of batches produced in each vessel, and a breakdown of conforming and nonconforming (defective) batches.

   (a) State the hypothesis test that the team should use. (Answer: Chi-sq Tables test)
   (b) State the null and alternate hypotheses.
   H_0: Quality of batches is independent of fermentation vessel used
   H_a: Quality of batches is dependent on fermentation vessel used
   (c) Run the test using MINITAB. State the p-value and both the statistical and practical conclusions.

   Data

	Fermentation vessel		
	A	B	C
Number of acceptable batches	98	113	82
Number of discarded batches	5	12	14

MINITAB Analysis
Chi-Square Test: Vessel A, Vessel B, Vessel C

Expected counts are printed below observed counts
Chi-Square contributions are printed below expected counts

	Vessel A	Vessel B	Vessel C	Total
1	98	113	82	293
	93.15	113.04	86.81	
	0.253	0.000	0.267	
2	5	12	14	31
	9.85	11.96	9.19	
	2.392	0.000	2.524	
Total	103	125	96	324

Chi-Sq = 5.436, DF = 2, P-Value = 0.066

Fail to reject H_0.

Conclusions. Testing at an alpha level of .05, the quality of production runs, as measured by number of accepted or discarded batches, does not differ among the three fermentation vessels. Quality is independent of fermentation vessel.

4. A quality improvement team is investigating whether different approaches to document development yield different document development times. For ten technical writers, the team measured the work time (in hours) necessary to complete finished documents for each of two different approaches (design options). Determine if there is a difference among the approaches.

Data

Technical Writer	Option 1	Option 2
A	6.7	5.8
B	5.4	5.0
C	5.1	4.0
D	3.2	2.7
E	5.8	5.6
F	6.1	5.4
G	4.9	5.3
H	4.7	4.2
I	5.5	4.1
J	6.8	5.1

MINITAB Analysis
Normal data, equal variance assumptions ok.

Paired T-Test and CI: Option 1, Option 2

Paired T for Option 1 - Option 2

	N	Mean	StDev	SE Mean
Option 1	10	5.42000	1.05283	0.33293
Option 2	10	4.72000	0.95545	0.30214
Difference	10	0.700000	0.607362	0.192065

95% CI for mean difference: (0.265519, 1.134481)
T-Test of mean difference = 0 (vs not = 0): T-Value = 3.64
P-Value = 0.005

Conclusion. Assuming an alpha level of .05, there is a significant difference in document development time between the two approaches (Option 1 and Option 2).

5. A clothing manufacturer was interested in the strength of cotton fiber provided by four different suppliers. Twenty random samples were taken and their breaking strength measured in grams. Is there a difference in fiber strength among the suppliers?

Data

Supplier 1	Supplier 2	Supplier 3	Supplier 4
25.6	29.0	29.1	20.3
24.7	26.6	21.3	18.6
26.6	26.8	21.9	21.7
23.2	28.7	27.8	22.6
23.3	24.3	25.2	22.8
29.5	24.7	24.0	21.8
23.8	26.1	22.4	19.4
26.9	29.6	26.8	18.7
23.1	26.8	23.8	20.8
22.7	27.6	24.9	19.6
24.0	29.1	25.1	21.2
28.0	28.0	26.4	24.4
24.8	29.0	25.6	22.4
27.5	25.7	27.8	21.5
24.2	28.5	25.0	21.1
22.1	27.6	22.3	17.7
19.0	26.3	26.6	27.4
24.2	32.6	28.2	25.4
20.6	27.0	28.6	23.3
19.6	28.0	23.2	22.0

MINITAB Analysis
One-way ANOVA: C1, C2, C3, C4

```
Source     DF         SS         MS         F          P
Factor      3     367.72     122.57     22.10      0.000
Error      76     421.43       5.55
Total      79     789.15

S = 2.355  R-SQ = 46.60%  R-SQ(adj) = 44.49%
```

```
                                    Individual 95% CIs For Mean
                                    based on Pooled StDev
Level     N      Mean    StDev      -------+------+------+------+-
C1       20    24.164    2.720             (----*----)
C2       20    27.592    1.897                           (---*---)
C3       20    25.304    2.361                 (---*---)
C4       20    21.638    2.369      (----*---)
                                    -------+------+------+------+-
                                        22.5    25.0   27.5   30.0

Pooled StDev = 2.355
```

Conclusions. There is a significant difference in fiber strength among the four suppliers.

SUMMARY

- Estimation is the process of analyzing a sample result to predict the corresponding value of the population parameter.
- The point estimate is a single value used to estimate the population parameter. The confidence interval is a range of values that include (with a preassigned probability called a confidence level) the true value of a population parameter.
- A hypothesis is an assertion made about a population. A test of hypothesis is a test of the validity of the assertion and is carried out by analyzing a sample of data.
- In evaluating a hypothesis, two types of errors can be made: type I errors (reject the hypothesis when it is true) and type II errors (fail to reject the hypothesis when it is false).
- The statistical design of experiments provides plans for conducting experiments from which valid statistical analyses can be made.
- Randomization is the assignment of specimens to treatments in a purely chance manner.
- Replication is the repetition of an observation or measurement.
- Regression analysis is the study of the relationship between two or more variables with continuous data.

PROBLEMS

Note: The specific questions have purposely been stated in nonstatistical language to provide the student with some practice in choosing techniques and making assumptions. When required, use a type I error of .05 and a confidence level of 95%. State any other assumptions needed.

18.1. In the casting industry, the pouring temperature of metal is important. For an aluminum alloy, past experience shows a standard deviation of 15°. During a particular day, five temperature tests were done during the pouring time.
   (a) If the average of these measurements was 1650°, make a statement about the average pouring temperature.
   (b) If you had taken 25 measurements and obtained the same results, what effect would this have on your statement? Make such a revised statement.

18.2. At the casting firm in problem 18.1, a new aluminum alloy is being poured. During the first day of pouring, five pouring temperature tests were done, with these results:

   1705°   1725°   1685°   1690°   1715°

   Make a statement about the average pouring temperature of this metal.

18.3. A manufacturer pressure tests gaskets for leaks. The pressure at which this gasket leaked on nine trials was (in psi)

4000	3900	4500
4200	4400	4300
4800	4800	4300

   Make a statement about the average "leak" pressure of this gasket.

18.4. The acceptable time for answering the phone at a call center is 15 seconds. In a sample of 500 calls, 427 were answered within 15 seconds. Make a statement concerning the true proportion that would be acceptable.

18.5. In a meat-packing firm, out of 600 pieces of beef, 420 were found to be Grade A. Make a statement about the true proportion of Grade A beef.

18.6. A specification requires that the average breaking strength of a certain material be at least 180 psi. Past data indicate that the standard deviation of individual measurements is 5 psi. How many tests are necessary to be 99% sure of detecting a lot that has an average strength of 170 psi?

18.7. Tests are to be run to estimate the average life of a product. Based on past data on similar products, it is assumed that the standard deviation of individual units is about 20% of the average life.
(a) How many units must be tested to be 90% sure that the sample estimate will be within 5% of the true average?
(b) Suppose funds are available to run only 25 tests. How sure would we be of obtaining an estimate within 5%?

Answer: (a) 44. (b) 78.8%.

18.8. A manufacturer of needles has a new method of controlling a diameter dimension. From many measurements of the present method, the average diameter is 0.076 cm with a standard deviation of 0.010 cm. A sample of 25 needles from the new process shows that the average was 0.071. If a smaller diameter is desirable, should the new method be adopted? (Assume that the standard deviation of the new method is the same as that for the present method.)

18.9. In the garment industry, the breaking strength of cloth is important. A heavy cotton cloth must have an average breaking strength of at least 200 psi. From one particular lot of this cloth, these five measurements of breaking strength (in psi) were obtained:

206
194
203
196
192

Does this lot of cloth meet the requirement of an average breaking strength of 200 psi?

Answer: $t = -0.67$.

18.10. In a drug firm, the variation in the weight of an antibiotic from batch to batch is important. With our present process, the standard deviation is 0.11 g. The research department has developed a new process that it believes will produce less variation. The following weight measurements (in grams) were obtained with the new process:

7.47
7.49
7.64
7.59
7.55

Does the new process have less variation than the old?

18.11. A paper manufacturer has a new method of coating paper. The lower the variation in the weight of this coating, the more uniform and better the product. The following 10 sample coatings were obtained by the new method:

Coating weights (in weight/unit area × 100)	
223	234
215	229
220	223
238	235
230	227

If the standard deviation in the past was 9.3, is this proposed method any better? Should the company switch to this method?

Answer: $\chi^2 = 5.43$.

18.12. A manufacturer of rubber products is trying to decide which "recipe" to use for a particular rubber compound. High tensile strength is desirable. Recipe 1 is cheaper to mix, but he is not sure if its strength is about the same as that of recipe 2. Five batches of rubber were made by each recipe and tested for tensile strength. These are the data collected (in psi):

Recipe 1	Recipe 2
3067	3200
2730	2777
2840	2623
2913	2044
2789	2834

Which recipe would you recommend?

18.13. Test runs with five models of an experimental engine showed that they operated, respectively, for 20, 18, 22, 17, and 18 min with 1 gal of a certain kind of fuel. A proposed specification states that the engine must operate for a mean of at least 22 min.
(a) What can we conclude about the ability of the engine to meet the specification?
(b) What is the probability that the sample mean could have come from a process whose true mean is equal to the specification mean?
(c) How low would the mean operating minutes (of the engine population) have to be to have a 50% chance of concluding that the engine does not meet the specification?

Answer: (a) $t = -3.4$. (b) Approx. 0.03. (c) 20.1.

18.14. A manufacturer claims that the average length in a large lot of parts is 2.680 in. A large amount of past data indicates that the standard deviation of individual lengths is 0.002 in. A sample of 25 parts shows an average of 2.678 in. The manufacturer

says that the result is still consistent with his claim because only a small sample was taken.
(a) State a hypothesis to evaluate the claim.
(b) Evaluate his claim using the standard hypothesis testing approach.
(c) Evaluate his claim using the confidence limit approach.

18.15. An engineer wants to determine whether the type of test oven or temperature has a significant effect on the average life of a component. She proposes the following design of experiment:

	Oven 1	Oven 2	Oven 3
550°	1	0	1
575°	0	1	1
600°	1	1	0

The numbers in the body of the table represent the number of measurements to be made in the experiment. What are two reasons that interaction cannot be adequately evaluated in this design?

18.16. The molding department in a manufacturing plant has been making too many defectives. There are many opinions about the reasons. One opinion states that the molding time per record has a cause-and-effect relationship with the number of defectives produced per 100 units. Several trial lots of 100 units each were made with various mold times. The results were as follows:

Time, s	Number defective
2	16
4	13
5	8
7	8
10	4
11	6
13	5
17	3
17	5
20	3

Plot the data and graphically estimate the Y intercept and the slope.

Answer: The least-squares estimates are a Y intercept of 13.54 and a slope of -0.6076.

18.17. Assemble a team of three people to prepare a proposal for a designed experiment. Select a problem on a specific performance characteristic in a service industry. Using the 11 points in Section 18.16, develop an outline for the proposal.

REFERENCES

Agresti, A. (2007). *An Introduction to Categorical Data Analysis* (2nd ed.), John Wiley and Sons, New York, NY.

Barton, R. R. (1997). "Pre-Experiment Planning for Designed Experiments: Graphical Methods," *Journal of Quality Technology,* July, pp. 307–316.

Bisgaard, S. (1999). "Quality Quandries," *Quality Engineering,* vol. 11, no. 4, pp. 645–650.

Box, G. E. P., J. S. Hunter, and W. G. Hunter (2005). *Statistics for Experimenters,* John Wiley and Sons, Hoboken, NJ.

Box, G. E. P. (1999). "Statistics as a Catalyst to Learning by Scientific Method, Part II—A Discussion," *Journal of Quality Technology,* vol. 31, no. 1, pp. 16–29.

Box, G. E. P. and P. Y. T. Liu (1999). "Statistics as a Catalyst to Learning by Scientific Method, Part I—An Example," *Journal of Quality Technology,* vol. 31, no. 1, pp. 1–15.

Carter, C. W. (1996). "Sequenced-Level Experimental Designs," *Quality Engineering,* vol. 8, no. 1 (pp. 181–188), no. 2 (pp. 361–366), no. 3 (pp. 499–504), no. 4 (pp. 695–698).

Coleman, D. E. and D. C. Montgomery (1993). "A Systematic Approach to Planning for a Designed Experiment" (with discussion), *Technometrics,* vol. 35, pp. 1–27.

DeVor, R. E., T.-H. Chang, and J. W. Sutherland (1992). *Statistical Quality Design and Control,* Prentice-Hall, Upper Saddle River, NJ.

Emanuel, J. T. and M. Palanisamy (2000). "Sequential Experimentation Using Two-Level Fractional Factorials," *Quality Engineering,* vol. 12, no. 3, pp. 335–346.

Finn, L. M. (1995). "Getting Statistical Tools to the Masses: Common Problems in Training Design and Execution," *Annual Quality Congress Proceedings,* ASQ, Milwaukee, pp. 356–364.

Juran, J. M. and Joseph A. DeFeo (2010). *Juran's Quality Handbook* (6th ed.), McGraw-Hill, New York.

Ross, P. J. (1996). *Taguchi Techniques for Quality Engineering,* McGraw-Hill, New York.

Sprent, P. and N. C. Smeeton (2007). *Applied Nonparametric Statistical Methods* (4th ed.), Chapman & Hall/CRC, Boca Raton, FL.

Wadsworth, H. M., Jr. (1998). *Handbook of Statistical Methods for Engineers and Scientists,* 2nd ed., McGraw-Hill, New York.

CHAPTER 19

Statistical Tools to Design for Quality

➔ 19.1 THE STATISTICAL TOOLKIT FOR DESIGN

Statistical tools for quality in the design and development process include techniques such as graphical summaries of data, probability distributions, theorems of probability, confidence limits, tests of hypotheses, design of experiments, and regression analysis. These are explained in Chapter 17, "The Role of Statistics and Probability," and Chapter 18, "Tools for Analyzing Data."

To supplement these techniques, this chapter explains some statistical tools for reliability and availability and tools for setting specification limits on product characteristics.

We proceed to examine some statistical tools that are useful in design.

➔ 19.2 FAILURE PATTERNS FOR COMPLEX PRODUCTS

Methodology for quantifying reliability was first developed for complex products. Suppose that a piece of equipment is placed on test, is run until it fails, and the failure time is recorded. The equipment is repaired and again placed on test, and the time of the next failure is recorded. The procedure is repeated to accumulate the data shown in Table 19.1. The failure rate is calculated, for equal time intervals, as the number of failures per unit of time. When the failure rate is plotted against time, the result (Figure 19.1) often follows a familiar pattern of failure known as the *bathtub curve*. Three periods are apparent. These periods differ in the frequency of failure and in the failure causation pattern:

1. *The infant mortality period.* This period is characterized by high failure rates that show up early in use (see the lower half of Figure 19.1). These failures can be the

TABLE 19.1
Failure history for a unit of electronic ground support equipment

Time of failure, infant mortality period		Time of failure, constant-failure-rate period		Time of failure, wear-out period	
1.0	7.2	28.1	60.2	100.8	125.8
1.2	7.9	28.2	63.7	102.6	126.6
1.3	8.3	29.0	64.6	103.2	127.7
2.0	8.7	29.9	65.3	104.0	128.4
2.4	9.2	30.6	66.2	104.3	129.2
2.9	9.8	32.4	70.1	105.0	129.5
3.0	10.2	33.0	71.0	105.8	129.9
3.1	10.4	35.3	75.1	106.5	
3.3	11.9	36.1	75.6	110.7	
3.5	13.8	40.1	78.4	112.6	
3.8	14.4	42.8	79.2	113.5	
4.3	15.6	43.7	84.1	114.8	
4.6	16.2	44.5	86.0	115.1	
4.7	17.0	50.4	87.9	117.4	
4.8	17.5	51.2	88.4	118.3	
5.2	19.2	52.0	89.9	119.7	
5.4		53.3	90.8	120.6	
5.9		54.2	91.1	121.0	
6.4		55.6	91.5	122.9	
6.8		56.4	92.1	123.3	
6.9		58.3	97.9	124.5	

result of blunders in design or manufacture, misuse, or misapplication. Usually, once corrected, these failures do not occur again, e.g., an oil hole that is not drilled. Sometimes it is possible to "debug" the product by a simulated use test or by overstressing (in electronics this is known as burn-in). The weak units still fail, but the failure takes place in the test rig rather than in service. O'Connor (1995), Chapter 9, explains the use of burn-in tests and environmental screening tests.

2. *The constant-failure-rate period.* Here the failures result from the limitations inherent in the design, excess variability in the production process, changes in the environment, and accidents caused by use or maintenance. The accidents can be held down by good control of operating and maintenance procedures. However, a reduction in the failure rate requires structured quality improvement, which may lead to a wide variety of different remedies from basic redesign of the product to changes in the assembly process or production of components and raw materials.

3. *The wear-out period.* These failures are due to old age, e.g., a metal becomes embrittled or insulation dries out. A reduction in failure rates requires preventive replacement of these dying components before they result in catastrophic failure. Designing future models of the product with greater durability can offer a competitive advantage.

The top portion of Figure 19.1 shows the corresponding Weibull plot when $\alpha = 2.6$ was applied to the original data (see Section 17.12, "The Weibull Probability Distribution"). The values of the shape parameter, β, were approximately 0.5, 1.0, and

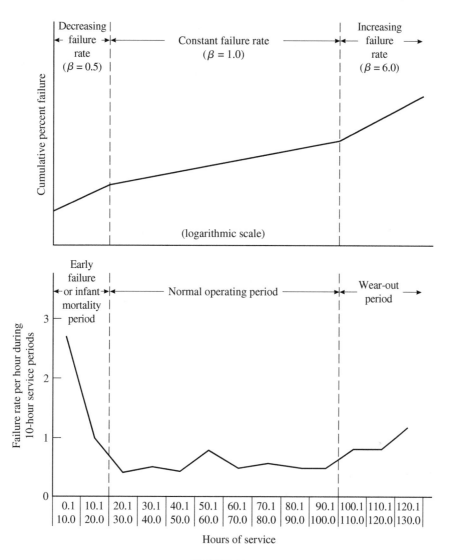

FIGURE 19.1
Failure rate versus time.

6.0, respectively. A shape parameter less than 1.0 indicates a decreasing failure rate, a value of 1.0 a constant failure rate, and a value greater than 1.0 an increasing failure rate (see Figure 19.1).

The Distribution of Time between Failures

Users are concerned with the length of time that a product will run without failure. Thus, for repairable products, the *time between failures* (TBF) is a critical characteristic.

The variation in time between failures can be studied statistically. The corresponding characteristic for nonrepairable products is usually called the *time to failure.*

When the failure rate is constant, the distribution of time between failures is distributed exponentially. Consider the 42 failure times in the constant-failure-rate portion of Table 19.1. The time between failures for successive failures can be tallied, and the 41 resulting TBFs can be formed into the frequency distribution shown in Figure 19.2a. The distribution is roughly exponential in shape, indicating that when the underlying failure rate is constant with observed failures subject to random variation, the distribution of time between failures (not *mean time* between failures) is exponential. This distribution is the basis of the *exponential formula for reliability.*

→ 19.3 THE EXPONENTIAL FORMULA FOR RELIABILITY

The distribution of TBF indicates the chance of failure-free operation for the specified time period. The chance of obtaining failure-free operation for a specified time period *or longer* can be shown by changing the TBF distribution to a distribution showing the number of intervals equal to or greater than a specified time length (Figure 19.2b). If the frequencies are expressed as relative frequencies, they become estimates of the probability of survival. When the failure rate is constant, the probability of survival (or reliability) is

$$P_s = R = e^{-t/\mu} = e^{-t\lambda}$$

where $P_s = R$ = probability of failure-free operation for a time period equal to or greater than t
e = 2.718
t = specified period of failure-free operation
μ = mean time between failures (the mean of TBF distribution)
λ = failure rate (the reciprocal of μ)

Note that this formula is simply the exponential probability distribution rewritten in terms of reliability.

EXAMPLE 19.1. A washing machine requires 30 min to clean a load of clothes. The mean time between failures of the machine is 100 h. Assuming a constant failure rate, what is the chance of the machine completing a cycle without failure?

$$R = e^{-t/\mu} = e^{-0.5/100} = 0.995$$

There is a 99.5% chance of completing a washing cycle.

How about the assumption of a constant failure rate? In practice, sufficient data are usually not available to evaluate the assumption. However, experience suggests that this assumption is often true, particularly when (1) infant mortality types of failures have been eliminated before delivery of the product to the user and (2) the user replaces the product or specific components before the wear-out phase begins.

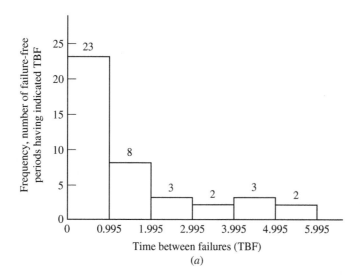
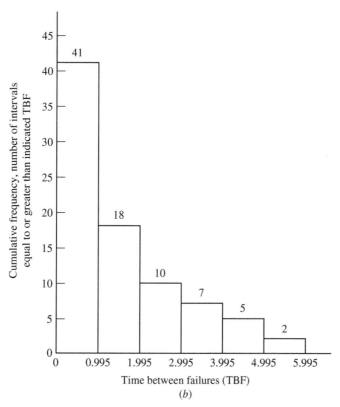

FIGURE 19.2
(*a*) Histogram of TBF. (*b*) Cumulative histogram of TBF.

The Meaning of *Mean Time between Failures*

Confusion surrounds the meaning of *mean time between failures* (MTBF). Further explanation is warranted:

1. The *MTBF* is the mean (or average) time between successive failures of a product. This definition assumes that the product in question can be repaired and placed back into operation after each failure. For nonrepairable products, the term *mean time to failure* (MTTF) is used.
2. If the failure rate is constant, the probability that a product will operate without failure for a time equal to or greater than its MTBF is only 37%. This outcome is based on the exponential distribution. (R is equal to 0.37 when t is equal to the MTBF.) This result is contrary to the intuitive feeling that there is a 50–50 chance of exceeding an MTBF.
3. MTBF is not the same as "operating life," "service life," or other indexes, which generally connote overhaul or replacement time.
4. An increase in an MTBF does not result in a proportional increase in reliability (the probability of survival). If $t = 1$ hour, the following table shows the MTBF required to obtain various reliabilities:

MTBF	R
5	0.82
10	0.90
20	0.95
100	0.99

A fivefold increase in MTBF from 20 to 100 hours is necessary to increase the reliability by 4 percentage points compared with a doubling of the MTBF from 5 to 10 hours to get an 8 percentage point increase in reliability.

MTBF is a useful measure of reliability, but it is *not* correct for all applications. Section 12.5, "Designing for Time-Oriented Performance (Reliability)," includes a list of other reliability indexes.

➔ 19.4 THE RELATIONSHIP BETWEEN PART AND SYSTEM RELIABILITY

It is often assumed that system reliability (i.e., the probability of survival P_s) is the product of the individual reliabilities of the n parts within the system:

$$P_s = P_1 P_2 \ldots P_n$$

For example, if a communications system has four subsystems with reliabilities of 0.970, 0.989, 0.995, and 0.996, the system reliability is 0.951. The formula assumes that (1) the failure of any part causes failure of the system and (2) the reliabilities of the parts are independent of one another, i.e., the reliability of one part does not depend on the functioning of another part.

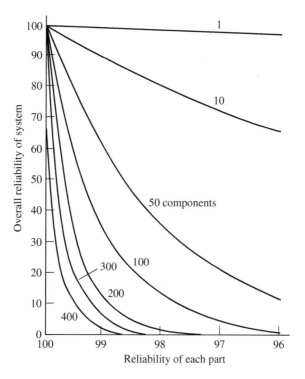

FIGURE 19.3
Relationship of part and system reliability.

These assumptions are *not* always true, but in practice, the formula serves two purposes. First, it shows the effect of increased complexity of equipment on overall reliability. As the number of parts in a system increases, the system reliability decreases dramatically (see Figure 19.3). Second, the formula is often a convenient approximation that can be refined as information on the interrelationships of the parts becomes available.

Pyzdek (1994) discusses the design of an ambulance service to handle emergency calls. The reliability block diagram for the rescue system is shown in Figure 19.4. For this example, the reliability of the system is $(0.998)(0.999) \ldots (0.994) = 0.9195$, i.e., the system would succeed in its mission about 92% of the time.

When it can be assumed that (1) the failure of any part causes system failure, (2) the parts are independent, and (3) each part follows an exponential distribution, then

$$P_s = e^{-t_1\lambda_1}e^{\lambda_1}e^{-t_2\lambda_2}\ldots e^{-t_n\lambda_n}$$

Further, if t is the same for each part,

$$P_s = e^{-t\Sigma\lambda}$$

Thus when the failure rate is constant (and therefore the exponential distribution can be applied), the reliability of a system can be predicted based on the addition of the part failure rates (see Section 19.5).

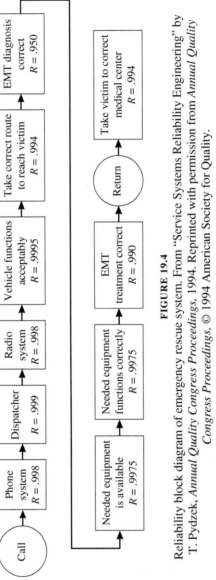

FIGURE 19.4
Reliability block diagram of emergency rescue system. From "Service Systems Reliability Engineering" by T. Pydzek, *Annual Quality Congress Proceedings*, 1994. Reprinted with permission from *Annual Quality Congress Proceedings*, © 1994 American Society for Quality.

Sometimes designs are planned with redundancy so that the failure of one part will not cause system failure. Redundancy is an old design technique invented long before the advent of reliability prediction techniques. However, the designer can now predict the effect of redundancy on system reliability in *quantitative* terms.

Redundancy is the existence of more than one element for accomplishing a given task, where all elements must fail before there is an overall failure of the system. In *parallel redundancy* (one of several types of redundancy), two or more elements operate at the same time to accomplish the task, and any single element is capable of handling the job itself in case of failure of the other elements. When parallel redundancy is used, the overall reliability is calculated as follows:

$$P_s = 1 - (1 - P_1)^n$$

where P_s = reliability of the system
P_1 = reliability of the individual elements in the redundancy
n = number of identical redundant elements

EXAMPLE 19.2. Suppose that a unit has a reliability of 99.0% for a specified mission time. If two identical units are used in parallel redundancy, what overall reliability will be obtained?

$$R = 1 - (1 - 0.99)(1 - 0.99) = 0.9999, \text{ or } 99.99\%$$

➔ 19.5 PREDICTING RELIABILITY DURING DESIGN

In Section 12.5 we introduced the reliability prediction method. Reliability prediction methods are still evolving. Several methods will be discussed in this chapter. Ireson et al. (1996) provide an extensive discussion of reliability prediction.

The following steps make up a reliability prediction method:

1. *Define the product and its functional operation.* The system, subsystems, and units must be precisely defined in terms of their functional configurations and boundaries. This precise definition is aided by preparation of a functional block diagram that shows the subsystems and lower level products, their interrelationships, and the interfaces with other systems.

   Given a functional block diagram and a well-defined statement of the functional requirements of the product, the conditions that constitute failure or unsatisfactory performance can be defined.

2. *Prepare a reliability block diagram.* For systems in which there are redundancies or other special interrelationships among parts, a reliability block diagram is useful. This diagram is similar to a functional block diagram, but the reliability block diagram shows exactly what must function for successful operation of the system. The diagram shows redundancies and alternative modes of operation. The reliability block diagram is the foundation for developing the probability model for reliability. O'Connor (1995) provides further discussion.

3. *Develop the probability model for predicting reliability.* A simple model may add only failure rates; or a complex model can account for redundancies and other conditions.

634   Quality Management and Analysis

4. *Collect information relevant to parts reliability.* The data include information such as parts function, parts ratings, stresses, internal and external environments, and operating time. Many sources of failure-rate information state failure rates as a function of operating parameters. For example, failure rates for fixed ceramic capacitors are stated as a function of (1) expected operating temperature and (2) the ratio of the operating voltage to the rated voltage. Such data show the effect of derating (assigning a part to operate below its rated voltage) on reducing the failure rate.
5. *Select parts reliability data.* The required parts data consist of information on catastrophic failures and on tolerance variations with respect to time under known operating and environmental conditions. Acquiring these data is a major problem for the designer because there is no single reliability data bank comparable to handbooks such as those for physical properties of materials. Instead, the designer must build a data bank by securing reliability data from a variety of sources:
   - Field performance studies conducted under controlled conditions.
   - Specified life tests.
   - Data from parts manufacturers or industry associations.
   - Customers' parts qualification and inspection tests.
   - Government agency data banks such as the Government Industry Data Exchange Program (GIDEP) and the Reliability Analysis Center (RAC).
6. *Combine all of the preceding to obtain the numerical reliability prediction.* An example of a relatively simple reliability prediction method is shown in Section 12.5. Other prediction methods are based on various statistical distributions, as explained in the following sections.

   The concept of reliability growth is a technique for predicting future performance from tests and field data. The concept assumes that a product is undergoing continuous improvements in design and refinements in operating and maintenance procedures. Thus the product performance will improve ("grow") with time. See the following discussion.

Ireson et al. (1996) and O'Connor (1995) are excellent references for reliability prediction. Included are the basic methods of prediction, repairable versus nonrepairable systems, electronic and mechanical reliability, reliability testing, and software reliability. *JQH5,* Section 48, provides extensive discussion of reliability data analysis, including topics such as censored life data (not all test units have failed during the test) and accelerated life test data analysis. Dodson (1999) explains how the use of computer spreadsheets can simplify reliability modeling using various statistical distributions.

Reliability prediction techniques based on component failure data to estimate system failure rates have generated controversy. Jones and Hayes (1999) present a comparison of predicted and observed performance for five prediction techniques using parts count analyses. The predictions differed greatly from observed field behavior and from each other. ANSI/IEC/ASQ D60300-3-1 (1997) compares five analysis techniques—FMEA/FMECA, fault tree analysis, reliability block diagram, Markov analysis, and parts count reliability prediction. It continues to be the international standard.

The reliability of a system evolves during design, development, testing, production, and field use. The concept of *reliability growth* assumes that the causes of product

TABLE 19.2
Example of mechanical parts and subsystem failure rates

Part description	Quantity	Generic failure rate per million hours	Total failure rate per million hours
Heavy-duty ball bearing	6	14.4	86.4
Brake assembly	4	16.8	67.2
Cam	2	0.016	0.032
Pneumatic hose	1	29.28	29.28
Fixed displacement pump	1	1.464	1.464
Manifold	1	8.80	8.80
Guide pin	5	13.0	65.0
Control valve	1	15.20	15.20
Total assembly failure rate			273.376

MTBF = 1/0.000273376 = 3657.9 h
Source: Adapted from Ireson et al., p. 19.9.

failures are discovered and action is taken to remove the causes, thus resulting in improved reliability of future units ("test, analyze, and fix"). Reliability growth models provide predictions of reliability due to such improvements. For elaboration, see O'Connor (1995). Also, ANSI/IEC/ASQ D1164-1997 describes several methods of estimating reliability growth.

➔ 19.6 PREDICTING RELIABILITY BASED ON THE EXPONENTIAL DISTRIBUTION

When the failure rate is constant and when study of a functional block diagram reveals that all parts must function for system success, then reliability is predicted to be the simple total of failure rates. An example of a subsystem prediction is shown in Table 19.2. The prediction for the subsystem is made by adding the failure rates of the parts; the MTBF is then calculated as the reciprocal of the failure rate.

For further discussion of reliability prediction, including an example for an electronic system, see Section 12.5.

➔ 19.7 PREDICTING RELIABILITY BASED ON THE WEIBULL DISTRIBUTION

Prediction of overall reliability based on the simple addition of component failure rates is valid only if the failure rate is constant. When this assumption cannot be made, an alternative approach based on the Weibull distribution can be used.

1. Graphically, use the Weibull distribution to predict the reliability R for the time period specified. $R = 100 - \%$ failure. Do this for each component. (See Section 17.12 under "The Weibull Probability Distribution.")

2. Combine the component reliabilities using the product rule and/or redundancy formulas to predict system reliability.

Predictions of reliability using the exponential distribution or the Weibull distribution are based on reliability as a function of time. Next we consider reliability as a function of stress and strength.

19.8 RELIABILITY AS A FUNCTION OF APPLIED STRESS AND STRENGTH

Failures are not always a function of time. In some cases, a part will function indefinitely if its strength is greater than the stress applied to it. The terms *strength* and *stress* here are used in the broad sense of inherent capability and operating conditions applied to a part, respectively.

For example, operating temperature is a critical parameter, and the maximum expected temperature is 145°F (63°C). Further, capability is indicated by a strength distribution having a mean of 172°F (78°C) and a standard deviation of 13°F (7°C) (see Figure 19.5). With knowledge of only the maximum temperatures, the safety margin is

$$\frac{172 - 145}{13} = 2.08$$

The safety margin says that the average strength is 2.08 standard deviations above the maximum expected temperature of 145°F (63°C). The normal probability distribution on your computer or Table A in Appendix I can be used to calculate a reliability of 0.981 [the area beyond 145°F (63°C)]. Note that normality is assumed and is widely applicable, but if data are available to check that assumption, they should be applied because various skewed distributions might give different results.

This calculation illustrates the importance of *variation* in addition to the *average* value during design. Designers have always recognized the existence of variation by using a *safety factor* in design. However, the safety factor is often defined as the ratio of average strength to the worst stress expected.

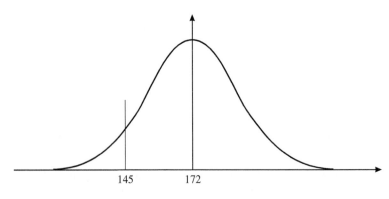

FIGURE 19.5
Distribution of strength.

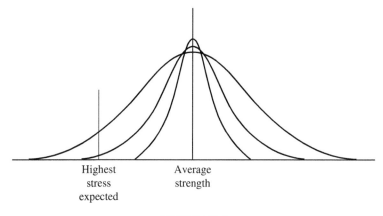

FIGURE 19.6
Variation and safety factor.

Note that in Figure 19.6, all designs have the same safety factor. Also note that the reliability (probability of a part having a strength greater than the stress) varies considerably. Thus the uncertainty often associated with this definition of safety factor is in part due to its failure to reflect the *variation* in both strength and stress. Such variation is partially reflected in a safety margin, defined as

$$\frac{\text{Average strength} - \text{Worst stress}}{\text{Standard deviation of strength}}$$

This recognizes the variation in strength but is conservative because it does not recognize a variation in stress.

The recent emphasis of the six sigma approach relates to a fascinating historical event. Almost half a century ago, Lusser (1958) proposed the use of safety margins as a way to design-in high reliability for critical products such as guided missiles. He suggested safety margins for both strength and stress. Specifically, the reliability boundary (maximum stress) is defined as six standard deviations (of stress) above the average stress. He also proposed that the average strength should be at least five standard deviations (of strength) above the reliability boundary. High reliability is provided because it is unlikely that a high stress (with extremely low probability of occurrence) would combine with a low strength (also with low probability of occurrence). Perhaps Lusser was an early pioneer of the six sigma approach.

➔ 19.9 AVAILABILITY

Availability has been defined as the probability that a product, when used under given conditions, will perform satisfactorily when called upon. Availability considers the operating time of the product and the time required for repairs. Idle time, during which the product is not needed, is excluded.

Availability is calculated as the ratio of operating time to operating time plus downtime. However, downtime can be viewed in two ways:

1. *Total downtime.* This period includes active repair (diagnosis and repair time), preventive maintenance time, and logistics time (time spent waiting for personnel, spare parts, etc.). When total downtime is used, the resulting ratio is called *operational availability* (A_0).
2. *Active repair time.* This always includes diagnosis and actual cycle time of repairing, as earlier. Be clear on the definition, however, because some also include post-repair testing and logistics time. The resulting ratio is called *intrinsic availability* (A_i). Under certain conditions, availability can be calculated as

$$A_0 = \frac{\text{MTBF}}{\text{MTBF} + \text{MDT}} \quad \text{and} \quad A_i = \frac{\text{MTBF}}{\text{MTBF} + \text{MTTR}}$$

where MTBF = mean time between failures
MDT = mean downtime
MTTR = mean time to repair

This is known as the steady-state formula for availability.

Garrick and Mulvihill (1974) present data on certain subsystems of a mechanized bulk mail system (see Table 19.3). If reliability and maintainability can be estimated during the design process, availability can be evaluated before the design is released for production.

The steady-state formula for availability has the virtue of simplicity. However, the formula is based on several assumptions that are not always met in the real world. The assumptions are

- The product is operating in the constant-failure-rate period of the overall life. Thus the failure-time distribution is exponential.
- The downtime or repair-time distribution is exponential.
- Attempts to locate system failures do not change the overall system failure rate.
- No reliability growth occurs. (Such growth might be due to design improvements or through debugging of bad parts.)
- Preventive maintenance is scheduled outside the time frame included in the availability calculation.

TABLE 19.3
Availability data for mail system equipment

Equipment	MTBF, h	MTTR, h	Availability (%)
Sack sorter	90	1.620	98.2
Parcel sorter	160	0.8867	99.4
Conveyor, induction	17,900	1.920	100.0
Deflector, traveling	3,516	3.070	99.9

More precise formulas for calculating availability depend on operational conditions and statistical assumptions. These formulas are discussed by Ireson et al. (1996).

19.10 SETTING SPECIFICATION LIMITS

A major step in the development of physical products is the conversion of product *features* into dimensional, chemical, electrical, and other *characteristics* of the product. Thus a heating system for an automobile will have many characteristics for the heater, air ducts, blower assembly, engine coolant, etc.

For each characteristic, the designer must specify (1) the desired average (or "nominal value") and (2) the specification limits (or "tolerance limits") above and below the nominal value that individual units of product must meet. These two elements relate to parameter design and tolerance design discussed in Section 12.4, "Designing for Basic Functional Requirements."

The specification limits should reflect the functional needs of the product, manufacturing variability, and economic consequences. These three aspects are addressed in the next three sections.

Anand (1996) describes several cases of the role of statistics in determining specification limits.

19.11 SPECIFICATION LIMITS AND FUNCTIONAL NEEDS

Setting tolerances around key values in a product is, unfortunately, too often an act of informed guesswork rather than science. If, however, the design has been based on proper statistical analysis and strong experiments, as described Chapter 18, then tolerances can be set in a much more robust fashion, typically called "realistic" tolerances. The name may seem to imply some element of ad hoc judgment, but in fact this is a robust statistical method.

One of the performance features of a car radio is its ability to distinguish different stations in part based on the differences in signal strength. One product feature that affects this performance is the spacing of elements on the antenna array. See Figure 19.7 for the estimated functional relationship between these two variables based on a regression line fit. The graph shows the individual observations along with the regression equation. Also added is the "prediction interval," which shows the 95% confidence interval for individual values predicted with this equation. The performance specification is that the radio should be able to distinguish differences between 36 and 42 arbitrary strength units. Note that in order to meet this performance with 95% certainty, the array spacing must be between 79.6 and 84.9 μm. These limits are set by where the performance limits intersect the upper and lower prediction intervals, not the regression line. This method captures, then, not only the functional relationship between the two variables, but also the statistical variation in that relationship. Using the regression line would have set much wider tolerances and resulted in poorer performance.

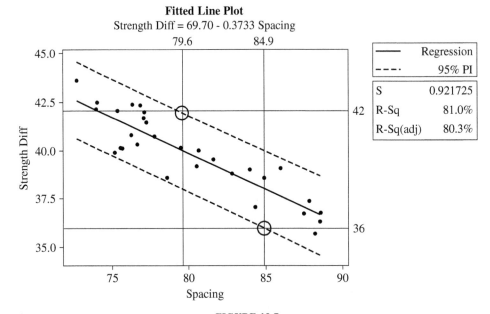

FIGURE 19.7
Realistic tolerancing of radio antenna array. *Source:* Juran Institute, 2012.

O'Connor (1995), Chapter 8, discusses how this approach can be related to the Taguchi approach to robust design.

→ 19.12 SPECIFICATION LIMITS AND MANUFACTURING VARIABILITY

Too often, designers do not consider information on process capability. The best approach is to observe the "realistic" tolerancing methods described in Section 19.11. So long as we invest in generating the data, the same methods can be used to relate tolerances on production variables to the product tolerances as were used to set product tolerances based on functional tolerances.

If they are unable to meet this stronger approach, designers minimally need to obtain a sample of data from the process, calculate the limits that the process can meet, and compare these to the limits they were going to specify to see if the process is capable of meeting those tolerances.

Statistically, the problem is to predict the limits of variation of individual items in the total population based on a sample of data. For example, suppose that a product characteristic is normally distributed with a population average of 5.000 in (12.7 cm) and a population standard deviation of 0.001 in (0.00254 cm). Limits can then be calculated to include any given percentage of the population. Figure 19.8 shows the location of the 99% limits. Table A in Appendix I indicates that ±2.575 standard

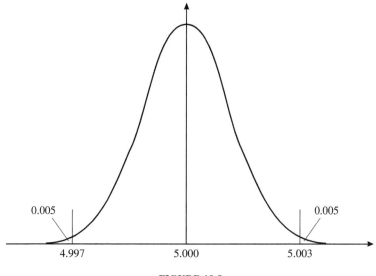

FIGURE 19.8
Distribution with 99% limits.

deviations will include 99% of the population. Thus, in this example, a set of tolerance limits would be

$$5.000 \pm 2.575(0.001) = \begin{matrix} 5.003 \\ 4.997 \end{matrix}$$

Ninety-nine percent of the individual pieces in the population will have values between 4.997 and 5.003.

In practice, the average and standard deviation of the population are not known but must be estimated from a sample of product from the process. We, of course, have seen how to do this in general, but this particular problem is a little more complex because the setting of these tolerances will simultaneously have uncertainty around both the mean and the standard deviation. As a first approximation, tolerance limits are sometimes set at

$$\overline{X} \pm 3s$$

Here the average $\overline{X}$ and standard deviation s of the sample are used directly as estimates of the population values. If the true average and standard deviation of the population happen to be equal to those of the sample and if the characteristic is normally distributed, then 99.73% of the pieces in the population will fall within the limits calculated. These limits are frequently called *natural tolerance limits* (limits that recognize the actual variation of the process and therefore can likely be met). This approximation ignores the possible error in both the average and standard deviation as estimated from the sample.

Methodology has been developed for validating tolerance limits in a more precise manner. For example, formulas and tables are available for determining tolerance limits based on a normally distributed population. Table H in Appendix I provides factors for calculating tolerance limits that recognize the uncertainty in the sample mean and sample standard deviation. The tolerance limits are determined as

$$\overline{X} \pm Ks$$

The factor K is a function of the confidence level desired, the percentage of the population to be included within the tolerance limits, and the number of data values in the sample.

For example, suppose that a sample of 10 resistors from a process yielded an average and standard deviation of 5.04 Ω and 0.016 Ω, respectively. The tolerance limits are to include 99% of the population, and the tolerance statement is to have a confidence level of 95%. Referring to Table H in Appendix I, the value of K is 4.433, and tolerance limits are then calculated as

$$5.04 \pm 4.433(0.016) = \begin{matrix} 5.11 \\ 4.97 \end{matrix}$$

We are 95% confident that at least 99% of the resistors in the population will have resistance between 4.97 Ω and 5.11 Ω. Tolerance limits calculated in this manner are often called statistical tolerance limits. This approach is more rigorous than the $\pm 3s$ natural tolerance limits, but the two percentages in the statement are a mystery to those without a statistical background.

For products in some industries (e.g., electronics), the number of units outside of specification limits is stated in terms of parts per million (ppm). Thus if limits are set at ± 3 standard deviations, 2700 ppm (100% − 99.73%) will fall outside the limits. For many applications (e.g., a personal computer with many logic gates), such a level is totally unacceptable. Table 19.4 shows the ppm for several standard deviations. These levels of ppm assume that the process average is constant at the nominal specification. A deviation from the nominal value will result in a higher ppm value. To allow for modest shifts in the process average, some manufacturers follow a guideline for setting specification limits at $\pm 6\sigma$. See Section 20.13 under "Six Sigma Concept of Process Capability."

TABLE 19.4
Standard deviations and PPM (centered process)[1]

Number of standard deviations	Parts per million (ppm)
$\pm 3\sigma$	2700
$\pm 4\sigma$	63
$\pm 5\sigma$	0.57
$\pm 6\sigma$	0.002

---

[1] If the process is not centered and the mean shifts by up to 1.5σ, then $\pm 6\sigma$ will be 3.4 ppm.

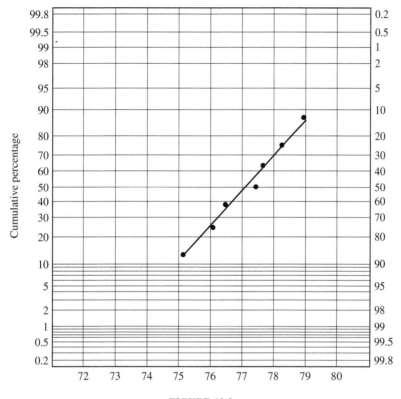

FIGURE 19.9
Probability plot of development data.

Designers must often set tolerance limits with only a few measurements from the process (or more likely from the development tests conducted under laboratory conditions). In developing a paint formulation, the following values of gloss were obtained: 76.5, 75.2, 77.5, 78.9, 76.1, 78.3, and 77.7. A group of chemists was asked where they would set a minimum specification limit. Their answer was 75.0—a reasonable answer for those without statistical knowledge. Figure 19.9 shows the data on a normal probability plot. If the line is extrapolated to 75.0, the plot predicts that about 11% of the population will fall below 75.0—even though all of the sample data exceed 75.0. Of course, a larger sample size is preferred and further statistical analyses could be made, but the plot provides a simple tool for evaluating a small sample of data.

All methods of setting tolerance limits based on process data assume that the sample of data represents a process that is sufficiently stable to be predictable. In practice, the assumption is often accepted without any formal evaluation. If sufficient data are available, the assumption should be checked with a control chart.

Statistical tolerance limits are sometimes confused with other limits used in engineering and statistics. Table 19.5 summarizes the distinctions among five types of limits.

TABLE 19.5
Distinctions among limits

Name of limit	Meaning
Tolerance	Set by the design function to define the minimum and maximum values allowable for the product to work properly
Statistical tolerance	Calculated from process data to define the amount of variation that the process exhibits; these limits will contain a specified proportion of the total population, but will not necessarily meet the customer's need
Realistic tolerance	Either a product feature tolerance calculated based on performance tolerance limits or production feature tolerances calculated based on product tolerances; incorporates both the functional relationship between the two tolerances and the natural statistical variation around that relationship
Prediction	Calculated from process data to define the limits which will contain all of k future observations
Confidence	Calculated from data to define an interval within which a population parameter lies
Control	Calculated from process data to define the limits of chance (random) variation around some central value

➔ 19.13 SPECIFICATION LIMITS AND ECONOMIC CONSEQUENCES

In setting traditional specification limits around a nominal value, we assume that there is no monetary loss for product falling within specification limits. For product falling outside the specification limits, the loss is the cost of replacing the product.

Another viewpoint holds that *any* deviation from the nominal value causes a loss. Thus there is an ideal (nominal) value that customers desire, and any deviation from this ideal results in customer dissatisfaction. This loss can be described by a loss function (Figure 19.10).

Many formulas can predict loss as a function of deviation from the target. Taguchi proposes the use of a simple quadratic loss function:

$$L = k(X - T)^2$$

where L = loss in monetary terms
k = cost coefficient
X = value of quality characteristic
T = target value

Ross (1996) provides an example to illustrate how the loss function can help to determine specification limits. In automatic transmissions for trucks, shift points are designed to occur at a certain speed and throttle position. Suppose it costs the producer $100 to adjust a valve body under warranty when a customer complains of the shift point. Research indicates that the average customer would request an adjustment if

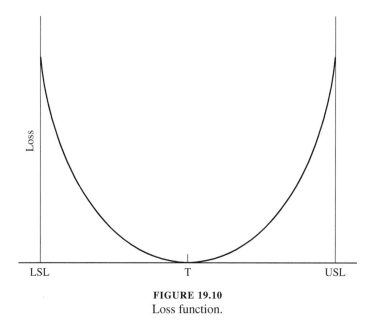

FIGURE 19.10
Loss function.

the shift point is off from the nominal by 40 rpm transmission output speed on the first-to-second gear shift. The loss function is then,

$$\text{Loss} = k(X - T)^2$$
$$100 = k(40)^2$$
$$k = \$0.0625$$

This adjustment can be made at the factory at a lower cost, about $10. The loss function is now used to calculate the specification limits:

$$\$10 = 0.0625(X - T)^2$$
$$(X - T) = \pm 12.65 \text{ or } \pm 13 \text{ rpm}$$

The specification limits should be set at ±13 rpm around the desired nominal value. If the transmission shift point is more than 13 rpm from the nominal, adjustment at the factory is less expensive than waiting for a customer complaint and making the adjustment under warranty in the field. Ross discusses how the loss function can be applied to set one-sided specification limits, e.g., a minimum value or a maximum value.

↦ 19.14 SPECIFICATION LIMITS FOR INTERACTING DIMENSIONS

Interacting dimensions mate or merge with other dimensions to create a final result. Consider the simple mechanical assembly shown in Figure 19.11. The lengths of components A, B, and C are interacting dimensions because they determine the overall assembly length.

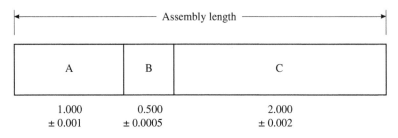

FIGURE 19.11
Mechanical assembly.

Suppose the components were manufactured to the specifications indicated in Figure 19.11. A logical specification for the assembly length would be 3.500 ± 0.0035, giving limits of 3.5035 and 3.4965. This logic may be verified from the two extreme assemblies:

Maximum	Minimum
1.001	0.999
0.5005	0.4995
2.002	1.998
3.5035	3.4965

The approach of adding component tolerances is arithmetically correct but is statistically deceptive, and often too conservative. Suppose that about 1% of the pieces of component A are expected to be below the lower tolerance limit for component A and suppose the same for components B and C. If a component A is selected at random, there is, on average, 1 chance in 100 that it will be on the low side, and similarly for components B and C. The key point is this: If assemblies are made at random and if the components are manufactured independently, then the chance that an assembly will have all *three* components simultaneously below the lower tolerance limit is

$$\frac{1}{100} \times \frac{1}{100} \times \frac{1}{100} = \frac{1}{1,000,000}$$

There is only about one chance in a million that all three components will be too small, resulting in a small assembly. Thus setting component and assembly tolerances based on the simple addition formula is conservative in that it fails to recognize the extremely low probability of an assembly containing all low (or all high) components.

The statistical approach is based on the relationship between the variances of a number of independent causes and the variance of the dependent or overall result. This may be written as

$$\sigma_{result} = \sqrt{\sigma^2_{cause\ A} + \sigma^2_{cause\ B} + \sigma^2_{cause\ C} + \cdots}$$

In terms of the assembly example, the formula is

$$\sigma_{assembly} = \sqrt{\sigma^2_A + \sigma^2_B + \sigma^2_C}$$

Now suppose that, for each component, the tolerance range is equal to ±3 standard deviations (or any constant multiple of the standard deviation). Because σ is equal to T divided by 3, the variance relationship may be rewritten as

$$\frac{T}{3} = \sqrt{\left(\frac{T_A}{3}\right)^2 + \left(\frac{T_B}{3}\right)^2 + \left(\frac{T_C}{3}\right)^2}$$

or

$$T_{\text{assembly}} = \sqrt{T_A^2 + T_B^2 + T_C^2}$$

Thus the squares of tolerances are added to determine the square of the tolerance for the overall result. This formula compares to the simple addition of tolerances commonly used.

The effect of the statistical approach is dramatic. Listed here are two possible sets of component tolerances that will yield an assembly tolerance equal to ±0.0035 when used with the previous formula.

Component	Alternative 1	Alternative 2
A	±0.002	±0.001
B	±0.002	±0.001
C	±0.002	±0.003

With alternative 1, the tolerance for component A has been doubled, the tolerance for component B has been quadrupled, and the tolerance for component C has been kept the same as the original component tolerance based on the simple addition approach. If alternative 2 is chosen, similar significant increases in the component tolerances may be achieved. This formula, then, may result in a larger component tolerance with *no* change in the manufacturing processes and *no* change in the assembly tolerance.

The risk of this approach is that an assembly may fall outside the assembly tolerance. However, this probability can be calculated by expressing the component tolerances as standard deviations, calculating the standard deviation of the result, and finding the area under the normal curve outside the assembly tolerance limits. For example, if each component tolerance is equal to 3σ, then 99.73% of the assemblies will be within the assembly tolerance, i.e., 0.27%, or about 3 assemblies in 1000 taken at random would fail to meet the assembly tolerance. The risk can be eliminated by changing components for the few assemblies that do not meet the assembly tolerance.

The tolerance formula is not restricted to outside dimensions of assemblies. Generalizing, the left side of the equation contains the dependent variable or *physical result,* and the right side of the equation contains the independent variables of *physical causes.* If the result is placed on the left and the causes on the right, the formula always has *plus* signs under the square root—even if the result is an internal dimension (such as the clearance between a shaft and hole). The causes of variation are *additive* wherever the physical result happens to fall.

The formula has been applied to a variety of mechanical and electronic products. The concept may be applied to several interacting variables in an engineering relationship. The nature of the relationship need not be additive (assembly example) or subtractive (shaft-and-hole example). The tolerance formula can be adapted to predict the variation of results that are the product and/or the division of several variables.

Assumptions of the Formula

The formula is based on several assumptions:

- The component dimensions are independent and the components are assembled randomly. This assumption is usually met in practice.
- Each component dimension should be normally distributed. Some departure from this assumption is permissible.
- The actual average for each component is equal to the nominal value stated in the specification. For the original assembly example, the actual averages for components A, B, and C must be 1.000, 0.500, and 2.000, respectively. Otherwise, the nominal value of 3.500 will not be achieved for the assembly, and tolerance limits set at about 3.500 will not be realistic. Thus it is important to control the average value for interacting dimensions. Consequently, process control techniques are needed using variables measurement.

Use caution if any assumption is violated. Reasonable departures from the assumptions may still permit applying the concept of the formula. Notice that in the example, the formula resulted in the doubling of certain tolerances. This much of an increase may not even be necessary from the viewpoint of process capability.

Bender (1975) has studied these assumptions for some complex assemblies and concluded, based on a "combination of probability and experience," that a factor of 1.5 should be included to account for the assumptions:

$$T_{result} = 1.5\sqrt{T_A^2 + T_B^2 + T_C^2 + \cdots}$$

Graves (1997) suggests developing different factors for initial versus mature production, high versus low volume production, and mature versus developing technology and measurement processes.

Finally, variation simulation analysis is a technique that uses computer simulation to analyze tolerances. This technique can handle product characteristics having either normal or nonnormal distributions.

Dodson (1999) describes the use of simulation in the tolerance design of circuits; Gomer (1998) demonstrates simulation to analyze tolerances in engine design.

SUMMARY

- Complex products typically experience three periods of product life: infant mortality, constant failure rate, and wear out.
- When the failure rate is constant, the distribution of time between failures is exponential. This distribution is the basis of the exponential formula for reliability.

- If the failure of any part causes the failure of the system and if the reliabilities of the parts are independent, then the system reliability is the product of the parts reliabilities.
- Redundancy is the existence of more than one element for accomplishing a given task.
- For time-oriented products, reliability can be predicted during design by using the exponential or Weibull distribution.
- For non-time-oriented products, reliability can be predicted as a function of stress and strength.
- Availability is the probability that a product, when used under given conditions, will perform satisfactorily.
- Specification limits must be based on the functional needs of the product and the variability of the manufacturing process. Statistical analysis quantifies the relationship of functional requirement with product tolerance and the relationship of product tolerance with process feature. It also incorporates process variability around that relationship for the designer.
- Specification limits for interacting dimensions should recognize the probabilistic features of forming the final result.

WORKED EXAMPLES USING MINITAB

1. A design engineer of small power tools is studying the clearance between an internal fan and the housing surrounding it. Initial tooling has been constructed for the plastic fan, and the process for creating the housing diameter has been set up. The following data are provided to the design engineer for evaluation.
   (a) What is the standard deviation of the distribution of interference values.
   (b) If the clearance is less than 0.002, there is likely to be interference between the fan and the housing when the parts warm up. What portion of the product may have this problem?

   (Filename: Fan-Housing Clearance.mpj)

Fan Diameter		Housing Diameter	
2.628	2.630	2.655	2.650
2.642	2.622	2.650	2.651
2.624	2.621	2.646	2.652
2.632	2.620	2.653	2.653
2.632	2.626	2.652	2.651
2.638	2.619	2.651	2.650
2.627	2.621	2.649	2.655
2.639	2.622	2.651	2.656
2.638	2.624	2.653	2.650
2.620	2.627	2.654	2.652
2.634	2.626	2.646	2.652
2.638	2.617	2.651	2.657
2.616	2.639	2.653	2.654
2.631	2.609	2.652	2.650
2.624	2.630	2.654	2.651

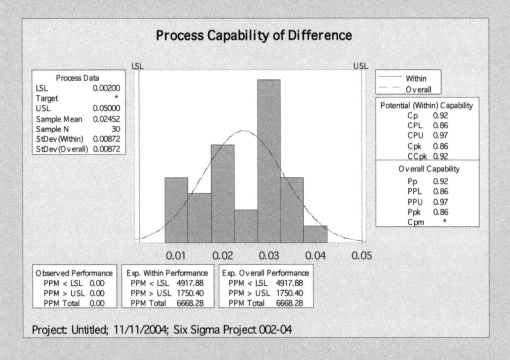

Project: Untitled; 11/11/2004; Six Sigma Project 002-04

Answer:
(a) (MINITAB Calc>Calculator to take differences between housing and fan; then Stat> Basic Stats>Descriptive Statistics to get:

Variable	N	Mean	StDev
Difference	30	0.02452	0.00865

(b) (MINITAB: Stat>Quality Tools> Capability Analysis>Normal on differences column) From the capability chart from MINITAB, the portion of mated components in a finished product that may have an interference problem (difference < 0.002) is 4917 ppm (0.4917%).

2. In studying the cycle time of an order fulfillment operation at a small distributor, the team focused on one type of order that had very similar properties. Looking at the five major steps to filling an order, the cycle times were measured and recorded. Upon examination, the distribution of cycle times could be taken as normally distributed. What would be the average and variation of overall cycle time?

Step	Average	Range of Variation
1	1.5 h	± 20 min
2	1 h	± 10 min
3	2.5 h	± 30 min
4	0.5 h	± 5 min
5	2 h	± 20 min

Answer: The average is the sum of the above averages = 7.5 h. The variation can be calculated by taking the square of each variation measurement, summing them together, and then taking the square root. Range of variation = 42.7 min.

3. A design engineer was considering the stress–strength relationship between two critical structural components. He took some stress data and determined that data were distributed normally with a mean value of 1200 lb with a standard deviation of 50 lb. The strength data on the component revealed it also was normally distributed and had a mean of 1500 lb with a standard deviation of 40 lb. If a safety factor of 300 lb is considered appropriate, to what degree has this design achieved that goal? (What is the probability that the safety factor will be less than 300 lb?)

Answer: Calculated Z for this is $[(1500 - 1200)/(50^2 + 40^2)^{1/2}] = 4.68$. Using MINITAB or a table of standard normal values returns the probability of 1.5 ppm.

PROBLEMS

19.1. A radar set has a MTBF of 240 h based on an exponential distribution. Suppose that a certain mission requires failure-free operation of the set for 24 h. What is the chance that the set will complete a mission without failure?

Answer: 0.91.

19.2. A piece of ground support equipment for a missile has a specified MTBF of 100 h. What is the reliability for a mission time of 1 h? 10 h? 50 h? 100 h? 200 h? 300 h? Graph these answers by plotting mission time versus reliability. Assume an exponential distribution.

19.3. The average life of subassembly A is 2000 h. Data indicate that this life characteristic is exponentially distributed.
   (a) What percentage of the subassemblies in the population will last at least 200 h?
   (b) The average life of subassembly B is 1000 h, and the life is exponentially distributed. What percentage of the subassemblies in the population will last at least 200 hours?
   (c) These subassemblies are independently manufactured and then are connected in series to form the total assembly. What percentage of assemblies in the population will last at least 200 hours?

Answer: (a) 90.5%. (b) 81.9%. (c) 74.1%.

19.4. It is expected that the average time to repair a failure on a certain product is 4 h. Assume that repair time is exponentially distributed. What is the chance that the time for a repair will be between 3 h and 5 h?

19.5. The following table summarizes basic failure-rate data on components in an electronic subsystem:

Component	Quantity	Failure rate per hour
Silicon transistor	40	74.0×10^{-6}
Film resistor	100	3.0×10^{-6}
Paper capacitor	50	10.0×10^{-6}

Estimate the MTBF. (Assume an exponential distribution. All components are critical for subsystem success.)

Answer: 267 h.

19.6. A system consists of subsystems A, B, and C. The system is used primarily on a certain mission that lasts 8 h. The following information has been collected:

Subsystem	Required operating time during mission, h	Type of failure distribution	Reliability information
A	8	Exponential	50% of subsystems will last at least 14 h
B	3	Normal	Average life is 6 h with a standard deviation of 1.5 h
C	4	Weibull with $\beta = 1.0$	Average life is 40 h

Assuming independence of the subsystems, calculate the reliability for a mission.

19.7. A hydraulic subsystem consists of two subsystems in parallel, each has the following components and characteristics:

Components	Failures/10^6 h	Number of components
Pump	23.4	1
Quick disconnect	2.4	3
Check valve	6.1	2
Shutoff valve	7.9	1
Lines and fittings	3.13	7

The components within each subsystem are all necessary for subsystem success. The two parallel subsystems operate simultaneously, and either can perform the mission. What is the mission reliability if the mission time is 300 h? (Assume an exponential distribution.)

19.8. The following estimates, based on field experience, are available on three subsystems:

Subsystem	Percent failed at 1000 mi	Weibull β value
A	0.1	2.0
B	0.2	1.8
C	0.5	1.0

If these estimates are assumed to be applicable for similar subsystems that will be used in a new system, predict the reliability (in terms of percentage successful) at the end of 3000 mi, 5000 mi, 8000 mi, and 10,000 mi.

19.9. It is desired that a power plant be in operating condition 95% of the time. The average time required for repairing a failure is about 24 h. What must the MTBF be for the power plant to meet the 95% objective?

Answer: 456 h.

19.10. A manufacturing process runs continuously 24 h per day and 7 days a week (except for planned shutdowns). Past data indicate a 50% probability that the time between successive failures is 100 h or more. The average repair time for failures is 6 h. Failure times and repair times are both exponentially distributed. Calculate the availability of the process.

19.11. The following table summarizes data on components in a hydraulic system:

Component	Quality	Failure rate per hour
Relief valve	1	200×10^{-6}
Check valve	1	150×10^{-6}
Filter	1	100×10^{-6}
Cylinder	1	50×10^{-6}

Assume that all components must operate for system success and that the system is used continuously throughout the 8760 h in a year with no shutdowns except for failures. Repair time varies with the type of failure, but 50% of the repairs require 3 h or more. The following average cost estimates apply:

$$\text{Material cost/failure} = \$400.00$$

$$\text{Repair labor cost/hour} = \$20.00$$

Assume that failure time and repair time are exponentially distributed. Calculate the average cost of repairing failures per year.

Answer: $2029.

19.12. Measurements were made on the bore dimension of an impeller. A sample of 20 from a pilot run production showed a mean value of 25.038 cm and a standard deviation of 0.000381 cm. All the units functioned properly, so it was decided to use the data to set specification limits for regular production.
  (a) Suppose it was assumed that the sample estimates were exactly equal to the population mean and standard deviation. What specification limits should be set to include 99% of production?
  (b) There is uncertainty that the sample and population values are equal. Based on the sample of 20, what limits should be set to be 95% sure of including 99% of production?
  (c) Explain the meaning of the difference in (a) and (b).
  (d) What assumptions were necessary to determine both sets of limits?

19.13. A circuit contains three resistors in series. Past data show these data for resistance:

Resistor	Mean, Ω	Standard deviation, Ω
1	125	3
2	200	4
3	600	12

  (a) What percentage of circuits would meet the specification of total resistance of $930 \pm 30\Omega$?
  (b) Ask a local distributor whether it is reasonable to assume that the resistance of a resistor is normally distributed.

19.14. A manufacturer of rotary lawn mowers received numerous complaints concerning the effort required to push its product. Studies soon found that the small clearance between the wheel bushing and shaft was the cause. Designers chose to make the clearance large enough for easy rotation of the wheel (or for a heavy grease coating) but still "tight enough" to prevent wobbling. Because of a large inventory of wheels and shafts, a decision was made to ream the bushings to a larger inside diameter (i.d.) and retain the shafts. The following specifications were proposed:

$$\text{Shaft diameter} = 0.800 \pm 0.002 \text{ in}$$

$$\text{New clearance} = 0.800 \pm 0.003 \text{ in}$$

$$\text{Bushing i.d.} = 0.800 \pm 0.001 \text{ in}$$

Production people claimed that they could not economically hold the tolerance on the bushing i.d. What comment would you make about this claim?

Answer: i.d. tolerance of ± 0.0022 could be allowed.

19.15. An assembly consists of two parts (A and B) that mate together "end to end" to form an overall length, C. It is desired that the overall length, C, meet a specification of 3.000 ± 0.005 cm. The nominal specification on A is 2.000, and on B it is 1.000. The manufacturing process for B has much more variability than the process for A. Specifically, the tolerance for part B should be twice as large as for part A. Assemblies are to be made at random, and parts A and B are independently manufactured. Assuming that we want only a small risk of not meeting the specification on C, what tolerances should be set on A and B?

19.16. A canning factory decided to set tolerance limits for filling cans of a new product by sampling a pilot run of 30 cans. The results of that run yielded an average of 446 g with a standard deviation of 1.25 g. What tolerance limits would be 95% certain of including 99% of production? The label on the can states that it contains 453 g of the product. How many grams of product should the average can contain if the company is to be 95% certain that 99% of production contains at least 453 g?

REFERENCES

Anand, K. N. (1996). "The Role of Statistics in Determining Product and Part Specifications: A Few Indian Experiences," *Quality Engineering,* vol. 9, no. 2, pp. 187–193.

ANSI/IEC/ASQ D60300-3-1-1997 (1997). *Dependability Management Part 3: Application Guide—Section 1: Analysis Techniques for Dependability: Guide on Methodology,* ASQ, Milwaukee.

Bender, A. (1975). "Statistical Tolerancing as It Relates to Quality Control and the Designer," *Automotive Division Newsletter of ASQC,* April, p. 12.

Dodson, B. (1999). "Reliability Modeling with Spreadsheets," *Proceedings of the Annual Quality Congress,* ASQ, Milwaukee, pp. 575–585.

Garrick, J. B. and R. J. Mulvihill (1974). "Reliability and Maintainability of Mechanized Bulk Mail Systems," *Proceedings of Annual Reliability and Maintainability Symposium,* Institute of Electrical and Electronics Engineers, New York.

Gomer, P. (1998). "Design for Tolerance of Dynamic Mechanical Assemblies," *Annual Quality Congress Proceedings,* ASQ, Milwaukee, pp. 490–500.

Graves, S. B. (1997). "How to Reduce Costs Using a Tolerance Analysis Formula Tailored to Your Organization," *Report No. 157,* Center for Quality and Productivity Improvement, University of Wisconsin, Madison.

Ireson, W. G., C. F. Coombs, Jr., and R. Y. Moss (1996). *Handbook of Reliability Engineering and Management,* 2nd ed., McGraw-Hill, New York.

Jones, J. and J. Hayes (1999). "A Comparison of Electronic Reliability Prediction Models," *IEEE Transactions on Reliability,* vol. 48, no. 2, pp. 127–134.

Lusser, R. (1958). *Reliability Through Safety Margins,* United States Army Ordnance Missile Command, Redstone Arsenal, AL.

O'Connor, P. D. T. (1995). *Practical Reliability Engineering,* 3rd ed. rev., John Wiley and Sons, New York.

Pyzdek, T. (1994). "Service Systems Reliability Engineering," *Proceedings of the Annual Quality Congress,* ASQ, Milwaukee, pp. 174–187.

Ross, P. J. (1996). *Taguchi Techniques for Quality Engineering,* McGraw-Hill, New York.

CHAPTER 20

Tools to Maintain the Control of Quality

➔ 20.1 DEFINITION AND IMPORTANCE OF QUALITY CONTROL

Dateline 1950: I was installing control charts as part of a statistical process control system to control the variation in the weight of packages of dry soup mix at the Lipton Company. We set up the system and quietly tried it out on one filling line by having the operator weigh samples of packages and plot the average and range of the samples. Within one day, operators on the other lines enthusiastically asked us to show them how to plot and use the charts. Their message to us was clear: They wanted to be better able to control their process (self-control), and they wanted to have fun plotting the data. Moral: Statistical process control is not all numbers.

(Dr. Joseph M. Juran)

We define *statistical process control* as the application of statistical methods to the measurement and analysis of variation in a process. This technique applies to both in-process parameters and end-of-process (product) parameters. Let the reader be aware, however, that the term *statistical process control* has also assumed other definitions, even including some that involve little or no use of statistical analysis!

A process is a collection of activities that converts inputs into outputs or results. More specifically, a process is a unique combination of machine, tools, methods, materials, and people that attain an output in goods, software, or services—a circuit chip, a computer program, or answers provided on a consumer hot line.

In the current chapter, we examine the significance of variation, the use of control charts in analyzing and minimizing variation, the quantification of process capability, the statistical basis of the six sigma approach, and the relation of these concepts to other techniques for process improvement.

The evidence is clear that these fascinating techniques can make an important contribution to achieving quality objectives. For most organizations, statistical process control techniques are essential. To help ensure successful and continued application of these concepts in the reality of lean operating budgets, statistical process control techniques must not become an end in themselves.

Pragmatic operating managers correctly demand that each potential application show a tangible opportunity for significant benefits—a situation that quality professionals should never forget.

➔ 20.2 ADVANTAGES OF DECREASING PROCESS VARIABILITY

Reducing the variation in a process leads to some great benefits:

- Lower variability may result in improved product performance that is discernible by the customer. Sullivan (1984) describes two Sony plants making the same television set (Figure 20.1). The San Diego plant had no product outside specifications, but the distribution was virtually rectangular, with a large percentage of product close to the specification limits. In contrast, the plant in Japan did have some product outside of specification limits, but the distribution was normal and was concentrated around the target value. Field experience revealed that product near the specification limits generated complaints from customers. This and other reasons led to a higher loss per unit at San Diego even though that plant was superior in meeting the specification. The higher internal loss due to complaints would, of course, likely result in lower future sales.
- Lower variability of a component characteristic may be the only way to compensate for high variability in other components and thereby meet performance requirements in an assembly or system. To meet these requirements may also require strict control

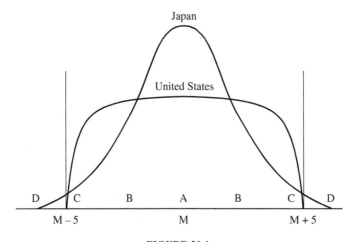

FIGURE 20.1
Uniformity and production quality of television sets produced in Japan and the United States. (*From Sullivan, 1984.*)

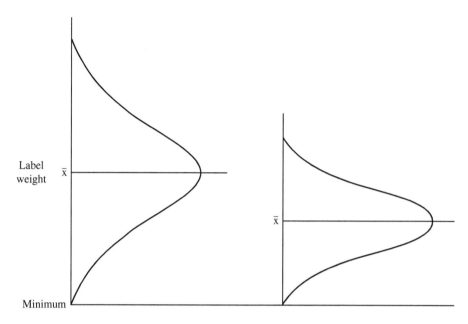

FIGURE 20.2
Reducing average overfill by reducing the standard deviation.

of the average values of each component, as was the case in the design and manufacture of undersea cable.
- For some characteristics such as weight, lower variability may provide the opportunity to change the process average. Reducing the standard deviation of fill content in a food package permits a reduction in the *average* fill, thereby resulting in cost reduction (Figure 20.2). Imagine the cumulative cost reduction over millions of packages!
- Lower variability results in less need for inspection. In the extreme, if there were *no* variability, inspection of only one unit of product would tell the whole story.
- Lower variability may command a premium price for a product. Some electronic components have traditionally been priced as a function of the amount of variability.
- Lower variability may be a competitive factor in determining market share. Increasingly, meeting specification limits is no longer sufficient. Industrial customers in particular realize that high variability of purchased material and components often requires frequent (and costly) adjustment to their own processes to compensate for the variability of purchased products. The result is that these customers compare suppliers on variability of important product characteristics.

> **EXAMPLE 20.1.** The marketing manager of a commodity chemicals manufacturer describes two scenarios—old and new—between a customer and a salesperson.
>
> Old scenario:
> Customer: "Your product quality is no good."
> Salesperson: "I'll lower the price."
> Customer: "Good, that's what I wanted to hear."

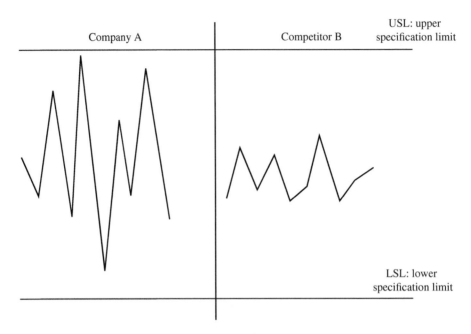

FIGURE 20.3
Variation and competition.

New scenario:
    Customer: "Your product quality is no good."
    Salesperson: "I'll lower the price."
    Customer: "The price is acceptable; I said your quality is no good."
    Salesperson: "Was some of the product out of specification?"
    Customer: "No."
    Salesperson: "I don't understand."
    Customer: "Look at these data." (See Figure 20.3.)
    Customer: "It's not enough to meet the specifications. Your competitor meets the same specification with less variability."

➔ 20.3 STATISTICAL CONTROL CHARTS—GENERAL

A *statistical control chart*, or simply *control chart*, compares process performance data to computed "statistical control limits," drawn as limit lines on the chart. The process performance data usually consist of groups of measurements *(rational subgroups)* from the regular sequence of production while preserving the order of the data.

A prime objective of a control chart is detecting *special* (or assignable) causes of variation in a process—by analyzing data from both the past and the future. Knowing the meaning of "special causes" is essential to understanding the control chart concept (see Table 6.5).

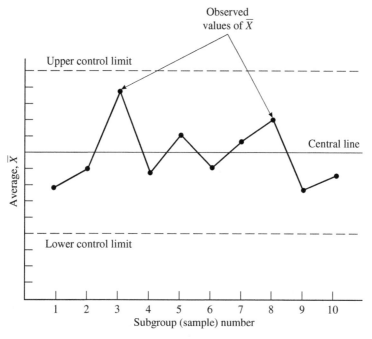

FIGURE 20.4
Generalized control chart for averages.

Process variations have two kinds of causes: (1) common (or random or chance), which are inherent in the process as currently designed, and (2) special (or assignable), which arise new and create abnormal variation. Ideally, only common causes should be present in a process because they represent a stable and predictable process. A process that is operating without special causes of variation is said to be "in a state of statistical control" because the variation is predictable and consistent with historical performance. The control chart for such a process has all of the data points within the statistical control limits. The object of a control chart is not to achieve a state of statistical control as an end in itself but to reduce variation and prevent deterioration of the process by catching new special causes and flagging them for elimination.

The control chart distinguishes between common and special causes of variation through the choice of control limits (Figure 20.4). When the variation *exceeds* the statistical control limits, it is a signal that special causes have entered the process and the process should be investigated to identify these causes of excessive variation. Random variation *within* the control limits means that only common (random) causes are present; the amount of variation has stabilized, and minor process adjustments should be avoided. Note that a control chart detects the presence of a special cause but does not *find* the cause—that task must be handled by a subsequent investigation of the process.

Also note that even when a process is in statistical control, it may still be performing poorly from the point of view of the customer. The resulting product of the process may still have an abundance of defects that need to be eliminated. Control charts are of little use in addressing the chronic defects and waste of excessive common cause variation. See Chapters 3 and 18 for the methods and tools of quality improvement that reduce chronic common-cause variation.

➔ 20.4 ADVANTAGES OF STATISTICAL CONTROL

A state of statistical control exists when only common causes of variation exist in a process. This condition provides several important advantages:

- The process is stable, which makes it possible to predict its behavior, at least in the near term.
- The process has an identity in terms of a given set of conditions that are necessary for predictions. An analogy to baseball is useful. If we say that a player has a batting capability of about .250, we mean that, on average, he will get one hit in four times at bat. Of course, this statement assumes that the player is in a normal state of health, i.e., free of any assignable or special causes that would detract from batting capability. Predictions of any kind must be related to an assumed set of conditions. For manufacturing and business processes, these conditions are represented by a state of statistical control. Ideally, drawing conclusions from any set of data should be preceded by an analysis to see whether the data come from a process in statistical control.
- A process in statistical control operates with less variability than a process having special causes. Lower variability has become an important tool of competition (see Section 5.11, "Planning for Product Quality to Generate Sales Income").
- A process having special causes is unstable, and the excessive variation may hide the effect of changes introduced to achieve improvement. Also, the removal of some special causes and subsequent replotting of the control chart may reveal that additional special causes exist that were masked earlier.
- Knowing that a process is in statistical control is helpful to the workers running a process. It says that when data fall within the statistical control limits, adjustments should *not* be made. Making such adjustments will add to the variability, not decrease it. Conversely, a control chart helps to avoid underadjustment because out-of-control points will signal the presence of special causes that must be addressed.
- Knowing that a process is in statistical control provides direction to those who are trying to make a long-term reduction in process variability. To reduce process variability, the process system must be analyzed and changed rather than having management expect the workers who run the process to reduce the variability by themselves.
- An analysis for statistical control, which includes the plotting of data in order of production, will easily identify trends over time that are hidden by other summarizations of data such as histograms.

- A stable process (as verified by statistical control) that also meets product specifications (as verified by process capability studies) provides evidence that the process has conditions that, if maintained, will result in an acceptable product. Such evidence is needed *before* a process is transferred from the planning stage to full production.

These advantages of statistical control feed into the main objectives of statistical process control, which are to reduce process variation and prevent problems. Real prevention is achieved by having capable processes that are maintained in a state of statistical control.

✣ 20.5 STEPS IN DEVELOPING CONTROL CHARTS

Setting up a control chart requires the following steps:

1. Choose the characteristic to be charted.
    - Give high priority to characteristics that are currently running with a high defective rate, especially those with high cost or customer impact. A Pareto analysis can establish priorities.
    - Identify the process variables and conditions that contribute to the end-product characteristics, in order to define potential charting applications from raw materials through processing steps to final characteristics. For example, the pH, salt concentration, and temperature of a plating solution are process variables contributing to consistent plating thickness. Systematic analysis and experiments as described in Chapter 18 can pinpoint the vital few operational causes of defects. These are the *dominant* variables and should be the subject of control plans.
    - Verify that the measurement process has sufficient accuracy and precision (see Section 10.8) to provide data that does not obscure variation in the manufacturing or service process. The observed variation in a process reflects the *combined* variations in the actual operations process and in the measurement processes. Anthis et al. (1991) describe how the measurement process was a roadblock to improvement by hiding important clues to the sources of variation in a manufacturing process. Dechert et al. (2000) explain how large measurement variation can be controlled and result in effective statistical process control methods.
    - Determine the earliest point in the production process at which testing can be done to get information on assignable causes so that the chart serves as an effective early-warning device to prevent defectives.
2. Choose the type of control chart. Table 20.1 compares three basic control charts. Schilling (1990) provides additional guidance in choosing the type of control chart.
3. Decide on the central line to be used and the basis of calculating the limits. The parameter for the central line will generally be determined by the type of variable being measured. It will be some average of past data. The choice of historical period should reflect a significant duration (typically at least 20 different times without substantial shifts in the level). The limits are usually set at $\pm 3\sigma$, but other multiples may be chosen for different statistical risks.

TABLE 20.1
Comparison of some control charts

Statistical measure plotted	Average $\bar{X}$ and range R	Percentage nonconforming (p)	Number of nonconformities per unit (u)
Type of data required	Variable data (measured values of a characteristic)	Attribute data (number of defective units of product)	Attribute data (number of defects per unit of product)
General field of application	Control of individual characteristics	Control of overall fraction defective of a process	Control of overall number of defects per unit
Significant advantages	Provides maximum use of information available from data	Data required are often already available from inspection records	Same advantages as p chart but also provides a measure of defectiveness
	Provides detailed information on process average and variation for control of individual dimensions	Easily understood by all personnel Provides an overall picture of quality	
Significant disadvantages	Not understood unless training is provided; can cause confusion between control limits and specification limits	Does not provide detailed information for control of individual characteristics	Does not provide detailed information for control of individual characteristics
	Cannot be used with go/no go type of data	Does not recognize different degrees of defectiveness in units of product	
Sample size	Usually four or five, but more is better	Use given inspection results of samples of at least 25, but 50 or 100 better	Any convenient unit of product such as 100 feet of wire or five television sets

4. Choose the "rational subgroup." Each point on a control chart represents a subgroup (or sample) consisting of several units of product. For process control, subgroups should be homogeneous as to time and space. That is, they should all be produced within a short period of time using exactly the same facilities and workers chosen so that the units *within* a subgroup have the greatest chance of being alike and the units *between* subgroups have the greatest chance of being different.
5. Provide a system for collecting the data. If the control chart is to serve as a day-to-day shop tool, it must be simple and convenient to use. Measurement must be simplified and kept free of error. Indicating instruments must be designed to give prompt, reliable readings. Better yet, instruments should be designed that can record as well as indicate. Recording of data can be simplified by skillful design of data or tally sheets. Most modern workplaces now provide networked computing, including wireless, which greatly facilitates data collection and processing. Working conditions are, of course, a factor and may require creative solutions.

TABLE 20.2
Control chart limits—attaining a state of control

Chart for	Central line	Lower limit	Upper limit
Averages $\overline{X}$	$\overline{\overline{X}}$	$\overline{\overline{X}} - A_2 \overline{R}$	$\overline{\overline{X}} + A_2 \overline{R}$
Ranges R	$\overline{R}$	$D_3 \overline{R}$	$D_4 \overline{R}$
Proportion nonconforming p	$\overline{p}$	$\overline{p} - 3\sqrt{\dfrac{\overline{p}(1-\overline{p})}{n}}$	$\overline{p} + 3\sqrt{\dfrac{\overline{p}(1-\overline{p})}{n}}$
Nonconformities per unit (u)	$\overline{u} = \dfrac{\Sigma x_i}{\Sigma n_i}$	$\overline{u} - 3\sqrt{\dfrac{\overline{u}}{n_i}}$	$\overline{u} + 3\sqrt{\dfrac{\overline{u}}{n_i}}$

6. Calculate the control limits and provide specific instructions for the interpretation of the results and the actions that various production personnel are to take (discussed later). Control limit formulas for the three basic types of control charts are given in Table 20.2. These formulas are based on $\pm 3\sigma$ and use a central line equal to the average of the data used in calculating the control limits. Values of the A_2, D_3, and D_4 factors used in the formulas are given in Table I in Appendix I. For some technical statistical reasons, small sample estimates of standard deviations are slightly biased downward. Most software applications have added unbiasing constants in front of estimates of s for increased precision. Each year, *Quality Progress* magazine publishes a directory that includes software for calculating sample parameters and control limits and for plotting the data.
7. Plot the data and interpret the results.
8. Provide assignments and procedures for acting on any out-of-control results.

The control chart is a powerful statistical concept, but its use should be kept in perspective. The ultimate purpose of an operations process is to make product that is fit for use—not to make product that simply meets statistical control limits. Once the charts have served their purpose, many should be taken down and the effort shifted to other characteristics needing improvement. Schilling (1990) traces the life cycle of control chart applications.

✦ 20.6 CONTROL CHART FOR VARIABLES DATA

For variables data (or continuous data), the control chart for sample averages and sample ranges provides a powerful technique for analyzing process data.

A small sample (e.g., five units) is periodically taken from the process, and the average ($\overline{X}$) and range (R) are calculated for each sample. A total of at least 50 individual measurements (e.g., 10 samples of five each) should be collected before the control limits are calculated. The control limits are set at $\pm 3\sigma$ for sample averages and sample ranges. The $\overline{X}$ and R values are plotted on separate charts against their $\pm 3\sigma$ limits.

Standard deviations are readily computed by modern calculators and software, but calculations can be avoided by using shortcuts in the absence of software.

The shortcut formulas for the upper control limit (UCL) and lower control limit (LCL) for sample averages are

$$\text{UCL} = \bar{\bar{X}} + A_2 \bar{R}$$

$$\text{LCL} = \bar{\bar{X}} - A_2 \bar{R}$$

where $\bar{\bar{X}}$ = grand average = average of the sample averages
$\bar{R}$ = average of the sample ranges
A_2 = constant found from Table I in Appendix I

The shortcut consists of (1) computing the range (difference between largest and smallest) of the individuals for each sample, (2) averaging the ranges thus obtained, and then (3) multiplying the average range by a conversion factor to get the distance from the expected average to the limit line. The central line is merely the average of all individual observations.

The shortcut formulas for control limits on sample ranges are

$$\text{UCL} = D_4 \bar{R}$$

$$\text{LCL} = D_3 \bar{R}$$

where D_3 and D_4 are constants found in Table I in Appendix I.

A partial tabulation of the A_2, D_3, and D_4 factors is reproduced in Table 20.3 for the convenience of the reader in following the text.

Consider the data for machines N-5 and N-7 in Figure 20.5. For each machine, the data consist of 10 samples (with six units each) plotted in time order of production (sample number). Figure 20.5 shows the $\bar{X}$ and R charts for each machine. The upper part of the figure displays the individual observations.

For machine N-5, the UCL and LCL are calculated as

AVERAGES:

$$\text{UCL} = \bar{\bar{X}} + A_2 \bar{R} = 9.59 + 0.483(6.6) = 12.77$$

$$\text{LCL} = \bar{\bar{X}} - A_2 \bar{R} = 9.59 - 0.483(6.6) = 6.40$$

TABLE 20.3
Constants for $\bar{X}$ and R chart

n	A_2	D_3	D_4	d_2
2	1.880	0	3.268	1.128
3	1.023	0	2.574	1.693
4	0.729	0	2.282	2.059
5	0.577	0	2.114	2.326
6	0.483	0	2.004	2.534
7	0.419	0.076	1.924	2.704
8	0.373	0.136	1.864	2.847
9	0.337	0.184	1.816	2.970
10	0.308	0.223	1.777	3.078

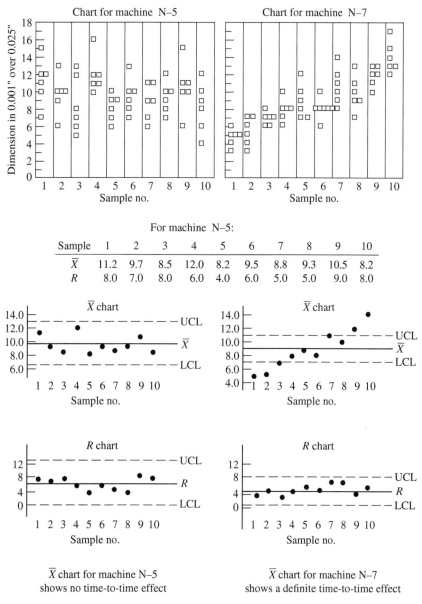

FIGURE 20.5
$\bar{X}$ and R charts confirm suggested machine differences.

RANGES:

$$\text{UCL} = D_4\bar{R} = 2.004(6.6) = 13.23$$
$$\text{LCL} = D_3\bar{R} = 0(6.6) = 0$$

Because all points fall within the control limits, it is concluded that the process is free of assignable causes of variation.

Control limits for a chart of averages represent three standard deviations of sample *averages* (not individual values). Because specification limits usually apply to *individual* values, the control limits *cannot* be compared to specification limits, averages inherently vary less than the individual measurements going into the averages (see Figure 18.1). Therefore, specification limits should *not* be placed on a control chart for averages. Sample averages, rather than individual values, are plotted because averages are more sensitive to detecting process changes than individual values.

Another example is presented in Figure 20.5 for machine N-7. This machine has both within-sample variation shown by the range chart and between-sample variation, as illustrated by the chart for sample averages. The $\bar{X}$ chart indicates that some factor such as tool wear is present that results in larger values of the characteristic with the passing of time (note the importance of preserving the order of the measurements). In such cases, measures of process capability should reflect both sources of variation.

Interpretation of Charts

Place the charts for $\bar{X}$ and R (or s if the rational subgroup is > 8) one above the other so the average and range for any one subgroup are on the same vertical line. Observe whether either or both indicate lack of control for that subgroup.

$\bar{X}$'s outside the control limits are evidence of a general change affecting all pieces after the first out-of-limits subgroup. The log kept during data collection, the operation of the process, and the worker's experience should be studied to discover a variable that could have caused the out-of-control subgroups. Typical causes are a change in material, personnel, machine setting, tool wear, temperature, or vibration.

R's outside control limits are evidence that the uniformity of the process has changed. Typical causes are a change in personnel, increased variability of material, or excessive wear in the process machinery. In one case, a sudden increase in R warned of an impending machine accident.

A single out-of-control R can be caused by a shift in the process that occurred while the subgroup was being taken.

Look for unusual patterns and nonrandomness. Nelson (1984, 1985) provides eight tests to detect such patterns on control charts using 3σ control limits (Figure 20.6). Each of the zones shown is 1σ wide.

Nelson's helpful rules for identifying additional types of nonrandom variation have been widely automated. As a result, automated flagging of out-of-control conditions are quick and easy. Unfortunately, the consequences of implementing such automated multiple rules have been poorly understood. Most users are comfortable with the basic Shewhart construct of the $\pm 3s$ control limit flag (rule 1) that will give a false alarm only about 1 time in 370. The other seven rules have similar, but not identical probabilities of false alarms. But, because the rules are at least somewhat independent, if one scans for out-of-control conditions using automated application of multiple rules, the likelihood of a false alarm rises.

The probability of a false alarm for rule 1 is easily computed and widely known: 0.27%. The probability of a false alarm for most of the other rules can also be

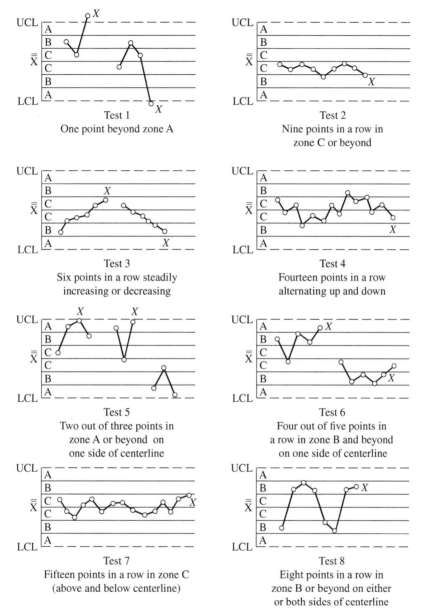

FIGURE 20.6
Illustrations of tests for special causes applied to $\bar{X}$ control charts. (*From Nelson, 1984.*)

computed with a little work, but rules 3 and 4 cannot. Also, joint probabilities among the rules cannot be readily calculated. Table 20.4 contains the calculated probabilities along with the results of a Monte Carlo simulation with 1 million runs calculating the probability of individual false alarms and all the joint probabilities.

TABLE 20.4
Probabilities for false alarms from Nelson's eight rules

	Probabilities in ppm									
	Random Signal		Simulated Joint Probability of Random Signal							
	Calculated	Simulated	Rule 1	Rule 2	Rule 3	Rule 4	Rule 5	Rule 6	Rule 7	Rule 8
Rule 1	2700	2778		6	4	11	120	48	X	—
Rule 2	3906	3983	6		2	12	24	133	7	—
Rule 3	*	364	4	2		X	4	5	1	—
Rule 4	*	2943	11	12	X		6	14	8	—
Rule 5	3058	3031	120	24	4	6		205	X	—
Rule 6	5532	5397	48	133	5	14	205		X	32
Rule 7	3261	3129	X	7	1	8	X	X		X
Rule 8	103	97	—	—	—	—	—	32	X	
Any of 8		21,099								

*Could not find a simple exact algorithm. X = not theoretically possible.
Source: John F. Early, "Probabilities of False Alarms from Multiple Out-of-Control Rules," *Juran Institute, 2013.*

None of the rules yields an unusually high false-alarm rate, although two are very small. The two-way joint probabilities are mostly vanishingly small. The cumulative effect for false alarms from applying multiple rules simultaneously, therefore, rises rapidly. If all eight rules are applied simultaneously, the probability of a false alarm rises to 2.1%—seven times the familiar rule 1 rate. These results are a good caution that with the use of multiple rules, the potential for false alarms increases substantially, especially if one is monitoring multiple characteristics at the same time. More time will then likely be spent chasing special causes that are not there.

While most of the rules are designed to catch trends or shifts in the process centering, three of the rules can identify deficiencies in the control system itself. Rule 7 (15 points within one standard deviation of the mean) suggests a failure in the selection of homogeneous rational subgroups. If the subgroups are not homogeneous, then the within-group variation will be large, thereby creating large standard deviations and wide control limits. In addition, if the subgroups are nonhomogeneous in time, they will capture much of the between-group variation, thereby compressing the individual points closer to the centerline.

Rule 8 (eight points in a row in one standard deviation or beyond on either or both sides of centerline) can also be triggered by nonhomogeneous subgroups. If the subgroups are selected across time from different operating conditions but are homogeneous within each point in time, then the within variation will yield relatively small standard deviations and relatively tight control limits. But the individual observations will not be close to the centerline because they each have a different mean, which averages out to the grand mean. As a result, observations from some subgroups are likely to be consistently above the grand mean and others systematically below. Rule 4 (14 points alternating up and down) could also be consistent with a specific variation of that same faulty design or execution of the rational subgroups in which alternating samples are not taken under identical conditions—e.g., alternating among machines, shifts, workers, and other conditions.

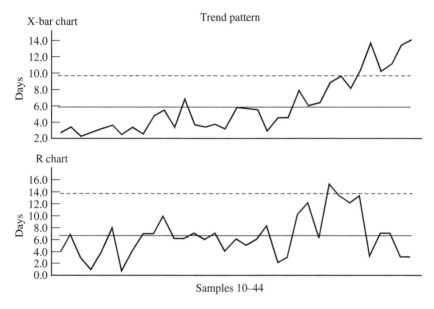

FIGURE 20.7
Average and range control charts.

AT&T (1990) provides an example from the service industry:

EXAMPLE 20.2. A manager of a personnel database at AT&T collected data on the average time to implement an employee change notification.

The average and range control chart is shown in Figure 20.7. The manager learned from the chart that the average time to implement began in the 2- to 3-day range and, over the course of 30 days, continuously increased to the 10- to 12-day range. The increase began gradually and then accelerated, suggesting that the process had changed. Investigation revealed that more and more change notification requests arrived with incomplete information, requiring workers to call the originating organization and increasing the time to implement.

Ott and Schilling (1990) provide a useful text on analysis after the initial control charts by presenting an extensive collection of cases with innovative statistical analysis clearly described.

Introducing Control Charts

To quality specialists, control charts are sensitive devices for detecting process changes; to operating forces, the charts represent a major change from the traditional "law of the shop," i.e., the specification limits. In introducing control charts, it is essential to prevent confusion about the role of control limits versus specification limits. Workers react to nonconforming product because specification limits have been the law of the shop; they do not react to control limits in the same way because the legitimacy of control limits may not have been fully established and clarified. For example, what is a worker to do if a control chart is frequently out of control but the product is well within specification limits? See Section 20.14.

Chart for Individuals

An alternative to the $\bar{X}$ and R chart is the chart for individual X values. This chart, often called a run chart, is a plot of individual observations against time. See Table 20.5 for more details. Individual values charts are usually accompanied by an MR—moving range—chart. The *moving range* is the difference between the maximum and minimum values over a range of observations—frequently, but not always, only two. The MR can be used to then calculate estimates of the standard deviation to plot the $\pm 3s$ control limits of individual values.

A chart of individual values is not as sensitive as the X-bar chart. Sometimes individual charts are necessary because of the nature of the process and its timing. But sometimes operations will adopt individual charts simply because of the difficulty, expense, or even just inconvenience of measuring several times. If, in fact, the measurement is very difficult or expensive—e.g., requiring destructive testing—consideration should be given to sampling less frequently and then drawing two to four samples each time so that an X-bar chart can be used. Each individual case will have different technical and economic consequences, but the alternative deserves evaluation.

EXAMPLE 20.3. Tom Pohlen, a quality engineering specialist, provides a personal example of a chart for individual values (Pohlen, 1999). His wife is diabetic, and she needed to control the variability of her blood glucose level. For nondiabetics, the normal glucose range is 70 to 120 milligrams/deciliter; the level for Pohlen's wife sometimes varied from under 70 to over 300 within a 24-hour period—that's variability. High levels can be caused by a variety of factors, including different foods, exercise, illness, infections, and emotional stress.

Figure 20.8 shows the control chart for several months of glucose readings. With the help of their physician and the chart, the couple learned about the causes of high blood glucose and successfully developed a "control strategy" for reducing the variability. This case is not an avocation of an engineer—any diabetic person will verify the seriousness of this matter. The reference makes fascinating reading and gives us a testimonial to the power of understanding variation—and the determination of Pohlen's wife.

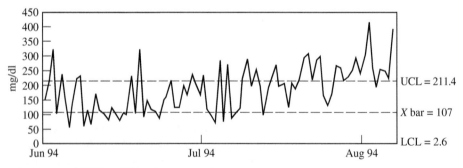

Limits based on 11/29/94 to 1/3/95—a future period of good control

FIGURE 20.8

Blood glucose—summer 1994. From "Statistical Thinking: A Personal Application" by T. Pohlen, *Annual Quality Congress Proceedings,* 1999. Reprinted with permission from *Annual Quality Congress Proceedings,* © 1999 American Society for Quality.

TABLE 20.5
Types of control charts

Common Control Charts for Continuous (or Variable) Data:

- **X-bar and R chart:** Also called "average and range" chart. The "x-bar" means the average of a sample or subgroup. It measures the central tendency of the response variable over time. R is the range or difference between the highest and lowest values in each subgroup. R charts measure the gain or loss of uniformity within a subgroup which represents the variability in the response variable over time.
- **X-bar and S chart:** The average chart is similar, but the standard deviation (instead of the range) is used in the S chart to track the variability within the subgroup. Requires a minimum of 9 observations per subgroup.
- **X-mR Chart** (also known as **I-mR chart**): Also known as an individual and moving range chart. Instead of charting the subgroup average and range over time, this chart plots each individual reading (subgroup size = 1) and a moving range.
- **Z-mR chart:** Standardized individuals and moving range chart. This is used for short runs with different nominal values for the control variable. Individual values are coded or standardized (Z transformation) so that the performance of the process can be continuously monitored across different products produced by that process. $Z_i = (X_i - \overline{X})/s$

Common Control Charts for Attribute (or Categorical) Data:

- **P Chart:** Also called proportions chart. It tracks the proportion or percent of nonconforming units (or percent defective) in each sample over time.
- **nP Chart:** A chart used to track the number of nonconforming units (or defective units) in each sample over time.
- **C Chart:** Used to track the number of nonconformities (i.e., defects, not defective units). Especially useful when a single unit (or length or area) of product may have infinite possibilities for defects. For example, the number of defects on a car.
- **U Chart:** This is a variation of the c chart. This chart tracks the number of nonconformities (or defects) per unit in a sample of n units.

A chart of individual measurements can be useful when the normal process measurements are spaced some time apart, e.g., one measurement per day from a chemical process or a single weekly measurement from accounting data.

◆ 20.7 PRE-CONTROL

PRE-Control is a statistical technique for detecting process conditions and changes that may cause defects (rather than changes that are statistically significant). PRE-Control focuses on controlling conformance to specifications, rather than statistical control. PRE-Control starts a process centered between specification limits and detects shifts that might result in making some of the parts outside a specification limit. PRE-Control requires no plotting and no computations, and it needs only three measurements to give control information. The technique uses the normal distribution curve to determine significant changes in either the aim or the spread of a production process that could result in increased production of defective work.

The principle of PRE-Control is demonstrated by assuming the worst condition that can be accepted from a process capable of quality production, i.e., when the natural tolerance is the same as the specification allows and when the process is precisely centered and any shift would result in some defective work.

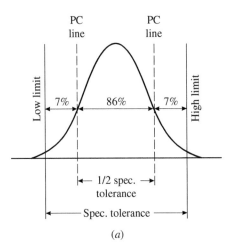
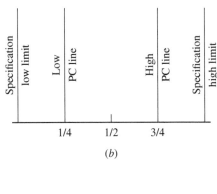

FIGURE 20.9
(a) Assumptions underlying PRE-Control. *(b)* Location of PRE-Control lines.

If we draw two PRE-Control (PC) lines, each one-fourth of the tolerance width in from each specification limit (Figure 20.9), it can be shown that, in round numbers, 86% of the parts will be inside the PC lines, with 7% in each of the outer sections. In other words, 7%, or 1 part in 14, will occur outside a PC line under normal circumstances.

The chance that two measurements in a row will fall outside a PC line is 1/14 times 1/14, or 1/196. Therefore, only once in about every 200 measurements should we expect to get two in a row in a given outer band. When two in a row do occur, the chance that the process has shifted unfavorably is much greater (195/196). It is therefore advisable to reset the process to the center. It is equally unlikely to get a measurement beyond one given PC line and the next outside the other PC line. In this case, the indication is not that the process has shifted, but that some factor has been introduced that has widened the pattern to an extent that defective pieces are almost inevitable. An immediate remedy of the cause of the trouble must be made before the process can safely continue.

The zone within the PC lines is the green zone; between the PC lines and the specification limits is the yellow zone; outside the specification limits is the red zone.

To qualify a process for PRE-Control:

1. Take consecutive individual measurements on a characteristic until five consecutive measurements fall within the green zone.
2. If one yellow occurs, restart the count.
3. If two consecutive yellows occur, adjust the process.
4. Anytime an adjustment or other process change occurs, requalify the process.

When the process is qualified, the following PRE-Control rules are applied to running the process:

1. Use a sample of two consecutive measurements, A and B. If A is green, continue running the process. If A is yellow, take a second measurement, B.
2. If A and B are both yellow, stop the process and investigate.

If a red occurs during either the qualification or running stages, stop the process and investigate.

Most processes require periodic adjustments to remain within specifications. Six A, B pairs of measurements between adjustments are viewed as sufficient to provide virtually no out-of-specification product. Thus if a process typically requires an adjustment about every two hours, then an A, B pair of measurements should be taken every 20 minutes.

PRE-Control is an example of a concept known as "narrow-limit gauging." The broader concept provides sampling procedures (sample size, location of the narrow limits, and allowable number of units outside the narrow limits) to meet predefined risks of accepting bad product. Narrow-limit gauging is discussed by Ott and Schilling (1990, Chapter 7).

The relative simplicity of PRE-Control versus statistical control charts can have important advantages in many applications. The concept, however, has generated some controversy. For a comparison of PRE-Control versus other approaches and the most appropriate applications of PRE-Control, see Ledolter and Swersey (1997) and Steiner (1997). For a complete story, also see the references in both of these papers.

20.8 ATTRIBUTES CONTROL CHARTS

Control charts for $\bar{X}$, R, S, MR, Z, and X require actual numerical measurements, e.g., line width from a photoresist process. Control charts for attribute data require only a count of observations of a characteristic, e.g., the number of nonconforming items in a sample.

Examples of Attribute Charts

The fraction of nonconforming units (p) chart can be illustrated with data on magnets used in electrical relays. For each four weeks, three samples of magnets were inspected during each day. The number of magnets inspected and the number of nonconforming magnets were recorded. For the first two weeks, 500 were inspected in each sample. After that, the sample was doubled. While use of the average sample sizes makes preparation easier and the resulting chart more pleasing visually, if the samples sizes vary substantially, the use of the average does increase the risk of either missing of falsely chasing a special cause. Modern software makes it unnecessary to use averages in most cases.

The resulting control chart is shown in Figure 20.10. Note that the last sample is below the LCL, indicating a significantly low fraction nonconforming. Although this sample might mean that some assignable cause is resulting in better quality, such points can also be (1) the result of an inspector's accepting some nonconforming units in error or (2) if an average sample size were used in calculation, a sample size quite different from the average used to calculate the limits. Note that three points

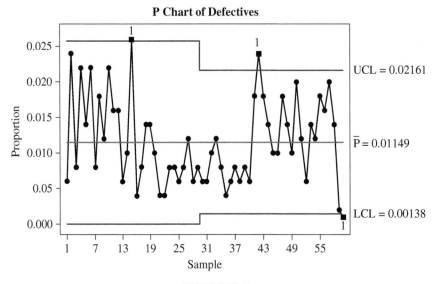

FIGURE 20.10
p chart for permanent magnets.

are beyond the control limits even though the data were included in calculating the control limits. Control limits do have the ability to detect the presence of (at least some) special causes, even though the control limits were influenced by those causes. Nevertheless, control limits used during operations should not include points with proven special causes in their calculation.

Leonard (1986) reports on the application of a *p* chart to recruiting new employees at the Rogers Corporation. Figure 20.11 shows a plot of the percentage of open job requisitions filled during calendar quarters. All of the points fall within the control limits, indicating that the variation is due to a system of common causes. Reducing that variation requires action "across the board" (as the team called it) rather than analyzing the cause of a low value such as 3% in the fourth quarter of 1981. Note that the control limits vary for each quarter. Instead of using an *average* sample size to calculate one set of control limits, the exact sample size is substituted in the formula to obtain the precise limits for each quarter. The price we pay for this precision is the difficulty in explaining why the control limits vary.

Heyes (1988) presents a vivid discussion of a *p* chart and a regression study. The analysis graphically convinced people that machines were aging, and the regression study quantified the age–performance relationship.

The *u* chart will be illustrated with Figure 20.12.

Four times per day, a textile plant removes 400 linear inches of fabric from production and subjects it to careful examination for a variety of different flaws. The exigencies of production, however, do not always allow for getting all 400 inches, so the amount of fabric actually used varies. Hence, the different control limits reflect the actual sample size used.

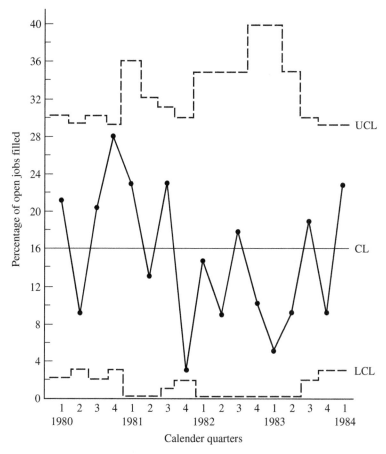

FIGURE 20.11
p chart of jobs filled as a percentage of jobs open by quarter
(January 1, 1980–April 2, 1984). (*From Leonard, 1986.*)

➔ 20.9 SPECIAL CONTROL CHARTS

The previous paragraphs presented the basic types of control charts for variable and attribute data—the ones needed for most applications. Sometimes other types of control charts are employed to address special needs. Such special control charts have ingenious aspects, and several types are mentioned to encourage the reader to explore further.

The *cumulative sum* (CUMSUM) *control chart* is a chronological plot of the cumulative sum of deviations of a sample statistic from a reference value, typically the nominal or target specification. By definition, the CUMSUM chart focuses on a target value rather than on the actual average of process data. Note that the centerline is zero because the plotted values are deviations from the target.

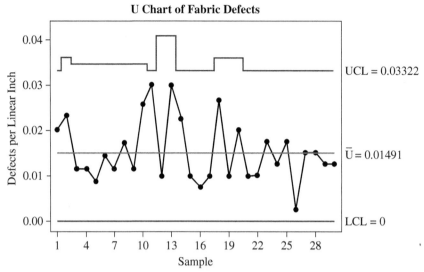

FIGURE 20.12
Control chart for fabric defects per linear inch.

Each point plotted contains information from all preceding observations (i.e., a *cumulative* sum). Note that there are two lines on this chart: one to cumulate the positive changes and a second to cumulate the negative. CUMSUM charts are particularly useful in detecting small shifts in the process average (say 0.5σ to 2.0σ). In our example, the chart flags an alert for a decline in water absorption. Calculations for constructing the chart shown in Figure 20.13 are given in *JQH6*, pages 565–657.

Another special chart is the *moving average chart.* This chart is a chronological plot of the moving average, which is calculated as the average value updated by dropping the oldest individual measurement and adding the newest individual measurement. Thus a new average is calculated with each individual measurement. A further refinement is the *exponentially weighted moving average* (EWMA) *chart.* In the EWMA chart, the observations are weighted, and the highest weight is given to the most recent data. Moving average charts are effective in detecting small shifts, highlighting trends, and using data in processes in which it takes a long time to produce a single item.

Still another chart is the Box-Jenkins manual adjustment chart. The average and range, CUMSUM, and EWMA charts for variables focus on *monitoring* a process and reducing variability due to special causes of variation identified by the charts. Box-Jenkins charts have a different objective: to analyze process data to *regulate* the process after each observation and thereby minimize process variation. For elaboration on this advanced technique, see Box and Luceno (1997).

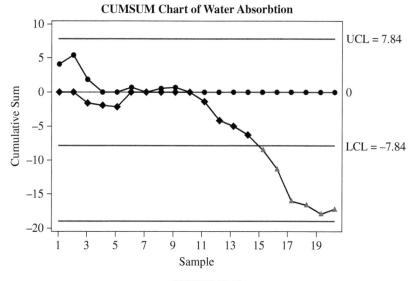

FIGURE 20.13
Cumulative sum control chart. (*From JQH6, page 657.*)

Finally, we consider the concept of multivariate control charts. When there are two or more quality characteristics on a unit of product, these could be monitored independently with separate control charts. Then the probability that a sample average on either control chart exceeds three sigma limits is 0.0027. But the joint probability that both variables exceed their control limits simultaneously when they are both in control is (0.0027)(0.0027) or 0.00000729, which is much smaller than 0.0027. The situation becomes more distorted as the number of characteristics increases. For this and other reasons, monitoring several characteristics independently can be misleading. Multivariate control charts address this issue. See Bersimis et al. (2007) for a good overview and extensive references to the research.

➔ 20.10 PROCESS CAPABILITY

In planning the quality aspects of operations, nothing is more important than advance assurance that the processes will meet the specifications. In recent decades, a concept of *process capability* has emerged to provide a quantified prediction of process adequacy. This ability to predict quantitatively has resulted in widespread adoption of the concept as a major element of quality planning.

Process capability is the measured, inherent variation of the product turned out by a process.

Basic Definitions

Each key word in this definition must itself be clearly defined because the concept of capability has an enormous extent of application and nonscientific terms are inadequate for communication within the industrial community.

- *Process* refers to some unique combination of machine, tools, methods, materials, *and people* engaged in production. It is often feasible and illuminating to separate and quantify the effect of the variables entering this combination.
- *Capability* refers to an ability, based on tested performance, to achieve measurable goals or targets.
- *Measured capability* refers to the fact that process capability is quantified from data that, in turn, are the results of measurement of work performed by the process.
- *Inherent capability* refers to the product uniformity resulting from a process that is in a state of statistical control, i.e., in the absence of time-to-time "drift" or other assignable causes of variation. "Instantaneous reproducibility" is a synonym for inherent capability.
- The characteristics of the *product* are measured because product variation is the end result.

Uses of Process Capability Information

Process capability information serves multiple purposes:

1. Predicting the extent of variability that processes will exhibit. Such capability information, when provided to designers, provides important information in setting realistic specification limits.
2. Choosing from among competing processes that are most appropriate to meet the tolerances.
3. Planning the interrelationship of sequential processes. For example, one process may distort the precision achieved by a predecessor process, as in hardening of gear teeth. Quantifying the respective process capabilities often points the way to a solution.
4. Providing a quantified basis for establishing a schedule of periodic process control checks and readjustments.
5. Assigning machines to classes of work for which they are best suited.
6. Testing theories of causes of defects during quality improvement programs.
7. Serving as a basis for specifying the quality performance requirements for purchased machines.

These purposes account for the growing use of the process capability concept.

Standardized Formula

The most widely adopted formula for process capability is

$$\text{Process capability} = \pm 3\sigma \text{ (a total of } 6\sigma\text{)}$$

where σ = the standard deviation of the process under a state of statistical control, i.e., under no drift and no sudden changes. See Section 20.13, however, for the six sigma concept of process capability.

If the process is centered at the nominal specification and follows a normal probability distribution, 99.73% of production will fall within $\pm 3\sigma$ of the nominal specification.

Some industrial processes operate under a state of statistical control. For such processes, the computed process capability of 6σ can be compared directly to specification limits, and judgments of adequacy can be made. However, most industrial processes exhibit both drift and sudden changes. These departures from the ideal are a fact of life, and the practitioner must deal with them.

Nevertheless, there is great value in standardizing on a formula for process capability based on a state of statistical control. In this state, product variations are the result of numerous small variables (rather than the effect of a single large variable) and, hence, have the character of random variation. It is most helpful for planners to have such limits in quantified form.

Relationship to Product Specifications

A major reason for quantifying process capability is to compute the ability of the process to hold product specifications. For processes that are in a state of statistical control, a comparison of the variation of 6σ to the specification limits permits ready calculation of percentage defective by conventional statistical theory.

Planners try to select processes with the 6σ process capability well within the specification width, usually called *tolerance*. A measure of this relationship is the capability ratio:

$$C_p = \text{Capability ratio} = \frac{\text{Specification range}}{\text{Process capability}} = \frac{\text{USL} - \text{LSL}}{6s}$$

where USL = upper specification limit
LSL = lower specification limit

Note that $6s$ is used as an estimate of 6σ.

Some companies define the ratio as the reciprocal. Some industries now express defect rates in terms of parts per million. A defect rate of one part per million requires a capability ratio (specification range over process capability) of about 1.63.

Figure 20.14 shows four of many possible relations between process variability and specification limits and the likely courses of action for each. Note that, in all of these cases, the average of the process is at the midpoint between the specification limits.

Table 20.6 shows selected capability ratios and the corresponding level of defects, assuming that the process average is midway between the specification limits. A process that is just meeting specification limits (specification range = $\pm 3\sigma$) has a C_p of 1.0. The criticality of many applications and the reality that the process average will not remain at the midpoint of the specification range suggest that C_p should be at least 1.33.

Note that the C_p index measures whether the process variability can fit within the specification range. It does not indicate whether the process is actually running within

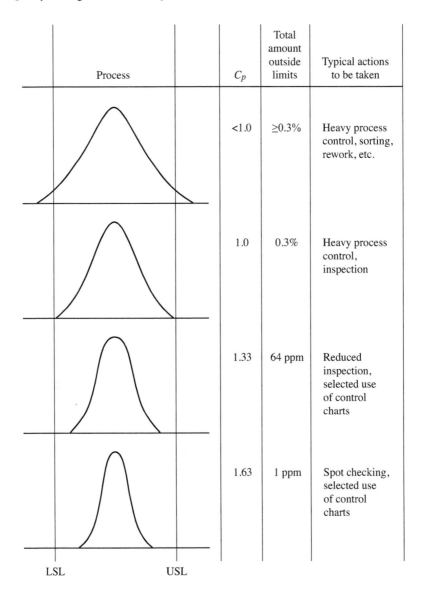

FIGURE 20.14
Four examples of process variability.

the specification because the index does not include a measure of the process average (this issue is addressed by another measure, C_{pk}).

Three capability indexes commonly in use are shown in Table 20.7. The higher the value of any indexes, the lower the amount of product outside the specification limits.

Note that the "capability" indexes use a short-term measure of the standard deviation and the "performance" indexes use a long-term standard deviation. The same distinction applies here as for the within and between for control charts. The short term

TABLE 20.6
Process capability index (C_p) and product outside specification limits

Process capability index (C_p)	Total product outside two-sided specification limits*
0.5	133,614 ppm
0.67	44,431 ppm
1.00	2700 ppm
1.33	664 ppm
1.63	1 ppm
2.00	0.002 ppm

*Assuming that the process is centered midway between the specification limits.

is the instantaneous variation in a rational subgroup, while the long term incorporates both the instantaneous variation and the between-subgroup variation. ANOVA can be used to estimate these properly from samples including subgroups.

Because capability indexes are estimated using a sample estimate for the standard deviation, it is often prudent to calculate the confidence interval for the standard deviation and use the endpoints of the interval to compute confidence intervals for the capability indexes. The following formula applies the chi-square distribution to find the confidence interval for the standard deviation:

$$s\sqrt{\frac{n-1}{\chi^2_{n-1,1-\alpha/2}}} \leq \sigma \leq s\sqrt{\frac{n-1}{\chi^2_{n-1,\alpha/2}}}$$

Bothe (1997) provides a comprehensive reference book that includes extensive discussion of the mathematical aspects of capability indexes. These references explain how to calculate confidence bounds for various process capability indexes.

TABLE 20.7
Process capability and process performance indexes

Process capability	Process performance
$C_p = \dfrac{USL - LSL}{6s_{st}}$	$P_p = \dfrac{USL - LSL}{6s_{Lt}}$
$C_{pk} = \min\left[\dfrac{USL - \mu}{3s_{st}}, \dfrac{\mu - LSL}{3s_{st}}\right]$	$P_{pk} = \min\left[\dfrac{USL - \bar{X}}{3s_{Lt}}, \dfrac{\bar{X} - LSL}{3s_{Lt}}\right]$
$C_{pm} = \dfrac{USL - LSL}{6\sqrt{s_{st}^2 + (\mu - T)^2}}$	$P_{pm} = \dfrac{USL - LSL}{6\sqrt{s_{Lt}^2 + (\bar{X} - T)^2}}$

s_{st} = short-term standard deviation
s_{Lt} = long-term standard deviation
T = target, midpoint between LSL and USL

The C_{pk} Capability Index

Process capability, as measured by C_p, refers to the variation in a process about the average value. This concept is illustrated in Figure 20.15. The two processes have equal capabilities (C_p) because 6σ is the same for each distribution, as indicated by the widths of the distribution curves. The process aimed at μ_2 is producing defectives because the aim is off center, not because of the inherent variation about the aim (i.e., the capability).

Thus the C_p index measures *potential* capability, assuming that the process average is equal to the midpoint of the specification limits and the process is operating in statistical control; because the average is often not at the midpoint, it is useful to have a capability index that reflects both variation and the location of the process average. Such an index is C_{pk}.

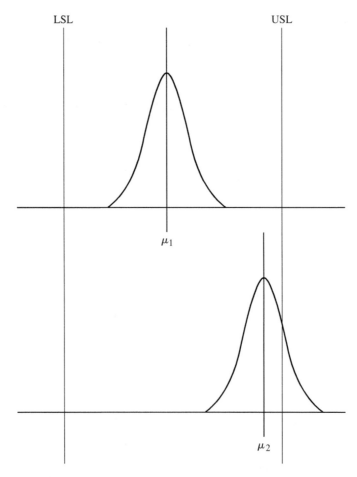

FIGURE 20.15
Process with equal C_p but different aim.

C_{pk} reflects the current process mean's proximity to either the USL or LSL. C_{pk} is estimated by

$$\hat{C}_{pk} = \min\left[\frac{\overline{X} - \text{LSL}}{3s}, \frac{\text{USL} - \overline{X}}{3s}\right]$$

In an example from Kane (1986),

$$\text{USL} = 20 \quad \overline{X} = 16$$
$$\text{LSL} = 8 \quad s = 2.$$

The standard capability ratio is estimated as

$$\frac{\text{USL} - \text{LSL}}{6\sigma} = \frac{20 - 8}{12} = 1.0,$$

which implies that *if* the process were centered between the specification limits (at 14), then only a small proportion (about 0.27%) of product would be defective.

However, when we calculate C_{pk}, we obtain

$$\hat{C}_{pk} = \min\left[\frac{16 - 8}{6}, \frac{20 - 16}{6}\right] = 0.67,$$

which indicates that the process mean is *currently* nearer the USL. (Note that, if the process were centered at 14, the value of C_{pk} would be 1.0.) An acceptable process will require reducing the standard deviation and/or centering the mean.

Note that, if the *actual* average is equal to the midpoint of the specification range, then $C_{pk} = C_p$.

The higher the value of C_p, the lower the amount of product outside specification limits. In certifying suppliers, some organizations use C_{pk} as one element of certification criteria. In these applications, the value of C_{pk} desired from suppliers can be a function of the type of commodity purchased.

A capability index can also be calculated around a target value rather than the actual average. This index, called C_{pm} or the Taguchi index, focuses on reduction of variation from a target value rather than reduction of variability to meet specifications.

Originally, capability indexes assumed that the quality characteristic is normally distributed. Readily available software such as MINITAB now make it easy to identify and fit distributions other than normal. From a proper definition of the distribution, comparable indexes can be calculated for any population.

Two types of process capability studies are as follows:

1. *Study of process potential.* In this study, an estimate is obtained of what the process *can* do under certain conditions, i.e., variability under short-run defined conditions for a process in a state of statistical control. The C_p index estimates the potential process capability.
2. *Study of process performance.* In this study, an estimate of capability provides a picture of what the process *is* doing over an extended period of time. A state of statistical control is also assumed. The C_{pk} index estimates the performance capability.

The Assumption of Statistical Control and Its Effect on Process Capability

All statistical predictions assume a stable population. In a statistical sense, a stable population is one that is repeatable, i.e., a population that is in a state of statistical control. The statistician rightfully insists that this be the case before predictions can be made. The manufacturing engineer also insists that the process conditions (feeds, speeds, etc.) be fully defined.

In practice, the original control chart analysis will often show that the process is out of statistical control. (It may or may not be meeting product specifications.) In theory, process capability should not be predicted until the process is in statistical control. However, in practice, some kind of comparison of capability to specifications is needed. The danger in delaying the comparison is that the assignable causes may never be eliminated from the process. The resulting indecision will thereby prolong interdepartmental bickering on whether "the specification is too tight" or "manufacturing is too careless."

A good way to start is by plotting individual measurements against specification limits. This step may show that the process can meet the product specifications even with assignable causes present. If a process has assignable causes of variation but is able to meet the specifications, usually no economic problem exists. The statistician can properly point out that a process with assignable variation is unpredictable. This point is well taken, but in establishing priorities of quality improvement efforts, processes that are meeting specifications are seldom given high priority.

It is important to distinguish between a process that is in a state of statistical control and a process that is meeting specifications. A state of statistical control does not necessarily mean that the product from the process conforms to specifications. Statistical control limits on sample averages *cannot* be compared to specification limits because specification limits refer to individual units. For some processes that are not in control, the specifications are being met and no action is required; other processes are in control, but the specifications are not being met, and action is needed (see Section 20.14).

In summary, we need processes that are both stable (in statistical control) and capable (meeting product specifications).

The increasing use of capability indexes has also led to the failure to understand and verify some important assumptions that are essential for statistical validity of the results. Five key assumptions are

1. *Process stability.* Statistical validity requires a state of statistical control with no drift or oscillation.
2. *Underlying of the characteristic being measured.* Normality is needed to draw statistical inferences about the population using normal assumptions. If another distribution can be fit to the data, then capability can be computed using that distribution. If no distribution can be fit, then care is required in interpretation based on approximations.
3. *Sufficient data.* Sufficient data are necessary to minimize the sampling error for the capability indexes. A good rule of thumb is a minimum of 40 data points. Applying the confidence interval for the population standard deviation to the calculation of

the capability indexes will yield confidence intervals on the indexes and quantify the effects of sample size.
4. *Representativeness of samples.* Samples must be random draws from the full population.
5. *Independence of measurements.* Consecutive measurements cannot be correlated.

These assumptions are not theoretical refinements—they are important conditions for properly applying capability indexes.

20.11 MEASURING PROCESS PERFORMANCE

A process performance study collects data from a process that is operating under typical conditions but includes normal changes in material batches, workers, tools, or process settings. This study, which spans a longer term than the process potential study, also requires that the process be in statistical control.

The capability index for a process performance study is

$$C_{pk} = \min\left[\frac{\overline{X} - \text{LSL}}{3s}, \frac{\text{USL} - \overline{X}}{3s}\right].$$

EXAMPLE 20.4. Consider a pump cassette used to deliver intravenous solutions (Baxter Travenol Laboratories, 1986). A key quality characteristic is the volume of solution delivered in a predefined time. The specification limits are

$$\text{USL} = 103.5 \qquad \text{LSL} = 94.5$$

A control chart was run for one month, and no out-of-control points were encountered. From the control chart data, we know that

$$\overline{X} = 98.2 \quad \text{and} \quad s = 0.98$$

Figure 20.16 shows the process data and the specification limits.
The capability index is

$$C_{pk} = \min\left[\frac{98.2 - 94.5}{3(0.98)}, \frac{103.5 - 98.2}{3(0.98)}\right]$$

$$C_{pk} = 1.26.$$

For many applications, 1.26 is an acceptable value of C_{pk}.

Interpretation of C_{pk}

In using C_{pk} to evaluate a process, we must recognize that C_{pk} is a summary of two parameters—the average and the standard deviation. Such a summary can

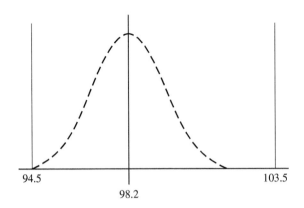

FIGURE 20.16
Delivered volume of solution. (*From Baxter Travenol Laboratories, 1986.*)

inadvertently mask important detail in these parameters. For example, Figure 20.17 shows that three extremely different processes can all have the same C_{pk} (in this case $C_{pk} = 1$).

Increasing the value of C_{pk} may require a change in the process average, the process standard deviation, or both. For some processes, increasing the value of C_{pk} by changing the average value (perhaps by a simple adjustment of the process aim) may be easier than reducing the standard deviation (by investigating the many causes of variability). The histogram of the process should always be reviewed to highlight both the average and the spread of the process.

Note that Table 20.7 also includes the capability index C_{pm}. This index measures the capability around a target value T rather than the mean value. When the target value equals the mean value, the C_{pm} index is identical to the C_p index.

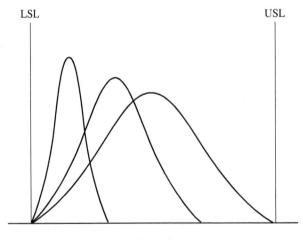

FIGURE 20.17
Three processes with $C_{pk} = 1$.

Attribute (or Categorical) Data Analysis

The methods discussed earlier assume that numerical measurements are available from the process. Sometimes, however, the only data available are in attribute or categorical form, i.e., the number of nonconforming units and the number acceptable.

The data in Table 4.5 on errors in preparing insurance policies also can be used to illustrate process capability for attribute data. The data reported 80 errors from six policy writers or 13.3 errors per writer—the current *performance.* The process *capability* can be calculated by excluding the abnormal performance identified in the study—type 3 errors by worker B, type 5 errors, and errors of worker E. The error data for the remaining five writers becomes 4, 3, 5, 2, and 5, with an average of 3.8 errors per writer. The process capability estimate of 3.8 compares with the original performance estimate of 13.3.

This example calculates process capability in terms of errors or mistakes rather than the variability of a process parameter. Hinckley and Barkan (1995) point out that in many processes, nonconforming product can be caused by excessive variability or by mistakes (e.g., missing parts, wrong parts, wrong information, or other processing errors). For some processes, mistakes can be a major cause of failing to meet customer quality goals.

Process Capability in Service Industries

Although the concept of process capability grew up in manufacturing, the concept also applies to service industry processes.

For certain parameters in service processes, process capability can be measured by the various capability indexes. For example, in a bank, the cycle time to complete the loan approval process is critical, and time data is readily available in quantitative form to calculate a capability index.

Other service processes may not have continuous data available. For example, a firm provides a service of guaranteeing checks written by customers at retail stores. The decision to guarantee is based on a process that employs an online evaluation of six factors. Unfortunately, a percentage of checks default ("bounce"), and this percentage could be viewed as a measure of the capability of the process for making the decision to guarantee. This approach uses attribute or categorical data as in the earlier example on errors in insurance policies.

✦ 20.12 PLANNING FOR THE PROCESS CAPABILITY STUDY

Capability studies are conducted for various reasons, e.g., to respond to a customer request for a capability index number or to evaluate and improve product quality. Prior to data collection, clarify the purpose for making the study and the steps needed to ensure that it is achieved.

In some cases, the capability study will focus on determining a histogram and capability index for a relatively simple process. Here the planning should ensure that process conditions (e.g., temperature, pressure) are completely defined and recorded. All other inputs must clearly be representative, i.e., specific equipment; material; and of course, personnel.

For more complex processes or when defect levels of 1 to 10 parts per million are desired, the following steps are recommended:

1. Develop a process description including inputs, process steps, and output quality characteristics. This description can range from simply identifying the equipment to developing a mathematical equation that shows the effect of each process variable on the quality characteristics.
2. Define the process conditions for each process variable. In a simple case, this step involves stating the settings for temperature and pressure. But for some processes, it means determining the optimum value or aim of each process variable. The statistical design of experiments provides the methodology. Also, determine the operating ranges of the process variables around the optimum because the range will affect the variability of the product results.
3. Make sure that each quality characteristic has at least one process variable that can be used to adjust it.
4. Decide whether measurement error is significant. This can be determined from a separate error of measurement study (see Section 10.8). In some cases, the error of measurement can be evaluated as part of the overall study.
5. Decide whether the capability study will focus only on variability or will also include mistakes or errors that cause quality problems.
6. Plan for the use of control charts to evaluate the stability of the process.
7. Prepare a data collection plan including adequate sample size that documents results on quality characteristics along with the process conditions (e.g., values of all process variables) and preserves information on the order of measurements so that trends can be evaluated.
8. Plan which methods will be used to analyze data from the study to ensure, before starting the study, that all necessary data for the analysis will be available. The analyses should include process capability calculations on variability and also analysis of attribute or categorical data on mistakes and analysis of data from statistically designed experiments built into the study.
9. Be prepared to spend time investigating interim results before process capability calculations can be made. These investigations can include analysis of optimum values and ranges of process variables, out-of-control points on control charts, or other unusual results. The investigations then lead to the ultimate objective, i.e., improvement of the process.

Note that these steps focus on improvement rather than just on determining a capability index.

➔ 20.13 SIX SIGMA CONCEPT OF PROCESS CAPABILITY

The term *six sigma* has been adopted as a label for quality improvement. Chapter 4 explains the five steps in the approach along with the specific tools involved. In the current chapter, we present only the statistical basis of the concept.

For some processes, shifts in the process average are so common that they should be recognized in setting acceptable values of C_p. In some industries, shifts in the process

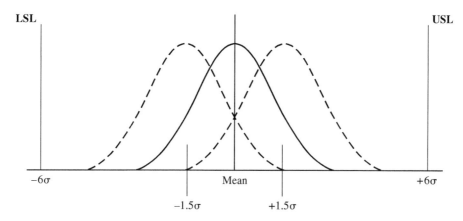

FIGURE 20.18
Six Sigma concept of process capability.

average of ±1.5 standard deviations (of individual values) are not unusual. To allow for such shifts, high values of C_p are needed. For example, if specification limits are at ±6σ (Figure 20.18) and if the mean shifts ±1.5σ, then 3.4 parts per million will be beyond specification limits. The Motorola Company's initiation of the six sigma approach adopted this typical shift in the process average as part of the definition of six sigma.

Table 20.8 shows capability indexes and defect levels in parts per million (ppm) for a centered process and for a process with the mean shifted 1.5σ.

Note that at the 3σ level with a centered process, the defect level is 2700 parts per million (or 0.27%); with a shift of 1.5σ, the defect level is 66,803 parts per million (or 6.68%). These calculations apply to a single product or process characteristic. In reality, many products or processes have numerous characteristics, and thus the overall yield decreases with this complexity. McFadden (1993) discusses this matter. For example, assuming a process with a 1.5σ shift, a process with 100 steps each operating at the three sigma level would have a yield of only 0.10%; at the five sigma level, the yield would still only be about 54%. This sobering outcome demonstrates the importance of low variability (i.e., a high sigma level), particularly for complex products or processes.

TABLE 20.8
Sigma levels and defect levels

Sigma level	Centered process		Shifted (±1.5σ) process	
	C_p	ppm*	C_{pk}	ppm
3	1	2,700	0.5	66,803
4	1.33	63	0.833	6,200
5	1.67	0.57	1.167	233
6	2	0.002	1.5	3.4

From "Six Sigma Quality Programs" by F. R. McFadden, *Quality Progress*, June, 1993. Reprinted with permission from the *Quality Progress*, © 1993 American Society for Quality.
*ppm = parts per million.

We must emphasize that the six sigma approach to improvement is a strategy that assumes process shifts of 1.5σ (in practice a realistic assumption for many processes). This use of the term six sigma is quite different from the classical standardized formula for process capability as $\pm 3\sigma$ or a total of 6σ.

✣ 20.14 STATISTICAL PROCESS CONTROL AND QUALITY IMPROVEMENT

In Chapter 1, we distinguished between control and improvement. The control process detects and takes action on *sporadic* quality problems; the improvement process identifies and takes action on *chronic* quality problems.

In the control process, statistical control charts detect the existence of special causes of variation that result in sporadic problems. The charts show sample data falling beyond statistical control limits, i.e., the process is "out of statistical control." Conversely, when a chart shows that a process is "in statistical control," the process is in a state of stability, and variation is due to a set of common causes inherent in the process. Statistical control means stability, but stability does not always mean customer satisfaction with the result. Unfortunately, a process in statistical control can have serious quality problems. Because the process is stable, the problems will continue (become chronic) unless a basic change in the system of common causes is made. Such a change, which typically affects the average or variation, is the job of improvement. "Removal of a special cause of variation, to move toward statistical control, important though it may be, is not improvement of the process" (Deming, 1986, page 338).

Process improvement is directed at several problems:

1. *The process average is misdirected.* Table 20.9 shows possible corrective action.
2. *The inherent variability of the process is too large.* Table 20.9 provides some of the approaches available to reduce variability.
3. *The instrumentation is inadequate.* See Section 10.8, "Errors of Measurement."
4. *A process drift exists.* Here the need is to quantify the amount of drift in a given period of time and to provide a means of resetting the process to compensate for this drift.
5. *Cyclical changes in the process exist.* The need is to identify the underlying cause and either remove it or reduce the effect on the process.
6. *The process is erratic.* Sudden changes can take place in processes. As the capability studies quantify the size of these changes and help to discover the reasons for them, appropriate planning action can be taken:
   - *Temporary phenomena* (e.g., a cold machine coming up to operating temperature) can be dealt with by scheduling warming periods plus checks at the predicted time of stability.
   - *More enduring phenomena* (e.g., changes due to new materials) can be dealt with by specifying setup reverification when such changes are introduced.

The statistical design of experiments is an essential analytical tool for improvement that goes far beyond the investigation of out-of-control points on a statistical control chart. This tool, when combined with the knowledge of those who plan and run the processes, replaces intuitive decision making with a scientific basis. See

TABLE 20.9
Approaches to process improvement

Changing the average	Reducing variability
Adjust settings on the process equipment.	Investigate work methods and equipment factors. This includes identifying the process variables that affect the product result.
Change selected product design parameters so that the design is more robust to manufacturing conditions.	Change selected product design parameters so that the design is more robust to manufacturing conditions.
Identify the process variables that affect the product results, determine the optimum values for the variables, and set the process to these values.	Identify and reduce the causes of variability due to human inputs. The concept of system versus worker-controllable input and the concept of self-control are useful guides.
Employ automated process controls to continually measure, analyze, and adjust process variables that affect the average.	Reduce the variability of inputs to the process through an improvement program with internal and external suppliers.
	Employ automated process controls to continually measure, analyze, and adjust process variables that affect variability.

Section 18.11, "The Design of Experiments," and Section 12.4 under "Parameter Design and Robust Design."

Now, how do these matters relate to customer needs?

Clearly, in using statistical process control and taking subsequent actions, the focus must be on meeting customer needs. One definition—which is far from perfect—is given by specification limits. Limits on statistical control charts are different from specification limits. In some situations, a process is not in statistical control but may not require action because the product specifications are easily met. In other situations, a process is in statistical control, but the product specifications are not being met.

If a product *does not* meet specifications, then some type of action is needed—changing the average value, reducing the variability, doing both, changing the specifications, sorting the product, etc. If a product *does* meet the specifications, the alternatives are different—taking no action, using a less precise process, or reducing the variability further (see later discussion for reasons). Table 20.10 shows the more usual permutations encountered and provides suggestions on the type of action to be taken.

Our goal for processes is clear: Be in statistical control and be capable of meeting product specifications.

Note that this chapter focuses on the variability of product characteristics and its relationship to specification limits. A broader issue of improvement is the ability of existing product features to meet customer needs (see Chapter 11).

⇥ 20.15 SOFTWARE FOR STATISTICAL PROCESS CONTROL

The availability of software has contributed substantially to the use of statistical process control techniques. Software has made it practical to collect large quantities of data and perform complex analyses.

694   Quality Management and Analysis

**TABLE 20.10
Action to be taken**

	Product meets specifications	
	Process variation small relative to specifications*	**Process variation large relative to specifications***
Process is in control	Consider value in marketplace of tighter specifications. Reduce inspection.	Continue tight controls on process average.
Process is out of control	Process is erratic and unpredictable and may be heading for trouble. Investigate causes of lack of control.	
	Product does not meet specifications	
Process is in control	Process is "misdirected" to wrong average. Generally easy to correct permanently.	Process may be misdirected and also too scattered. Correct misdirection. Consider economics of more precise process versus wider specifications versus sorting the product.
Process is out of control	Process is misdirected or erratic or both. Correct misdirection. Discover cause of lack of control. Consider economics of more precise process versus wider specifications versus sorting the product.	

*As a rule of thumb, a process variation (sometimes called natural tolerance = 6σ) less than one-third of the specification range is small; more than two-thirds is large.

Statistical process control software will calculate sample statistics, calculate control limits, and plot a control chart. Additional summaries and analyses can be provided such as a listing of raw data, out-of-specification values, histograms, checks for runs and other patterns on a control chart, Pareto charts, and process capability calculations.

While specialized software is required for many statistical calculations, control charts are sufficiently simple that they can readily be incorporated into spreadsheets installed on most personal computers.

✧ 20.16 WORKED EXAMPLE USING MINITAB®

The U.S. FDA establishes standards regarding allowable levels of insect fragments within foods. A chocolate manufacturer collected data from its product line over time; with 100 gram samples and subgroup sizes of 6 taken once each day for 60 days. Is the number of insect parts in statistical control? Comment on any patterns. Assuming a food defect action level of 60 or more insect fragments per 100 grams of chocolate (average of subgroups), is the process meeting specification?

(Gryna 022805.MPJ; Columns C19-20, Subgroup, Fragments)

The number of insect parts detected is in statistical control, and no clear patterns are evident as seen in Figure 20.19. However, from Figure 20.20 the process is not capable, as evidenced by the C_{pk} of 0.28. To ensure product more consistently meets FDA standards, the process should be improved to reduce variation in insect parts.

Tools to Maintain the Control of Quality 695

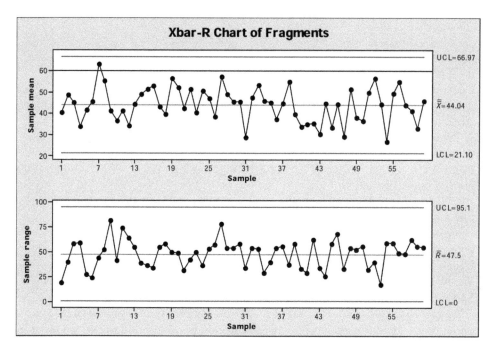

FIGURE 20.19

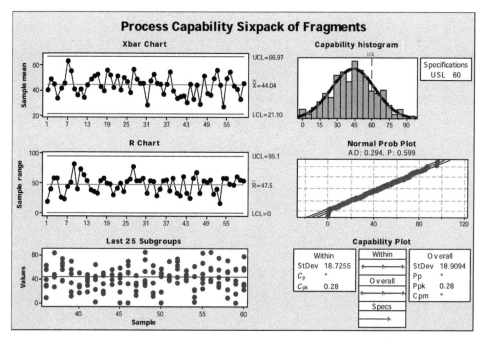

FIGURE 20.20

SUMMARY

- Statistical process control is the application of statistical methods to the measurement and analysis of any process.
- Decreased process variability has important advantages.
- A statistical control chart is a graphic comparison of process performance data to computed statistical control limits drawn as limit lines on the chart.
- A control chart distinguishes between common causes and special causes of variation.
- A state of statistical control has important advantages for any process.
- Control charts come in many types for both variable or continuous data and attribute or categorical data.
- PRE-Control is a statistical technique for detecting process conditions and changes that may cause defects.
- Process capability is the measured, inherent reproducibility of a product turned out by a process.
- Capability ratios help to quantify process capability.
- We want processes that are in statistical control and capable of meeting specification limits.
- The six sigma approach is a strategy for improvement. The goal is for process variation so small that the specification limits are 6σ above and below the process average.

PROBLEMS

20.1. A large manufacturer of watches makes some of its own parts and buys other parts from a supplier. The supplier submits lots of parts that meet the specifications of the horologist. The supplier thus wishes to keep a continuous check on its production of watch parts. One gear has been a special problem. A check of 25 samples of five pieces gave the following data on a key dimension:

$$\bar{\bar{X}} = 0.3175 \text{ cm} \quad \bar{R} = 0.00508 \text{ cm}$$

What criterion should be set up to determine when the process is out of control? How should this criterion compare with the specification? What are some alternatives if the criterion is not compatible with the specification?

20.2. A manufacturer of dustless chalk is concerned with the density of the product. Previous analysis has shown that the chalk has the required characteristics only if the density is between 4.4 g/cm² and 5.04 g/cm². If a sample of 100 pieces gives an average of 4.8 and a standard deviation of 0.2, is the process aimed at the proper density? If not, what should the aim be? Is the process capable of meeting the density requirements? Calculate C_p and C_{pk}.

20.3. The head of an automobile engine must be machined so that both the surface that meets the engine block and the surface that meets the valve covers are flat. These surfaces must also be 4.875 ± 0.001 in. apart. Presuming that the valve-cover side of the head is finished correctly, compare the capability of two processes for performing the finishing of the engine-block side of the head. A broach set up to do the job gave an average thickness of 4.877 in. with an average range of 0.0005 in. for 25 samples of four each. A milling machine gave an average of 4.875 in. and average range of 0.001 in. for 20 samples of four each. For each machine, calculate C_p and C_{pk}.

Answer: Broach $\pm 3s = \pm 0.00072$, milling machine $\pm 3s = 0.00144$.

20.4. A critical dimension on a double-armed armature has been causing trouble, and the designer has decided to change the specification from 0.033 ± 0.005 in. to 0.033 ± 0.001 in. To evaluate the proposed change, the manufacturing planning department has obtained the following coded data from the process:

Time	Left arm			Right arm			Comments
8:00	331	330	331	329	330	328	
8:30	332	331	329	327	331	329	
9:00	330	329	329	330	329	327	
9:30	332	330	331	331	332	328	
10:00	333	332	333	326	331	326	
10:30	332	331	332	329	330	331	
11:00	333	331	331	330	326	327	
11:30	332	332	333	327	326	329	
12:00	331	332	334	337	328	337	Adjustment
12:30	335	334	336	326	325	325	
1:00	333	332	332	329	332	330	
1:30	336	331	330	331	328	329	
2:00	332	334	329	332	330	329	
2:30	336	336	330	329	329	327	
3:00	329	335	338	333	330	331	Adjustment
3:30	341	333	330	329	331	332	

Comment on the proposal to change the specification.

20.5. A company manufactures an expensive chemical. The net package weight has a minimum specification value of 25.0 lb. Data from a control chart analysis show (based on 20 samples of five each):

$$\bar{\bar{X}} = 26.0 \quad \bar{R} = 1.4$$

The points on both the average and range chart are all in control.
(a) Draw conclusions about the ability of the process to meet the specification.
(b) What action, if any, would you suggest on the process? If any action is suggested, are there any disadvantages to the action?

20.6. Samples of four units were taken from a manufacturing process at regular intervals. The width of a slot on a part was measured, and the average and range were computed for each sample. After 25 samples of four, the following coded results were obtained:

	Averages	Ranges
Upper control limit	626	37.5
Average value	$\bar{\bar{X}} = 614$	$\bar{R} = 16.5$
Lower control limit	602	0

All points on the $\overline{X}$, and R charts fell within control limits. The specification requirements are 610 ± 15. If the width is normally distributed and the distribution is centered at $\overline{X}$, what percentage of product would you expect to fall outside the specification limits?

Answer: 9.4%.

20.7. (a) What conclusion about a process can be made from an average and range control chart that cannot be made from a histogram?
(b) What conclusion can be made from a histogram that cannot be made from an average and range control chart?

20.8. The percentage of water absorption is an important characteristic of common building brick. A certain company occasionally measured this characteristic of its product but never kept records. Management decided to analyze the process with a control chart. Twenty-five samples of four bricks each yielded these results:

Sample number	$\overline{X}$	R	Sample number	$\overline{X}$	R
1	15.1	9.1	14	9.8	17.5
2	12.3	9.9	15	8.8	10.5
3	7.4	9.7	16	8.1	4.4
4	8.7	6.7	17	6.3	4.1
5	8.8	7.1	18	10.5	5.7
6	11.7	9.1	19	9.7	6.4
7	10.2	12.1	20	11.7	4.6
8	11.5	10.8	21	13.2	7.2
9	11.2	13.5	22	12.5	8.3
10	10.2	6.9	23	7.5	6.4
11	9.6	5.0	24	8.8	6.9
12	7.6	8.2	25	8.0	6.4
13	7.6	5.4			

Plot the data on an average and range control chart with control limits. Comment.

Answer: 15.8 and 3.97; 18.4 and 0.

20.9. The specification for a certain dimension is 3.000 ± 0.004 in. A large sample from the process indicates an average of 2.998 in. and a standard deviation of 0.002 in. Suppose that controls are instituted to shift the process average to the nominal specification of 3.000. Each part outside specification limits results in a loss of $5. In a lot of 1000 parts, how much money would be saved by shifting the average versus keeping it at 2.998?

Answer: $573.

20.10. A statistical control chart for averages and ranges has been used to help control a manufacturing process. The sample data are consistently within control limits, and the control limits are inside the engineering tolerance limits. The supervisor is confused because a high percentage of product is outside the tolerance limits, even though the process is within control limits. What is your explanation?

20.11. The following data represent the number of defects found on each sewing machine cabinet inspected:

Sample number	Number of defects	Sample number	Number of defects
1	8	14	6
2	10	15	4
3	7	16	7
4	7	17	5
5	8	18	8
6	6	19	6
7	9	20	4
8	8	21	5
9	4	22	7
10	7	23	4
11	9	24	5
12	6	25	5
13	5		

Plot a control chart with control limits. Comment on the chart.

20.12. A sample of 100 electrical connectors was inspected each shift. Three characteristics were inspected on each connector, but each connector was classified simply as defective or acceptable. The results follow:

Sample number	Defective, %	Sample number	Defective, %
1	4	14	4
2	3	15	4
3	5	16	5
4	6	17	3
5	7	18	0
6	5	19	3
7	4	20	2
8	2	21	1
9	5	22	3
10	6	23	4
11	4	24	2
12	3	25	2
13	3		

(a) Plot a control chart with control limits. Comment on the chart.
(b) If the inspection results had been recorded in sufficient detail, what other type of chart could also have been plotted?

Answer: (a) 9.2%, 0.

20.13. An average and range chart based on a sample size of five has been run with the following results:

	Averages	Ranges
Upper control limit	78.0	8.0
Average value	75.8	3.8
Lower control limit	73.6	0

How large an increase in the overall process average would have to occur to have a 30% chance that a sample average will exceed the upper control limit?

Answer: 1.83.

20.14. As part of a quality improvement program, a textile manufacturer decides to use a control chart to monitor the number of imperfections in bolts of cloth. The data from the last 25 inspections are recorded in the following table. From these data, compute control limits for an appropriate type of control chart. Plot the chart.

Bolt of cloth	Number of imperfections	Bolt of cloth	Number of imperfections
1	14	14	22
2	5	15	1
3	10	16	6
4	19	17	14
5	0	18	8
6	6	19	6
7	2	20	9
8	9	21	7
9	8	22	1
10	7	23	5
11	3	24	12
12	12	25	4
13	1	Total	191

Answer: Limits are 15.93 and 0.

20.15. Control chart data were collected for the softening point (in degrees) in a polymerization process. Based on 25 samples of four each, the following control limits were calculated:

	Averages	Ranges
Upper control limit	12.9	13.4
Average value	9.4	5.9
Lower control limit	5.9	0

Suppose that the population average shifts to 12.4. How large a sample is necessary to have a 25% probability that the control chart for averages will signal out of control?

20.16. A p chart is to be used to analyze the September record for 100% inspection of certain radio components. The total number inspected during the month was 2196, and the total number of defectives was 158. Compute $\bar{p}$ Compute control limits for the following 3 days and state whether the percentage defective falls within the control limits for each day.

Date	Number inspected	Number of defectives
Sept. 14	54	8
Sept. 15	162	24
Sept. 16	213	3

20.17. A control chart with upper and lower control limits can detect when a process is out of control. It cannot, however, identify the causes. Select a process from a service industry and describe an approach to identify the variable(s) that caused an out-of-control process.

20.18. What is the main objective of statistical control for evaluating applications for mortgage loans at a financial institution?

REFERENCES

Anthis, D. L., R. F. Hart, and R. J. Stanula (1991). "The Measurement Process: Roadblock to Product Improvement," *Quality Engineering,* vol. 3, no. 4, pp. 461–470.

AT&T (1990). *Analyzing Business Process Data: The Looking Glass,* AT&T Customer Information Center, Indianapolis, IN.

Baxter Travenol Laboratories (1986). *Special Process Control Guideline,* Deerfield, IL, p. 17.

Bersimis, S., S. Psarakis, and J. Panaretos (2007). "Multivariate Statistical Process Control Charts: An Overview," *Quality and Reliability Engineering International,* vol. 23, pp. 517–543.

Bothe, D. R. (1997). *Measuring Process Capability,* McGraw-Hill, New York.

Box, G. E. P. and A. Luceno (1997). *Statistical Control by Monitoring and Adjustment,* 3rd ed., John Wiley and Sons, New York.

Dechert, J., K. E. Case, and T. L. Kautiainen (2000). "Statistical Process Control in the Presence of Large Measurement Variation," *Quality Engineering,* vol. 12, no. 3, pp. 417–423.

Deming, W. E. (1986). *Out of the Crisis,* Massachusetts Institute of Technology, Center for Advanced Engineering Study, Cambridge, MA.

Heyes, G. B. (1988) "Do We Need New Machines? A p-Chart and Regression Study," *Quality Engineering,* vol. 1, no. 1, pp. 13–28.

Hinckley, C. M. and P. Barkan (1995). "The Role of Variation, Mistakes, and Complexity in Producing Nonconformities," *Journal of Quality Technology,* vol. 27, no. 3, pp. 242–249.

Kane, V. E. (1986). "Process Capability Indices," *Journal of Quality Technology,* vol. 18, no. 1, pp. 41–52.

Ledolter, J. and A. Swersey (1997). "An Evaluation of PRE-Control," *Journal of Quality Technology,* vol. 29, no. 2, pp. 163–171.

Leonard, J. F. (1986). "Quality Improvement in Recruiting and Employment," *Juran Report Number Six,* Winter, Juran Institute, Inc., Wilton, CT, pp. 111–118.

McFadden, F. R. (1993). "Six Sigma Quality Programs," *Quality Progress,* June, pp. 37–42.

Nelson, L. S. (1984). "The Shewhart Control Chart-Tests for Special Causes," *Journal of Quality Technology,* vol. 16, no. 4, October, pp. 237–239.

Nelson, L. S. (1985). "Interpreting Shewhart Charts," *Journal of Quality Technology,* vol. 17, no. 2, pp. 114–116.

Ott, E. R. and E. G. Schilling (1990). *Process Quality Control,* McGraw-Hill, New York.

Pohlen, T. (1999). "Statistical Thinking: A Personal Application," *Proceedings of the Annual Quality Congress Proceedings,* ASQ, Milwaukee, pp. 230–235.

Schilling, E. G. (1990). "Elements of Process Control," *Quality Engineering,* vol. 2, no. 2, p. 132. Reprinted by courtesy of Marcel Dekker, Inc.

Steiner, S. H. (1997). "PRE-Control and Some Simple Alternatives," *Quality Engineering,* vol. 10, no. 1, pp. 65–74.

Sullivan, L. P. (1984). "Reducing Variability: A New Approach to Quality," *Quality Progress,* July, pp. 15–21.

Appendix I

Tables

Table A: Normal distribution
Table B: Exponential distribution values of $e^{-X/\mu}$ for various values
Table C: Poisson distribution
Table D: Distribution of t
Table E: Distribution of χ^2
Table F: Ninety-five percent confidence belts for population proportion
Table G: Distribution of F
Table H: Tolerance factors for normal distributions (two sided)
Table I: Factors for $\bar{X}$ and R control charts; factors for estimating s from R

TABLE A
Normal distribution

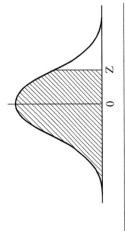

Proportion of total area under the curve from $-\infty$ to $Z = \dfrac{X - \mu}{\sigma}$. To illustrate: when $Z = 2$, the probability is .9773 of obtaining a value equal to or less than X.

Z	0.09	0.08	0.07	0.06	0.05	0.04	0.03	0.02	0.01	0.00
-3.0	.00100	.00104	.00107	.00111	.00114	.00118	.00122	.00126	.00131	.00135
-2.9	.0014	.0014	.0015	.0015	.0016	.0016	.0017	.0017	.0018	.0019
-2.8	.0019	.0020	.0021	.0021	.0022	.0023	.0023	.0024	.0025	.0026
-2.7	.0026	.0027	.0028	.0029	.0030	.0031	.0032	.0033	.0034	.0035
-2.6	.0036	.0037	.0038	.0039	.0040	.0041	.0043	.0044	.0045	.0047
-2.5	.0048	.0049	.0051	.0052	.0054	.0055	.0057	.0059	.0060	.0062
-2.4	.0064	.0066	.0068	.0069	.0071	.0073	.0075	.0078	.0080	.0082
-2.3	.0084	.0087	.0089	.0091	.0094	.0096	.0099	.0102	.0104	.0107
-2.2	.0110	.0113	.0116	.0119	.0122	.0125	.0129	.0132	.0136	.0139
-2.1	.0143	.0146	.0150	.0154	.0158	.0162	.0166	.0170	.0174	.0179
-2.0	.0183	.0188	.0192	.0197	.0202	.0207	.0212	.0217	.0222	.0228
-1.9	.0233	.0239	.0244	.0250	.0256	.0262	.0268	.0274	.0281	.0287
-1.8	.0294	.0301	.0307	.0314	.0322	.0329	.0336	.0344	.0351	.0359
-1.7	.0367	.0375	.0384	.0392	.0401	.0409	.0418	.0427	.0436	.0446
-1.6	.0455	.0465	.0475	.0485	.0495	.0505	.0516	.0526	.0537	.0548
-1.5	.0559	.0571	.0582	.0594	.0606	.0618	.0630	.0643	.0655	.0668
-1.4	.0681	.0694	.0708	.0721	.0735	.0749	.0764	.0778	.0793	.0808
-1.3	.0823	.0838	.0853	.0869	.0885	.0901	.0918	.0934	.0951	.0968
-1.2	.0985	.1003	.1020	.1038	.1057	.1075	.1093	.1112	.1131	.1151
-1.1	.1170	.1190	.1210	.1230	.1251	.1271	.1292	.1314	.1335	.1357

Z	0.00	0.01	0.02	0.03	0.04	0.05	0.06	0.07	0.08	0.09
−1.0	.1379	.1401	.1423	.1446	.1469	.1492	.1515	.1539	.1562	.1587
−0.9	.1611	.1635	.1660	.1685	.1711	.1736	.1762	.1788	.1814	.1841
−0.8	.1867	.1894	.1922	.1949	.1977	.2005	.2033	.2061	.2090	.2119
−0.7	.2148	.2177	.2207	.2236	.2266	.2297	.2327	.2358	.2389	.2420
−0.6	.2451	.2483	.2514	.2546	.2578	.2611	.2643	.2676	.2709	.2743
−0.5	.2776	.2810	.2843	.2877	.2912	.2946	.2981	.3015	.3050	.3085
−0.4	.3121	.3156	.3192	.3228	.3264	.3300	.3336	.3372	.3409	.3446
−0.3	.3483	.3520	.3557	.3594	.3632	.3669	.3707	.3745	.3783	.3821
−0.2	.3859	.3897	.3936	.3974	.4013	.4052	.4090	.4129	.4168	.4207
−0.1	.4247	.4286	.4325	.4364	.4404	.4443	.4483	.4562	.1562	.4602
−0.0	.4641	.4681	.4721	.4761	.4801	.4840	.4880	.4920	.4960	.5000

Z	0.00	0.01	0.02	0.03	0.04	0.05	0.06	0.07	0.08	0.09
+0.0	.5000	.5040	.5080	.5120	.5160	.5199	.5239	.5279	.5319	.5359
+0.1	.5398	.5438	.5478	.5517	.5557	.5596	.5636	.5675	.5714	.5753
+0.2	.5793	.5832	.5871	.5910	.5948	.5987	.6026	.6064	.6103	.6141
+0.3	.6179	.6217	.6255	.6293	.6331	.6368	.6406	.6443	.6480	.6517
+0.4	.6554	.6591	.6628	.6664	.6700	.6736	.6772	.6808	.6844	.6879
+0.5	.6915	.6950	.6985	.7019	.7054	.7088	.7123	.7157	.7190	.7224
+0.6	.7257	.7291	.7324	.7357	.7389	.7422	.7454	.7486	.7517	.7549
+0.7	.7580	.7611	.7642	.7673	.7704	.7734	.7764	.7794	.7823	.7852
+0.8	.7881	.7910	.7939	.7967	.7995	.8023	.8051	.8079	.8106	.8133
+0.9	.8159	.8186	.8212	.8238	.8264	.8289	.8315	.8340	.8365	.8389
+1.0	.8413	.8438	.8461	.8485	.8508	.8531	.8554	.8577	.8599	.8621
+1.1	.8643	.8665	.8686	.8708	.8729	.8749	.8770	.8790	.8810	.8830
+1.2	.8849	.8869	.8888	.8907	.8925	.8944	.8962	.8980	.8997	.9015
+1.3	.9032	.9049	.9066	.9082	.9099	.9115	.9131	.9147	.9162	.9177
+1.4	.9192	.9207	.9222	.9236	.9251	.9265	.9279	.9292	.9306	.9319
+1.5	.9332	.9345	.9357	.9370	.9382	.9394	.9406	.9418	.9429	.9441

(*continued*)

TABLE A (continued)

Z	0.00	0.01	0.02	0.03	0.04	0.05	0.06	0.07	0.08	0.09
+1.6	.9452	.9463	.9474	.9484	.9495	.9505	.9515	.9525	.9535	.9545
+1.7	.9554	.9564	.9573	.9582	.9591	.9599	.9608	.9616	.9625	.9633
+1.8	.9641	.9649	.9656	.9664	.9671	.9678	.9686	.9693	.9699	.9706
+1.9	.9713	.9719	.9726	.9732	.9738	.9744	.9750	.9756	.9761	.9767
+2.0	.9773	.9778	.9783	.9788	.9793	.9798	.9803	.9808	.9812	.9817
+2.1	.9821	.9826	.9830	.9834	.9838	.9842	.9846	.9850	.9854	.9857
+2.2	.9861	.9864	.9868	.9871	.9875	.9878	.9881	.9884	.9887	.9890
+2.3	.9893	.9896	.9898	.9901	.9904	.9906	.9909	.9911	.9913	.9916
+2.4	.9918	.9920	.9922	.9925	.9927	.9929	.9931	.9932	.9934	.9936
+2.5	.9938	.9940	.9941	.9943	.9945	.9946	.9948	.9949	.9951	.9952
+2.6	.9953	.9955	.9956	.9957	.9959	.9960	.9961	.9962	.9963	.9964
+2.7	.9965	.9966	.9967	.9968	.9969	.9970	.9971	.9972	.9973	.9974
+2.8	.9974	.9975	.9976	.9977	.9977	.9978	.9979	.9979	.9980	.9981
+2.9	.9981	.9982	.9983	.9983	.9984	.9984	.9985	.9985	.9986	.9986
+3.0	.99865	.99869	.99874	.99878	.99882	.99886	.99889	.99893	.99896	.99900

"Table A—Normal Distribution" from *Statistical Quality Control, 4E* by Eugene L. Grant and Richard S. Leavenworth. © 1972 by The McGraw-Hill Companies, Inc. Reprinted by permission of The McGraw-Hill Companies, Inc.

And we can keep going:

Z	0.00
+4.0	.9999683
+5.0	.9999997133
+6.0	.9999999990

TABLE B
Exponential distribution values of $e^{-X/\mu}$ for various values

Fractional parts of the total area (1.000) under the exponential curve greater than X. To illustrate: if X/μ is 0.45, the probability of occurrence for a value greater than X is 0.6376.

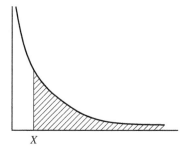

$\frac{X}{\mu}$	0.00	0.01	0.02	0.03	0.04	0.05	0.06	0.07	0.08	0.09
0.0	1.000	.9900	.9802	.9704	.9608	.9512	.9418	.9324	.9231	.9139
0.1	.9048	.8958	.8860	.8781	.8694	.8607	.8521	.8437	.8353	.8270
0.2	.8187	.8106	.8025	.7945	.7866	.7788	.7711	.7634	.7558	.7483
0.3	.7408	.7334	.7261	.7189	.7118	.7047	.6977	.6907	.6839	.6771
0.4	.6703	.6637	.6570	.6505	.6440	.6376	.6313	.6250	.6188	.6126
0.5	.6065	.6005	.5945	.5886	.5827	.5769	.5712	.5655	.5599	.5543
0.6	.5488	.5434	.5379	.5326	.5273	.5220	.5169	.5117	.5066	.5016
0.7	.4966	.4916	.4868	.4819	.4771	.4724	.4877	.4630	.4584	.4538
0.8	.4493	.4449	.4404	.4360	.4317	.4274	.4232	.4190	.4148	.4107
0.9	.4066	.4025	.3985	.3946	.3906	.3867	.3829	.3791	.3753	.3716

$\frac{X}{\mu}$	0.0	0.1	0.2	0.3	0.4	0.5	0.6	0.7	0.8	0.9
1.0	.3679	.3329	.3012	.2725	.2466	.2231	.2019	.1827	.1653	.1496
2.0	.1353	.1225	.1108	.1003	.0907	.0821	.0743	.0672	.0608	.0550
3.0	.0498	.0450	.0408	.0369	.0334	.0302	.0273	.0247	.0224	.0202
4.0	.0183	.0166	.0150	.0130	.0123	.0111	.0101	.0091	.0082	.0074
5.0	.0067	.0061	.0055	.0050	.0045	.0041	.0037	.0033	.0030	.0027
6.0	.0025	.0022	.0020	.0018	.0017	.0015	.0014	.0012	.0011	.0010

Source: Adapted from S. M. Selby, ed., *CRC Standard Mathematical Tables,* 17th ed., CRC Press, Cleveland, OH, 1969, pp. 201–207.

TABLE C
Poisson distribution

$1000 \times$ probability of r or fewer occurrences of event that has average number of occurrences equal to np.

np	0	1	2	3	4	5	6	7	8	9
0.02	980	1000								
0.04	961	999	1000							
0.06	942	998	1000							
0.08	923	997	1000							
0.10	905	995	1000							
0.15	861	990	999	1000						
0.20	819	982	999	1000						
0.25	779	974	998	1000						
0.30	741	963	996	1000						
0.35	705	951	994	1000						
0.40	670	938	992	999	1000					
0.45	638	925	989	999	1000					
0.50	607	910	986	998	1000					
0.55	577	894	982	998	1000					
0.60	549	878	977	997	1000					
0.65	522	861	972	996	999	1000				
0.70	497	844	966	994	999	1000				
0.75	472	827	959	993	999	1000				
0.80	449	809	953	991	999	1000				
0.85	427	791	945	989	998	1000				
0.90	407	772	937	987	998	1000				
0.95	387	754	929	984	997	1000				
1.00	368	736	920	981	996	999	1000			
1.1	333	699	900	974	995	999	1000			
1.2	301	663	879	966	992	998	1000			
1.3	273	627	857	957	989	998	1000			
1.4	247	592	833	946	986	997	999	1000		
1.5	223	558	809	934	981	996	999	1000		
1.6	202	525	783	921	976	994	999	1000		
1.7	183	493	757	907	970	992	998	1000		
1.8	165	463	731	891	964	990	997	999	1000	
1.9	150	434	704	875	956	987	997	999	1000	
2.0	135	406	677	857	947	983	995	999	1000	
2.2	111	355	623	819	928	975	993	998	1000	
2.4	091	308	570	779	904	964	988	997	999	1000
2.6	014	267	518	736	877	951	983	995	999	1000
2.8	061	231	469	692	848	935	976	992	998	999
3.0	050	199	423	647	815	916	966	988	996	999
3.2	041	171	380	603	781	895	955	983	994	998
3.4	033	147	340	558	744	871	942	977	992	997
3.6	027	126	303	515	706	844	927	969	988	996
3.8	022	107	269	473	668	816	909	960	984	994
4.0	018	092	238	433	629	785	889	949	979	992

TABLE C (*continued*)

np \ r	0	1	2	3	4	5	6	7	8	9
4.2	015	078	210	395	590	753	867	936	972	989
4.4	012	066	185	359	551	720	844	921	964	985
4.6	010	056	163	326	513	686	818	905	955	980
4.8	008	048	143	294	476	651	791	887	944	975
5.0	007	040	125	265	440	616	762	867	932	968
5.2	006	034	109	238	406	581	732	845	918	960
5.4	005	029	095	213	373	546	702	822	903	951
5.6	004	024	082	191	342	512	670	797	886	941
5.8	003	021	072	170	313	478	638	771	867	929
6.0	002	017	062	151	285	446	606	744	847	916

np \ r	10	11	12	13	14	15	16			
2.8	1000									
3.0	1000									
3.2	1000									
3.4	999	1000								
3.6	999	1000								
3.8	998	999	1000							
4.0	997	999	1000							
4.2	996	999	1000							
4.4	994	998	999	1000						
4.6	992	997	999	1000						
4.8	990	996	999	1000						
5.0	986	995	998	999	1000					
5.2	982	993	997	999	1000					
5.4	977	990	996	999	1000					
5.6	972	988	995	998	999	1000				
5.8	965	984	993	997	999	1000				
6.0	957	980	991	996	999	999	1000			
6.2	002	015	054	134	259	414	574	716	826	902
6.4	002	012	046	119	235	384	542	687	803	886
6.6	001	010	040	105	213	355	511	658	780	869
6.8	001	009	034	093	192	327	480	628	755	850
7.0	001	007	030	082	173	301	450	599	729	830
7.2	001	006	025	072	156	276	420	569	703	810
7.4	001	005	022	063	140	253	392	539	676	788
7.6	001	004	019	055	125	231	365	510	648	765
7.8	000	004	016	048	112	210	338	481	620	741
8.0	000	003	014	042	100	191	313	453	593	717
8.5	000	002	009	030	074	150	256	386	523	653
9.0	000	001	006	021	055	116	207	324	456	587
9.5	000	001	004	015	040	089	165	269	392	522
10.0	000	000	003	010	029	067	130	220	333	458

(*continued*)

TABLE C (*continued*)

np \ r	10	11	12	13	14	15	16	17	18	19
6.2	949	975	989	995	998	999	1000			
6.4	939	969	986	994	997	999	1000			
6.6	927	963	982	992	997	999	999	1000		
6.8	915	955	978	990	996	998	999	1000		
7.0	901	947	973	987	994	998	999	1000		
7.2	887	937	967	984	993	997	999	999	1000	
7.4	871	926	961	980	991	996	998	999	1000	
7.6	854	915	954	976	989	995	998	999	1000	
7.8	835	902	945	971	986	993	997	999	1000	
8.0	816	888	936	966	983	992	996	998	999	1000
8.5	763	849	909	949	973	986	993	997	999	999
9.0	706	803	876	926	959	978	989	995	998	999
9.5	645	752	836	898	940	967	982	991	996	998
10.0	583	697	792	864	917	951	973	986	993	997

np \ r	20	21	22
8.5	1000		
9.0	1000		
9.5	999	1000	
10.0	998	999	1000

"Table C—Poisson Distribution" from *Statistical Quality Control, 4E* by Eugene L. Grant and Richard S. Leavenworth. © 1972 by The McGraw-Hill Companies, Inc. Reprinted by permission of The McGraw-Hill Companies, Inc.

TABLE D
Distribution of t

Value of t corresponding to certain selected probabilities (i.e., tail areas under the curve). To illustrate: the probability is .975 that a sample with 20 degrees of freedom would have $t = +2.086$ or smaller.

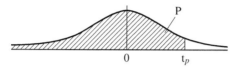

DF	$t_{.60}$	$t_{.70}$	$t_{.80}$	$t_{.90}$	$t_{.95}$	$t_{.975}$	$t_{.99}$	$t_{.995}$
1	0.325	0.727	1.376	3.078	6.314	12.706	31.821	63.657
2	0.289	0.617	1.061	1.886	2.920	4.303	6.965	9.925
3	0.277	0.584	0.978	1.638	2.353	3.182	4.541	5.841
4	0.271	0.569	0.941	1.533	2.132	2.776	3.747	4.604
5	0.267	0.559	0.920	1.476	2.015	2.571	3.365	4.032
6	0.265	0.553	0.906	1.440	1.943	2.447	3.143	3.707
7	0.263	0.549	0.896	1.415	1.895	2.365	2.998	3.499
8	0.262	0.546	0.889	1.397	1.860	2.306	2.896	3.355
9	0.261	0.543	0.883	1.383	1.833	2.262	2.821	3.250
10	0.260	0.542	0.879	1.372	1.812	2.228	2.764	3.169
11	0.260	0.540	0.876	1.363	1.796	2.201	2.718	3.106
12	0.259	0.539	0.873	1.356	1.782	2.179	2.681	3.055
13	0.259	0.538	0.870	1.350	1.771	2.160	2.650	3.012
14	0.258	0.537	0.868	1.345	1.761	2.145	2.624	2.977
15	0.258	0.536	0.866	1.341	1.753	2.131	2.602	2.947
16	0.258	0.535	0.865	1.337	1.746	2.120	2.583	2.921
17	0.257	0.534	0.863	1.333	1.740	2.110	2.567	2.898
18	0.257	0.534	0.862	1.330	1.734	2.101	2.552	2.878
19	0.257	0.533	0.861	1.328	1.729	2.093	2.539	2.861
20	0.257	0.5.33	0.860	1.325	1.725	2.086	2.528	2.845
21	0.257	0.532	0.859	1.323	1.721	2.080	2.518	2.831
22	0.256	0.532	0.858	1.321	1.717	2.074	2.508	2.819
23	0.256	0.532	0.858	1.319	1.714	2.069	2.500	2.807
24	0.256	0.531	0.857	1.318	1.711	2.064	2.492	2.797
25	0.256	0.531	0.856	1.316	1.708	2.060	2.485	2.787
26	0.256	0.531	0.856	1.315	1.706	2.056	2.479	2.779
27	0.256	0.531	0.855	1.314	1.703	2.052	2.473	2.771
28	0.256	0.530	0.855	1.313	1.701	2.048	2.467	2.763
29	0.256	0.530	0.854	1.311	1.699	2.045	2.462	2.756
30	0.256	0.530	0.854	1.310	1.697	2.042	2.457	2.750
40	0.255	0.529	0.851	1.303	1.684	2.021	2.423	2.704
60	0.254	0.527	0.848	1.296	1.671	2.000	2.390	2.660
120	0.254	0.526	0.845	1.289	1.658	1.980	2.358	2.617
∞	0.253	0.524	0.842	1.282	1.645	1.960	2.326	2.576

"Table D—Distribution of t" from *Introduction to Statistical Analysis, 3E* by W. J. Dixon and F. J. Massey, Jr.. © 1969 by The McGraw-Hill Companies, Inc. Reprinted by permission of The McGraw-Hill Companies, Inc.

TABLE E
Distribution of χ^2

Values of χ^2 corresponding to certain selected probabilities (i.e., tail areas under the curve). To illustrate: the probability is .95 that a sample with 20 degrees of freedom, taken from a normal distribution, would have $\chi^2 = 31.41$ or smaller.

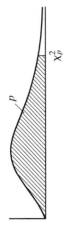

Values of χ_P^2 corresponding to P

DF	$\chi^2_{.005}$	$\chi^2_{.01}$	$\chi^2_{.025}$	$\chi^2_{.05}$	$\chi^2_{.10}$	$\chi^2_{.90}$	$\chi^2_{.95}$	$\chi^2_{.975}$	$\chi^2_{.99}$	$\chi^2_{.995}$
1	0.000039	0.00016	0.00098	0.0039	0.0158	2.71	3.84	5.02	6.63	7.88
2	0.0100	0.0201	0.0506	0.1026	0.2107	4.61	5.99	7.38	9.21	10.60
3	0.0717	0.115	0.216	0.352	0.584	6.25	7.81	9.35	11.34	12.84
4	0.207	0.297	0.484	0.711	1.064	7.78	9.49	11.14	13.28	14.86
5	0.412	0.554	0.831	1.15	1.61	9.24	11.07	12.83	15.09	16.75
6	0.676	0.872	1.24	1.64	2.20	10.64	12.59	14.45	16.81	18.55
7	0.989	1.24	1.69	2.17	2.83	12.02	14.07	16.01	18.48	20.28
8	1.34	1.65	2.18	2.73	3.49	13.36	15.51	17.53	20.09	21.96
9	1.73	2.09	2.70	3.33	4.17	14.68	16.92	19.02	21.67	23.59
10	2.16	2.56	3.25	3.94	4.87	15.99	18.31	20.48	23.21	25.19
11	2.60	3.05	3.82	4.57	5.58	17.28	19.68	21.92	24.73	26.76
12	3.07	3.57	4.40	5.23	6.30	18.55	21.03	23.34	26.22	28.30
13	3.57	4.11	5.01	5.89	7.04	19.81	22.36	24.74	27.69	29.82
14	4.07	4.66	5.63	6.57	7.79	21.06	23.68	26.12	29.14	31.32
15	4.60	5.23	6.26	7.26	8.55	22.31	25.00	27.49	30.58	32.80
16	5.14	5.81	6.91	7.96	9.31	23.54	26.30	28.85	32.00	34.27
18	6.26	7.01	8.23	9.39	10.86	25.99	28.87	31.53	34.81	37.16
20	7.43	8.26	9.59	10.85	12.44	28.41	31.41	34.17	37.57	40.00
24	9.89	10.86	12.40	13.85	15.66	33.20	36.42	39.36	42.98	45.56
30	13.79	14.95	16.79	18.49	20.60	40.26	43.77	46.98	50.89	53.67
40	20.71	22.16	24.43	26.51	29.05	51.81	55.76	59.34	63.69	66.77
60	35.53	37.48	40.48	43.19	46.46	74.40	79.08	83.30	88.38	91.95
120	83.85	86.92	91.58	95.70	100.62	140.23	146.57	152.21	158.95	163.64

"Table E—Distribution of χ^2" from *Introduction to Statistical Analysis, 3E* by W. J. Dixon and F. J. Massey, Jr., © 1969 by The McGraw-Hill Companies, Inc. Reprinted by permission of The McGraw-Hill Companies, Inc.

TABLE F
Ninety-five percent confidence belts for population proportion

Example In a sample of 10 items, 8 were defective ($X/n = 8/10$). The 95% confidence limits on the population proportion defective are read from the two curves (for $n = 10$) as 0.43 and 0.98.

"Table F—Ninety-five percent confidence belts for population proportion" from *Selected Techniques of Statistical Analysis—OSRD* by C. Eisenhart, M. W. Hastay, and W. A. Wallis. Copyright 1947 by The McGraw-Hill Companies, Inc. Reprinted by permission of The McGraw-Hill Companies, Inc.

TABLE G
Distribution of F

Values of F corresponding to certain selected probabilities (i.e., tail areas under the curve). To illustrate: the probability is .05 that the ratio of two sample variances obtained with 20 and 10 degrees of freedom in numerator and denominator, respectively, would have $F = 2.77$ or larger. For a two-sided test, a lower limit is found by taking the reciprocal of the tabulated F value for the degrees of freedom in reverse. For the above example, with 10 and 20 degrees of freedom in numerator and denominator respectively, F is 2.35 and $1/F$ is 1/2.35, or .43. The probability is .10 that F is .43 or smaller or 2.77 or larger.

n_2 \ n_1	1	2	3	4	5	6	7	8	9
					$F_{.95}(n_1, n_2)$				
1	161.4	199.5	215.7	224.6	230.2	234.0	236.8	238.9	240.5
2	18.51	19.00	19.16	19.25	19.30	19.33	19.35	19.37	19.38
3	10.13	9.55	9.28	9.12	9.01	8.94	8.89	8.85	8.81
4	7.71	6.94	6.59	6.39	6.26	6.16	6.09	6.04	6.00
5	6.61	5.79	5.41	5.19	5.05	4.95	4.88	4.82	4.77
6	5.99	5.14	4.76	4.53	4.39	4.28	4.21	4.15	4.10
7	5.59	4.74	4.35	4.12	3.97	3.87	3.79	3.73	3.68
8	5.32	4.46	4.07	3.84	3.69	3.58	3.50	3.44	3.39
9	5.12	4.26	3.86	3.63	3.48	3.37	3.29	3.23	3.18
10	4.96	4.10	3.71	3.48	3.33	3.22	3.14	3.07	3.02
11	4.84	3.98	3.59	3.36	3.20	3.09	3.01	2.95	2.90
12	4.75	3.89	3.49	3.26	3.11	3.00	2.91	2.85	2.80
13	4.67	3.81	3.41	3.18	3.03	2.92	2.83	2.77	2.71
14	4.60	3.74	3.34	3.11	2.96	2.85	2.76	2.70	2.65
15	4.54	3.68	3.29	3.06	2.90	2.79	2.71	2.64	2.59
16	4.49	3.63	3.24	3.01	2.85	2.74	2.66	2.59	2.54
17	4.45	3.59	3.20	2.96	2.81	2.70	2.61	2.55	2.49
18	4.41	3.55	3.16	2.93	2.77	2.66	2.58	2.51	2.46
19	4.38	3.52	3.13	2.90	2.74	2.63	2.54	2.48	2.42
20	4.35	3.49	3.10	2.87	2.71	2.60	2.51	2.45	2.39
21	4.32	3.47	3.07	2.84	2.68	2.57	2.49	2.42	2.37
22	4.30	3.44	3.05	2.82	2.66	2.55	2.46	2.40	2.34
23	4.28	3.42	3.03	2.80	2.64	2.53	2.44	2.37	2.32
24	4.26	3.40	3.01	2.78	2.62	2.51	2.42	2.36	2.30
25	4.24	3.39	2.99	2.76	2.60	2.49	2.40	2.34	2.28
26	4.23	3.37	2.98	2.74	2.59	2.47	2.39	2.32	2.27
27	4.21	3.35	2.96	2.73	2.57	2.46	2.37	2.31	2.25
28	4.20	3.34	2.95	2.71	2.56	2.45	2.36	2.29	2.24
29	4.18	3.33	2.93	2.70	2.55	2.43	2.35	2.28	2.22
30	4.17	3.32	2.92	2.69	2.53	2.42	2.33	2.27	2.21
40	4.08	3.23	2.84	2.61	2.45	2.34	2.25	2.18	2.12
60	4.00	3.15	2.76	2.53	2.37	2.25	2.17	2.10	2.04
120	3.92	3.07	2.68	2.45	2.29	2.17	2.09	2.02	1.96
∞	3.84	3.00	2.60	2.37	2.21	2.10	2.01	1.94	1.88

Note: n_1 = degrees of freedom for numerator; n_2 = degrees of freedom for denominator.

Adapted from E. S. Pearson & H. O. Hartley, eds, *Biometrika Tables for Statisticians*, 2E, Vol. I, 1958. Reprinted by permission of Oxford University Press.

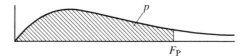

10	12	15	20	24	30	40	60	120	∞
				$F_{.95}(n_1, n_2)$					
241.9	243.9	245.9	248.0	249.1	250.1	251.1	252.2	253.3	254.3
19.40	19.41	19.43	19.45	19.45	19.46	19.47	19.48	19.49	19.50
8.79	8.74	8.70	8.66	8.64	8.62	8.59	8.57	8.55	8.53
5.96	5.91	5.86	5.80	5.77	5.75	5.72	5.69	5.66	5.63
4.74	4.68	4.62	4.56	4.53	4.50	4.46	4.43	4.40	4.36
4.06	4.00	3.94	3.87	3.84	3.81	3.77	3.74	3.70	3.67
3.64	3.57	3.51	3.44	3.41	3.38	3.34	3.30	3.27	3.23
3.35	3.28	3.22	3.15	3.12	3.08	3.04	3.01	2.97	2.93
3.14	3.07	3.01	2.94	2.90	2.86	2.83	2.79	2.75	2.71
2.98	2.91	2.85	2.77	2.74	2.70	2.66	2.62	2.58	2.54
2.85	2.79	2.72	2.65	2.61	2.57	2.53	2.49	2.45	2.40
2.75	2.69	2.62	2.54	2.51	2.47	2.43	2.38	2.34	2.30
2.67	2.60	2.53	2.46	2.42	2.38	2.34	2.30	2.25	2.21
2.60	2.53	2.46	2.39	2.35	2.31	2.27	2.22	2.18	2.13
2.54	2.48	2.40	2.33	2.29	2.25	2.20	2.16	2.11	2.07
2.49	2.42	2.35	2.28	2.24	2.19	2.15	2.11	2.06	2.01
2.45	2.38	2.31	2.23	2.19	2.15	2.10	2.06	2.01	1.96
2.41	2.34	2.27	2.19	2.15	2.11	2.06	2.02	1.97	1.92
2.38	2.31	2.23	2.16	2.11	2.07	2.03	1.98	1.93	1.88
2.35	2.28	2.20	2.12	2.08	2.04	1.99	1.95	1.90	1.84
2.32	2.25	2.18	2.10	2.05	2.01	1.96	1.92	1.87	1.81
2.30	2.23	2.15	2.07	2.03	1.98	1.94	1.89	1.84	1.78
2.27	2.20	2.13	2.05	2.01	1.96	1.91	1.86	1.81	1.76
2.25	2.18	2.11	2.03	1.98	1.94	1.89	1.84	1.79	1.73
2.24	2.16	2.09	2.01	1.96	1.92	1.87	1.82	1.77	1.71
2.22	2.15	2.07	1.99	1.95	1.90	1.85	1.80	1.75	1.69
2.20	2.13	2.06	1.97	1.93	1.88	1.84	1.79	1.73	1.67
2.19	2.12	2.04	1.96	1.91	1.87	1.82	1.77	1.71	1.65
2.18	2.10	2.03	1.94	1.90	1.85	1.81	1.75	1.70	1.64
2.16	2.09	2.01	1.93	1.89	1.84	1.79	1.74	1.68	1.62
2.08	2.00	1.92	1.84	1.79	1.74	1.69	1.64	1.58	1.51
1.99	1.92	1.84	1.75	1.70	1.65	1.59	1.53	1.47	1.39
1.91	1.83	1.75	1.66	1.61	1.55	1.50	1.43	1.35	1.25
1.83	1.75	1.67	1.57	1.52	1.46	1.39	1.32	1.22	1.00

TABLE H
Tolerance factors for normal distributions (two sided)

P \ N	$\gamma = 0.75$					$\gamma = 0.90$				
	0.75	0.90	0.95	0.99	0.999	0.75	0.90	0.95	0.99	0.999
2	4.498	6.301	7.414	9.531	11.920	11.407	15.978	18.800	24.167	30.227
3	2.501	3.538	4.187	5.431	6.844	4.132	5.847	6.919	8.974	11.309
4	2.035	2.892	3.431	4.471	5.657	2.932	4.166	4.943	6.440	8.149
5	1.825	2.599	3.088	4.033	5.117	2.454	3.494	4.152	5.423	6.879
6	1.704	2.429	2.889	3.779	4.802	2.196	3.131	3.723	4.870	6.188
7	1.624	2.318	2.757	3.611	4.593	2.034	2.902	3.452	4.521	5.750
8	1.568	2.238	2.663	3.491	4.444	1.921	2.743	3.264	4.278	5.446
9	1.525	2.178	2.593	3.400	4.330	1.839	2.626	3.125	4.098	5.220
10	1.492	2.131	2.537	3.328	4.241	1.775	2.535	3.018	3.959	5.046
11	1.465	2.093	2.493	3.271	4.169	1.724	2.463	2.933	3.849	4.906
12	1.443	2.062	2.456	3.223	4.110	1.683	2.404	2.863	3.758	4.792
13	1.425	2.036	2.424	3.183	4.059	1.648	2.355	2.805	3.682	4.697
14	1.409	2.013	2.398	3.148	4.016	1.619	2.314	2.756	3.618	4.615
15	1.395	1.994	2.375	3.118	3.979	1.594	2.278	2.713	3.562	4.545
16	1.383	1.977	2.355	3.092	3.946	1.572	2.246	2.676	3.514	4.484
17	1.372	1.962	2.337	3.069	3.917	1.552	2.219	2.643	3.471	4.430
18	1.363	1.948	2.321	3.048	3.891	1.535	2.194	2.614	3.433	4.382
19	1.355	1.936	2.307	3.030	3.867	1.520	2.172	2.588	3.399	4.339
20	1.347	1.925	2.294	3.013	3.846	1.506	2.152	2.564	3.368	4.300
21	1.340	1.915	2.282	2.998	3.827	1.493	2.135	2.543	3.340	4.264
22	1.334	1.906	2.271	2.984	3.809	1.482	2.118	2.524	3.315	4.232
23	1.328	1.898	2.261	2.971	3.793	1.471	2.103	2.506	3.292	4.203
24	1.322	1.891	2.252	2.950	3.778	1.462	2.089	2.480	3.270	4.176
25	1.317	1.883	2.244	2.948	3.764	1.453	2.077	2.474	3.251	4.151
26	1.313	1.877	2.236	2.938	3.751	1.444	2.065	2.460	3.232	4.127
27	1.309	1.871	2.229	2.929	3.740	1.437	2.054	2.447	3.215	4.106
30	1.297	1.855	2.210	2.904	3.708	1.417	2.025	2.413	3.170	4.049
35	1.283	1.834	2.185	2.871	3.667	1.390	1.988	2.368	3.112	3.974
40	1.271	1.818	2.166	2.846	3.635	1.370	1.959	2.334	3.066	3.917
100	1.218	1.742	2.075	2.727	3.484	1.275	1.822	2.172	2.854	3.646
500	1.177	1.683	2.006	2.636	3.368	1.201	1.717	2.046	2.689	3.434
1000	1.169	1.671	1.992	2.617	3.344	1.185	1.695	2.019	2.654	3.390
∞	1.150	1.645	1.960	2.576	3.291	1.150	1.645	1.960	2.576	3.291

"Table H—Tolerance factors for normal distributions" from *Selected Techniques of Statistical Analysis—OSRD* by C. Eisenhart, M. W. Hastay, and W. A. Wallis. Copyright 1947 by The McGraw-Hill Companies, Inc. Reprinted by permission of The McGraw-Hill Companies, Inc.

γ = confidence level
P = percentage of population within tolerance limits
N = number of values in sample

	$\gamma = 0.95$					$\gamma = 0.99$			
0.75	0.90	0.95	0.99	0.999	0.75	0.90	0.95	0.99	0.999
22.858	32.019	37.674	48.430	60.573	114.363	160.193	188.491	242.300	303.054
5.922	8.380	9.916	12.861	16.208	13.378	18.930	22.401	29.055	36.616
3.779	5.369	6.370	8.299	10.502	6.614	9.398	11.150	14.527	18.383
3.002	4.275	5.079	6.634	8.415	4.643	6.612	7.855	10.260	13.015
2.604	3.712	4.414	5.775	7.337	3.743	5.337	6.345	8.301	10.548
2.361	3.369	4.007	5.248	6.676	3.233	4.613	5.488	7.187	9.142
2.197	3.136	3.732	4.891	6.226	2.905	4.147	4.936	6.468	8.234
2.078	2.967	3.532	4.631	5.899	2.677	3.822	4.550	5.966	7.600
1.987	2.839	3.379	4.433	5.649	2.508	3.582	4.265	5.594	7.129
1.916	2.737	3.259	4.277	5.452	2.378	3.397	4.045	5.308	6.766
1.858	2.655	3.162	4.150	5.291	2.274	3.250	3.870	5.079	6.477
1.810	2.587	3.081	4.044	5.158	2.190	3.130	3.727	4.893	6.240
1.770	2.529	3.012	3.955	5.045	2.120	3.029	3.608	4.737	6.043
1.735	2.480	2.954	3.878	4.949	2.060	2.945	3.507	4.605	5.876
1.705	2.437	2.903	3.812	4.865	2.009	2.872	3.421	4.492	5.732
1.679	2.400	2.858	3.754	4.791	1.965	2.808	3.345	4.393	5.607
1.655	2.366	2.819	3.702	4.725	1.926	2.753	3.279	4.307	5.497
1.635	2.337	2.784	3.656	4.667	1.891	2.703	3.221	4.230	5.399
1.616	2.310	2.752	3.615	4.614	1.860	2.659	3.168	4.161	5.312
1.599	2.286	2.723	3.577	4.567	1.833	2.620	3.121	4.100	5.234
1.584	2.264	2.697	3.543	4.523	1.808	2.584	3.078	4.044	5.163
1.570	2.244	2.673	3.512	4.484	1.795	2.551	3.040	3.993	5.098
1.557	2.225	2.651	3.483	4.447	1.764	2.522	3.004	3.947	5.039
1.545	2.208	2.631	3.457	4.413	1.745	2.494	2.972	3.904	4.985
1.534	2.193	2.612	3.432	4.382	1.727	2.460	2.941	3.865	4.935
1.523	2.178	2.595	3.409	4.353	1.711	2.446	2.914	3.828	4.888
1.497	2.140	2.549	3.350	4.278	1.668	2.385	2.841	3.733	4.768
1.462	2.090	2.490	3.272	4.179	1.613	2.306	2.748	3.611	4.611
1.435	2.052	2.445	3.213	4.104	1.571	2.247	2.677	3.518	4.493
1.311	1.874	2.233	2.934	3.748	1.383	1.977	2.355	3.096	3.954
1.215	1.737	2.070	2.721	3.475	1.243	1.777	2.117	2.783	3.555
1.195	1.709	2.036	2.676	3.418	1.214	1.736	2.068	2.718	3.472
1.150	1.645	1.960	2.576	3.291	1.150	1.645	1.960	2.576	3.291

TABLE I
Factors for $\bar{X}$ and R control charts; factors for estimating s from $\bar{R}$†

$\begin{cases} \text{Upper control limit for } \bar{X} = \text{UCL}_{\bar{X}} = \bar{\bar{X}} + A_2 \bar{R} \\ \text{Lower control limit for } \bar{X} = \text{LCL}_{\bar{X}} = \bar{\bar{X}} - A_2 \bar{R} \end{cases}$

$\begin{cases} \text{Upper control limit for } R = \text{UCL}_R = D_4 \bar{R} \\ \text{Lower control limit for } R = \text{LCL}_R = D_3 \bar{R} \end{cases}$

$s = \bar{R}/d_2$

Number of observations in sample	A_2	D_3	D_4	Factor for estimate from $\bar{R}$ $d_2 = \bar{R}/s$
2	1.880	0	3.268	1.128
3	1.023	0	2.574	1.693
4	0.729	0	2.282	2.059
5	0.577	0	2.114	2.326
6	0.483	0	2.004	2.534
7	0.419	0.076	1.924	2.704
8	0.373	0.136	1.864	2.847
9	0.337	0.184	1.816	2.970
10	0.308	0.223	1.777	3.078
11	0.285	0.256	1.744	3.173
12	0.266	0.284	1.717	3.258
13	0.249	0.308	1.692	3.336
14	0.235	0.329	1.671	3.407
15	0.223	0.348	1.652	3.472

From *1950 ASTM Manual on Quality Control of Materials* and *ASTM Manual on Presentation of Data*, 1945, American Society for Testing and Materials. Copyright ASTM International. Reprinted with permission.

Name Index

Abetti, F., 144, 163
Allen, R.L., 25, 44
Anand, 90, 91
Anderson, D., 144, 163
Anderson, G.F., 140, 164
Arquette, Chris, 151
Atkinson, H., 55, 60, 103

Bailes, C.V., 25, 44
Bajaria, 113
Baker, C., 200, 215
Banks, J., 145, 163
Barnard, W., 16, 20, 53, 103, 113, 163
Batson, R.G., 144, 163, 228, 235
Betker, H.A., 108f, 109f, 163
Bhote, K.R., 132, 163
Bottome, R., 113, 163
Box, G.E.P., 142
Brache, 218
Brust, P.J., 13, 20
Bultmann, C., 175t, 191
Burns, R.K., 175, 191
Buzzell, R.D., 166t, 191

Campanella, J., 55, 60, 103
Camp, R.C., 33, 34, 45, 89, 229, 235
Carr, N.G., 229, 235
Champy, J., 230, 231, 235
Chua, Richard C.H., 113, 127, 156, 163, 164
Cokins, G., 58, 103
Coleman, J., 113, 123, 163
Coletti, O.J., 178, 191
Cooper, D.R., 200, 215
Crosby, Philip, 6, 7, 11

DeFeo, J.A., 16, 20, 30, 45, 48, 53, 62, 103, 113, 163
Dell, L.D., 140, 164
Deming, W. Edwards, 6, 7, 11

Early, 118
Early, J.F., 191, 200, 215

Emory, C.W., 200, 215
Endres, Al C., 183, 191, 215

Feigenbaum, A.V., 6, 7
Forsha, H.I., 12, 20, 123, 163

Gale, B.T., 166t, 167t, 168, 191, 192
Gitlow, H., 215
Godfrey, A.B., 140, 164
Gryna, F.M., 13, 20, 176, 192

Hacker, S.K., 224, 235
Hamburg, J., 55, 60, 103
Hammer, M., 230, 231, 235
Hanewinckel, E., 163
Hardaker, M., 220, 235
Harry, 113
Harshbarger, 224
Hartman, B., 116, 163
Henson, 113
Hildreth, S.W., 235
Hooyman, 224
Humphrey, Watts, 84

Imai, M., 215
Ishikawa, Kaoru, 6, 7
Ittner, C., 55, 60, 103

Janssen, A., 113, 163
Juran Institute, Inc., 34, 45, 47, 49f, 50f, 86f, 130, 133, 140, 141f, 156, 163, 179, 184, 185f, 211, 215, 218, 219f, 221, 222f, 223f, 226f, 228f, 229f, 230f, 235
Juran, Joseph M., 6–7, 20, 94, 95, 96, 107, 114, 163, 171, 178, 180f, 192

Kano, N., 197, 208f, 215
Kaplan, R.S., 40f, 45
Karlin, E.W., 163
Kaynak, H., 13, 20
Kearns, 89

Name Index

Keiningham, T.L., 66, 67*f*, 103, 192
Klavans, R., 167*t*, 192
Klenz, B.W., 121, 163
Koch, 116
Kodali, 90, 91
Kolarik, W.J., 140, 163
Kondo, Y., 197, 208*f*, 215

Lauter, B.E., 67, 103
Leger, E., 233*t*, 235
Lenhardt, L., 119, 163
Leo, R.J., 27*f*, 45
Link, Albert N., 15, 20
Lowenstein, 172
Luther, D.B., 39, 45
Lynch, J., 170, 171, 192

MacDorman, J.C., 233*t*, 235
McCain, C., 26*f*, 45
McCracken, M.J., 13, 20
McNamara, D.M., 232, 235
Mears, P., 140, 163
Mortimer, J., 62, 103

Nadler, 89
Nash, M., 170, 171, 192
Norton, D.P., 40*f*, 45

Oppenheim, A., 215
Oppenheim, R., 215

Pautz, S.J., 213, 214, 215
Perry, A.C., 32*f*, 45
Plsek, P.E., 140, 163
Porter, M.E., 28, 45

Reichheld, F.F., 170, 172, 173, 174, 192
Rethmeier, K.A., 160, 163
Roberts, H.V., 12, 20
Ross, B., 113, 123, 163
Rummler, 218
Rust, R.T., 66, 67f, 103, 168, 175, 192

Salvendy, G., 228, 235
Sanders, D., 113, 123, 163
Savage, P., 144, 163
Scanlan, P.M., 175*t*, 192
Schonberger, R.J., 198, 215
Schottmiller, J.C., 60, 103
Schroeder, 113
Scott, J.T., 15, 20
Selwitchka, R., 164
Sergesketter, B.F., 12, 20
Shaw, G.T., 233*t*, 235
Sink, 118
Slater, R., 35, 45, 112, 164
Smith, 175
Smith, G.F., 140, 164
Smith, W., 191
Snee, R.D., 227, 235
Stephens, K.S., 140, 164
Stepnick, L., 62, 103
Stoner, J.A.F., 62, 103
Strickland III, A.J., 22, 45

Taguchi, Genichi, 6, 11
Thompson, Jr., A.A., 22, 45

Utzig, L., 23, 45

Van Aken, 118
Van Aken, E.M., 224, 235

Wadsworth, H.M., 140, 164
Waltz, B., 170, 192
Ward, B.K., 220, 235
Werner, F.M., 62, 103
Williams, T.K., 144, 163, 228, 235
Wilson, P.F., 140, 164
Wong, D., 200, 215

Yun, J.Y., 113, 164

Zahorik, A.J., 66, 67f, 103, 192, 196
Zairi, M., 215

Subject Index

ABC. *See* activity-based costing
accountability
   absence, 43
   communication, relationship, 35
action plans, development, 33
activities network diagram (arrow diagram), 139
activity-based costing (ABC), usage, 58
actual performance, measurement, 214
ad-hoc interviews, items (examples), 52–53
ADLI. *See* approach, deployment, learning, and integration
Aetna Health Plans, simulation model, 144
affinity diagram, 139
agility, Baldrige Award core value, 69, 71
allowances, 56
American Red Cross, mission, 25
analysis
   Baldrige Award criterion, 69
   phase, assistance (techniques), 139
analysis of variance (ANOVA), 150
analyze (Six Sigma phase), 113, 125–140, 149–150
   data collection plan, 156
   theories, selection, 155
annual business plans
   improvement, goals, 112
   strategies, translation/deployment, 23
annual goal-setting process, 32f
annual goals, leadership team creation, 50
annual quality cost, examples, 59t, 177t
annual strategic plan review/refresh, 42f
ANOVA. *See* analysis of variance
ANSI/RAB accreditation program, 84
application, 74
appraisal costs, 57
   prevention costs, combination, 63
approach, deployment, learning, and integration (ADLI), 72–73
arrow diagram (activities network diagram), 139
AS 9100, 76
as is process, 224

assessment, 48–51
   collaborative assessment, 75
   international standards, 75–78
   plan, 51–53
asset descriptions, 88
assignable causes, 209
assignable causes of variation, random causes of variation (contrast), 210t
association searches, 133–134
AT&T, 232
   process quality management/improvement methodology, 233t
attrition, opportunity cost, 68f
audit, establishment, 182
audits, review, 39
Avis Rent-a-Car, total customer satisfaction, 25

balanced scorecard, review, 39
balancing, 57
Baldrige Model
   example (2013), 49f
   naming, 48
Baldrige National Quality Program, 74
bank, market research (example), 66t
baseline performance
   measurement, 118
   target/achieved (contrast), 159
bellwether project, usage, 111
benchmarkers, organization, 98–99
benchmarking
   business unit/site benchmarking, 92–93
   categorization, 92
   classification, 91–93
   competitive benchmarking, 33–34, 94–95
      best practices, 47
   conducting, 92
   consortium benchmarking, 95, 98
   database benchmarking, 95, 96
   defining process, 90
   explanation, 89–93
   external benchmarking, 93, 94

benchmarking (*Continued*)
  functional benchmarking, 92
  generic benchmarking, 92, 93
  industrial tourism, contrast, 89
  internal benchmarking, 93–94
  noncompetitive benchmarking, 93, 95
  objectives, 90–91
  one-to-one benchmarking, 95, 97–98
  participation, 90
  process, application, 33
  process benchmarking, 92
  projects benchmarking, 92, 93
  reasons, 91
  self-assessment benchmarking, 95, 97
  study, conducting, 97–98
  survey benchmarking, 95, 96–97
  two-phase process, 91
  type, 96
benchmark-quality ratings, provision, 85
benchmarks, 33t
  definition, 33, 168
  local telephone service, benchmark data, 66t
  partners, identification, 33
  recalibration, 33
  report, 88
  subjects, identification, 33
benchmark subjects, identification, 33
best fit design, 188, 189
best in class
  determination, 168
  focus, 179
best practices
  benchmarking, 85–88
  identification, 90
  learning, establishment, 88
bias, minimum, 127
  data collection, 127
big Q, 12, 13t, 22
billing process, macro-level process flow diagram, 225f
Black Belt, nomination, 186
blisters, defect rate, 152
blitz teams, 118
brainstorming technique, 122, 149
brand loyalties, cementing, 165
breakthrough, 16
  improvement, project-by-project approach, 107
  sequence, 107, 110
burn-in testing, 38
business
  excellence models, 92
  goals, 30
  measures, cost of poor quality (relationship), 61–62
  objectives/indicators, linkages, 228f
  performance, 5
  plans, strategies (translation/deployment), 23
  process
    measurements, 227f
    performance, 30
  risk categories, 31t
  system, 48
  unit/site benchmarking, 92–93
business process management, 217, 218–219
  example, 221–222
  impact, 232–234
  organization infrastructure, 222f
  planning phase, 222–232
  road map, 219f

Cadillac Motor Car Division
  strategies, identification, 28
  vision statement, 28
calamity response times, 93
capability
  meaning, 38–39
  requirements, 38–39
Capability Maturity Model (CMM), 84
  origination, 84
Capability Maturity Model Integration (CMMI) (CMMi)
  model, usage, 85
  software standard, 76
capacity
  meaning, 38
  requirements, 38–39
Carnegie-Mellon University (CMU), 84
Carolina Power and Light
  approach, 36
  middle managers, input source, 25
carry-over analysis, 187
case history, usage, 111
catch ball, elements, 36
cause, 119
cause-and-effect diagram, 123, 124f
cause-effect diagram, 139
  example, 153f
  fishbone diagram, 149
certification
  body, 79
  marketplace meaning, 79
certified quality engineers, employment, 30
cGMPs. *See* Current Good Manufacturing Practices
change
  management, 150
  resistance, 110, 145–147
checking, 57
check sheet, 139
chi-square contingency tables (test of independence), 150

chronic quality problems, 105, 106f
  project-by-project approach, 106
chronic quality-related problems, impact, 112
chronic waste, size (estimation), 110
classical control, 197t
cloning, 112
CMM. *See* Capability Maturity Model (CMM)
CMMI. *See* Capability Maturity Model Integration
coding error, 121
collaborative assessment, 75
collections process, macro-level process flow diagram, 225f
communication
  accountability, relationship, 35
  plan, establishment, 51
company internal costs, reduction, 169
Compaq Computers, vision statement, 26
comparisons, results evaluation factor, 73
competition, 7
  comparison, 24
  definition, 33
  market standing, assessment, 170
  meaning, 33
competitive advantage, achievement, 28, 165
competitive benchmarking, 33–34, 92, 94–95, 168–169
  best practices, 47
  importance, 34
competitive gap, determination, 33
competitiveness, organization effectiveness (contrast), 29
competitive performance, 30
complaints
  adjustments, 56
  analysis, 172
  handling, 171
Complexity Factor (CF), 87
complexity factor assessments, 88
compliance systems, 193–194
computer-aided design (CAD), usage (resistance), 146
computer software/firmware, incorporation, 83
confidence intervals, 150
conformance, quality, 10
conformance to specifications, 11
consortium benchmarking, 95, 98
constraint, 140
consumer products, life-cycle costs, 177t
contamination
  defect rate, 152
  prevention, 81
contingency planning matrix, 212f
continuous improvement (CI)
  focus, maintenance, 155–160
  kaizen, 105
  practices, 14f

continuous process regulation, 211
contract
  management, 222
  process, 223f
  special-contract management process, flowchart, 226f
control
  charts, 132, 139
    example, 209f
  classical control, 197t
  expedite order entry control plan, example, 202–203
  goals, 205t
  importance, understanding, 198–199
  limits, computation, 209
  process, feedback loop (requirement), 193
  stations, 207
    review, 206
  subjects, 199t, 205t
  systems, 193–194
control (Six Sigma phase), 113, 147–148
  charts, 132, 150
  design, 147–148
  development, 182, 183
  final process capability, determination, 148
  measurement system, validation, 148
  plans, development, 182
  process (improvement), documentation, 147–148
controllable variables, 113
COP3. *See* cost of poor quality
COPQ. *See* cost of poor quality
core competencies
  examples, 29t
  leveraging, 28–29
core values/concepts, role, 71f
corporate mission/beliefs, commitment, 25
corporate quality
  goals, examples, 30
  policies/values, preparation, 37
correlation analysis, 133
cost of appraisal, 63
cost of poor quality (COPQ) (COP3), 48, 53–55, 150
  business measures, relationship, 61–62
  chart, 54f
  estimation, 47, 64
    reasons, 54
  hidden cost, 61f
  improvement, 30
  interpretation, 61–62
  knowledge, importance, 54
  measurement, 54
  reduction, 62, 155
  sales review, relationship, 62
  study, 24, 114–115

cost of poor service, calculation, 60
cost of quality
   optimum, 62–64
   term, usage, 59
cost reduction, opportunities (identification), 54
costs
   appraisal costs, 57
   categories, 55–61
   decrease, 105
   external failures costs, 56–57
   failure costs, 63
   hidden quality costs, 60–61
   internal failure costs, 55–56
   optimum costs, model, 63f
   prevention costs, 57–60
   quality, relationship, 13–15
   reduction, 111
count, 220
creative thinking, 150
criteria selection matrix, 150
criticality analysis, 188
critical success factors, 220f
critical-to-quality (CTQ) elements, 181
critical-to-quality (CTQ) requirements, 149
   necessity, 186
   needs, translation, 187
   prioritized list, 187
cross-functional processes, 218, 221
   emphasis, 232
   teams, focus, 234
cross-functional project teams, requirement, 112
cross-industry noncompetitive benchmarking, 95
CTO. *See* critical-to-quality
cumulative plots, comparison, 131
Current Good Manufacturing Practices (cGMPs), 75, 80–81
   FDA enforcement, 80
   importance, 81
   regulations
      differences, 82
      organization compliance, FDA determination process, 81–82
current performance, 220
current process, defining, 222, 224
customer-based measurements, 24t
customer care center (CCC), 205
customer defections
   financial impact, 172
   rate, 173–174
   reasons, determination, 172
   reduction, profit (impact), 171
customer-driven excellence, Baldrige Award core value, 69
customer-focused organization, 7

customer loyalty, 30, 169–172
   analysis, results, 172
   drivers, determination, 171
   economic worth, 173–174
customer needs, 22
   assessment/translation, 170
   discovery, 181, 183, 224
   meeting, failure, 55–57
   prioritized list, 187
   spreadsheets, 187
customer-related measurements, emphasis, 195
customer retention, 30, 169–172
   rate (increase), profit (impact), 174f
   satisfaction level, 175–176
   tracking, 170
customer(s)
   customer-based measurements, 24t
   customer-driven excellence, Baldrige Award core value, 69
   customer-focused organization, 7
   customer-related measurements, emphasis, 195
   defections, 57
   disloyalty drive map, 173f
   dissatisfaction, opportunities (identification), 54
   emphasis, categories, 178t
   expectation
      levels, increase, 7–8
      surpassing, 35t
   focus
      Baldrige Award criterion, 69
      group, response value, 199
   loyalty, data collection, 67
   organization (second party), 78
   problems, 176
   requirements, meeting (failure), 55–57
   retention, data collection, 67
   satisfaction, 175
   segmentation, initial assessment, 186
   service, 22f
   spectrum, 177–178
customer satisfaction, 30
   approach, 179
   customer loyalty, contrast, 169t
   increase, 114
Customer Satisfaction Index survey, 66
customer service representation (CSR), 144
cycle time, quality (relationship), 13–15

data
   analysis, 87, 126, 128
      worker-to-worker differences, 138
   cumulative data plots, 131–132
   mining, example, 116–117
   normalization, 98–100
   sources, 95–98

database benchmarking, 95, 96
data collection, 128, 214
    bias, minimum, 127
    method, determination, 33
    objectives, establishment, 126–127
    plan, 118, 121–123, 125–128
        example, 127f, 156
        flowchart, 126f
    planning, 150
defect (disconnect), 119
defect-concentration analysis, 132–133
defection rate, 173–174
defect level, reduction, 110
deficiencies, 10
    chronic sources, elimination, 211
    sporadic sources, elimination, 211
define (Six Sigma phase), 113, 114–118, 148–149
    steps, 114
define (step), 186
define-discover-design-develop-deliver, 184
Define, Measure, Analyze, Design, Verify (DMADV), 11, 17, 113, 184
Define, Measure, Analyze, Improve, Control (DMAIC), 11, 17, 148, 184
    analyze, 113, 125–140
    control, 113, 147–148
    define, 113, 114–118
    improve, 113, 140–147
    measure, 113, 118–125
    project, Six Sigma example, 151–155
deliver results (step), 189–190
Deming cycle, 197f
departmental goals, 24
departmental plans, development, 24
deployment, 48–51, 72
    meaning, 34
    process evaluation factor, 72
design (step), 187–188
    documents, recording, 190
    matrices, usage, 189f
    parameters, 188
    scorecard, 188f
        update, 189
design for assembly (DFA), 188
design for manufacturability (DFM), 188
Design for Six Sigma (DFSS), 184–190
design of experiments (DOE), 141–142, 150
design, quality, 10
design verification test (DVT), 189
develop (step), 188–189
development plan, example, 41f
DFA. *See* design for assembly
DFSS. *See* Design for Six Sigma
diagnostic experiments, types, 141t

diagnostic journey
    requirement, 119
    steps, 120
dirt, defect rate, 152
disciplines, 18t
discover (step), 186–187
disloyalty drive map, 173f
dissection
    process dissection, 129–130
    products dissection, 129–130
    simultaneous dissection, 132
distribution, example, 129f
DMADV. *See* Define, Measure, Analyze, Design, Verify
DMAIC. *See* Define, Measure, Analyze, Improve, Control
document review, 57
DOE. *see* design of experiments
Dow Corning, process knowledge management, 229
drift, 131
driving satisfaction, performance/importance (contrast), 67f
Drug Manufacturing Inspections Program (systems-based inspectional program), 82
DVT. *See* design verification test

economic significance, 210
economies of scale, achievement, 165
effort, chronicle, 151
EFQM. *See* European Framework for Quality Management
EIU. *See* environmental impact unit
electronic commerce, 8
engineering, internal customer, 9
Enterprise strategic plan, quality (integration), 21
environmental analysis, 26
environmental impact unit (EIU), 99
equal variances, tests, 150
equipment, unplanned downtime, 56
errors
    coding error, 121
    error-proofing, 139
    inadvertent errors, 137–138
    matrix, example, 136t
    Pareto diagram, example, 122f
    patterns, interrelationship, 137t
EU. *See* European Union
European Framework for Quality Management (EFQM) excellence model, 48f
European Union (EU), 77
events, symbols (usage), 228
evidence-based input, provision, 90
evolutionary operations (EVOP), 142
    plan, 143f
        production experimentation structure, 143

excess inventory reserves, 115
    Pareto analysis, 116t
executive leadership, accountability, 21
executive owner, 221
Executive summary, 151
exhortation approach, success, 43
expedite order entry control plan, example, 202–203
experimentation, consideration, 142
experiments, design, 140–142
exploratory experiments (screening experiments), 140–141
external benchmarking, 93, 94
external customers
    categories, 9
    components, 9
    proposal, format/content, 222
external failure costs, 56–57

fail-safing, 198
failure analysis, 172
failure costs, 63
failure mode and effects analysis (FMEA), 148
    update, 150
fault tree analysis, 187
feedback loops
    action, 211–212
    development, 182
    elements, 197
    flowchart, 194f
    requirement, 193
    usage, 147–148
Fidelity Investments, 171
field failures, 131
field studies, examples, 64–67
filler ingredient, usage, 133
final inspection/test, 57
final yield, 206, 206f
financial analysis, 187
financial control, 16t
financial goals, 24
financial improvement, 16t
financial management, accomplishment, 15–16
financial performance, quality
    impact, 7–8
    relationship, 166–167
financial planning, 16t
financial processes, 16t
first party (supplier), 78
first-party audits, 78
first-time yield, 206, 206f
fishbone diagram (cause-effect diagram), 149
fitness for use, 11
    usages, 9
5S method, 198

flag diagram, 207
    example, 208f
flow diagram, 128
    analysis, 230f
    symbols, usage, 228
FMEA. *See* failure mode and effects analysis
focus groups, 74–75
    items, targeting, 52–53
Ford Taurus/Mustang, quality goals, 181
foundry equipment, annual quality cost, 177t
functional activities, impact, 8
functional benchmarking, 92
functional departments (silos), 217
    policies, requirement, 37–38
functional goals, establishment, 33
functional organization, work flow, 218f
functional process management, business process management (contrast), 217–218
future, focus (Baldrige Award core value), 71
future performance, projection, 33
fuzzy logic concepts, 145

gains, holding (institute controls), 110
gap analysis, 88
GATT. *See* General Agreement on Tariffs and Trade
Genentech (Error-Proofing efforts), Six Sigma/Lean (usage), 113
General Agreement on Tariffs and Trade (GATT), 77
General Electric (GE)
    Capital Mortgage Insurance Corporation, process control system (example), 213
    projects, increase, 112
    Six Sigma
        elements, identification, 113
        goal, 35
    strategic planning study, 23
generic benchmarking, 92, 93
global benchmarking consortium, 95
GMP. *See* Good Manufacturing Practice
goals
    achievement, obstacles, 41, 43
    annual goal-setting process, 32f
    definition, 30
    deployment, 32, 34–35
    formulation
        competitive benchmarking, importance, 34
        inputs, 31
    macrostrategies, 25
    requirement, JQH5 identification, 30
    stretch goals, 32
    subdivision, 21
Good Manufacturing Practice (GMP) regulations, 80

hardware, redesign, 55
health, safety, and environmental (HSE) processes, 95

heavy equipment study, example, 65t
hidden quality costs, 60–61
high-level design, 187
    capability, results, 188
    concept, 188
    process, design matrices (usage), 189f
high-level flow diagram
    construction, 227
    example, 120f
histograms, 139
    comparison, 131f
Hoshin Kanri
    goals, deployment, 34–35
    strategic planning, 23f
Hoshin planning, 23
HSE. *See* health, safety, and environmental
human attention, dependence (reduction), 138
human behavior, understanding, 7
human error, theories test, 136–140

improve (Six Sigma phase), 113, 140–147
    alternative remedies, evaluation, 140
    change
        introduction, rules, 146–147
        resistance, 145–147
    experiments, design, 140–142
    production experiments, 142–143
    remedy, design, 145
    simulation experiments, 143–145
    written plan, 142
improvement, 16. *See* quality improvement
    action planning/implementation, 88
    approach, 112
    example, 158
    focus, maintenance, 159
    goals, 112
    infrastructure, 159
    project
        gains, maintenance, 194
        Six Sigma, example, 107–110
        tasks, 107
    radical forms, pursuit, 159
    tools, application, 160
Inadvertent errors, 137–138
    remedies, 138
individual-contract approach, 78
industrial tourism, benchmarking (contrast), 89
industry-specific processes, 94
inefficient processes, cost, 56
information
    appearance, 231
    generation, 126
    importance, 194–196
    loss, 55
    revolution, 8
    sources, 95–98
    technology, 12
innovation, management (Baldrige Award core
        value), 71
in-process inspection/test, 57
in-process quality data analysis technique,
        development/implementation, 30
input variables
    definition, 143
    distribution data, 144
    process outputs, relationship, 113
in-sourcing, 22f
institutionalized learning, 8
insurance policy writers, error matrix (example),
        136t
insurance product, example, 171f
integrated quality system, definition, 9–11
integration
    process evaluation factor, 72–73
    results evaluation factor, 73
interlocking spreadsheets, QbD backbone, 185f
intermediate stock inventory (net), 115
internal benchmarking, 93–94
internal costs, reduction, 169
internal customers, 9
    categories, 9
    requirements, nonconformance, 222
internal failure costs, 55–56
    Pareto analysis, 115t
internal functions/processors, 9
internal support operations, rework, 56
International Organization for Standardization (ISO)
    ISO 176 Technical Committee, 75
    ISO 9000, 50, 75
        adoption/implementation, 76–77
        industry-specific adoptions/extensions,
            79–85
        management system, 78
        quality management system standard,
            76–78
    ISO 9000-3:1991, ISO/TC176 development/
        publication, 83
    ISO 9001
        application, issues, 83
        clauses/structure, 77t
        ISO 9003, combination, 83
    ISO 9002:1987 requirements, 80
    ISO 14000, 75
    ISO/TC176 technical committee, 83
    ISO/TC210 technical committee, 82
    quality definition, 11
    Quality Management System, 5, 18
    Quality Standards, 82
international system standards, assessment, 47
interrelationship diagraph (relations diagram), 139

Subject Index

interval scale, 200
interview, 74
invoice errors, stratification (example), 130f
Ishikawa cause-and-effect diagram, 109f, 123, 124f
ISO. *See* International Organization for Standardization

Japanese tools, 139t
Juran
   Complexity Factor (CF), 87
   Organization Health Check, 50
   organization quality, 52t
   Pareto principle, 114–115
   Quality Handbook fifth edition (JQH5)
      goal requirement identification, 30
      Section 6, 220
      Section 47, 141
   Quality Handbook sixth edition (JQH6), 7
      Section 13, 22
   Quality Index, 86, 87
   Quality Management System (QMS), 5, 18
   Quality Risk Assessment, usage, 48
   risk check scoring criteria, 52t
   7-Step Benchmarking Process, flowchart, 86f
   Trilogy
      diagram, 17f
      processes, interrelationship, 16
Juran Complexity Factor (CF), 87
   development, 99

Kaiser-Permanente, vision statement, 28
Kaizen, 105
   event, 150
kaizen (continuous improvement), 105
Kelly Services, 25
   strategic planning process, 26f
key-factor analysis, 117
key performance indicators (KPIs), 87, 88
   usage, 99
key process input variables (KPIVs), 119
key process output variables (KPOVs), 119
Kilo of Lines of Code (KLOC), 199
knack, 138
knowledge management, Baldrige Award criterion, 69
KPIs. *See* key performance indicators
KPIVs. *See* key process input variables
KPOVs. *See* key process output variables

leadership
   absence, 41
   Baldrige Award criterion, 68
   strategy implementation, 36–39
Lean event, 150

Lean Six Sigma
   breakthrough improvement, 112–113
   model, 105–106
   summary, 148–155
Lean Six Sigma (LSS), 18
learning
   potential, 94
   process evaluation factor, 72
levels, trends, comparisons, and integration (LeTCI), 73
life-cycle costs, 176–177
   consumer products, 177t
little q, 12, 13t
L.L. Bean Company, Xerox selection, 33, 89
local telephone service, benchmark data, 66t
Long John Silver's vision statement, 28
long-range company goals, 30
long-term goal, defining, 21
long-term strategic goals, development, 21
long-term strategies, development, 28–29
long-term variation, 125
loss to society, 11
lost sales, poor quality (impact), 174–175
loyalty. *See* customer loyalty
   information, tracking, 170
LSS. *See* Lean Six Sigma

macro-level process flow diagram, 225f
macrostrategies, 25
Malcolm Baldrige National Award for Excellence, 5
Malcolm Baldrige National Quality Award, 68
   concepts, role, 71f
   core values, 69, 71–72
      role, 71f
   criteria, 68–69
   scoring system, 72–73
   usage, system assessment tool, 73–75
management by fact, Baldrige Award core value, 71
management controllable, 196–197
management-controllable problems, theories (testing), 128–132
management-controllable, term (usage), 128
management, languages, 62t
management system (ISO 9000 standards), 78
management team, Black Belt nomination, 186
managing-for-results process (Xerox), 27f
Managing the Software Process (Humphrey), 84
MAP, 76
market
   focus, Baldrige Award criterion, 69
   initial assessment, 186
   leadership, requirement, 168
   research
      results, graphing, 66
      study, 64
   standing, assessment, 170

marketing, 22f
marketplace
   assessment/standing, 64–67
   bank, market research (example), 66t
   field studies, examples, 64–67
   heavy equipment study, example, 65t
   local telephone service, benchmark data, 66t
   multiattribute study, example, 65t
   superiority, achievement, 167–169
market share
   flowchart, 11f
   increase, 165
matrix data analysis (prioritization matrix), 139
matrix diagram, 139
matrix-ranking technique, 133
measure (Six Sigma phase), 113, 118–125, 148, 151
   steps, 118
measurement
   Baldrige Award criterion, 69
   decision, 127
   drivers, 195f
   establishment, 213
   importance, 194–196
   intermediate operation stages, 135
   noncontrolled operations, 135
   scales, 200
   system
      creation, 199
      validation, 123–124
measurement system analysis (MSA), 149
medical device industry, ISO 9000 standards, 80
metrics, conversion process, 99
middle managers, input source, 25
Minitab (MINITAB), 124, 142
mission
   development, 25–28
   statement, 25
      basis, 117
      projects, 117
   verification, 107
mistake proofing, 150
MSA. *See* measurement system analysis
multiattribute study, example, 65t
multivari chart, example, 132f

NAFTA. *See* North American Free Trade Association
national performance standards/awards, usage, 47, 68–72
national quality awards, importance, 68
net present value (NPV), calculation, 173–174
new-products review, 58
new quality program, skepticism, 43
new windows, cutting, 135
nominal scale, 200

noncompetitive benchmarking, 92, 93, 95
   cross-industry noncompetitive benchmarking, 95
noncontrolled operations, measurement, 135
nonlinear models, handling, 188
nonparametric tests, 150
non-value-added activities, 56
normality, test, 150
normalization, 98–99. *See also* Data normalization
   data, analysis, 100
   method, efficacy (testing), 100
North American Free Trade Association (NAFTA), 77
NPV. *See* net present value
numerical reliability, defining, 30

objectives, defining, 51
obsolete reserves, 115
occurrences, 200
one-to-one benchmarking, 95, 97–98
operational control subjects, sensors (usage), 201
operational effectiveness, 28
operational goals, criteria, 205
operational plans, development, 21
operational results, review, 52–53
operation, intermediate stages (measurement), 135
operations, remedy (transfer), 147
opportunity cost of attrition, 68f
opportunity, defining, 179–181, 183
Optima!, 224
optimum costs, model, 63f
Oracle, 170
order forms errors, Pareto diagram (example), 122f
ordinal scale, 200
organization
   benchmarking studies, 114
   business process management, impact, 232–234
   effectiveness, competitiveness (contrast), 29
   example, 180f
   forms, changes, 8
   leadership, impact, 49
   measurements, combination, 39
   quality (Juran), 52t
organizational learning, Baldrige Award core value, 69
organizational vision/mission, 49
Otis Elevator, reliability, 25
out-of-control points, elimination, 110
output variable(s), definition, 143
out-sourcing, 22f
ownership, total cost, 176

Pareto analysis, 139
   example, 108f
   excess inventory reserves, 116t
   internal failure costs, 115t
   slide, example, 152

Pareto concept, key-factor analysis, 117
Pareto principle (Juran), 114–115
   application, 121
   entries, 136
Pareto Priority Index (PPI), usage, 116, 116t
partners, valuing, 69
payoff matrix, 150
PDSA. *See* Plan, Do, Study, Act
perfection, economic goal, 64
performance
   assessment, reasons, 47
   baseline, assessment, 51
   current performance, 220
   data, analysis, 100
   determination, 91
   differences, consideration, 100–101
   excellence, 8
      criteria, item listing, 70f
   gaps
      quantification, 90
      understanding, 91
   improvement, 8, 94
      foundations (building), learning (enablement), 91
   levels, determination, 90
   maintenance, quality control, 193
   measurement, 206–207, 214
      establishment, 199–205
   reasons, evaluation, 90
   regulation, 196
   standards, 205
      comparison, 207–210
      economic significance, 210
      establishment, 193, 199–205, 214
      statistical significance, 207–210
   working practices, knowledge (sharing), 91
personal learning, Baldrige Award core value, 69
Pharmaceutical CGMPs for the 21st Century Initiative (FDA announcement), 81–82
phase-by-phase deliverables/tools, 148–151
piece-to-piece variation, 132
pilot activities, learning, 43
pilot plant, setup, 142
PIMS. *See* Profit Impact of Market Strategies
pitted castings, problem (analysis), 133
plan, do, check, act (PDCA), 197
Plan, Do, Study, Act (PDSA), 105, 107, 197
planning
   phase, business process management, 222–232
   process, 16–17
   requirement, 51–53
policy
   clarity, 36–37
   definition, 36
   deployment, 23

   issue, sensitivity, 38
   requirement, 37–38
poor quality. *See* Cost of poor quality
   hidden costs, 61f
   penalties, 56
   revenue loss, 59t
population, measurement decision, 127
practices, measuring (continuous process), 33
pre-assessment, completion, 51
premium prices, securing, 165
prevention costs, 57–60
   appraisal costs, combination, 63
   compilation, importance, 58
price, quality (relationship), 167t
primary process, 217
Primer flash, wind tunnel, 152
primer, gram weight settings, 152
prioritization matrix (matrix data analysis), 139
problems
   diagnosis, 211–212
   identification, 211
   statement, 151
   troubleshooting, 211–212
process capability
   demonstration, 182
   example, 125f
   measurement, 118, 124–125
   studies, 187
process control
   development, 213–214
   plans, 150, 190
   system, example, 213–214
process(es)
   benchmarking, 92
   business process management, 217, 218–219
   characteristics, variation, 56
   computer simulation, 228
   continuous process regulation, 211
   control, 58
   controllability, demonstration, 182
   cross-functional processes, 221
   current performance, 220
   current process, defining, 222, 224
   data, analysis, 222, 227–229
   decision program chart, 139
   definition, 217
   descriptions, 88
   design, 229–232
      details, 189
      finalization, 182
   development, 182, 183
   dissection, 129–130
   documentation, 118–121
   equipment capacity, excess, 60
   evaluation, factors, 72–73

final process capability, determination, 148
flowcharting, 224
flow diagram (process map), 119
FMEA, 149
goals, 193, 220
improvement
    documentation, 147–148
    implementation/monitoring, 148
improvement, addressing, 159
knowledge, management, 229
management, 35t, 232
    Baldrige Award criterion, 69
map, 224
meaning, 72
measurement, establishment, 222, 224, 227
measures, linkages, 228f
mission, 220
outputs, input variables, relationship, 113
planning, 58
primary process, 217
process-capability analysis, 128–129
procession type, 129
properties, measurement, 135
quality management/improvement methodology (AT&T), 233t
redesign, 229–232
repository, 229
selection, 219–220
self-managed process, 196–198
team, organization, 221
transfer, 232
variation
    analysis, promotion, 194
    sigma, denotation, 112–113
Process Maturity Framework, 85
production
    experiments, 142–143
    scrap, 115
productivity
    definition, 13
    quality, relationship, 13–15
products
    attributes, 24t
    characteristics, variability, 56
    deficiencies, units of measure, 200
    design, 181–183
    dissection, 129–130
    evaluation, 37
    features, 9–10
    functionality, 187
    goals, 193
    impact, 79
    liability lawsuits, growth, 114
    meaning, 9
    measuring, continuous process, 33
    performance, 30
        data, 31
    properties, measurement, 135
    quality audits, 57
    salability, 182
    standards, 78
    suppliers, change, 175t
profit, impact, 171
Profit Impact of Market Strategies (PIMS), 166
profound knowledge system, parts, 7
progress assessments, review, 39
project-by-project approach, 106, 107
    lessons, learning, 111–112
project need, verification, 107
projects
    benchmarking, 92, 93
    evaluation, 114, 116–117
    example, 148–155
    identification, 114–117
    management, 150
    mission statement, 117
    need, verification, 118–119
    nomination, 114
        review, 116
    problem, 117
    results, 159
    scope/charter, defining, 107
    selection, 107, 114, 117
project teams
    meetings, requirements, 118
    organization/launching, 107
    selection/launching, 117–118
proportions test, 150
Pugh matrix, 150

QbD. *See* quality by design
QFD. *See* quality function deployment
QMS. *See* Juran
QS 9000, 75
quality
    activities, scope, 12
    annual costs, example, 59t
    annual quality cost, examples, 59t, 177t
    assurance requirements, 78
    attainment, costs, 53
    audits, 58
    business performance, 5
    concepts, application, 12
    contribution, 165–166
    culture, creation, 36
    customer emphasis, categories, 178t
    cycle time, relationship, 15
    data warehouse, 121
    defining, 124, 165
    definition, 7

Subject Index

quality (*Continued*)
  department, role, 8
  disciplines, 17
  financial performance, relationship, 166–167
  flowchart, 11f
  function, 11–13, 17
  impact, 7–8
  improvement, 13, 105
    diagnostic journey, 119
    projects, 6
  infrastructure, absence, 41, 43
  initiative, management approval, 110–111
  integration, 21
  internal results, 24
  ISO definition, 11
  Japanese revolution, 6
  Juran comprehensive assessment, 47–48
  loss approach, 60
  management, measurement (impact), 194–195
  meanings, 5
  measurements, development, 195–196
  national quality awards, 68
  organizationwide assessment, 47
  parameters, 37
  planning, 57–58
    road map, 179f
    steps, 178, 183
  policies/values, 36–38
  price, relationship, 167t
  procedures manual, development, 30
  processes, 194
  productivity, relationship, 13
  progress, spiral, 12f
  relationship, 13–15
  risk assessment, 47–48, 50
    criteria, 51
  scorecard, 213f
  strategic plan, 21–23
  strategies, examples, 29t
  task, responsibility, 37
  value, relationship, 15
  views, 19t
quality by design (QbD), 11, 165, 184–190
  backbone, 185f
  example, 183
  road map, 178–184
  spreadsheets, 185f
    construction, 184f
  steps, 179–183
quality control, 193
  subjects, 199
    alignment/linkage, 198
quality costs
  categories, 55–61
  distribution, exploration, 63
  example, 59t
  optimum, 62–64
  reduction, 30
  relationship, 13, 15
quality function deployment (QFD), 183
quality goals
  achievement, 43
  defining, alternative criteria, 32
  examples, 30
  formulation, 31–33
Quality Handbook (Juran), 7
quality improvement, 30
  business case, 110–111
  cost, 30
  cost decrease, 105
  methods, 105
  program, launching (steps), 107
quality management. *See* Strategic quality
    management
  guideline standard (ISO 9004), 76
  universal principles, 15–17
  universal processes, 16t
quality of conformance, 10
quality of design, 10
quality problems
  analysis, 143–144
  chronic quality problems, 105, 106f
  costs, 53–55
    estimation, 47
  management/systems controllable, 135–136
  revenue loss, 59t
  size, quantification, 54
  sporadic quality problems, 105, 106f
quality-related losses, estimation, 110
quality-related problems, 139
quality system
  certification/registration, 78–79
  certification/registration-level activities, 79
  standards, 78
Quality System Guidance Development working
    group (QS working group), 82
quantitative analyses, usage, 100

random causes of variation, assignable causes of
    variation (contrast), 210t
ranking, 133. *See also* theory
rate of return, quality (impact), 167f
rational subgroups, 209
ratio scale, 200
recognition, absence, 43
reconciliation, 57
reengineering
  aspects, 231
  contribution, 231
  defining, 230

registrar, 79
registration, marketplace meaning, 79
reinspection, 55
relations diagram (interrelationship diagraph), 139
reliability/lifetime analysis results, 189
remediation, requirement, 212
remedy, 119
   alternative remedies, evaluation, 140
   design, 145
   effectiveness, 145
   final evaluation, 145
   preliminary evaluation, 145
   selection matrix, 141t
   transfer, 147
report development, 101
repurchase intent, 176t
repurchasing (likelihood), complaints (handling), 171
requirements, conformance, 7
resistance, impact, 145, 147
resistance to change, 146
resources
   assignment, 32
   requirement, 24
   underestimation, 43
response surface methodology (RSM), 143
responsibility, 194
results
   delivery, steps, 182–183
   evaluation, factors, 73
   focus (Baldrige Award core value), 73
retention. *See* customer retention
returned material, 56
return on assets (ROA), calculation, 62
return on investment (ROI), 24
   calculation, 111
   data, 166
   flowchart, 11f
   market share, impact, 166t
   quality, impact, 166t
return on sales (ROS), 166
return rate, quality (impact), 167f
review, 57
reward, absence, 43
rework, 55, 57
   allowances, 60
   loops, flow diagrams, 229f
   reduction, 13–14
   time standards, allowances, 60
risk analysis, 187
   results, 188
risk check scoring criteria (Juran), 52t
risks, consideration, 31
ROA. *See* return on assets

ROI. *See* return on investment
ROS. *See* return on sales
RSM. *See* response surface methodology

safety stock, 115
sales
   income, quality (contribution), 165–166
   increase, quality by design (impact), 165
   loss, poor quality (impact), 174–175
   revenue, opportunities (loss), 57
   satisfaction, relationship, 175t
sample, measurement decision, 127
Samsung Electronics
   quality, mantra, 6
   Six Sigma success, 113
satisfaction. *See* customer satisfaction
   sales, relationship, 175t
savings, estimation, 110–111
scope/criteria, defining, 51
scrap, 55
   allowances, 60
   time standards, allowances, 60
screening experiments (exploratory experiments), 140–141
SDSA. *See* standardize, do, study, act
second party (customer organization), 78
second-party audits, 78
self-assessment benchmarking, 95, 97
self-control, 197t
   achievement, 194
   analysis, 151
   concept, 196
   criteria, 196
   guidelines, 190
self-managed process, 196–198
sensor, 199, 201, 205
service performance, 30
services
   measurement, continuous process, 33
   quality by design, example, 183
7-Step Benchmarking Process
   data analysis, 87–88
   data collection/validation, 87
   development, 86
   flowchart, 86f
   improvement action planning/implementation, 88
   institutionalized learning, 88
   learning, establishment, 88
   preparation/planning, 87
   reporting (benchmark report), 88
share-of-wallet concept, 170
shelf-life data, stratified scatter diagram (example), 134f
shipping shortages, data collection plan (example), 127f

short-range financial goal, 24
short-range goals, 30
short-term annual goals, development, 30–34
short-term strategic goals, development, 21
sigma, denotation, 112–113
simulation technique, 143–144
simultaneous dissection, 132
SIPOC. *See* supplier-input-process-output-customer
Six Sigma. *See* Lean Six Sigma
   causes, diagnosis, 108
   change, resistance, 110
   Control step, 110
   Define step, 107
   example, 125f
   improvement project, example, 107–110
   Improve step, 110
   Ishikawa cause-and-effect diagram, 109f
   measure/analyze steps, 108
   mission, verification, 107
   PDCA, 105
   phases, 113
   project need, verification, 107
   remedy, provision/effectiveness, 108
   temperature/time relationship, 109f
Six Sigma DMAIC project
   cause-effect diagram, 153f, 154f
   effort, chronicle, 151–152
   example, 151–155
   executive summary, 151
   improvement/results, 152, 155
   lessons, 155
   problem statement, 151
   results, 155
societal responsibility, Baldrige Award core value, 71–72
society, loss, 11
software development
   cGMPs, importance, 81
   CMM origination, 84
   ISO 9001, application (issues), 83
Software Engineering Institute (SEI), CMU
   creation, 84–85
software quality, 82–84
software redesign, 55
sorting inspection, 55
source inspection, 198
special-contract management process, flowchart, 226f
specifications, 166
   conformance, 11
sporadic quality problems, 105, 106f
spreadsheets, 185f, 187, 204f
   construction, 184f
stakeholders, 9
   analysis, 150

Standard CMMI Appraisal Method for Process Improvement (SCAMPI), 85
standardize, do, study, act (SDSA), 197–198
standard operating procedures, 151
standards of performance, 205
   comparison, 207–210
   establishment, 193, 199–205, 214
standing, assessment, 47
starting small, failure, 43
statistical control charts, out-of-control points (elimination), 110
statistical process control, 151, 198
statistical significance, 207–210
statistical variation, understanding, 7
stocks, evaluation, 57
storyboarding, 122
strategic alignment, 48–51
strategic goals, 30
   development, 21
strategic management, elements, 21
strategic plan
   annual strategic plan review/refresh, 42f
   projects, 35t
   quality, relationship, 21–23
   review/refresh, 39–41
strategic planning
   Baldrige Award criterion, 68
   model, 23–25
strategic quality management, 21
   approaches, 22
strategy (strategies)
   definition, 28
   implementation requirements, 36
   leadership implementation, 36–39
   progress, review, 39, 41
   translation, 40f
stratification
   analysis, 139
   concept, usage, 131
   example, 130f
stratified scatter diagram, example, 134f
stream-to-stream analysis, 130–131
strengths, weaknesses, opportunities, and threats (SWOT), 51
   analysis, 22, 31
      strategic planning phase, 28–29
   example, 22f
stretch goals, 32
success, system plan, 18
Suez Canal, 93
Sun Microsystems, loyalty driver classification, 171
supplier-input-process-output-customer (SIPOC) diagram, 149

suppliers
    cost of poor quality, 61
    first party, 78
    quality evaluation, 58
    written statement, 38
support operations
    cost of errors, 61
    revenue losses, 57
    rework, 57
survey, 73–74
    benchmarking, 95, 96–97
        drawbacks, 97
    process, 97
sustainability, 85–88
SWOT. *See* strengths, weaknesses, opportunities, and threats
symptom, 119
    description, 121
    quantification, 121
system assessment tool, Malcolm Baldrige National Quality Award (usage), 73–75
systematic change process, 16
systematic diagram (data diagram), 139
system controllable, 196–197
system-controllable, term (usage), 128
system model, development/translation, 144
system plan, development, 18
systems approach, 7
systems-based inspectional program, 82
systems perspective, Baldrige Award core value, 72
systems viewpoint, 219

task observations, items (examples), 52–53
TAT. *See* turnaround time
team
    defining, 51
    members, responsibilities (clarification), 194
technical assistance, provision, 37
Technique errors, 138–140
Texas Instruments, customer needs (satisfaction), 25
theory (theories), 119
    arrangement, 122–123
    cause-effect diagram, 154f
    formulation, 122–123
    generation, 122
    orderly arrangement, 123t
    selection, 123
    tabular arrangement, 122–123
    test, examples, 157
    testing, 128–132
        human error, involvement, 136–140
        new data, collection, 135–136
        ranking, usage (example), 134t
theory of constraints (TOC), 140
third-party consultants, 96

three-year development plan, 41f
TickIT, 83
time standards, allowances, 60
time-to-time analysis, 131–132
time-to-time drift, 131
time-to-time variations, analysis, 131
time, underestimation, 43
TOC. *See* theory of constraints
total company costs, 37
total cost of ownership, 176
total quality management (TQM), 18
trade-off analysis, 187, 188
training, 58
tree diagram (systematic diagram), 139
trends, results evaluation factor, 73
troubleshooting (firefighting), 211–212
    effectiveness, 212
true values (master values), 124
T-tests, 150
turnaround time (TAT), average, 144
21st Century Initiative
    CGMPs, impact, 82
    Pharmaceutical CGMPs, 81
two-way communication, 36

uniformity, predictable degree, 11
United States Postal Service, Customer Satisfaction Index survey, 66
units
    quality, defining, 124
    usage, 113
units of measure, 199
    calculation, 200
    examples, 201t
upper level managers, quality improvement, 6
upper management
    leadership, 22
        absence, 41
        importance, 36
    members, progress review, 39–40
use, fitness, 11

value
    analysis, 187
    calculation, 15
    creation, Baldrige Award core value, 72
    quality, relationship, 13–15
value-driven leadership, 35t
values, precision, 36–37
variation
    minimization, continuous process regulation, 211
    random causes/assignable causes (contrast), 210t
Visio, 224

vision, 30
   achievement, 21
   components, 50
   deployment, 34f
   development, 25–28
   statement, 26
   translation, 40f
visionary leadership, Baldrige Award core value, 69
voice of customer (VOC) analysis, 149

warranty charges, 56
weight distribution, example, 129f
Wong-Baker FACES pain rating scale, 200
wording, imprecision, 121
work activities, focus, 234
worker
   attentiveness, helping, 138
   methods, study, 135
   motivation, 137
   performance, difference, 138
   worker-controllable, term (usage), 128
   worker-to-worker differences, 138

worker error
   random pattern, 137
   subspecies, error pattern (interrelationship), 137t
work flow (functional organization), 218
workforce
   change, 8
   focus, Baldrige Award criterion, 69
   members/partners, valuation (Baldrige Award core value), 69
   valuing, 69
working owner, 221
working practices, knowledge (sharing), 91
work practices, standardization (concept), 197
written responses, 73

Xerox Corp.
   benchmarking activities, 89
   managing-for-results process, 27f
   operations, evaluation, 33
   warehouse orders, benchmark, 168